Applied Probability
for Engineers and Scientists

Applied Probability for Engineers and Scientists

Ephraim Suhir

*Bell Laboratories, Lucent Technologies,
Physical Sciences and Engineering Research Division*

McGraw-Hill
New York San Francisco Washington, D.C. Auckland Bogotá
Caracas Lisbon London Madrid Mexico City Milan
Montreal New Delhi San Juan Singapore
Sydney Tokyo Toronto

Library of Congress Cataloging-in-Publication Data

Suhir, Ephraim.
  Applied probability for engineers and scientists / Ephraim Suhir.
    p. cm.
  Includes bibliographical references (p. 529–533) and index.
  ISBN 0-07-061860-7
  1. Probabilities.   I. Title.
OA273.S773   1997
519.2′02462--dc21   96-48238
                                                                   CIP

McGraw-Hill
A Division of The McGraw-Hill Companies

Copyright © 1997 by The Bell Laboratories, Lucent Technologies, Inc. All rights reserved. Printed in the United States of America. Except as permitted under the United States Copyright Act of 1976, no part of this publication may be reproduced or distributed in any form or by any means, or stored in a data base or retrieval system, without the prior written permission of the publisher.

1 2 3 4 5 6 7 8 9   Doc/Doc   9 0 2 1 0 9 8 7

ISBN 0-07-061860-7

The sponsoring editor for this book was Harold B. Crawford, the editing supervisor was Patricia V. Amoroso, and the production supervisors were Donald F. Schmidt and Suzanne W. B. Rapcavage. It was set in Century Schoolbook by Santype International.

Printed and bound by R. R. Donnelley & Sons Company.

Information contained in this work has been obtained by The McGraw-Hill Companies, Inc., ("McGraw-Hill") from sources believed to be reliable. However, neither McGraw-Hill nor its authors guarantee the accuracy or completeness of any information published herein and neither McGraw-Hill nor its authors shall be responsible for any errors, omissions, or damages arising out of use of this information. This work is published with the understanding that McGraw-Hill and its authors are supplying information but are not attempting to render engineering or other professional services. If such services are required, the assistance of an appropriate professional should be sought.

McGraw-Hill books are available at special quantity discounts to use as premiums and sales promotions, or for use in corporate training programs. For more information, please write to the Director of Special Sales, McGraw-Hill, 11 West 19th Street, New York, NY 10011. Or contact your local bookstore.

Chance governs all
        John Milton, Paradise Lost

There are things in this world, far more important than the most splendid discoveries—it is the methods by which they were made
        Gottfried Leibnitz

To my family, Raisa, Elena, and Eugene Suhir, to my mother Betty, and to the memory of my father, Leonid Suhir

Contents

Preface xiii
List of Basic Symbols xv
Introduction xix

Chapter 1. Random Events   1

1.1 Frequency and Probability   1
1.2 Axioms of the Probability Theory   11
1.3 Algebra of Events   15
1.4 Procedures for Counting the Sample Points   23
1.5 "Geometric" Probabilities   34
1.6 Bayes Formula   39

Chapter 2. Discrete Random Variables   47

2.1 Random Variables   47
2.2 Frequently Encountered Discrete Distributions   52
    2.2.1 Bernoulli trials and the binomial distribution   52
    2.2.2 Poisson distribution: flow of events   60
    2.2.3 Geometric distribution   69
    2.2.4 Hypergeometric distribution   70
2.3 Generating Functions   72

Chapter 3. Continuous Random Variables   79

3.1 Characteristics of Continuous Random Variables   79
3.2 Bayes Formula for Continuous Random Variables   82
3.3 Frequently Encountered Continuous Distributions   84
    3.3.1 Uniform distribution   84
    3.3.2 Exponential distribution   85
    3.3.3 Normal (Gaussian) distribution   88
    3.3.4 Distributions associated with the normal law   93
    3.3.5 Rayleigh distribution   95

3.3.6	Rice distribution	98
3.3.7	Weibull distribution	99
3.3.8	χ^2-distribution. Gamma distribution. Erlang distribution	101
3.3.9	χ-distribution	103
3.3.10	Cauchy distribution	103
3.3.11	Beta distribution	104
3.3.12	Student distribution	105

Chapter 4. Systems of Random Variables — 107

- 4.1 System of Two Random Variables — 107
- 4.2 System of Many Random Variables — 113
- 4.3 Complex Random Variables — 114

Chapter 5. Functions of Random Variables — 117

- 5.1 Characteristics of Functions of Random Variables — 117
- 5.2 Distributions of Functions of Random Variables. Convolution of Distributions — 121
- 5.3 Limit Theorems of the Probability Theory — 144

Chapter 6. Entropy and Information — 149

- 6.1 Uncertainty of an Experiment — 149
- 6.2 Entropy — 152
- 6.3 Information — 156
- 6.4 Communication systems — 160

Chapter 7. Random Processes: Correlation Theory — 167

- 7.1 Random Functions and Processes — 167
- 7.2 Characteristics of Random Processes — 169
- 7.3 Stationary Random Processes — 175
- 7.4 Crossing the Given Level — 184
- 7.5 Duhamel Integral: Transformation of Random Processes by Linear Dynamic Systems — 186

Chapter 8. Random Processes: Spectral Theory — 195

- 8.1 Theory of Spectra — 195
  - 8.1.1 Fourier series — 195
  - 8.1.2 Fourier integral — 197
  - 8.1.3 Fourier transforms: Spectrum of a function — 198
  - 8.1.4 Parseval's formula — 200
  - 8.1.5 Solution of ordinary differential equations using the Fourier integral — 201
  - 8.1.6 The complex frequency characteristic as a spectrum of the impulse response of a dynamic system — 204
  - 8.1.7 The duration of a process and the width of its spectrum — 206
- 8.2 Wiener–Khinchin's Formulas — 208
- 8.3 "White Noise." Wiener's Process. Narrow Band Processes — 211

8.4	Spectral Theory of the Transformation of Stationary Random Processes by Linear Dynamic Systems	214
8.5	Response of a Dynamic System to Dual Inputs	219
8.6	Information Capacity of a Communication Channel	221
8.7	Characteristic Functions	223

Chapter 9. Extreme Value Distributions — 227

9.1	Extreme Value Statistics	227
9.2	Exceeding the Given Level by Normal Processes	230
9.3	Extreme Value Distribution for Stationary Normal Processes	232
9.4	Hermite Polynomials	246
9.5	Application of Extreme Value Distributions for Reducing Casualties at Sea	248
	9.5.1 Ship passing a shallow waterway	249
	9.5.2 Helicopter undercarriage strength when landing on a ship deck	256
	9.5.3 Probability of occurrence of ship slamming	260

Chapter 10. Reliability — 265

10.1	Major Definitions	265
10.2	Deterministic and Probabilistic Approaches in Reliability Engineering	266
10.3	Dependability. Reliability Function	271
10.4	Experimental Evaluation of Dependability and the Failure Rate	279
10.5	Goodness-of-Fit Criteria	280
	10.5.1 Goodness-of-fit tests	280
	10.5.2 Pierson's criterion	281
	10.5.3 Kolmogorov's criterion	284
10.6	Reliability of Repairable Items	287
10.7	Choosing the Appropriate Reliability Indices	289
10.8	Updating Reliability. Conjugate Distributions	295
10.9	Confidence Intervals	297
10.10	Accelerated Testing	299
10.11	Technical Diagnostics	302
	10.11.1 Objective of technical diagnostics	302
	10.11.2 Statistical methods of recognition	303
	10.11.2.1 Application of the Bayes formula	303
	10.11.2.2 Wald's method	308
	10.11.2.3 Methods of statistical decisions	311
10.12	Structural Reliability	316
	10.12.1 Ultimate and fatigue strength	316
	10.12.2 Dependability of a structural element	319
	10.12.3 Solder glass attachment in a ceramic electronic package	323
	10.12.4 Integrated circuit in a "smart card" design	327
	10.12.5 Kinetic approach to material failure	333

Chapter 11. Markovian Processes — 337

11.1	Flows of Events	337
11.2	Markovian Processes and Chains	339

Contents

- 11.3 Some Often Encountered Markovian Processes — 346
  - 11.3.1 Random telegraph signal — 346
  - 11.3.2 Shot noise — 347
  - 11.3.3 Random pulses — 348
  - 11.3.4 Poisson impulse process — 348
  - 11.3.5 Gaussian process — 349
  - 11.3.6 Brownian process — 350
  - 11.3.7 Flows of failures and restorations. Availability of an item — 352
  - 11.3.8 Linear filtering. Wiener–Hopf equation — 354
- 11.4 Queues — 358
- 11.5 Continuous Markovian Processes. Smoluchowski's Equation — 371
- 11.6 Fokker–Planck Equation — 371
- 11.7 Kolmogorov Backward Equation — 380
- 11.8 Barriers — 381
- 11.9 Method of Stochastic Differential Equations — 386

Chapter 12. Random Fatigue — 389

- 12.1 Characteristics of Fatigue Failure — 389
- 12.2 Linear Accumulation of Damages — 391
- 12.3 Probabilistic Assessment of Fatigue Life — 392

Chapter 13. Random Vibrations — 399

- 13.1 Vibrations of Elastic Systems — 399
  - 13.1.1 Equations of nonlinear vibrations of rectangular plates — 399
  - 13.1.2 Steady-state vibrations of linear elastic systems — 402
  - 13.1.3 Free vibrations with random initial conditions — 410
- 13.2 Nonlinear Vibrations — 415
  - 13.2.1 Methods for analyzing nonlinear random vibrations — 415
  - 13.2.2 Statistical linearization — 416
  - 13.2.3 Fokker–Planck equation for an elastic system — 422
  - 13.2.4 Fokker–Planck equation for a one-degree-of-freedom nonlinear system — 424
- 13.3 Vibrations Caused by Periodic Impulses. Stochastic Instability — 431
  - 13.3.1 Dynamic response of a one-degree-of-freedom system to a train of instantaneous periodic impulses — 431
  - 13.3.2 Linear vibrations — 432
  - 13.3.3 Nonlinear steady-state undamped vibrations — 436
  - 13.3.4 Stochastic instability — 440
  - 13.3.5 Vibrations of an elongated plate due to periodic impulses — 446

Chapter 14. Geometric Tolerance — 451

- 14.1 Random Geometry — 451
- 14.2 Confidence Intervals — 454

Chapter 15. Random Loads and Responses in Some Engineering Systems — 459

- 15.1 Cars — 459
  - 15.1.1. Loads — 459
  - 15.1.2. Responses — 461

	15.2	Ships		465
		15.2.1.	Loads	465
		15.2.2.	Responses	474
	15.3	Aircrafts		478
		15.3.1.	Loads	478
		15.3.2.	Responses	481
	15.4	Earthquakes		483
		15.4.1.	Loading	483
		15.4.2.	Responses	486

Chapter 16. Processing of Random Data — 491

	16.1	Evaluation of the Moments of Random Variables		491
	16.2	Fiducial Probabilities and Confidence Intervals		494
	16.3	Least-Square Method		499
	16.4	Monte Carlo Simulation		503
	16.5	Statistical Methods of Quality Control		508
		16.5.1.	Single sampling	508
		16.5.2.	Double sampling	510
		16.5.3.	Chain sampling	511
	16.6	Delphi Method		517

Bibliography — 529

Appendices — 535

A	Poisson Distribution Tables	535
B	Error (Laplace) Function Tables	541
C	Gamma Function Table	555
D	χ^2-Distribution Tables	557
E	Student Distribution Tables	567
F	Hermite Polynomils	573
G	Knots and Coefficients for Numerical Integration Using Hermite Polynomials	585

Index — 589

Preface

> *"Education is man's going forward from cocksure ignorance to thoughtful uncertainty."*
> DON CLARK "SCRAPBOOK"

> *"The only real voyage to discovery consists not in seeking new landscapes, but in having new eyes."*
> MARCEL PROUST

> *"All the general theories stem from examination of specific problems."*
> RICHARD COURANT

This book provides an introduction to probabilistic methods used in engineering and some other areas of applied science. Its main objective is to be suitable for young engineers and engineering students of different majors who, with some background in calculus, ordinary and partial differential equations, and general engineering disciplines, would like to gain an initial insight into the substance and use of probabilistic methods. The book describes the major concepts, methods, and approaches of the applied theory of probability and provides numerous examples from various areas of engineering: mechanical, structural, civil, earthquake, industrial, reliability, materials, ocean, aerospace, electronic, telecommunications, etc. Many applied physics problems and everyday life situations are also addressed. It is these examples (God is in details!) which is the book's forte.

Although the overall structure of the book is shaped with a particular audience in mind, it is the author's belief that it will prove of interest to a rather broad audience of engineers and applied scientists who, in one way or another, are encountered with, or are interested in, uncertainties in engineering analyses: practicing and research engineers, technical managers, college professors and students, instructors of various continuing education programs, etc. The book is also well-suited as a guide for self-study. I have tried to provide a combination of a clear, reasonably detailed, palatable, and, to an extent possible, entertaining introductory manual, and a useful reference source on the basics of the Applied Probability. I hope that some chapters will be useful also for experimental physicists, biologists, chemists, and even economists and market analysts. The bibliography at the end of the book, though not meant to be exhaustive, is rather extensive and will be useful for further study.

Ephraim Suhir
IEEE Fellow, ASME Fellow
Lucent Technologies, Bell Laboratories
Physical Science and Engineering Research Division
Murray Hill, New Jersey.

List of Basic Symbols

A	area
a, b, c	constants
$C(n, m)$	combination of n elements taken m at a time
D, d	diameter
$D_x = D(X) = \text{var}(X) = \langle (x - \langle x \rangle)^2 \rangle$	variance of the random variable X
$\sqrt{D_x} = \sigma_x$	standard deviation of the random variable X
E	modulus of elasticity, characteristic function
$E(X) = \langle x \rangle = m_x$	mean (expected) value of the random variable X
$F(t)$	time-to-failure distribution function
$F(x)$	cumulative distribution function of the random variable X
$F(x, y)$	joint cumulative distribution function of the random variables X and Y
$f(t)$	probability density function for the time-to-failure
$f(x) = \dfrac{dF(x)}{dx}$	probability density function of the random variable X
$f(x, y)$	joint density of the random variables X and Y
G	shear modulus (or elasticity)
$G(s), G_x(s)$	generating function, spectrum of the random process $X(t)$
g	acceleration due to gravity
H	hypothesis
h	height
$i = \sqrt{-1}$	"imaginary" unit
J	Jacobian

K_{xy}	correlation function (covariance) of the random variables X and Y
$K_x(\tau)$	(auto) correlation function of the stationary random process $X(t)$
$M(t)$	moment generating function
$P(A)$	probability of the event A
$P(A/B)$	conditional probability of event A given that event B has occurred
$\bar{P}(A) = 1 - P(A)$	complement of the probability of the event A
$P(n, m)$	permutation of n elements taken m at a time
$R(t) = P(t \geq T)$	reliability function
r	damping coefficient, correlation coefficient
$S_x(\omega)$	spectrum of the random process $X(t)$
T	time-to-failure (random variable)
t	time
$\langle t \rangle$	mean time-to-failure (MTTF)
$\langle t_r \rangle$	mean time-between-failures (MTBF)
$v_x = \sqrt{D_x}/\langle x \rangle$	coefficient of variation of the random variable X
X, Y, Z	random variables
x, y, z	actual values (realizations) of the corresponding random variables
α	producer's (seller's) risk (probability that a good lot or item will be rejected)
β	consumer's (buyer's) risk (probability that a bad lot or item will be accepted)
$\delta(t)$	Dirac delta function
λ	failure rate
σ	stress, standard deviation
$\langle \sigma \rangle$	mean (expected) value of stress
σ_f	fatigue limit
$\Phi(x)$	Laplace function
$\Phi(i\omega)$	complex frequency characteristic
$\phi(t)$	characteristic function
Ω	sample space
ω	elementary event, frequency
$\varnothing$	empty set

Applied Probability for Engineers and Scientists

Introduction

> *"We see that the theory of probability is at heart only common sense reduced to calculation; it makes us appreciate with exactitude what reasonable minds feel by a sort of instincts, often without being able to account for it.... The most important questions of life are, for the most part, really only problems of probability."*
> PIERRE SIMON, MARQUIS DE LAPLACE

> *"Coincidences, in general, are great stumbling blocks in the way of that class of thinkers who have been educated to know nothing of the theory of probabilities; that theory to which the most glorious objects of human research are indebted for the most glorious of illustration."*
> EDGAR ALLAN POE,
> *The Murders in the Rue Morgue*

> *"If you bet on a horse, that's gambling. If you bet you can make three spades, that's entertainment. If you bet the structure will survive for a hundred years, that's engineering. See the difference?"*
> UNKNOWN ENGINEER

Engineering products must have a worthwhile lifetime and operate during this time and under stated conditions successfully to match the user's expectations. To achieve this, the engineer must understand the ways in which the useful life-in-service of the product can be evaluated and to incorporate this understanding in the design. His tasks are to analyze, design, test, manufacture, operate, and maintain the product, system, or structure at all the stages of its creation and use, from the moment of the conception and substantiation of the initial idea to the moment of writing off and wrapping the product, so that it does not fail during the service period. When solving these problems, the engineer inevitably encounters variability in the employed materials, loads, manufacturing processes, testing techniques, and applications.

The traditional engineering approaches, when dealing with these problems, are referred to as *deterministic*, i.e., do not pay sufficient attention to the variability of the parameters and criteria used. Such approaches are acceptable and can be justified in many cases, when the deviations ("fluctuations") from the mean values are small, when the design parameters are known or can be predicted with reasonable accuracy, and when the processes and procedures that the engineer deals with are "stochastically stable," i.e., when "small causes" result in "small effects."

There are, however, numerous situations in which the "fluctuations" from the anticipated (mean) values are significant and in which the variability, change, and uncertainty play a vital role. In such situations the product will most likely fail if these uncertainties are ignored. Therefore understanding the role and significance of the "laws of chance," and the causes and effects of variability in materials properties, dimensions, tolerances, bearing clearances, loading conditions, stresses and strength, applications and environments, and a multitude of other design parameters, is critical for the creation and successful operation of a viable and reliable product or structure. In numerous practically important cases, the random nature and various uncertainties in the design characteristics and parameters can be described on the basis of the methods of the *theory of probability*.

Probabilistic methods proceed from the fact that various uncertainties is an inevitable and essential feature of the nature of an engineering system or design and provide ways of dealing with quantities whose values cannot be predicted with absolute certainty. Unlike deterministic methods, probabilistic approaches address more general and more complicated situations, in which the behavior of the given characteristic or parameter cannot be determined with certainty in each particular experiment or a situation, but, for products manufactured in large quantities and for experiments, which are repeated many times in identical conditions, follow certain, "statistical," relationships. These manifest themselves as trends in a large number of random events.

Probabilistic models reflect the actual physics of phenomena and the inevitable variability of the characteristics of an engineering system much better than the deterministic theories. Probabilistic models enable one to establish the scope and the limits of the application of deterministic theories, and provide a solid basis for a substantiated and goal-oriented accumulation, and effective use of empirical data. It would not be an exaggeration to say that all the fundamental theories and approaches of modern physics and engineering are probabilistic, and contain the corresponding deterministic models as first approximations. Realizing the fact that the probability of failure of an engineering product is never zero, the probabilistic methods enable one to quantitatively assess the degree of uncertainty in various factors, which determine the performance of the product, and to design on this basis a product with a low probability of failure. These methods, underlying modern methods of forecasting and decision making, allow one to extend the accumulated experience on new products and designs which differ from the existing ones by type, dimensions, materials, operating conditions, etc.

Probabilistic models are able to account for a substantially larger number of different practically important factors than the deterministic methods. It should be emphasized that the use of the probabilistic methods and approaches is due not so much to the fact that the available information is insufficient for a deterministic analysis, but, first of all, to the fact that the variability and uncertainty are inherent in the very nature of many physical phenomena, materials properties, engineering designs, and application conditions.

At the same time, it should be pointed out that probabilistic methods should not be viewed as a sort of panacea that cures all engineering troubles, and definitely should not be expected to perform miracles. Such methods have their limitations. For instance, they cannot be applied in situations where the conditions of an experiment or a trial are not reproducable or when the events are very rare. Was it possible to predict the outcome of the O. J. Simpson trials? Or was it possible to forecast the explosions in the World Trade Center or in the government building in Oklahoma City? Quite often a serious obstacle for applying probabilistic methods is also the difficulty of obtaining the necessary input information. In such cases the design has to be based on the worst case, i.e., in effect, on a "deterministic" situation, rather than on the application of probabilistic methods, or could be based on a combination of probabilistic and deterministic approaches. However, when the application of probabilistic methods is possible, justified, and is supported by reliable input information, these methods provide a powerful, effective, and well-substantiated means for engineering analyses and designs.

As it is clear from the book's title, it provides an introduction to probabilistic methods used in engineering and applied science. The book is written by an engineer and for engineers, and is not intended to be mathematically rigorous. The book is, first of all, a collection of examples (problems) and solutions. It is a "problem book," and although it provides numerous sets of mathematical tools for attacking problems, presenting well-structural theoretical developments of various mathematical techniques is not among the book's objectives. Such tools are presented, in many cases, in the process of solving a particular problem. Although such a style might strike some pedantic readers as being unreasonable and lacking organization, it has been taken deliberately to keep the discussions descriptive, lively, friendly and interesting. The author realizes also that his judgements on what the reader does and does not know, as far as various mathematical techniques are concerned, are to a great extent subjective, although are based on the author's engineering and teaching experience. I assumed, for instance, that the reader is comfortable with multivariable calculus, partial differential equations and even with some special functions (such as Dirac delta-function or elliptic functions and integrals), but is not familiar with the theory of spectra or the theory of Markovian processes. Finally, the author realizes that some important topics (e.g. martingales) are not addressed in the book at all and that the presentation of others (e.g. queueing) can be made more interesting and consistant.

The author acknowledges, with thanks, the support of Dr. J. W. Mitchell and useful comments made by Dr. A. Weiss and Dr. L. Shepherd. He would also like to thank Mrs. C. Martin for her skill, patience, and care in typing the manuscript.

Chapter 1

Random Events

"The concept of chance enters into the very first steps of scientific activity, by virtue of the fact that no observation is absolutely correct. I think that chance is a more fundamental concept than causality, for whether a cause-effect relationship exists in a concrete case, can only be judged by applying the laws of chance to the observations."
MAX BORN,
Natural Philosophy of
Cause and Chance

"We must believe in luck. For how else can we explain the success of those we don't like?"
JEAN COCTEAU

"If at first you succeed, don't take any more stupid chances."
STREET WISDOM

1.1. Frequency and Probability

The theory of probability studies *random phenomena*. These can be defined as phenomena that do not nearly yield the same outcomes in repeated observations under identical conditions. Any of the possible *outcomes* of an *experiment* (*trial*) or result of an observation is an *event*. An event is the simplest random phenomenon. Two dots appearing on the top face of a die, device failure, loss of structural integrity, buckling of a compressed element, an unacceptable distortion in a message transmitted over a communication channel, or an inability to serve a customer are all examples of events. The substance of the theory of probability is that each event has an associated quantity, which characterizes the *objective likelihood* of the occurrence of this event. This quantity is called the *probability* of the event.

The classical (opposite to the axiomatic) approach to the concept of probability is as follows. If the event will inevitably occur as the result of an experiment it is a *certain event*. An event that is known in advance as not being able to occur as a result of an experiment is an *impossible event*. The overwhelming majority of actual events are neither certain nor impossible, and may or may not occur during an experiment. These are *random events*.

Let us assume that the conditions of the experiment can be repeatedly reproduced, so that, in principle, a series of identical and independent trials can be conducted, and the random event (outcome) A does or does not occur in each of these trials. If the outcome A occurred m times in n equally likely and independent trials, then the ratio

$$P^*(A) = \frac{m}{n} \tag{1.1}$$

is called the *frequency*, or *statistical probability*, of the event. Obviously, $P^*(A) = 1$ for a certain event and $P^*(A) = 0$ for an impossible event. For a random event A, $0 \leq P^*(A) \leq 1$. When the number of experiments is small, the frequency $P^*(A)$ is an unstable quantity. It stabilizes with an increase in the number of trials. Numerous experiments indicate that the observed frequencies $P^*(A)$ in different runs of trials are close to each other and group around a certain value $P(A)$. This value is called the *probability* of the event A.

If an event is expected to occur as a result of numerous repeatedly reproduced experiments, in many cases there is no need to carry out actual experiments to determine the frequency of this event: this frequency can be predicted on the basis of the calculated probability. Such a probability is computed as the ratio of the number m of outcomes *favorable* to the occurrence of the given event to the total number n of trials. The content of the applied probability theory is, in effect, a collection of different ways, methods, and approaches that enable one to evaluate the probabilities of occurrence of complex events from the known probabilities of simple ("elementary") events. The probabilities of elementary events are either intuitively obvious or can be experimentally established. In practice, the elementary events can be often found simply by using common sense, e.g., from the consideration of the symmetry of the experiment (for a symmetric die, for instance, it is natural to assume that the appearance of each face is equipossible). The evaluation of the "elementary" events is the subject of special engineering or physical disciplines.

The following problems can be solved in a rather straightforward way ("direct evaluation of probabilities"), each requiring simple logic and common sense.

Example 1.1 Find the probability of drawing an ace from a well-shuffled deck of 52 cards.

Solution The number of favorable outcomes out of the $n = 52$ equally likely outcomes is $m = 4$. Therefore the sought probability is $P = 4/52 = 1/13$. It is understood that if one were to repeat the process a large number of times, an ace would appear in a number of trials which is 13 times smaller than the total number of trials.

Example 1.2 A box contains four standard elements for replacement. Two of these elements are sound and two are faulty. Two elements are taken at random from the box. What is the probability that by doing this one separates the sound elements from the faulty ones?

Solution Let us number all the elements from 1 to 4. The total number of outcomes when taking two elements out of four elements is equal to six:

$$1, 2; \quad 1, 3; \quad 1, 4; \quad 2, 3; \quad 2, 4; \quad 3, 4$$

The sound elements can be separated from the faulty ones in either of the following two cases: (a) both elements taken at random are sound; (b) both elements taken at random are faulty. Let the elements 1 and 2 be the sound ones and the elements 3 and 4 be the faulty ones. There is only one outcome (1, 2) when both elements are sound and only one outcome (3, 4) when both elements are faulty. Therefore the total number of favorable outcomes is equal to two. The sought probability is therefore $P = 2/6 = 1/3$.

Example 1.3 There are two boxes containing standard elements for replacement. The first box contains a sound elements and b faulty ones, and the second box contains c sound elements and d faulty ones. One element is drawn at random from each box. What is the probability that both elements are sound? What is the probability that both elements are faulty? What is the probability that both elements are different in quality?

Solution Since each element from the first box can be combined with each element from the second box, the total number of cases is

$$n = (a + b)(c + d)$$

The number of cases favorable for the event "both elements are sound" is $m = ac$, and therefore the probability of this event is

$$P = \frac{ac}{(a + b)(c + d)}$$

Similarly, the probability that both elements are faulty can be found as

$$P = \frac{bd}{(a + b)(c + d)}$$

The probability that the two elements are different in quality is

$$P = \frac{ad + bc}{(a + b)(c + d)}$$

Example 1.4 *De Méré's paradox.* Chevalier de Méré noticed that when three dice are cast simultaneously, a total score of 11 points is more likely to occur than a score of 12 points. From his viewpoint, however, both combinations should have the same probability. De Méré's reasoning was as follows. A score of 11 points can be obtained in six ways:

$$6+4+1, \quad 6+3+2, \quad 5+5+1, \quad 5+4+2, \quad 5+3+3, \quad 4+4+3$$

and the same is the number of ways by which 12 points can be obtained:

$$6+5+1, \quad 6+4+2, \quad 6+3+3, \quad 5+5+2, \quad 5+4+3, \quad 4+4+4$$

Since the number of favorable outcomes is the same, the corresponding probabilities have to be equal as well. The actually observed frequencies were, however, different. What was de Méré's mistake?

Solution The mistake of de Méré was explained by Blaise Pascal, a famous French mathematician and physicist and one of the founders of the classical probability theory. He paid attention to the fact that the above elementary outcomes are not equipossible, i.e., they do not have the same probabilities by virtue of the symmetry of the conditions of the experiment relative to these events. Indeed, when evaluating the elementary probabilities, one has to consider not only the number of points but also the dice on which these points appeared. After enumerating the dice and writing down the points in the proper sequence, one can notice that the combination 6, 4, 1 occurs in six outcomes:

$$6, 4, 1; \quad 6, 1, 4; \quad 4, 6, 1; \quad 4, 1, 6; \quad 1, 6, 4; \quad 1, 4, 6$$

while the combination 4, 4, 4 occurs only in one outcome. The equipossible outcomes in this experiment are those described by three numbers (a, b, c), where a is the number of points on the first die, b is the number of points on the second die, and c is the number of points on the third die. The total number of equipossible outcomes is $N = 216$. The number of outcomes favorable for the event "the sum of all points is equal to 11" is $N_1 = 27$, and the number of the outcomes that are favorable for the event "the sum of points is equal to 12" is only $N_2 = 25$. Therefore the probabilities of these events are also different: $P_1 = N_1/N = 0.1250$ and $P_2 = N_2/N = 0.1157$.

Example 1.5 A combination lock contains n disks on a common axis. Each of the disks is divided into m sections carrying different numbers. The lock can be opened only if every disk occupies a certain position with respect to the lock's enclosure. An arbitrary combination of the disk's positions has been set up. What is the probability that the lock will be open?

Solution The sought probability is, obviously, $P = 1/m^n$. If, for instance, $n = 4$ and $m = 10$, then the probability that the lock will be open is $P = 0.0001 = 0.01$ percent.

Example 1.6 A mechanism fails when at least one of its three parts breaks down. The probabilities of failure of these parts are 0.03, 0.04, and 0.06. What is the probability that the mechanism will not fail? How will this probability change if the first part is absolutely reliable?

Solution The sought probability is

$$P = 0.97 \times 0.96 \times 0.94 = 0.8753$$

If the first part is absolutely reliable, then

$$P = 0.96 \times 0.94 = 0.9024$$

Example 1.7 An engine consists of n blocks. The probability of failure of the kth block during the time T is p_k, $k = 1, 2, \ldots, n$. All the n blocks operate independently. What is the probability that at least one of the engine blocks will fail during the time T?

Solution The probability that the kth block will not fail is $q_k = 1 - p_k$. The probability that none of the n blocks will fail during the time T can be found as a product:

$$P_n = \prod_{k=1}^{n} q_k$$

The probability that at least one block will fail is

$$P = 1 - P_n = 1 - \prod_{k=1}^{n} q_k$$

Example 1.8 A batch of 100 machine parts is subjected to random inspection. The entire batch is rejected if there is at least one faulty part in a randomly taken sample of five parts. What is the probability that the batch be rejected, if it contains 5 percent of faulty parts?

Solution Since the total number of parts is 100, and 5 of them are faulty, the probability that the first taken-at-random part is sound is 95/100. If this event has really happened, only 94 sound parts will remain in the batch of 99 parts, and therefore the probability that the second taken-at-random part is sound is 94/99. Similarly, the probabilities that the third, the fourth, and the fifth parts are sound are 93/98, 92/97, and 91/96, respectively. Then the probability that all five parts are sound is

$$P_s = \frac{95}{100} \frac{94}{99} \frac{93}{98} \frac{92}{97} \frac{91}{96} = 0.7696$$

The probability that at least one part of the sample will be faulty (i.e., the probability that the batch is rejected) can be computed as

$$P = 1 - P_s = 0.2304$$

This example deals, in effect, with "conditional probabilities." These will be introduced and discussed in detail in Sec. 1.3.

Example 1.9 A heavy nondeformable beam is fastened to a ceiling by means of two parallel brittle rods. The weight G of the beam is distributed evenly between the rods if none of them failed, and is transmitted completely to one of the rods if the other one failed. Determine the probability of non-failure and the probability of failure of the entire structure.

Solution The structure will fail if one of the rods is unable to withstand a load equal to $G/2$, and the other rod will turn out to be unable to withstand a load equal to G. Let the probabilities of these events be $P_{1/2}$ and P_1, respectively. The structure will not fail (a) if none of the rods will fail under the action of the load $G/2$ [the probability of this event is $(1 - P_{1/2})^2$], (b) if the first rod will fail under the action of the load $G/2$ and the second rod will not fail under the action of the load G [the probability of this event is $P_{1/2}(1 - P_1)$], or (c) if the second rod will fail under the action of the load $G/2$ and the first rod will not fail under the action of the load G [the probability of this event is also $P_{1/2}(1 - P_1)$]. Thus, the probability of the nonfailure of the structure can be calculated as

$$P = (1 - P_{1/2})^2 + 2P_{1/2}(1 - P_1) = 1 + P_{1/2}^2 - 2P_{1/2}P_1$$

The probability of failure of the entire structure can be computed as

$$Q = 1 - P = 2P_{1/2}P_1 - P_{1/2}^2$$

Let, for instance, the probability that the rod will fail under the action of the load G be

$$P_1 = 1 - e^{-G^2/G_0^2}$$

where G_0 is the most likely value of this load. Then we have

$$P_{1/2} = 1 - e^{-G^2/4G_0^2} = 1 - (1 - P_1)^{1/4}$$

and the probability of the nonfailure of the structure is

$$P = 2(1 - P_1)[1 - (1 - P_1)^{1/4}] + (1 - P_1)^{1/2}$$

The probability of failure can be found as

$$Q = 1 - P = [1 - (1 - P_1)^{1/4}][2P_1 + (1 - P_1)^{1/4} - 1]$$

Clearly, if $P_1 = 1$, then $P = 0$ and $Q = 1$, and if $P_1 = 0$, then $P = 1$ and $Q = 0$. If $P_1 = 1/2$, then $P = 0.8662$ and $Q = 0.1338$.

Example 1.10 The probability that event A will occur during the time interval T is equal to P. The event is equally likely to occur at any moment of time within the time T. It is known that this event did not occur during the time $t < T$. What is the probability P^* that it will occur during the remaining time $T - t$?

Solution The probability P that event A will occur during the time T can be represented as the sum of the probability $(t/T)P$ that the event occurs during the time t and the probability P_1 that this event will occur during the remaining time $T - t$, provided that it did not occur during the time t. The probability of the latter event is $1 - (t/T)P$, and therefore

$$P_1 = \left(1 - \frac{t}{T}P\right)P^*$$

where P^* is the sought probability that the event will occur during the time $T - t$. Thus, we have the following equation:

$$P = \frac{t}{T}P + \left(1 - \frac{t}{T}P\right)P^*$$

Solving this equation for the probability P^*, we obtain

$$P^* = P\frac{1 - t/T}{1 - tP/T} \tag{1.2}$$

Let, for instance, the dependability of a device be such that specified probability of its failure during the time of $T = 5$ years is $P = 0.01$. If the device was successfully operated during the time of $t = 4$ years, then the probability that the failure will occur during the remaining year is $P^* = 0.002016$. The probability of the failure-free operation of the device during this year is $P_s = 1 - P^* = 0.998$.

Example 1.11 *"Gambler's ruin"*. A gambler at the casino starts with m dollars and gambles with his bank manager (croupier) on the following conditions. The gambler (or the bank manager) tosses a coin repeatedly: if the coin comes up "heads," then the manager pays him one dollar, but if it comes up "tails," then the gambler pays the manager one dollar. The game is played until one of two events occurs: the gambler runs out of money (becomes bankrupt) or he achieves his goal by winning n dollars and stops the game. What is the probability that the gambler will ultimately become bankrupt?

Solution This problem was solved by Abraham de Moivre, one of the founders of the classical theory of probability. Examine first a more general problem, when the probability that the gembler wins a single trial is p and the probability that he loses it is $q = 1 - p$. Let the gambler be in the state of holding k dollars at the beginning of the game. After the next trial, if he wins, he will hold $k + 1$ dollars, but if he loses the trial, he will hold only $k - 1$ dollars. This situation can be described by the following difference equation for the probability Q_k of eventually losing the game:

$$Q_k = pQ_{k+1} + qQ_{k-1} \tag{1.3}$$

This equation has the following solution:

$$Q_k = C_1 + C_2\left(\frac{q}{p}\right)^k$$

One can make sure that this is indeed the solution to Eq. (1.3) by introducing it into this equation and considering the fact that $p + q = 1$.

The constants of integration C_1 and C_2 can be found from the following boundary conditions:

$$Q_0 = 1, \quad Q_{m+n} = 0$$

The first condition indicates that if the gambler has nothing at the beginning of the game, the probability of winning the game is zero and the probability of losing it is equal to unity. The second condition states that if at the start the gambler has all the capital, both the one he put at stake and the one he hopes to win, he has already won the game, and the probability of losing the game is zero. With these boundary conditions we obtain:

$$C_1 + C_2 = 1, \qquad C_1 + C_2 \left(\frac{q}{p}\right)^{m+n} = 0$$

so that

$$C_1 = -\frac{(q/p)^{m+n}}{1 - (q/p)^{m+n}}, \qquad C_2 = \frac{1}{1 - (q/p)^{m+n}}$$

and the probability Q_k of eventually losing the game with the stake of k dollars is

$$Q_k = \frac{(q/p)^k - (q/p)^{m+n}}{1 - (q/p)^{m+n}} \tag{1.4}$$

At the start of the game, $k = m$, so that the probability of losing the game with the initial capital m is

$$Q_m = \frac{(q/p)^m - (q/p)^{m+n}}{1 - (q/p)^{m+n}} = \left(\frac{q}{p}\right)^m \frac{1 - (q/p)^n}{1 - (q/p)^{m+n}} \tag{1.5}$$

The probability of winning the game can be found as the complement

$$P_m = 1 - Q_m = \frac{1 - (q/p)^m}{1 - (q/p)^{m+n}} \tag{1.6}$$

When the bank manager's resources are large compared to the resources of the gambler (n is significantly larger than m), we obtain

$$Q_m = \left(\frac{q}{p}\right)^m$$

In the case $p = q = 1/2$ of equal chances to win or to lose a single trial, the gambler will inevitably become bankrupt. This result can also be obtained in a different way. When $p = q = 1/2$, the formulas (1.5) and (1.6), using L'Hôpital's rule, yield

$$P_m = \frac{m}{m+n}, \qquad Q_m = \frac{n}{m+n}$$

If n is large, ultimate bankruptcy becomes very likely.

The examined problem can be also formulated as a problem of gambling between two players, one of which starts his game with m dollars and the other with n dollars. When $m = n$ and $p = q = 1/2$, the chances to win or to lose are the same for both gamblers: $P = Q = 1/2$. However, if the initial resources of one of the gamblers are twice as large as the initial resources of the other ($n = 2m$), then the formulas (1.7) yield $P = 1/3$, $Q = 2/3$, i.e., the chances to lose are twice as lage for the gambler with half the initial resources of the other gambler.

Let us assume now that the probability for one of the gamblers to win a single trial is twice as large as his chance to lose it ($p = 2q = 2/3$, $q = 1/3$), while his initial resources are only half the resources of the other gambler ($n = 2m$). In such a case, formula (1.6) yields

$$P_m = \frac{1 - (1/2)^m}{1 - (1/2)^{3m}} = \frac{1}{1 + (1/2)^m + (1/2)^{2m}}$$

If, for instance, $m = 1$, then $P_1 = 4/7$. This probability is larger than 1/2. Hence, it is better to have a two times higher probability to win a trial than to possess a two times higher initial resource. The above formula indicates also that the probability to win increases with an increase in the initial resource m, although the resource n of the second gambler will increase as well.

From the previous analysis it is clear that if the initial resource n is significantly larger than the resource m, for a "fair" game, with $P_m = Q_m = 1/2$, the probability p to win a single trial must be larger than the probability q to lose it. From formula (1.5), putting $Q_m = 1/2$ and $n \to \infty$, we find that the ratio of these probabilities is expressed as $p/q = 2^{1/m}$. For $m = 1$, the probability p is twice as large as the probability q, so that $p = 2/3$ and $q = 1/3$. When m increases, the two probabilities become closer.

The examined problem can be used in modeling various practical situations, such as, for instance, the number of customers lining up during the given time intervals (say, the number of cars at a toll booth), the Dow–Jones index value each Monday morning, etc. Although such situations are not exactly reflected by the conditions in the gambler's ruin problem, it may not be too bad a guess that the current reading differs from the previous one by a random quantity that is independent of the previous jumps but has the same probability distribution. These kinds of situations are often referred to as *simple random walk*—the motion of a particle that makes at each discrete moment of time either one step to the right with the probability p or one step to the left with the probability $q = 1 - p$. The state of this system (*random walk with absorbing barriers*) is described by the difference equation (1.3), and the probabilities Q_m and P_m are, in this case, the probabilities that a particle, initially located at point m, will be eventually absorbed at the origin ($k = m$) or at the boundary $k = n$, respectively. Such systems contain the *Markov property* (see Chapter 11): conditional upon knowing the value of the function (process) at the kth step, the value of the function (process) after this step is independent from the previous steps. In other words, conditional upon the present, the future of the process is independent from its past.

Example 1.12 A drunkard is standing at the distance of one step from the edge of a cliff. The probability that he makes a step towards this edge is 1/3 and the probability that he makes a step in the opposite direction is 2/3. What are the drunkard's chances to survive after a large number of steps is made?

Solution The drunkard's chances to survive can be calculated by formula (1.6), assuming $p = 2/3$, $q = 1/3$, $m = 1$, and $n \to \infty$:

$$P = 1 - \left(\frac{q}{p}\right)^m = 1 - \frac{1}{2} = \frac{1}{2}$$

Another approach to this problem is as follows. The probability to fall down the cliff after the first step is $P_1 = 1/3$. After the third step $(1 \to 2 \to 1 \to 0)$ this probability is

$$P_3 = \frac{1}{3} + \frac{2}{3} \times \frac{1}{3} \times \frac{1}{3} = \frac{11}{27} = 0.4074$$

After five steps $(1 \to 2 \to 1 \to 2 \to 1 \to 0$ or $1 \to 2 \to 3 \to 2 \to 1 \to 0)$ the probability to fall down the cliff is

$$P_5 = \frac{1}{3} + \frac{2}{3} \times \frac{1}{3} \times \frac{2}{3} \times \frac{1}{3} \times \frac{1}{3} + \frac{2}{3} \times \frac{2}{3} \times \frac{1}{3} \times \frac{1}{3} \times \frac{1}{3} = \frac{107}{243} = 0.4403$$

and so on. As one can see, the sought probability tends to 1/2. The schematics of random motions of the drunk can be seen from Table 1.

Example 1.13 *"Choosy bride"*. Imagine a choosy bride who either accepts a proposal—and in this case the process of selecting the best match is over—or rejects it, and then the groom is irretrievably lost. The bride follows a simple rule: not to accept a proposal from a groom who is worse than his predecessors. Obviously, there is a risk involved each time when a proposal is made: the bride can reject the absolutely best proposal in the hope of receiving a better one later. The question that the bride asks herself is: what is the probability that the kth groom is the best out of the entire "set" of m grooms, both those having been rejected and the future ones? Although we describe the problem that a choosy bride deals with, a business executive, a job hunter, or a gambler encounter with a similar situation. It is an example of a decision-making problem, in which one has to determine whether the extra information justifies the expense of obtaining it.

Solution We will provide, without derivation, an optimal strategy the bride should choose. She should turn down the proposals of the first k bride-

TABLE 1.1 The probabilities to fall down the cliff after each step

	Distance from the edge (steps)						
No. of steps	0	1	2	3	4	5	6
0		1					
1	1/3		2/3				
2		2/9		4/9			
3	2/27		8/27		8/27		
4		8/81		24/81		16/81	
5	8/243		40/243		64/243		32/243

grooms and then accept the proposal of the first bridegroom who is better than the previous ones. The number k could be determined from the double inequality

$$\frac{1}{k+1}+\frac{1}{k+2}+\cdots+\frac{1}{m-1}\leq 1<\frac{1}{k}+\frac{1}{k+1}+\cdots+\frac{1}{m-1}$$

With such a strategy, the probability that the best bridegroom is chosen out of m candidates can be calculated as

$$p=\frac{k}{m}\left(\frac{1}{k}+\frac{1}{k+1}+\cdots+\frac{1}{m-1}\right)$$

If, for instance, the bride can choose the best match out of 10 bridegrooms, then the above inequality yields $k = 3$. Hence, the choosy bride should reject the proposals of the first three bridegrooms and then accept the proposal of the first bridegroom who will turn out better than all his predecessors. The probability that the bride will select the best party is

$$p=\frac{3}{10}\left(\frac{1}{3}+\frac{1}{4}+\cdots+\frac{1}{9}\right)\simeq 0.399$$

If the number m is large ($m \to \infty$), the probability that the best bridegroom will be chosen can be found as $p \simeq 1/e = 0.368$.

Another, more serious, example is as follows. During a job fair, a manager has to hire two people out of 10 candidates. The manager has the first priority to choose. If he decides not to hire a certain candidate, the candidate will go to another manager and will never come back. How should the right strategy be chosen? This problem is somewhat more complicated than the bride's problem, since two, not one, best candidates are to be selected.

1.2. Axioms of the Probability Theory

Modern mathematical probability theory, unlike the classical one, is based on an axiomatic approach. This was first introduced by a Russian mathematician, A. N. Kolmogorov, and uses the concepts of *set theory*.

A *set* is a collection of objects called *elements* of the set. Depending on how many elements a set has, it may be *finite* or *infinite*. An infinite set is *countable* if all of its items can be enumerated; otherwise it is *uncountable*. Two sets, A and B, *coincide* (or are equivalent) if they consist of the same elements. The coincidence of sets is expressed as $A = B$. The notation $a \in A$ means that an object a is an element of a set A, or, in other words, a *belongs* to A. The notation $a \notin A$ means that an object a is not an element of a set A. An *empty set* is a set with no elements. It is designated as $\varnothing$. Clearly, all empty sets are equivalent. A set B is said to be a *subset* of set A if all the elements of B also

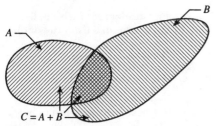

Figure 1.1 The union $C = A + B$ of the sets A and B.

belong to A. The notation is $B \subseteq A$ (or $A \supseteq B$). An empty set is a subset of any set A: $\emptyset \subseteq A$. The *union* (logical sum) of the sets A and B is a set $C = A + B$. A union of two sets is a collection of elements belonging to *at least one* of them. The union of two sets A and B is shown in Fig. 1.1. This is a *Venn diagram*—a symbolic representation of the subsets of a sample space. The name is given after an English mathematician John Venn (1834–1923). The shaded area in Fig. 1.1 is $A + B$. One can similarly define a union of any number of sets. The *intersection* (logical product) of two sets A and B is the set $D = AB$, which consists of the elements that belong to both A and B. An intersection of two sets A and B is shown in Fig. 1.2. One can similarly define an intersection of any number of sets. Two sets A and B are said to be *disjoint* (nonintersecting) if their intersection is an empty set: $AB = \emptyset$, i.e., no element belongs to both A and B.

In modern probability theory it is considered that an *elementary event* ω belongs to the *sample space* Ω: $\omega \in \Omega$. A sample space is a set of all possible outcomes of an experiment. Any subset of the set Ω is an event (or a random event), and any event A is a subset of the set Ω: $A \subseteq \Omega$. Let, for instance, an experiment involve casting a pentahedral die. Then the space of elementary events is $\Omega = \{1, 2, 3, 4, 5\}$. If an event A is an occurrence of an even score, then $A = \{2, 4\}$: $A \subseteq \Omega$. If a coin is tossed and the two possible outcomes are "heads" (H) and "tails" (T), then the sample space is $\Omega = \{H, T\}$. Two mutually dis-

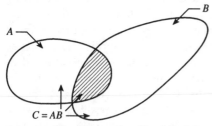

Figure 1.2 The intersection $C = AB$ of the sets A and B.

joint events A and B (such that $AB = \varnothing$) are said to be *incompatible*. The occurrence of one of the incompatible events precludes the occurrence of the other. Several events $A_1, A_2, \ldots, A_n$ are said to be *pairwise incompatible* (or simply incompatible), or *two-by-two mutually exclusive events*, if the occurrence of one of them precludes the occurrence of each of the others. Several events $A_1, A_2, \ldots, A_n$ form a *complete group*, if $\sum_{i=1}^{n} A_i = \Omega$, i.e., if their sum is a certain event (in other words, if at least one of these events will certainly occur as a result of the experiment). Let, for instance, an experiment consist of tossing a die. Then the events $A = \{1, 2\}$, $B = \{2, 3, 4\}$, and $C = \{4, 5, 6\}$ form a complete group: $A + B + C = \{1, 2, 3, 4, 5, 6\} = \Omega$.

Every event A is associated with a number $P(A)$ which is called the *probability* of this event. In modern probability theory, the probabilities should satisfy the following three basic axioms:

1. The probability of an event A is a number that always falls between zero and unity:

$$0 \leq P(A) \leq 1$$

2. If the events A and B are mutually exclusive, then the following *probability addition rule* takes place:

$$P(A + B) = P(A) + P(B)$$

For a finite number n of events, this axiom can be generalized as follows:

$$P\left(\sum_{i=1}^{n} A_i\right) = \sum_{i=1}^{n} P(A_i)$$

3. For an infinite sequence of mutually exclusive events $A_1, A_2, \ldots, A_n$, the following probability addition rule takes place:

$$P\left(\sum_{i=1}^{\infty} A_i\right) = \sum_{i=1}^{\infty} P(A_i)$$

These axioms can be used to calculate the probability of complex events (the subsets of Ω) from the probabilities of a finite or countable number of elementary events.

Several events $A_1, A_2, \ldots, A_n$ are said to be *equipossible* if they have the same probabilities by virtue of the symmetry of the conditions of the experiment relative to these events: $P(A_1) = P(A_2) = \cdots = P(A_n)$. If, in an experiment, one can represent the sample space Ω as a complete group of disjoint (independent) and equipossible events $\omega_1, \omega_2, \ldots, \omega_n$, then the events are called *cases* (*chances*) and the experiment is said to reduce to the *urn model*. If this is the case, the calculation of the probabilities can be carried out in a straightforward fashion (see Sec. 2.2.1). A case ω_i is said to be *favorable* to an event A if it is an element of a set $A : \omega_i \in A$.

Because the cases $\omega_1, \omega_2, \ldots, \omega_n$ form a complete group of events, then

$$\sum_{i=1}^{n} \omega_i = \Omega$$

Since the elementary events $\omega_1, \omega_2, \ldots, \omega_n$ are incompatible, it follows, from the probability addition rule, that

$$P\left(\sum_{i=1}^{n} \omega_i\right) = P(\Omega) = \sum_{i=1}^{n} P(\omega_i) = 1$$

Because the elementary events $\omega_1, \omega_2, \ldots, \omega_n$ are equipossible, their probabilities are the same and are equal to $1/n$:

$$P(\omega_1) = P(\omega_2) = \cdots = P(\omega_n) = \frac{1}{n}$$

This leads to the *classical formula* for the probability of a random event (see Sec. 1.1): if an experiment reduces to an urn model, then the probability of the event A in this experiment can be calculated as

$$P(A) = \frac{m}{n} \tag{1.7}$$

where m is the number of cases favorable to the event A and n is the total number of cases. Thus, the formula that is accepted as the definition of probability in the "classical" theory is just a corollary of the probability addition rule when the modern axiomatic approach is used.

Example 1.14 Three white and four black balls are thoroughly stirred in an urn and a ball is drawn from the urn at random. Construct the sample space and find the probability of the event A that "a white ball is drawn."

Solution Let us label the balls 1 to 7 inclusive. The first three balls are white and the last four are black. Hence, $\Omega = \{1, 2, 3, 4, 5, 6, 7\}$ and $A = \{1, 2, 3\}$. Since the experimental conditions are symmetric with respect to all the balls (a ball is drawn at random), the elementary events are equipossible. Since all the events are incompatible and form a complete group, the probability of the event A can be found as $P(A) = 3/7$, as suggested by formula (1.8). When the experiment is symmetric with respect to the possibility of the outcomes, formula (1.8) makes it possible to calculate the probabilities of events directly from the conditions of the experiment.

1.3. Algebra of Events

The *algebra of events* enables one to express an event in terms of other events. Here are the major rules of the algebra of events:

1. The sum $A + B$ of two events A and B is an event C consisting of the occurrence of at least one of these events. Similarly, the sum of several events $A_1, A_2, \ldots, A_n$ is the event $B = \sum_{i=1}^{n} A_i$ consisting of the occurrence of at least one of them. One can also form a sum of an infinite (countable) number of events $A_1, A_2, \ldots, A_n, \ldots$ as $B = \sum_{i=1}^{\infty} A_i$.
2. The product AB of two events A and B is the event C, consisting of a simultaneous realization of both events. The product of several events $A_1, A_2, \ldots, A_n$ is an event $B = \prod_{i=1}^{n} A_i$ consisting of a simultaneous realization of all the events. One can also form a product of an infinite (countable) number of events: $B = \prod_{i=1}^{\infty} A_i$.
3. It follows from the definitions of the sum and the product of events that

$$A + A = A, \quad A + \Omega = \Omega, \quad A + \emptyset = A,$$

$$AA = A, \quad A\Omega = A, \quad A\emptyset = \emptyset$$

If $A \subseteq B$, then $A + B = B$ and $AB = A$.

4. The operations of addition and multiplication of events possess the following properties:
   (a) Commutativity: $A + B = B + A$ and $AB = BA$;
   (b) Associativity: $(A + B) + C = A + (B + C)$ and $(AB)C = A(BC)$;
   (c) Distributivity: $A(B + C) = AB + AC$.
5. An "*opposite*," or *complementary*, event of A is an event $\bar{A}$ of non-occurrence of the event A. A domain $\bar{A}$ is the complement of the domain A with respect to the complete space Ω. As follows from the definition of the complementary event,

$$\bar{\bar{A}} = A, \quad \bar{\Omega} = \emptyset, \quad \bar{\emptyset} = \Omega$$

It is easy to make sure that if $B \subseteq A$, then $\bar{A} \subseteq \bar{B}$. The probabilities of compound events can be calculated from the probabilities of simpler events. This can be done of the basis on the following rules of the probability theory (rules of addition and multiplication of events):

1. The probability of the sum of two mutually exclusive events is equal to the sum of the probabilities of these events, i.e., if

$AB = \varnothing$, then

$$P(A + B) = P(A) + P(B)$$

This rule is, in effect, one of the axioms of the theory of probability. It can be easily generalized for an arbitrary number n of incompatible events: if $A_i A_j = \varnothing$ for $i \neq j$, then

$$P\left(\sum_{i=1}^{n} A_i\right) = \sum_{i=1}^{n} P(A_i)$$

This rule can be generalized also to the case of an infinite (countable) number of events: if $A_i A_j = \varnothing$ for $i \neq j$, then

$$P\left(\sum_{i=1}^{\infty} A_i\right) = \sum_{i=1}^{\infty} P(A_i)$$

As follows from the rule of the addition of probabilities, the sum of the probabilities of disjoint events $A_1, A_2, \ldots, A_n$ that form a complete group is equal to unity. In other words, if

$$\sum_{i=1}^{n} A_i = \Omega, \quad A_i A_j = \varnothing \text{ for } i \neq j$$

then

$$\sum_{i=1}^{n} P(A_i) = 1$$

In particular, since two opposite events A and $\bar{A}$ are mutually exclusive and form a complete group, the sum of their probabilities is equal to unity:

$$P(A) + P(\bar{A}) = 1$$

2. Let an experiment be repeated n times; on each occasion one observed the occurrences or nonoccurrences of two events A and B. Now, suppose that we only take an interest in those outcomes for which the event B occurs and disregard all the other experiments. The *conditional probability* of an event A, designated as $P(A/B)$, is defined as the probability of this event, calculated on the hypothesis that the event B has occurred.

With the concept of the conditional probability, the multiplication rule for probabilities can be formulated as follows. The probability of the product of two events A and B is equal to the probability of one of them multiplied by the conditional probability of the other, provided that the first event has occurred:

$$P(AB) = P(A)P(B/A) = P(B)P(A/B)$$

In the case of three events A, B, C, one can write this formula as

$$P(AH) = P(H)P(A|H)$$

and put first $H = BC$ and then apply the formula for the probability of two events again. Then one obtains

$$P(ABC) = P(C)P(B|C)P(A|BC)$$

This rule can be generalized for an arbitrary number of events as follows:

$$P(A_1 A_2 \cdots A_n) = P(A_1)P(A_2/A_1)P(A_3/A_1 A_2) \cdots P(A_n/A_1 A_2 \cdots A_{n-1})$$

Based on the formula for the probability of the product of two events, the conditional probability can be evaluated as

$$P(A/B) = \frac{P(AB)}{P(B)}$$

This relationship states that the conditional probability $P(A/B)$ of the events, on the hypothesis that the event B occurs, is equal to the probability $P(AB)$ of the product AB of the two event A and B divided by the probability $P(B)$ of the event B which is assumed to be realized.

Two events A and B are *independent* if the occurrence of one of them does not affect the probability of the occurrence of the other:

$$P(A/B) = P(A), \quad P(B/A) = P(B)$$

For two independent events A and B,

$$P(AB) = P(A)P(B)$$

For several independent events $A_1, A_2, \ldots, A_n$,

$$P(A_1 A_2 \cdots A_n) = P(A_1)P(A_2) \cdots P(A_n)$$

The following rule is a corollary of the rules of addition and multiplication for probabilities: if n independent trials are performed, and in each of these trials an event A occurs with the probability p, then the probability P_m of the event A occurring m times in this experiment is expressed by the formula

$$P_m = C(n, m)p^m q^{n-m}, \quad q = 1 - p$$

Here $C(n, m)$ is the number of combinations of n elements taken m at a time (see Sec. 1.4). The rule established by the above formula is known as *binomial distribution*. This important distribution will be addressed in detail in Sec. 2.2.1. If the binomial distribution is used, the probability of the event A occurring not less than m times in a series of n independent trials can be computed by the formula

$$R = \sum_{k=m}^{n} C(n, k) p^k q^{n-k}$$

or by the formula

$$R = 1 - \sum_{k=0}^{m-1} C(n, k) p^k q^{n-k}$$

Example 1.15 Solve the problem in Example 1.2, using the methods of the algebra of events.

Solution The sound elements will be separated from the faulty ones if one of the following two events takes place: (a) both elements taken at random are sound or (b) both elements taken at random are faulty. The probability that the first element taken at random is a sound one is equal to $2/4 = 1/2$. The probability that the second element is also a sound one is $1/3$. Then the probability that both elements are sound ones is $1/2 \times 1/3 = 1/6$. The probability that both elements taken at random are faulty ones is also $1/6$. Therefore the sought probability that the sound elements will be separated from the faulty elements by taking two elements out of four at random is $1/6 + 1/6 = 2/6 = 1/3$.

Example 1.16 Prove that if two events A and B are compatible, then

$$P(A + B) = P(A) + P(B) - P(AB)$$

Solution The event $A + B$ can be represented as a sum of three incompatible events: $A\bar{B}$ (A and not B), $\bar{A}B$ (B and not A), and AB (both A and B):

$$A + B = A\bar{B} + \bar{A}B + AB$$

On the other hand, the events A and B can be represented as

$$A = AB + A\bar{B}, \quad B = \bar{A}B + AB$$

In accordance with the addition rule for probabilities, we have:

$$P(A + B) = P(A\bar{B}) + P(\bar{A}B) + P(AB)$$

$$P(A) = P(A\bar{B}) + P(AB)$$

$$P(B) = P(\bar{A}B) + P(AB)$$

Adding the two last expressions together, we obtain

$$P(A) + P(B) = P(A\bar{B}) + P(\bar{A}B) + 2P(AB)$$

Subtracting the above expression for the probability $P(A + B)$ of the sum of the events A and B from this equation we have

$$P(A) + P(B) = P(A + B) + P(AB)$$

or

$$P(A + B) = P(A) + P(B) - P(AB)$$

This result can be easily obtained directly from the Venn diagram in Fig. 1.1. Similarly, for three joint events,

$$P(A + B + C) = P(A) + P(B) + P(C) - P(AB) - P(AC) - P(BC) + P(ABC)$$

This result can also be obtained from a Venn diagram for the events A, B, and C.

For any number of joint events, A_k, $k = 1, 2, \ldots, R$,

$$P\left(\sum_{k=1}^{n} A_k\right) = \sum_{k=1}^{n} P(A_k) - \sum_{k=1}^{n-1} \sum_{j=k+1}^{n} P(A_k A_j)$$
$$+ \sum_{k=1}^{n-2} \sum_{j=k+1}^{n} P(A_k A_j A_i - \cdots + (-1)^{n-1} P\left(\prod_{k=1}^{n} A_k\right) \quad (1.8)$$

For statistically independant events,

$$P\left(\sum_{k=1}^{n} A_k\right) = \sum_{k=1}^{n} P(A_k)$$

Example 1.17 The probability that a target is detected during one surveillance cycle of a radar unit is equal to p. Assuming that the detection of the target in each cycle is independent of other cycles, find the probability P that the target will be detected after n cycles of surveillance.

Solution The probability that the target will not be detected in one cycle is $q = 1 - p$. Then the probability that the target will not be detected in n cycles is q^n. Hence, the sought probability is

$$P = 1 - q^n$$

Example 1.18 There is a group of k targets, each of which can be detected by a radar unit, independently of the other targets, with the probability p. Each of m radar units tracks the targets independently of the other units. Find the probability that not all the targets in the group will be detected.

Solution The probability that a radar unit will miss a target is $q = 1 - p$. The probability that all the m radar units will miss a target is q^m. Then the probability that a target will be detected by any of the radar units is $1 - q^m$. Hence, the probability that all the targets will be detected is $(1 - q^m)^k$. Thus, the probability that not all the targets will be detected is $1 - (1 - q^m)^k$.

The situation examined in this problem is quite typical and is applicable to many other problems, encountered in engineering practice. Let, for instance, m possible failure modes considered in a reliability analysis of an electronic device are characterized by probabilities p_i, i, 2, ..., m. The probability of nonoccurrance of the i-th mode of failure is $q_i = 1 - p_i$. The probability that none of the modes of failure will take place (the probability of nonfailure of the device) is $\sum_{i=1}^{m} q_i$. The probability of failure of the device is therefore

$$P_f = 1 - \sum_{i=1}^{m} q_i$$

If, for instance, $p_i = 0.002$, $p_2 = 0.003$, $p_3 = 0.001$, $p_4 = 0.0008$, then $q_1 = 0.998$, $q_2 = 0.997$, $q_3 = 0.999$, $q_4 = 0.9992$, so that $P_f = 0.006784$.

Example 1.19 The probability that the engine gets started when the ignition is turned on is equal to p. Find: (a) the probabilty P that the engine gets started when the ignition is turned on for the second time and (b) the probability P_1 that it is necessary to turn the ignition on not more than two times to get the engine started.

Solution The probability that the engine gets started when the ignition is turned on for the second time can be evaluated as the product of the probability $q = 1 - p$ of the event "the engine did not get started for the first time" and the probability p that the engine did get started as a result of the second turning of the ignition key:

$$P = (1 - q)p$$

In order to determine the second probability, we determine first the probability that the engine did not pick up neither after the first attempt nor after the second attempt. This probability can be calculated as q^2. Then the probability of the opposite event, i.e. the probability that not more than two attempts will be needed to get the engine started, can be evaluated as

$$P_1 = 1 - q^2 = (2 - p)p$$

Example 1.20 Let $H_1, H_2, ..., H_n$ be a set of mutually exclusive events. One of these events will definitely occur. One is interested in a certain event A, and it is easier to calculate the conditional probability $P(A | H_j)$, when the event A takes place in combination with an event H_j, than to calculate the probability $P(A)$ of the event A of interest directly. Derive a formula for the evaluation of the probability $P(A)$ for the available probabilities $P(A | H_j)$.

Solution Since all the events $H_1, H_2 ..., H_j, ..., H_n$ are mutually exclusive, the events AH_j are mutually exclusive as well. Therefore the total probability of the events $AH_1, AH_2, ..., AH_j, ..., AH_n$ can be obtained by adding the probabilities of the particular events. Putting $H = H_j$ in the formula

$$P(AH) = P(H)P(A | H)$$

and forming a sum with respect to the indices j, we obtain

$$P(A) = \sum_{i=1}^{n} P(H_i)P(A | H_i)$$

This formula is known as the *total probability formula*. It can be useful when the assessment of the conditional probabilities $P(A\,|\,H_i)$ is easier than the direct evaluation of the probability $P(A)$.

Example 1.21 The probability that the first weather forecaster will predict the weather correctly is p_1 and the probability that the second forecaster will predict the weather correctly is p_2. The first forecaster predicted fair weather for the weekend and the second one predicted bad weather. What is the probability that the first forecaster is right?

Solution Let A be the event "the first forecaster is right" and B be the event "the second forecaster is right," so that $P(A) = p_1$ and $P(B) = p_2$. Since the two forecasters made different predictions, the event $A\bar{B} + \bar{A}B$ took place. The probability of this event can be computed as

$$P(A\bar{B} + \bar{A}B) = P(A\bar{B}) + P(\bar{A}B) = P(A)P(\bar{B}) + P(\bar{A})P(B)$$
$$= p_1(1 - p_2) + (1 - p_1)p_2$$

The weather will be fair if the event $A\bar{B}$ takes place. The probability of this event can be found as

$$P(A\bar{B}) = \frac{p_1(1 - p_2)}{p_1(1 - p_2) + (1 - p_1)p_2}$$

If, for instance, $p_1 = p_2 = 0.5$, then $P(A\bar{B}) = 0.5$ as well.

Example 1.22 The probability of receiving k telephone calls during the time T is equal to $P_T(k)$, $k = 0, 1, 2, \ldots$. What is the probability that s telephone calls will arrive during a time interval of duration $2T$?

Solution Let A_τ^k be the event "k calls arrive during the time τ." The event A_{2T}^s can be found as the sum of $s + 1$ mutually exclusive events "i calls, $i = 0, 1, 2, \ldots, s$, arrive during the time T" and "$s - i$ calls arrive during the next time interval of the same duration":

$$A_{2T}^s = A_T^0 A_T^s + A_T^1 A_T^{s-1} + \cdots + A_T^s A_T^0$$

Using the rule of addition of probabilities, we obtain

$$P(A_{2T}^s) = \sum_{i=0}^{s} P(A_T^i A_T^{s-i})$$

On the other hand, the rule of multiplication of probabilities of independent events yields

$$P(A_T^i A_T^{s-i}) = P(A_T^i)P(A_T^{s-i}) - P_T(i)P_T(s - i)$$

If we put

$$P_{2T}(s) = P(A_{2T}^s)$$

then

$$P_{2T}(s) = \sum_{i=0}^{s} P_T(i)P_T(s - i)$$

Later on we will see (Sec. 2.2.2) that for certain quite general conditions the following relationship is fulfilled (Poisson distribution):

$$P_T(k) = \frac{(aT)^k}{k!} e^{-aT}$$

where a is a constant. Then we obtain

$$P_{2T}(s) = \sum_{i=0}^{s} \frac{(aT)^s e^{-2aT}}{i!(s-i)!} = (aT)^s e^{-2aT} \sum_{i=0}^{s} \frac{1}{i!(s-i)!}$$

Since

$$\sum_{i=0}^{s} \frac{1}{i!(s-i)!} = \frac{1}{s!} \sum_{i=0}^{s} \frac{s!}{i!(s-i)!} = \frac{2^s}{s!}$$

then

$$P_{2T}(s) = \frac{(2aT)^s e^{-2aT}}{s!}, \qquad s = 0, 1, 2, \ldots$$

Thus, if the Poisson distribution is valid for the time interval T, it holds also for the time $2T$. Hence, this distribution holds also for any time intervals that are multiples of T.

Example 1.23 The probability that the given item is faulty is p_1. One takes n items at random for quality control. The probability that the defect is detected in a faulty item is p_2. Determine the probabilities of the events:

A: "no defects are detected,"
B: "the defects are detected in two items,"
C: "the defects are detected in at least two items."

Solution The probability that the defect will be detected in an item taken at random can be found as the product $p_1 p_2$ of the probability p_1 that the item is faulty and the probability p_2 that the defect will be detected, provided that the item is faulty. The probability that the defect is not found is $1 - p_1 p_2$. Then the probability that no defects are detected is

$$P(A) = (1 - p_1 p_2)^n$$

Two items can be taken out of n items in $C(n, 2) = n(n-1)/2$ ways. The probability that the defects are detected in two items can be computed as

$$P(B) = \frac{n(n-1)}{2} (p_1 p_2)^2 (1 - p_1 p_2)^{n-2}$$

The event $\bar{C}$, "the defects are detected in less than two items," can be represented as the sum $\bar{C} = \bar{C}_0 + \bar{C}_1$ of the event $\bar{C}_0$, "no defects are detected," and the event $\bar{C}_1$, "the defect is found in just one item." Using the rule of multiplication of probabilities we have

$$P(\bar{C}_0) = P(A) = (1 - p_1 p_2)^n$$

Applying the rules of multiplication and addition of probabilities we obtain

$$P(\bar{C}_1) = C(n, 1)p_1p_2(1 - p_1p_2)^{n-1} = np_1p_2(1 - p_1p_2)^{n-1}$$

Then we find

$$P(\bar{C}) = P(\bar{C}_0) + P(\bar{C}_1) = (1 - p_1p_2)^{n-1}(np_1p_2 + 1 - p_1p_2)$$

so that

$$P(C) = 1 - P(\bar{C}) = 1 - (1 - p_1p_2)^{n-1}[(n - 1)p_1p_2 + 1]$$

Example 1.24 The dependability (the probability of nonfailure) of a device is p. How many redundant devices should one employ, so that the dependability of the system would not be lower than P?

Solution The dependability of a system consisting of n devices is

$$P^* = 1 - (1 - p)^n$$

We require that this probability is not smaller than P:

$$1 - (1 - p)^n \geq P$$

Solving this inequality for n, we obtain

$$n \geq \frac{\log (1 - P)}{\log (1 - p)}$$

The number of the redundant devices can be computed as $n - 1$.

1.4. Procedures for Counting the Sample Points

In many actual problems, the sample space may be so complex and contain so many elements that it becomes a problem to construct. To simplify the solutions to various probability problems, two major procedures for counting the sample points can be applied in complicated experiments without constructing a sample space: permutations and combinations.

Permutation is an arrangement of a set of elements into a particular order. If, for instance, a set $\{1, 2, 3\}$ consists of three elements, six permutations can be made: 123, 132, 213, 231, 321, and 312.

A systematic way of determining the number of arrangements of the elements of a set is provided by the following *fundamental principle of counting*: if there are n_1 independent ways of doing one operation and after that there are n_2 independent ways of doing a second operation, then the total number of ways of doing the two operations is $N = n_1 \times n_2$. In general, for r operations,

$$N = n_1 n_2 \cdots n_r \tag{1.9}$$

Experimentalists often encounter such a situation in the design of experiments when choosing experimental matrices.

The formula (1.9) can also be interpreted as follows. Let there be r sets of elements: the first set contains n_1 elements, the second set contains n_2 elements, and so on; the rth set contains n_r elements:

$$a_{11}, a_{12}, \ldots, a_{1n_1}$$
$$a_{21}, a_{22}, \ldots, a_{2n_2}$$
$$\ldots\ldots\ldots\ldots\ldots\ldots$$
$$a_{r1}, a_{r2}, \ldots, a_{rn_r}$$

One makes arrangements in such a way that each arrangement contains only one element from every set. The total number of such arrangements is expressed by formula (1.9). Applying this principle to the example examined above, one can see that there are three positions of the first elements of the set that is selected, two positions for the second, and one for the third. Thus, there are $3 \times 2 \times 1$ permutations of the three elements of the set, when all the three elements are used in each permutation. In general, the total number of permutations of n distinct elements taken n at a time, denoted by $P(n, n)$ (read as "the permutations of n elements taken n at a time"), is $n!$ (read "n factorial"). The symbol $n!$, where n is a natural number, represents the product of n and each natural number less than n: $n! = 1 \times 2 \times 3 \cdots n$. Indeed, the possible number of arrangements of n elements taken n at a time is equivalent to choosing n different elements to fill n different positions. There are n choices for the first position, $n - 1$ choice for the second position, $n - 2$ choices for the third position, and so on, until one has $n - (n - 1) = 1$ elements to fill the $n - (n - 1) = 1$ position. Thus,

$$P(n, n) = n(n - 1)(n - 2) \cdots (n - n + 1) = n! \tag{1.10}$$

One is often interested in the number of permutations when a limited number of the total elements in the set are to be used. When such a situation exists, one refers to the permutation of n distinct elements taken r at a time, that is, $n > r$. For instance, the different permutations of two letters that can be made from the set $\{a, b, c, d\}$ are: ab, ac, ad; ba, bc, bd; ca, cb, cd; and da, db, dc.

When the set contains a large number of elements, listing the arrangements becomes tedious and prone to error. The need for a more economical method is obvious. A simple procedure can be established by the following reasoning. If there are four elements in the set and they are to be taken two at a time, then there are four choices for the first element. Since each arrangement can be used only once, there are three choices for the second element. The total arrangements are then easily computed as $4 \times 3 = 12$. In general, the total number of permutations of n distinct elements taken r at a time $(n > r)$ is given by

$$P(n, r) = n(n - 1)(n - 2) \cdots (n - r + 1)$$

Indeed, the total number of arrangements of n distinct elements taken r at a time is equal to the number of ways of choosing n different elements to fill r positions. There are n choices for the first position, $n - 1$ choices for the second position, $n - 2$ choices for the third position, and so on, until there are $n - (r1)$ choices for the rth position. Thus, the r positions can be filled in $n(n - 1)(n - 2) \cdots (n - r + 1)$ ways, so that

$$P(n, r) = n(n - 1)(n - 2) \cdots (n - r + 1)$$
$$= \frac{[n(n - 1)(n - 2) \cdots (n - r + 1)](n - r)!}{(n - r)!} = \frac{n!}{(n - r)!} \quad (1.11)$$

In the previous discussion, the permutations that were formed used each distinct element only once in each permutation. However, if repetitions are permitted, then the number of permutations that can be formed on n distinct elements taken r at a time $(n \geq r)$ is

$$P(n, r) = n! \quad (1.12)$$

When the elements in the set are not distinct, the number of permutations that can be made is affected. For example, in the word "reed" the letters are not distinct (there is no way of distinguishing between the two e's) and the number of permutations cannot be determined by $n!$ In such cases, the number of distinguishable permutations can be determined on the basis of the following rule: the number of permutations of n elements taken n at a time, when p are of one kind, q of another, and r of another kind, is

$$P\binom{p,\ q,\ r}{n,\ n} = \frac{n!}{p!\,q!\,r!} \quad (1.13)$$

Indeed, if the same elements appear more than once in the same permutation, the interchange of the elements will not produce a different permutation. Thus, if the set contains two elements that are alike, the number of distinguishable permutations that can be formed is found by dividing the total number of permutations by 2! If the set contains three identical elements, the total number of permutations is divided by 3! When the set contains p identical elements, q identical elements, and r identical elements, the number of distinguishable permutations is found by dividing the total number by $p!q!r!$

Sometimes one may wish to select elements regardless of the order in which the selections are made. The total number of different subsets each containing r elements that can be formed from a set of n elements is called *combinations* of n elements taken r at a time. The total number of combinations of n elements taken r at a time is given by

$$C(n, r) = \frac{n!}{(n-r)!r!} \tag{1.14}$$

Indeed, for any one of the $C(n, r)$ combinations consisting of r different elements, the r elements may be rearranged by permuting them in $r!$ different ways. Therefore for each combination of r elements there are $r!$ permutations. For all the possible combinations there are $C(n, r)r!$ different permutations. Since these are all possible permutations of n elements taken r at a time, we have $C(n, r)r! = P(n, r)$, or $C(n, r) = P(n, r)/r!$. Replacing $P(n, r)$ with $n!/(n-r)!$ we obtain the formula (1.11).

Even for not very large values of n, the evaluation of $n!$ by repetitive multiplication may become inconvenient. In such cases the following *Stirling formula* can be used:

$$n! \simeq \sqrt{2\pi n}\, n^n e^{-n} \tag{1.15}$$

or $$\ln(n!) \simeq \tfrac{1}{2} \ln(2\pi n) + n \ln n - n \tag{1.16}$$

These expressions result in errors smaller than 1 percent for n larger than 10.

Example 1.25 Five people enter a bus in which there are 10 vacant seats. In how many ways can they be seated?

Solution This is an arrangement of 10 seats taken five at a time, so that

$$P(10, 5) = \frac{10!}{(10-5)!} = \frac{10!}{5!} = \frac{5! \, 6 \times 7 \times 8 \times 9 \times 10}{5!} = 6 \times 7 \times 8 \times 9 \times 10$$

$$= 30\,240$$

Example 1.26 Five people enter a bus in which there are five vacant seats. In how many ways can they be seated?

Solution The solution is

$$P(5, 5) = 5! = 1 \times 2 \times 3 \times 4 \times 5 = 120$$

Example 1.27 How many two-digit numbers can be formed from the digits 4, 5, and 6?

Solution There are two positions and three possibilities for each position. Therefore, there are $3^2 = 9$ two-digit numbers that can be formed. These numbers are: 45, 46, 54, 56, 64, 65, 44, 55, and 66.

Example 1.28 How many four-letter words (which do not have to be all meaningful) can be formed from the letters of the word "reed"?

Solution Using formula (1.11) we obtain:

$$P = \frac{4!}{2!} = 12$$

Example 1.29 How many permutations can be formed of the 11 letters of the word Mississippi?

Solution In this word there are four i's, four s's, and two p's that are identical. Using formula (1.13) we have,

$$P\binom{4, 4, 2}{11, 11} = \frac{11!}{4!\,4!\,2!} = 34\,650$$

Example 1.30 In how many ways can a committee of 2 men and 3 women be selected from a group of 12 men and 10 women?

Solution Since the members of the committee can be selected without regard to order, the men can be selected in $C(12, 2)$ ways and the women in $C(10, 3)$ ways. Applying the "fundamental principle of counting" and using formula (1.14) for the number of combinations, we have:

$$C(12, 2)C(10, 3) = \frac{12!}{(12-2)!\,2!} \times \frac{10!}{(10-3)!\,3!} = \frac{12!}{10!\,2!} \times \frac{10!}{7!\,3!}$$

$$= \frac{12 \times 11}{2 \times 1} \times \frac{10 \times 9 \times 8}{3 \times 2 \times 1} = 6 \times 11 \times 10 \times 3 \times 4 = 7920$$

Example 1.31 A batch consists of 100 articles, 10 of which are faulty. A total of 10 articles are taken at random from the batch for inspection. If all the articles in the selected sample are sound, the batch is accepted, otherwise it is rejected. What is the probability that the batch is accepted?

Solution The total number of cases is the number of ways of selecting 10 articles from a batch of 100 articles. This can be found as

$$C(100, 10) = \frac{100!}{(100-10)!\,10!} = \frac{100!}{90!\,100!}$$

The event A "the batch is accepted" occurs if all the 10 articles in the sample are sound. Since the total number of sound articles is equal to 90, the number of favorable outcomes can be computed as

$$C(90, 10) = \frac{90!}{(90-10)!\,10!} = \frac{90!}{80!\,10!}$$

The sought probability is therefore

$$P = \frac{C(90, 10)}{C(100, 10)} = \frac{(90!)^2}{80!\,100!} = \frac{81 \times 82 \cdots 90}{91 \times 92 \cdots 100} = 0.3305$$

Example 1.32 A batch contains n sound articles and k faulty ones. A sample of m articles is taken at random for inspection. It turned out that the first l articles in this sample were sound. What is the probability that the next article will also be a sound one?

Solution The total number of articles that remained in the batch is $n + k - l$, and the number of sound articles is $n - l$. Therefore the sought probability is

$$P = \frac{n - l}{n + k - l} \tag{1.17}$$

Example 1.33 Each letter of the word "engineering" is written on a separate card. One takes at random six cards. What is the probability that the word "ginger" is obtained?

Solution Let A_1, A_2, A_3, A_4, A_5, and A_6 be the events which consist in the sequential drawing the letters g, i, n, g, e, and r. The corresponding probabilities are

$$P(A_1) = \frac{2}{11}$$

$$P(A_2 | A_1) = \frac{2}{10}$$

$$P(A_3 | A_1 A_2) = \frac{3}{9}$$

$$P(A_4 | A_1 A_2 A_3) = \frac{1}{8}$$

$$P(A_5 | A_1 A_2 A_3 A_4) = \frac{3}{7}$$

$$P(A_6 | A_1 A_2 A_3 A_4 A_5) = \frac{1}{6}$$

Hence, the sought probability is

$$P = \frac{2}{11} \times \frac{2}{10} \times \frac{3}{9} \times \frac{1}{8} \times \frac{3}{7} \times \frac{1}{6} = \frac{1}{9240} = 0.000\,108 = 0.0108 \text{ percent}$$

Example 1.34 Solve the previous problem by counting permutations.

Solution The total number of permutations of six cards forming the word "ginger" is

$$P(11, 6) = \frac{11!}{5!} = 11 \times 10 \times 9 \times 8 \times 7 \times 6 = 332\,640$$

Now consider how the permutations that spell the word "ginger" can be formed. The first position can be filled in two different ways (there are two g's in the word "engineering"), the second position can be filled also in two different ways (the word "engineering" contains two i's), the third position in three ways, the fourth position in only one way, the fifth position in three ways, and the sixth position is one way. Hence, the number of permutations that satisfy the stipulated requirement is

$$2 \times 2 \times 3 \times 1 \times 3 \times 1 = 36$$

Therefore the sought probability is

$$P = \frac{36}{332\,640} = \frac{1}{9240}$$

Example 1.35 If one writes n letters and then lists the addresses at random, what is the probability that at least one of the letter will reach the right address?

Solution The probability of the event A_k, "the kth envelope has the right address," can be calculated as

$$p = P(A_1 + A_2 + \cdots + A_k + \cdots + A_n)$$

where n is the total number of letters. Since,

$$P(A_k) = \frac{1}{n} = \frac{(n-1)!}{n!}$$

$$P(A_k A_i) = P(A_k)P(A_i/A_k) = \frac{(n-2)!}{n!}$$

$$P(A_k A_i A_j) = \frac{(n-3)!}{n!}$$

$$\cdots\cdots\cdots\cdots\cdots\cdots$$

$$P\left(\prod_{k=1}^{n} A_k\right) = \frac{1}{n!}$$

the sought probability, calculated on the basis of the formula (1.8) as the probability of the sum of n compatible events, is

$$p = C(n, 1)\frac{(n-1)!}{n!} - C(n, 2)\frac{(n-2)!}{n!} + C(n, 3)\frac{(n-3)!}{n!} - \cdots + (-1)^{n-1}\frac{1}{n!}$$

$$= 1 - \frac{1}{2!} + \frac{1}{3!} - \cdots + (-1)^{n-1}\frac{1}{n!}$$

For large n values,

$$p \simeq 1 - \frac{1}{e} = 0.632$$

Example 1.36 There are n particles, each of which can occupy each of $N(N > n)$ cells with the same probability $1/N$. Determine the probability that there will be one particle in each of the n cells.

Solution This problem plays an important role in statistical physics. A particular physical statistics (Boltzmann, Bose–Einstein, or Fermi–Dirac) is obtained depending on the way the complete group of equally probable events is formed.

In *Boltzmann statistics*, the distributions differ both by the number and by the "individuality" of the particles and all the possible distributions are equally probable. Each cell can accommodate any number of particles, from zero to n. The total number of possible distributions of n particles, each of which can be found in any of the N cells, is N^n. In other words, n particles can be distributed in the N cells in N^n different ways. Since the number of favorable cases (each of the n cells contains one particle) is $n!$, the sought probability is

$$P = \frac{n!}{N^n}$$

In *Bose–Einstein statistics*, the only important thing is how many particles there are in the given cell. The "individuality" of a particle is not important. All possible fillings of N cells with n particles can be computed as $(N + n - 1)!$ In order to find the number of favorable cases, we consider first identical permutations, in which each distribution among the cells is counted $(N - 1)!$ times. In addition, since we consider not only the number of particles in the given cell, but also the kind of particles in this cell, as well as their order, one can obtain $n!$ different distributions in each cell. The total number of favorable cases is $n!(N - 1)!$ and the sought probability is

$$P = \frac{(N + n - 1)!}{n!(N - 1)!}$$

In *Fermi–Dirac statistics*, each cell can accommodate one particle or none, and the "individuality" of a certain particle is not accounted for. In order to evaluate the total number of different particle distributions in cells, we notice that the first particle can be distributed in N different ways, the second particle can be distributed in $N - 1$ ways, the third particle in $N - 2$ ways, and so on. The nth particle can be distributed in $N - n + 1$ different ways. This number must be divided by $n!$ to eliminate particle "individuality." Therefore, the number of permutations in which n particles can be arranged in N cells is

$$\frac{N(N-1)\cdots(N-n+1)}{n!} = \frac{N!}{(N-n)!n!}$$

The sought probability can be found as

$$P = \frac{(N-n)!\,n!}{N!}$$

Example 1.37 The probability that each of n particles can occupy one of the N cells ($N > n$) is $1/N$. What is the probability that there will be one particle in each of n arbitrary cells? Give the answer for each of the statistics examined in Example 1.36.

Solution When Boltzmann statistics are used, the number of favorable cases is by a factor of $C(N, n)$ larger than in a situation when there is one particle in each of the definite cells. Therefore the sought probability is

$$P = C(N, n)\,\frac{n!}{N^n} = \frac{N!}{N^n(N-n)!}$$

When Bose–Einstein statistics are considered, this probability is

$$P = C_N^m\,\frac{n!(N-n)!}{(N+n-1)!} = \frac{N!(N-1)!}{(N-n)!(N+n-1)!}$$

In Fermi–Dirac statistics, $P = 1$.

Example 1.38 In the "choosy bride" problem (Example 1.13, Sec. 1.1), assess the probability that no better choices will appear after the kth one. Use the concept of the conditional probability (Sec. 1.3) and procedures for counting sample points.

Solution In effect, one would like to find out how the occurrence of event A (the kth object is the best out of the total m objects) is affected by event B (the last out of k examined objects is the absolute best). The probability $P(AB)$ that both events A and B occur can be calculated as

$$P(AB) = P(B)P(A\,|\,B)$$

where $P(B)$ is the probability of occurrence of the event B ("the last object is the best out of all the examined objects") and $P(A\,|\,B)$ is the conditional probability of the event A ("the last object is the absolute best"), provided that event B took place. The bride is interested in this latter probability:

$$P(A\,|\,B) = \frac{P(AB)}{P(B)}$$

Let us find the probabilities $P(AB)$ and $P(B)$. Since the event B contains the event A (if the object is the absolute best, it will be the best also among the first k objects), the event AB coincides with the event A. As to the probability of the event B, it can be calculated as a probability that the best out of k objects will be found at the fixed kth location as a result of a random permutation of k different objects. In accordance with the procedures for counting sample points, the probability $P(B)$ can be evaluated as

$$P(B) = \frac{(k-1)!}{k!} = \frac{1}{k}$$

Here $k!$ is the number of permutations for k objects and $(k-1)!$ is the number of permutations for $k-1$ objects, compatible with the condition that the best object is fixed at the kth location. The probability $P(A)$ can be obtained in a similar way, by calculating the probability that as a result of a random permutation of m different objects, the particular, the best, object will occupy the kth location:

$$P(A) = \frac{(m-1)!}{m!} = \frac{1}{m}$$

Thus, the sought probability that no better choices will appear after the kth one is

$$P(A \mid B) = \frac{P(AB)}{P(B)} = \frac{P(A)}{P(B)} = \frac{k}{m}$$

This result indicates that if the bride can wait, she should, since the probability to find the best match increases with the number k. If, for instance, the number m is 10 and the k value was found to be $k = 3$ (see Example 1.13, Sec. 1.1), then

$$P(A \mid B) = 0.3$$

(compare with the results obtained in Example 1.13).

Example 1.39 A batch of 36 bolts and nuts, containing 18 bolts and 18 nuts, is divided at random into two equal parts. What is the probability that both parts will have equal numbers of bolts and nuts?

Solution The total number of different ways to take 18 components from 36 is $C(36, 18)$. The favorable ways are those in which there will be 9 components taken from 18 bolts and 9 components from 18 nuts. Thus 9 bolts can be taken in $C(18, 9)$ different ways and 9 nuts also in $C(18, 9)$ different ways. If 9 particular bolts are taken, 9 nuts can be taken in $C(18, 9)$ different ways. Therefore the total number of favorable outcomes is $C(18, 9) \, C(18, 9)$. Thus, the sought probability is

$$P = \frac{C(18,\,9)C(18,\,9)}{C(36,\,18)} = \frac{(18!)^4}{36!\,(9!)^4}$$

Using Stirling's formula (1.15), we have

$$18! \simeq 18^{18} e^{-18} \sqrt{2\pi \times 18}$$

$$9! \simeq 9^{9} e^{-9} \sqrt{2\pi \times 9}$$

$$36! \simeq 36^{36} e^{-36} \sqrt{2\pi \times 36}$$

so that

$$P \simeq \frac{(\sqrt{2\pi \times 18} \times 18^{18} e^{-18})^4}{\sqrt{2\pi \times 36} \times 36^{36} e^{-36} (\sqrt{2\pi \times 9} \times 9^9 \times e^{-9})^4} \simeq \frac{2}{\sqrt{18\pi}} = 0.266$$

Example 1.40 There are n seats in the movie theater. All the tickets are numbered and so are the seats. The theater is sold out. The spectators are taking seats, however, in a random fashion. Find the probability that none of the spectators will take a designated seat.

Solution The total number of all the possible permutations is $n!$. The sought probability can be found as

$$P(n) = \frac{S(n)}{n!}$$

where $S(n)$ is the number of events that are favorable for the situation that none of the spectators will take a designated seat.

In order to find the $S(n)$ number one can use the following reasoning. Let us consider a situation where all the n spectators have indeed occupied the "wrong" seats and where a new, $n+1$, spectator has arrived and has been allowed to bring in an additional, $(n+1)$th, chair, so that this spectator is the only one who occupies the "right" seat. This situation can be changed, however, if one of the spectators that arrived earlier exchanges his seat with the newcomer. This leads to $nS(n)$ favorable permutations for $n+1$ seats. Consider now a situation where $n-1$ spectators out of the total number of n spectators have taken the "wrong" seats, and only one of them, the nth spectator, occupies the "right" seat. Let this spectator exchange his seat with the newcomer, occupying the $n+1$ seat. This also leads to $nS(n-1)$ favorable permutations in $n+1$ seats. Hence, the total number of favorable permutations for $n+1$ seats can be computed as

$$S(n+1) = nS(n) + nS(n-1)$$

Introducing the relationship

$$S(n) = n!P(n)$$

into this equation, we obtain

$$(n+1)!P(n+1) = n[n!P(n)] + n[(n-1)!P(n-1)]$$

or

$$(n+1)P(n+1) - nP(n) - P(n-1) = 0$$

This equation has the following solution:

$$P(n) = C \sum_{k=0}^{n} \frac{(-1)^k}{k!}$$

where the constant C can be found from the condition $P(2) = 1/2$. This condition yields $C = 1$, and therefore

$$P(n) = \sum_{k=0}^{n} \frac{(-1)^k}{k!}$$

When the number of seats is large ($n \to \infty$), this probability tends to $1/e$.

Example 1.41 Using the conditions of the previous problem, find the probability that m spectators ($m \leq n$) will occupy the "right" seats.

Solution The probability that particular m spectators will occupy the "right" seats is $(n-m)!/n!$. The probability that the remaining $n-m$ spectators will occupy the "wrong" seats is, in accordance with the solution to Example 1.40, as follows:

$$\sum_{k=0}^{n-m} \frac{(-1)^k}{k!}$$

The probability that m spectators will occupy the "right" seats and the remaining $n-m$ spectators will occupy the "wrong" seats can be obtained as a product:

$$\frac{(n-m)!}{n!} \sum_{k=0}^{n-m} \frac{(-1)^k}{k!}$$

Since m spectators can be selected out of n spectators in

$$C(m, n) = \frac{n!}{m!(n-m)!}$$

ways, the sought probability can be found as a product:

$$P = C(m, n) \frac{(n-m)!}{n!} \sum_{k=0}^{n-m} \frac{(-1)^k}{k!} = \frac{1}{m!} \sum_{k=0}^{n-m} \frac{(-1)^k}{k!}$$

1.5. "Geometric" Probabilities

A "geometrical" approach to the calculation of the probabilities of events is a generalization and extension of the direct calculation of probabilities in the urn model (see Sec. 1.2). Such an approach is helpful when the sample space Ω includes an uncountable set of elementary events $\omega \in \Omega$ and, because of symmetry, none of them is more likely than the others.

Assume that the sample space Ω is a domain on a plane and that the elementary events ω are points within this domain. If the experiment has a symmetry of possible outcomes (say, a "point" object is dropped at random in the interior of the domain), then all the elementary events are "equal in rights." It is natural to assume that the probabilities that the elementary event ω will be found within the domains of the same size are the same. Then the probability of any event $A \subseteq \Omega$ is equal to the ratio of the area S_A of the domain A to the area S_Ω of the entire domain Ω (Fig. 1.3):

$$P(A) = \frac{S_A}{S_\Omega} \qquad (1.18)$$

This formula is, a generalization of the basic formula (1.8) for the case of an uncountable set of elementary events. The symmetry of the

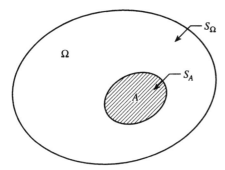

Figure 1.3 The probability of the event A can be calculated as the ratio of the area S_A of the domain A to the area S_Ω of the entire domain Ω.

experimental conditions with respect to the elementary outcomes ω is usually formulated using the words "at random." This is equivalent to the random choice of a ball in an urn.

Example 1.42 Two points with abscissas x_1 and x_2 are put at random within the interval from 0 to 2. What is the probability that the distance L between these points is less than unity?

Solution The elementary event ω is characterized by a pair of coordinates x_1 and x_2. The space of elementary events is a square in the $x_1 x_2$ plane with a side 2 (Fig. 1.4). Then we have: $L = |x_2 - x_1|$, and the occurrence of the event $A = \{|x_2 - x_1| < 1\}$ is associated with the shaded domain A in Fig. 1.4. Thus, the sought probability is

$$P(A) = \{|x_2 - x_1| < 1\} = S_A/S_\Omega = \frac{3}{4}$$

Example 1.43 A receiver acquires two signals during the time T. It is required that this time is such that the time interval $t_2 - t_1$ between the moments of acquisition be shorter than the given time t. If this requirement is not fulfilled, the receiver will be blocked. What is the probability that the receiver will be blocked?

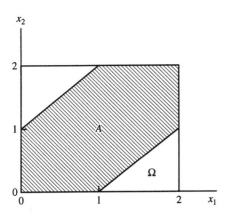

Figure 1.4 The space of elementary events: two points with abscissas x_1 and x_2 put at random within an interval (0, 2).

Solution The domain of the possible values of the moments of time t_1 and t_2 is a square whose sides are equal to T. The receiver will be blocked if $|t_2 - t_1| \leq \tau$. This domain is located between the straight lines $t_2 - t_1 = \tau$ and $t_2 - t_1 = -\tau$ (Fig. 1.5). The area of this domain is

$$S_A = S_\Omega - (T - \tau)^2$$

The sought probability can be found as

$$P = \frac{S_A}{S_\Omega} = 1 - \left(1 - \frac{\tau}{T}\right)^2$$

If the space of elementary events is three-dimensional, then one should replace the areas S_A and S_Ω by the volumes V_A and V_Ω. For a one-dimensional space, the lengths L_A and L_Ω of the corresponding segments of a straight line should be used.

The obtained solution can be used to solve the following "encounter problem." This can be formulated as follows. He and She have agreed to meet at a definite spot between O and T o'clock. The first one to come waits for τ minutes and then leaves. The arrival of Him and Her can occur at any moment at random during the time T. What is the probability that He and She will meet?

Denote the times of arrival of Him and Her by t_1 and t_2, respectively. The meeting will occur if $|t_1 - t_2| \leq \tau$ (Fig. 1.5). All possible outcomes can be described as points within a square with the side T and the favorable outcomes will lie in the shaded region. The sought probability is

$$P = 1 - \left(1 - \frac{\tau}{T}\right)^2$$

If, for instance, $T = 1$ hour $= 60$ min and $\tau = 20$ min, then $P = 5/9 = 0.5556$. The problem becomes more complicated in the case of more than two dates.

Example 1.44 A space satellite is moving on an orbit located between 60°N and 60°S and then falls to the Earth's surface. Any location in this region is equipossible. What is the probability that the satellite be found to the North of the 30°N?

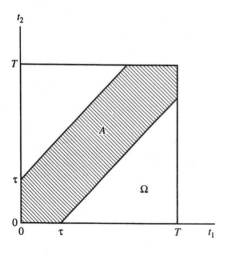

Figure 1.5 The space of elementary events: accession of two signals into a receiver.

Solution The sought probability can be calculated as

$$P = \frac{R^2 \int_0^{2\pi} \int_{\pi/6}^{\pi/3} \cos\phi \, d\phi \, d\psi}{2R^2 \int_0^{2\pi} \int_0^{\pi/3} \cos\phi \, d\phi \, d\psi} = \frac{\sqrt{3}-1}{2\sqrt{3}} = 0.2113$$

Example 1.45 A bar of unit length is broken into three parts whose lengths are a, b, and c. Find the probability that a triangle can be formed from the resulting parts.

Solution The elementary event ω is characterized by two parameters, a and b (since $c = 1 - a - b$). We depict the event by a point in the a, b plane (Fig. 1.6a). The values a and b must satisfy the conditions $a > 0$, $b > 0$, $a + b < 1$. The sample space is the interior of a right triangle with unit legs, that is, $S_\Omega = 1/2$. The condition requiring that a triangle could be formed from the segments a, b, and $c = 1 - a - b$ reduces to the following two conditions: (a) the sum of any two sides should be larger than the third side, (b) the difference between any two sides should be smaller than the third side. This condition is associated with the triangular domain A (Fig. 1.6b). Its area is

$$S_A = \frac{1}{2} \times \frac{1}{4} = \frac{1}{8}$$

The sought probability is, therefore,

$$P(A) = \frac{S_A}{S_\Omega} = \frac{1}{4}$$

Example 1.46 *Buffon's needle problem.* The eighteenth century French naturalist Buffon suggested the following way to determining the π number. Imagine a flat plane which is ruled with parallel straight lines spaced at the distance L from each other. A needle (a line segment) of the length $l < L$ is dropped at random on the plane. What is the probability that it will cross one of the lines?

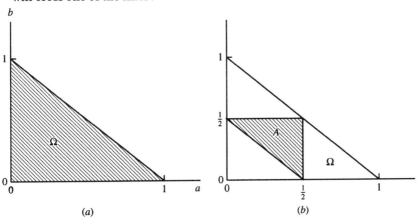

Figure 1.6 The spaces of elementary events: a bar of unit length broken into three parts.

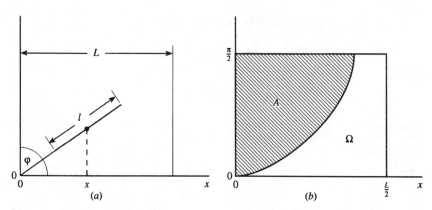

Figure 1.7 Buffon's needle problem.

Solution The outcome of the experiment (the position of the needle) can be defined by two numbers: the abscissa x of the center of the needle with respect to the nearest line on the left and by the angle ϕ that the needle forms with the direction of the lines (Fig. 1.7a). The fact that the needle is dropped on the plane at random means that all the values of x and ϕ are equipossible. One can evidently limit, without losing generality, all the possible values of x to the interval between 0 and $L/2$ and those of ϕ to the interval between 0 and $\pi/2$, and to consider the possibility of the needle crossing only one of the lines (the nearest line on the left). The sample space Ω is a rectangle with the sides $L/2$ and $\pi/2$ (Fig. 1.7b): $S_\Omega = (\pi/4)L$. The needle will cross the line if the abscissa x of the center of the needle is smaller than $(l/2) \sin \phi$. The event we are interested in is $A = \{x < (l/2) \sin \phi\}$. The domain A is shaded in Fig. 1.7b. The area of this domain is

$$S_A = \int_0^{\pi/2} \frac{l}{2} \sin \phi \, d\phi = \frac{l}{2}$$

so that

$$P(A) = \frac{S_A}{S_\Omega} = \frac{2l}{\pi L}$$

This formula, obtained by Buffon, was verified experimentally by many investigators.

One can carry out trials of tossing a needle and obtain an estimate of the π number. Since it is convenient to have a chance of "success" (the line is crossed) and "failure" (the line is not crossed) comparable, one can pick, say, $l/L \simeq 0.8$ and run the trials. The results are shown in the Table 1.2. It should be pointed out that these Monte Carlo runs (the Monte Carlo simulations will be examined in greater detail in Sec. 16.4) will not

TABLE 1.2 Estimates of π in Buffon's needle problem

"Successes"	43	87	142	192	246
Total number of trials	100	200	300	400	500
Estimate of π	3.721	3.678	3.380	3.333	3.252

produce an arbitrarily good estimate of π. It is more a test for a particular random number generator.

Nonetheless, a large number of actual needle-throwing experiments have been carried out by different experimentalists. A few are listed below: Wolf (1850) found $\pi = 3.1596$ with 5000 throws; Smith (1855) obtained $\pi = 3.1553$ as a result of 3204 throws; Fox arrived at $\pi = 3.1419$ after 1120 throws; Lazzarini calculated the π number as $\pi = 3.1415929$ after 3408 throws. The "exact" π number is $\pi = 3.14159265358$.

Example 1.47 *Bertrand's paradox.* A chord is chosen at random in a circle. What is the probability that its length will exceed the length of the side of the equilateral triangle inscribed in the circle?

Solution The solution depends on how we understand the words "at random." Examine the following three cases:

1. The orientation (direction) of the chord is specified beforehand, and the midpoint of the chord is uniformly distributed along the diameter perpendicular to the direction of the chord. It is obvious that only those chords that intersect the diameter in the interval between 1/4 and 3/4 of the diameter length will exceed the side of the equilateral triangle. The probability of this event is 1/2.

2. One end of the chord is fixed and the other end is uniformly distributed on the plane. The tangent of the circle at the point of the fixed end and two sides of the equilateral triangle with the vertex in this point form three 60° angles. Only those chords that fall in the middle angle are favorable to the conditions of the problem. The sought probability in this case is 1/3.

3. The midpoint of the chord is fixed. The chord will meet the condition of the problem if its midportion is located inside a circle concentric with the given one, but with one-half of its radius. The area of the circle is equal to one-fourth the area of the given circle. Therefore the sought probability is 1/4.

1.6. Bayes Formula

If n mutually exclusive hypotheses $H_1, H_2, \ldots, H_n$ can be made concerning the design of an experiment, and if an event A can occur only as a result of the realization of one of these hypotheses, then the probability of this event can be calculated by the following total probability formula (see Example 1.20, Sec. 1.3):

$$P(A) = \sum_{i=1}^{n} P(H_i) P(A/H_i) \qquad (1.19)$$

Here $P(H_i)$ is the probability of the hypothesis (event) H_i and $P(A/H_i)$ is the conditional probability of the event A, provided that the hypothesis H_i took place. If, before the experiment is conducted, the probabilities of the hypotheses $H_1, H_2, \ldots, H_n$ are $P(H_1), P(H_2) \ldots, P(H_n)$ and the experiment resulted in an event A, then the new (conditional) probabilities of the hypotheses can be calculated by the following *Bayes formula*:

$$P(H_i/A) = \frac{P(H_i)P(A/H_i)}{\sum_{i=1}^{n} P(H_i)P(A/H_i)}, \qquad i = 1, 2, \ldots, n \qquad (1.20)$$

This formula can be obtained by introducing the formula

$$P(AH) = P(H)P(A \mid H)$$

and the formula (1.19) into the formula

$$P(A \mid H) = \frac{P(AH)}{P(H)}$$

for the conditional probability (see Sec. 1.3). The formula (1.20) was suggested by an English mathematician Thomas Bayes in a paper published in 1763, three years after his death. In formula (1.20), $P(H_1)$, $P(H_2)$, ..., $P(H_n)$ are the *a priori* (*prior*) *probabilities* and $P(H_1/A)$, $P(H_2/A)$, ..., $P(H_n/A)$ are *a posteriori* (*posterior* or *inverse*) *probabilities*. Bayes formula makes it possible to "revise" the probabilities of the initial hypotheses based on the new experimental data. This formula indicates in a formal and an organized manner how new information can be used to update prior knowledge about random events.

If an experiment, resulting in an event A, is succeeded by an experiment, as a result of which an event B may occur, then the (conditional) probability of the event B can be calculated by the *total probability formula*, the "formula for the probabilities of future events"

$$P(B/A) = \sum_{i=1}^{n} P(H_i/A)P(B/H_i A) \qquad (1.21)$$

In this formula the new probabilities of the hypotheses $P(H_i/A)$ should be substituted for the former ones.

Example 1.48 The results of n medical analyses taken from a patient are fed into a diagnostic apparatus. Each analysis may prove to be erroneous with the probability p. The probability P of a correct diagnosis is a function of the number m of correct analyses: $P = P(m)$. During the time of functioning of the apparatus, diagnoses for k patients were made. What is the probability P^* that at least one erroneous diagnosis is made?

Solution The hypotheses concerning the number of the correct analyses made are as follows:

$$H_0 = \{\text{not a single correct analysis}\}$$
$$H_1 = \{\text{exactly one correct analysis}\}$$
$$\ldots\ldots\ldots\ldots\ldots\ldots\ldots$$

$H_m = \{\text{exactly } m \text{ correct analyses}\}$

. . . .

$H_n = \{n \text{ correct analyses}\}$

The probability of the event H_m for any m can be calculated by the following binomial distribution formula (see Sec. 2.2.1):

$$P_{n,m} = C(n, m)p^m(1-p)^{n-m}$$

in which p is replaced by $1 - p$:

$$P(H_0) = p^n$$
$$P(H_1) = n(1-p)p^{n-1}$$

. . . .

$$P(H_m) = C(n, m)(1-p)^m p^{n-m}$$

. . . .

$$P(H_n) = (1-p)^n$$

Using the total probability formula (1.21), we conclude that the probability that an erroneous diagnosis was made for a certain patient is

$$P_1 = \sum_{m=0}^{n} C(n, m)(1-p)^m p^{n-m} P(m)$$

The probability that no erroneous diagnosis was made for this patient is $1 - P_1$. The probability that none of the k diagnoses was erroneous is $(1 - P_1)^k$. Therefore the probability that at least one erroneous diagnosis was made is

$$P^* = 1 - (1 - P_1)^k$$

Example 1.49 The probability that the given article which a factory produces has a defect is equal to p. Each article is checked by a controller. The probability that he finds a defect, if any, is equal to p_1. Hence, the probability that the controller will not be able to detect a defect is $1 - p_1$. In addition, the controller can make another mistake: to reject a sound article. This occurs with the probability p_2. The controller checks n articles during one shift. What is the probability that at least one article will be assessed incorrectly (i.e., a sound article be rejected or a faulty one be accepted)?

Solution The hypotheses about a certain article are:

$$H_1 = \{\text{an article is faulty}\}$$
$$H_2 = \{\text{an article is sound}\}$$

Using the total probability formula, we find the probability that one article is assessed incorrectly as follows:

$$P = p(1 - p_1) + (1 - p)p_2$$

Then the probability that at least one article is assessed incorrectly is

$$R = 1 - (1 - P)^n$$

Note that the formula for the probability of error, i.e., for the probability that a given article is assessed incorrectly, can also be obtained on the basis of the following simple reasoning.

Let the given article be indeed faulty. The probability of this event is p. The probability that this article is accepted is $1 - p_1$, and therefore the probability of error in this case is $p(1 - p_1)$. If the given article is sound (the probability of this event is $1 - p$), but is erroneously rejected (the probability of this event is p_2), the probability of error is $(1 - p)p_2$. Therefore the total probability of error for one article is

$$P = p(1 - p_1) + (1 - p)p_2$$

Example 1.50 A machine consists of two units. Each unit must function for the machine to operate. The probability that the first unit will not fail during the time t is p_1. That for the second unit is p_2. The machine is tested during the time t and fails. Find the probability that only the first unit failed.

Solution Four hypotheses are possible prior to conducting the experiment:

$H_0 = \{$both units are sound$\}$

$H_1 = \{$the first unit failed and the second one is sound$\}$

$H_2 = \{$the first unit is sound and the second one failed$\}$

$H_3 = \{$both units failed$\}$

The probabilities of these hypotheses are

$$P(H_0) = p_1 p_2$$
$$P(H_1) = (1 - p_1)p_2$$
$$P(H_2) = p_1(1 - p_2)$$
$$P(H_3) = (1 - p_1)(1 - p_2)$$

The probability of the event A, "the machine failed," is

$$P(A/H_0) = 0, \quad P(A/H_1) = P(A/H_2) = P(A/H_3) = 1$$

By applying the Bayes formula (1.20) we obtain the following expression for the probability that only the first unit failed:

$$P(H_1/A) = \frac{(1 - p_1)p_2}{(1 - p_1)p_2 + p_1(1 - p_2) + (1 - p_1)(1 - p_2)} = \frac{(1 - p_1)p_2}{1 - p_1 p_2}$$

If the probability of failure is the same for both units ($p = p_1 = p_2$), then the probability that only one unit fails is

$$P(H_1/A) = \frac{p}{1 + p}$$

For very reliable equipment ($p \simeq 1$), the probability that only the first unit failed is 0.5. For very unreliable equipment ($p \simeq 0$), this probability is zero, i.e., it is quite likely that both units failed.

Example 1.51 A reliability engineer observes the operation of three groups of devices: 23 percent of all the devices belong to the first group, 31 percent belong to the second group, and 46 percent belong to the third group. The probabilities of failure in each of these groups are 0.0020, 0.0035, and 0.0055, respectively. What is the probability that a failure will occur in any of these groups? What is the probability that the observed failure occurred in the first, in the second, or in the third group of devices?

Solution The probability of the occurrence of a failure can be found as

$$P = 0.23 \times 0.0020 + 0.31 \times 0.0035 + 0.46 \times 0.0055 = 0.0041$$

The probabilities that the failure occurs in the first, in the second, or in the third groups of devices can be calculated, using the Bayes formula, as follows:

$$P_1 = \frac{0.23 \times 0.0020}{0.0041} = 0.11$$

$$P_2 = \frac{0.31 \times 0.0035}{0.0041} = 0.27$$

$$P_3 = \frac{0.46 \times 0.0055}{0.0041} = 0.62$$

Example 1.52 The probability that an engineering product, manufactured at the given factory, meets the requirement of the appropriate specification is 0.96. A series of control tests is carried out to determine whether the given batch of products meets the specification. The tests, however, are not completely certain. They give an affirmative answer with the probability 0.98 and a negative answer with the probability 0.05. What is the probability of the event A that the product that passed the tests meets the specification?

Solution The probability of the hypothesis H_1, "the product meets the specification," is $P(H_1) = 0.96$. The probability of the hypothesis H_2, "the product does not meet the specification," is $P(H_2) = 0.04$. The probability that the product passed the tests, provided that the first hypothesis takes place, is $P(A|H_1) = 0.98$. The probability that the product passed the tests, provided that the second hypothesis is fulfilled, is $P(A|H_2) = 0.05$. Then the posterior probability of the event A, "the product meets the specification," is

$$P(A) = \frac{0.96 \times 0.98}{0.96 \times 0.98 + 0.04 \times 0.05} = 0.998$$

Thus, if the product passed the tests, it might be faulty only in two cases out of a thousand (i.e., does not meet the specification).

Example 1.53 There are two identical boxes. The first one contains two sound devices and three faulty ones. The second box contains three sound devices and a faulty one. Two devices from the first box are taken out at random and put into the second box. Then one device from the second box is taken out at random. This device turns out to be sound. Find the most likely combination of the two devices taken from the first box and put into the second one: "sound–sound," "faulty–faulty," or "sound–faulty?"

Solution Let the event A be "the device taken from the second box is sound." This event can occur if one of the following hypotheses is fulfilled:

H_1 = both devices taken from the first box and put into the second box are sound

H_2 = both devices taken from the first box and put into the second box are faulty

H_3 = one of the devices taken from the first box and put into the second box is sound and the other one is faulty.

The probabilities of the above hypotheses are

$$P(H_1) = \frac{C(2, 2)}{C(5, 2)} = \frac{1}{10}$$

$$P(H_2) = \frac{C(3, 2)}{C(5, 2)} = \frac{3}{10}$$

$$P(H_3) = \frac{C(2, 1)C(3, 1)}{C(5, 2)} = \frac{6}{10}$$

The conditional probabilities can be calculated as

$$P(A \mid H_1) = \frac{3+2}{4+2} = \frac{5}{6}$$

$$P(A \mid H_2) = \frac{3}{6}$$

$$P(A \mid H_3) = \frac{3+1}{6} = \frac{4}{6}$$

In accordance with the Bayes formula,

$$P(H_i \mid A) = \frac{P(H_i)P(A \mid H_i)}{\sum_{k=1}^{n} P(H_k)P(A \mid H_k)} = \frac{P(H_i)P(A \mid H_i)}{P(A)}$$

we obtain the following values for the posterior probabilities:

$$P(H_1 \mid A) = \frac{P(H_1)P(A \mid H_1)}{\sum_{k=1}^{3} P(H_k)P(A \mid H_k)} = \frac{\frac{1}{10}\frac{5}{6}}{\frac{1}{10}\frac{5}{6} + \frac{3}{10}\frac{3}{6} + \frac{6}{10}\frac{4}{6}} = \frac{5}{38} = 0.132$$

$$P(H_2 \mid A) = \frac{\frac{3}{10}\frac{3}{6}}{\frac{38}{60}} = \frac{9}{38} = 0.237$$

$$P(H_3|A) = \frac{\dfrac{6}{10}\dfrac{4}{6}}{\dfrac{38}{60}} = \frac{12}{19} = 0.632$$

Comparing the obtained values of the posterior probabilities, we conclude that the probability that one of the devices was sound and the other one was faulty is the highest.

Example 1.54 Each of two boxes contains items of the same type. There are m items in the first box and one of these items is a faulty one. The second box contains n items and one of them is faulty. One item from the first box is taken at random and put into the second box. Then one item from the second box is taken out at random. What is the probability that this item is faulty?

Solution The probability that the item taken from the first box and put into the second box is faulty (hypothesis H_1) is $P(H_1) = 1/m$. The probability that this item is sound (hypothesis H_2) is $P(H_2) = (m-1)/m$. The conditional probability $P(A|H_1)$ of the event A, "the item taken from the second box is faulty," provided that the hypothesis H_1 took place, is $P(A|H_1) = 2/(n+1)$. The conditional probability $P(A|H_2)$ of this event, provided that the hypothesis H_2 was realized, is $P(A|H_2) = 1/(n+1)$. Then, in accordance with the formula for the complete probability,

$$P(A) = \sum_{i=1}^{n} P(H_i) P(A|H_i)$$

we find

$$P(A) = \sum_{i=1}^{2} P(H_i) P(A|H_i)$$
$$= \frac{1}{m}\frac{2}{n+1} + \frac{m-1}{m}\frac{1}{n+1} = \frac{m+1}{m(n+1)}$$

Example 1.55 The object of interest can be either available or unavailable (hypotheses H_1 and H_2). The prior probabilities of these states are $P(H_1) = 0.7$ and $P(H_2) = 0.3$. The actual condition of the object is judged based on the information obtained from two devices. These devices give contradictory information about the state of the object: the first device indicates that the object is unavailable, and the reliability of this prediction is 0.9; the second device indicates that the object is available, and the reliability of this prediction is 0.7. Find the posterior probabilities of the hypotheses H_1 and H_2.

Solution The conditional probabilities of the event A, "the first device gave the information H_2 and the second device gave the information H_1," can be found as $P(A|H_1) = 0.1 \times 0.7 = 0.07$ (the first device provided false information and the second one provided correct information) and $P(A|H_2) = 0.9 \times 0.3 = 0.27$ (the first device provided correct information and the second one provided false information).

In accordance with the Bayes formula we find

$$P(H_1|A) = \frac{0.7 \times 0.07}{0.7 \times 0.07 + 0.3 \times 0.27} = 0.377$$

$$P(H_2|A) = 1 - P(H_1|A) = 0.623$$

Thus, we conclude that the hypothesis H_2 is more likely: the object is unavailable.

Example 1.56 The probability that a plane be landed safely is p_1, provided that the weather is favorable and the pilot can see the runway. This probability does not change even in low cloud ceiling conditions, if the instruments needed for a blind landing function normally. If, however, these
instruments fail, the probability of safe landing is $p_2 < p_1$. It is known that in the k per cent of cases there is a low cloud ceiling at the given airport at the given time of the year. The probability of failure-free functioning of the instruments needed for a blind landing is p_*. What is the probability of safe landing? What is the probability that the pilot used the blind-landing instruments, if it is known that the plane landed safely?

Solution The hypotheses are:

$$H_1 = \text{the cloud ceiling is high}$$
$$H_2 = \text{the cloud ceiling is low}$$

We have:

$$P(H_1) = 1 - \frac{k}{100}, \quad P(H_2) = \frac{k}{100}$$

and the conditional probability $P(A/H_1)$ of the event A "the landing is safe" is $P(A/H_1) = p_1$. The conditional probability $P(A/H_2)$ can be found, on the basis of the total probability formula, as

$$P(A/H_2) = p_* p_1 + (1 - p_*) p_2$$

so that the sought probability of safe landing is

$$P(A) = \left(1 - \frac{k}{100}\right) p_1 + \frac{k}{100} [p_* p_1 + (1 - p_*) p_2]$$

$$= p_1 - \frac{k}{100} (p_1 - p_2)(1 - p_*)$$

If it is known that the pilot landed blind, then the cloud ceiling was low (hypothesis H_2 took place). Then we have:

$$P(H_2/A) = \frac{\frac{k}{100}[p_* p_1 + (1 - p_*)p_2]}{P(A)} = \frac{\frac{k}{100}[p_* p_1 + (1 - p_*)p_2]}{p_1 - \frac{k}{100}(p_1 - p_2)(1 - p_*)}$$

Future Reading (see *Bibliography*): 7, 20, 22, 30, 31, 32, 38, 40, 44, 47, 58, 59, 62, 84, 88, 98, 104, 110, 111, 122, 124, 130, 137.

Chapter 2

Discrete Random Variables

"Lord created integers. Everything else is due to man."
LEOPOLD KRONECKER

"The man who removes a mountain begins by carrying away small stones."
CHINESE PROVERB

"Any astronomer can predict just where every star will be at half past eleven tonight. He can make no such prediction about his daughter."
JAMES TRUSLOW ADAMS

2.1. Random Variables

A *random variable* is a variable which, as a result of an experiment with an unpredictable outcome, assumes a certain value that is unknown prior to the experiment. No matter how carefully a process is run, an experiment is performed, or a measurement is taken, there will be differences (variability) in repeatability due to the inability of an actual system to control or to predict completely all possible influences. Examples of random variables are: population of a town or a country, time to failure in a machine, stress level in a structure, acceleration of the ground due to an earthquake, height of an ocean wave, water level in a lake or a river, velocity of a wind gust, current or voltage in an electric circuit, gas pressure in a pipe line, etc. Random variables are usually denoted by capital letters from the end of the Latin alphabet, $X, Y, Z, \ldots$, and the possible values of these variables (*realizations*) by the corresponding small letters, $x, y, z, \ldots$. A random event can be viewed as a special case of a random variable, namely, as a random variable which assumes only two values— "unity" when this event takes place and "zero" when it does not.

In the set-theoretical interpretation of the probability concept, a random variable X is a function of an elementary event ω: $X = f(\omega)$, where $\omega \in \Omega$. The value of that function depends on which elementary event ω occurs as a result of an experiment.

The most general form to define a random variable is to establish the *law of probability distribution* for this variable. Such a law is, in effect, a rule which can be used to find the probability of the event related to the random variable. It could be a probability that the variable assumes a certain value or falls within a certain interval. The most general form of a probability distribution law is a *distribution function*—the probability that a random variable X assumes a value that is smaller than the given (nonrandom) value x:

$$F(x) = P\{X < x\}$$

Being a probability, this function possesses the following properties: $F(-\infty) = 0$ and $F(\infty) = 1$, so that $0 \le F(x) \le 1$. The function $F(x)$ does not decrease with an increase in x.

A random variable is *discrete* if it has a finite or countable set of possible values that can be enumerated. A discrete random variable X is defined if, for every value x_k, $k = 1, 2, \ldots$, of this variable, a corresponding probability p_k is given. This probability determines the objective likelihood with which this value can be realized. The simplest distribution for a discrete random variable X is an *ordered sample*, an *ordered series*, or a *frequency table*. This is a table whose top row contains the values x_k, $k = 1, 2, \ldots$, of the random variable in the ascending order and the bottom row contains the corresponding probabilities p_k, $k = 1, 2, \ldots$:

$$X: \begin{array}{|c|c|c|c|c|} \hline x_1 & x_2 & \cdots & x_k & \cdots \\ \hline p_1 & p_2 & \cdots & p_k & \cdots \\ \hline \end{array}$$

In this table, p_k is the probability that the random variable X assumes the value x_k:

$$p_k = P\{X = x_k\}$$

Clearly, the sum of all the probabilities p_k should be equal to unity:

$$\sum_i p_i = 1$$

This important property of probabilities is known as the *condition of normalization*.

A graphical representation of an ordered sample in the form of a bar chart is called a *frequency polygon* or a *histogram* (Fig. 2.1). The

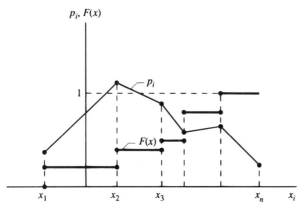

Figure 2.1 Frequency polygon p_i and a "jump" function $F(x)$.

frequency polygon provides a picture of the data from which one can judge the shape of the probability curve, the center of the distribution, and the spread of the data with respect to this center. An alternative way of representing statistical (experimental) data is by accumulating the number of observations less than, or equal to, a given value. Such a means of representing data leads to a *cumulative distribution (frequency) function*, or a "*jump*" *function* $F(x)$. The graphical representation of this function is shown in Fig. 2.1. The discontinuous steps ("jumps") in the distribution function $F(x)$ of a discrete random variable X correspond to the possible values x_k of the variable X and are equal to the probabilities p_k of these values. The function $F(x)$ is never decreasing, starts at zero, and goes to the total sample size (unity in fractional representation or 100 percent in percentage notation).

The *mean* value, or the *mathematical expectation*, of a discrete random variable X is the sum of the products of all its values x_k and their probabilities p_k:

$$E(X) = \langle x \rangle = \sum_k x_k p_k \tag{2.1}$$

The sample mean can be simply computed as the sum of the observations divided by the number of observations:

$$E(X) = \langle x \rangle = \frac{1}{n} \sum_{k=1}^{n} x_k$$

A *centered random variable* is the difference between the random variable X and its mean:

$$\mathring{X} = X - \langle x \rangle$$

The *variance* of a random variable X is the mean value of the square of the corresponding centered random variables:

$$D(X) = D_x = \text{var}(X) = \sum_k (x_k - \langle x \rangle)^2 p_k \qquad (2.2)$$

The sample variance can be calculated by the formula

$$D_x = \frac{1}{n-1} \sum_{k=1}^n (x_k - \langle x \rangle)^2$$

The $n-1$ term, not the n term, occurs because statistical theory shows that dividing by $n-1$ instead of n gives a better estimate of the population variance. This is due to the fact that a sample variance always underestimates the population variance, since the value of the sample mean is not usually the same as the value of the population mean. Because of this, the sum of the deviations of the values about the sample mean is usually smaller than the sum of the squares of the deviations about the population mean. The *mean square deviation*, or the *standard deviation*, of a random variable X is the square root of its variance:

$$\sigma_x = \sqrt{D_x} \qquad (2.3)$$

Unlike the variance, it is measured in the same units as the random variable itself.

In addition to the characteristics $\langle x \rangle$ and D_x, it is often useful to have measures of symmetry of the distribution of the random variable about the center and of how peaked the data are over the central region. These measures are called *skewness* and *kurtosis*, respectively, and are defined as the expected values of the third and the fourth moments about the mean:

$$\mu_3 = E[(x - \langle x \rangle)^3] = \langle (x - \langle x \rangle)^3 \rangle$$

$$\mu_4 = E[(x - \langle x \rangle)^4] = \langle (x - \langle x \rangle)^4 \rangle$$

A *unimodal* (single-peak) distribution with an extended right "tail" has a positive skewness and is referred to as *skewed right*. *Skewed left* implies a negative skewness and a corresponding extended left tail. The skewness of symmetric distributions is equal to zero. Kurtosis, on the other hand, characterizes the relative flatness (peakedness) of the distribution, and indicates how "heavy" the tails are. Quite often the normal distribution (see Sec. 3.3.3) is used as a measure of reference for the kurtosis of the given distribution. The kurtosis for the normal distribution is equal to $\mu_4 = 3$. If, for the given distribution, $\mu_4 < 3$, the distribution is *platykurtic* (mild peak), and if $\mu_4 > 3$, the distribution is *leptokurtic* (sharp peak). The magnitudes of the skewness and kurtosis enable one to check the validity of the assumed probability distributions.

Example 2.1 A constant nonrandom variable a is added to a random variable X. How does this affect its mean value, the variance, and the standard deviation?

Solution A term a is added to its mean value, while the variance and the standard deviation do not change.

Example 2.2 A random variable X is multiplied by a. How does this affect its mean value, the variance, and the standard deviation?

Solution The mean value is multiplied by a, the variance is multiplied by a^2, and the standard deviation is multiplied by $|a|$.

Example 2.3 The probability that an electric bulb will fail as a result of turning it on and off once is equal to p. The bulb is turned on and off many times. Examine a discrete random variable—the sequential number of turning the bulb on and off—and find the distribution of this variable.

Solution The probability that the bulb will not burn out after the kth turning on and off can be found as a product of the probability that it will not fail after the $k - 1$ turnings on and off, and will fail as a result of the kth turning on and off:

$$(1 - p)^{k-1} p$$

Hence, the possible values of the random variable $1, 2, \ldots, k, \ldots$ have the following probabilities:

$$p, (1 - p)p, \ldots, (1 - p)^{k-1} p, \ldots$$

The probability that the bulb will not fail after k turnings on and off is $(1 - p)^k$, and therefore the probability distribution function is

$$F(k) = 1 - (1 - p)^k$$

Let us make sure that

$$\sum_{k=1}^{\infty} p_k = 1$$

Indeed,

$$\sum_{k=1}^{\infty} (1 - p)^{k-1} p = \frac{p}{1-p} \sum_{k=1}^{\infty} (1 - p)^k = \frac{p}{1-p} \frac{1-p}{p} = 1$$

Here the formula for calculating the sum of a infinite geometric progression was used. This problem has been examined here as an example of the descrete random variable of general nature. It will be shown in Section 2.2.1, the sequential number of turning the bulb on the off is a random variable distributed in accordance with the binomial law (see Example 2.13, Sec. 2.2.1).

2.2. Frequently Encountered Discrete Distributions

2.2.1. Bernoulli trials and the binomial distribution

The Bernoulli trials, named after the famous Swiss mathematician Jacob Bernoulli (1654–1705), deal with a sequence of independent tests. Examine four typical situations:

1. *Urn problem* (we have encountered an urn model in Sec. 1.2). An urn contains white and black balls. The probability of taking out a white ball (event A) is $P(A) = p$. The probability of taking out a black ball (event $\bar{A}$) is $P(\bar{A}) = q = 1 - p$. One carries out n trials and, to make the trials independent, each time puts the drawn ball back. What is the probability P_m that the white ball will appear m times during the trials?

2. *Tossing of a coin.* A coin is tossed n times. The probability of the two possible outcomes ("heads" or "tails") is the same and is equal to 1/2. What is the probability that the "heads" will come up m times after n tossings?

3. *Newborn babies.* What is the probability that m out of n newborn babies are boys if the probability of an event "the newborn baby is a boy" is $p = 0.51$?

4. *Random motion* (we have encountered a similar situation in the "gambler's ruin" problem: Example 1.11, Sec. 1.1). A Brownian particle moves one unit to the right with the probability p and one unit to the left with the probability $q = 1 - p$. What is the probability that the particle will make m moves to the right if the total number of moves is equal to n?

These, and many other physical and engineering problems, including some of those examined in Chapter 1, can be formalized as follows. One carries out n independent trials, and during each of these trials one of the two opposite events, A ("success") or $\bar{A}$ ("failure"), takes place. What is the probability that the event A ("success") will occur m times? The solution to this problem is given by the binomial distribution (see Sec. 1.3).

A random variable X is *binomially distributed* if the probabilities of its values 0, 1, 2, ... are calculated by the formula:

$$P_m = P\{X = m\} = C(n, m)p^m q^{n-m}, \qquad q = 1 - p \qquad (2.4)$$

where $0 < p < 1$ and $m = 0, 1, 2, \ldots, n$. The binomial distribution depends on two parameters: the probability p of a particular success and the number n of trials. From the previously discussed theorem (rule) on repetition of trials it follows that the number X of the occurrence of an event in n independent trials has a binomial distribution. In order for the binomial distribution to be applied, the following conditions must take place: the outcome of each trials must be independent of all the other trials, only one of the two mutually exclusive outcomes ("success" or "failure") is possible, the probability of occurrence of "success" or "failure" in each trial is the same for all trials, and the order in which the "successes" and the "failures" occur is not important. The mean and variance of the binomial distribution are

$$\langle x \rangle = np, \qquad D_x = npq \tag{2.5}$$

Let us show how these formulas can be obtained. Examine a random variable ξ which is the total number m of "successes" in n trials. This variable can be written as a sum

$$\xi = \xi_1 + \xi_2 + \cdots + \xi_n$$

of n independent variables $\xi_1, \xi_2, \ldots, \xi_n$. These are defined as follows: $\xi_m = 1$ if the mth trial is successful and $\xi_m = 0$ if it is a failure. The mean and variance of each particular variable ξ_m are

$$\langle \xi_m \rangle = p, \qquad D_{\xi_m} = \langle \xi_m^2 \rangle - (\langle \xi_m \rangle)^2 = p - p^2 = p(1-p) = pq$$

Then the mean and variance of the random variable ξ can be obtained as

$$\langle \xi \rangle = np, \qquad D_\xi = npq$$

The generalization of the binomial distribution is a *polynomial distribution*. This distribution determines the probability that the events A_k, $k = 1, 2, \ldots, m$, characterized by the probabilities p_k, will occur n_k times during n independent trials, in each of which only one of the events A_k can take place. The polynomial distribution is as follows:

$$P_{n;\, n_1,\, n_2,\, \ldots,\, n_m} = \frac{n!}{n_1!\, n_2! \cdots n_m!} p_1^{n_1} p_2^{n_2} \cdots p_m^{n_m} \tag{2.6}$$

Clearly,

$$\sum_{k=1}^{m} n_k = n$$

Example 2.4 A number k of malfunctioning machines are tested to locate the faults. In each test a fault can be located with the probability p. When a fault is located, the machine is taken to a repair shop. The tests are terminated when the faults are located in all the machines. The inspection procedure allows not more than n tests $(n > k)$. What is the probability that the faults will be located in all k machines?

Solution The probability of the event A, "all the faults are located," (i.e., "the faults are located at least k times") can be calculated as

$$P(A) = \sum_{m=k}^{n} C(n, m) p^m q^{n-m}, \qquad q = 1 - p \tag{2.7}$$

When n, m, and k are large, the application of this formula becomes tedious and time consuming, and the following approximate *Moivre–Laplace formula* can be used:

$$P(A) = \frac{1}{\sqrt{\pi}} \int_a^b e^{-t^2} dt = \frac{1}{2} [\Phi(b) - \Phi(a)]$$

where

$$a = \frac{k - np}{\sqrt{2npq}}$$

$$b = \frac{n - np}{\sqrt{2npq}}$$

and

$$\Phi(\alpha) = \frac{2}{\sqrt{\pi}} \int_0^\alpha e^{-t^2} dt = \sqrt{\frac{2}{\pi}} \int_0^{\alpha\sqrt{2}} e^{-t^2/2} dt$$

is the *Laplace function* or the *probability integral* (see also Sec. 3.3.3 and Appendix B).

Example 2.5 A batch of n articles contains k faulty ones. A sample of m articles is taken at random from the batch for inspection. What is the probability that this sample contains l faulty articles? What is the probability that it contains at least one faulty article?

Solution The total number of ways by which one can take m articles out of n articles is $C(n, m)$. Favorable outcomes are those when l faulty articles are taken out of the total number k of faulty articles [this can be done in $C(k, l)$ ways], while the remaining $m - l$ articles, taken out of the total number of $n - k$ sound articles [this can be done in $C(n - k, m - l)$ ways], are sound ones. Therefore the total number of favorable cases is $C(k, l)C(n - k, m - l)$ and the sought probability is

$$P = \frac{C(k, l) C(n - k, m - l)}{C(n, m)} \tag{2.8}$$

The probability that there is only one faulty article in the sample can be obtained by putting $l = 1$ in this formula. Then the probability that there is at least one faulty article in the sample is

$$P = 1 - \frac{C(k, l)C(n - k, m - l)}{C(n, m)} \qquad (2.9)$$

Example 2.6 What is more likely to win when playing against an equally strong opponent: (a) three games out of four, or five games out of eight? (b) at least three games out of four, or at least five games out of eight?

Solution Since the opponents are equally strong, the probabilities of winning or losing a game are the same: $p = q = 1/2$. The probability to win three games out of four is

$$P(4, 3) = C(4, 1)\left(\frac{1}{2}\right)^4 = \frac{1}{4}$$

The probability to win five games out of eight is

$$P(8, 5) = C(8, 3)\left(\frac{1}{2}\right)^8 = \frac{7}{32}$$

which is less than 1/4.

The probability to win at least three games out of four is

$$R(4, 3) = P(4, 3) + P(4, 4) = \frac{1}{4} + \frac{1}{16} = \frac{5}{16} = 0.312$$

The probability to win at least five games out of eight is

$$R(8, 5) = P(8, 5) + P(8, 6) + P(8, 7) + P(8, 8)$$

$$= \frac{7}{32} + \left(\frac{8 \times 7}{2} + 8 + 1\right)\left(\frac{1}{2}\right)^8 = \frac{93}{256} = 0.363$$

which is greater than 5/16.

Example 2.7 A particle subjected to random repetitive impacts can move along the x axis one step right, with the probability p, or one step left, with the probability $q = 1 - p$. What is the probability $P_n(m)$ that the particle will be at the point m after n steps?

Solution Let k be the number of steps to the right from the initial location $k = 0$. Then the number of steps to the left is $k - m$. Since the total number of steps is n, that is, $k + (k - m) = n$, we conclude that the number of steps to the right is $k = (n + m)/2$. The number of steps to the left is $1 - k = (n - m)/2$. Thus, the sought event, "after n steps the particle will be located at the point m," will occur if the number of steps to the right will be equal to $(n + m)/2$ after the total number of n steps. Since the sequence in which the steps in different directions are made does not matter, the probability of such an event can be calculated as

$$P_n(m) = C\left(n, \frac{n+m}{2}\right) p^{(n+m)/2} q^{(n-m)/2} \qquad (2.10)$$

When the number n of steps increases, the variance of the variable m, i.e., the coordinate of the particle, increases as well. It could be shown that when the number n approaches infinity, the distribution of the variance $D_x = D_m = npq$ approaches the normal distribution (see Sec. 3.3.3).

Example 2.8 At a given segment of a ship's route, the wave climatology is such that two out of sixty days of the season are characterized by wave conditions that might be dangerous to her safety. Based on the ship's speed, it is envisioned that she will be in the dangerous region of the ocean for about half a day. What is the probability that the ship will encounter dangerous wave conditions?

Solution Since the ship can be found in the potentially dangerous region for half a day, this time can be taken as a time unit. Then the problem can be formalized as follows. The total number of time units is $n = 120$, and $k = 4$ of them can be dangerous to the ship's safety. The ship sails in the dangerous region of the ocean for $m = 1$ time units. What is the probability that $l = 1$ units ("out of $m = 1$") will be dangerous?

The solution to this problem is given by the formula (2.8):

$$P = \frac{C(4, 1)C(116, 0)}{C(120, 1)} = \frac{4!\,119!}{3!\,120!} = \frac{4}{120} = \frac{1}{30} = 0.0333$$

Example 2.9 A sinking vessel sends a rescue signal m times. The probability that this signal is received by one of the n other vessels in the given region of the ocean is equal to $p = 1/n$. What is the probability that the given vessel will be the one that will receive the rescue signal first?

Solution The random variable X (the number of vessels that can accept the rescue signal) can assume the following values: 0, 1, 2, ..., n. The probability $P(n, k)$ that this variable will assume the value k (i.e. the probability that the given vessel will receive the rescue signal k times) can be determined by the formula (2.4)

$$P(n, k) = C(m, k)p^k(1-p)^{m-k} = \frac{m!}{k!(m-k)!}\left(\frac{1}{n}\right)^k\left(1-\frac{1}{n}\right)^{m-k}$$

For $k = 1$, we obtain

$$P = C(m, 1)\left(\frac{1}{n}\right)\left(1-\frac{1}{n}\right)^{m-1} = \frac{m}{n}\left(1-\frac{1}{n}\right)^{m-1}$$

Example 2.10 A river dam has a projected life of 50 years. What is the probability that a hundred-year flood will occur once during the life of the dam?

Solution The probability of a hundred-year flood occurring in any one year is $p = 0.01$. From the formula (2.4), with $m = 1$ and $n = 50$, we obtain

$$P_1 = C(50, 1)(0.01)^1(0.99)^{49} = 0.31$$

Thus, the sought probability is approximately 31 percent (see also Example 2.26, Sec. 2.2.2).

Example 2.11 The probabilities that the diameter of a machine part is smaller than the allowable one, greater than the allowable one, and is found within the allowable limits are $p_1 = 0.05$, $p_2 = 0.10$, and $p_3 = 0.85$, respectively. The total number of trials is $n = 100$. Determine the probability that five of the parts will have a diameter that is smaller than the allowable one, and five will have a diameter that is larger than the allowable one.

Solution One has to find the probability p that the event A_1, "the part's diameter is too small," and the event A_2, "the part's diameter is too large," will occur five times each. Thus, $n_1 = n_2 = 5$ and $n_3 = 90$. The sought probability can be found in accordance with the formula (2.6) as follows:

$$p = P_{100;\,5,\,5,\,90} = \frac{100!}{5!\,5!\,90!}\, 0.05^5 \times 0.10^5 \times 0.85^{90} = 0.006$$

Example 2.12 One carries out n independent trials. The probability of occurrence of an event A is the same in all the trials and is equal to p. What is the most probable number m^* of the occurrence of the event A?

Solution Let us consider the case when $0 < m^* < n$. In this case the following two inequalities must be satisfied:

$$C(n, m^*)p^{m^*}q^{n-m^*} \geq C(n, m^* + 1)p^{m^*+1}q^{n-m^*-1}$$
$$C(n, m^*)p^{m^*}q^{n-m^*} \geq C(n, m^* - 1)p^{m^*-1}q^{n-m^*+1}$$

These are equivalent to the following inequalities:

$$(m^* + 1)q \geq (n - m^*)p; \qquad (n - m^* + 1)p \geq m^*q$$

Hence, m^* must be an integer satisfying the condition

$$(n + 1)p - 1 \leq m^* \leq (n + 1)p$$

It can be verified that this condition takes place also if $p < 1/(n + 1)$ ($m^* = 0$ in this case), as well as in another extreme case, $m^* = n$, in which $p > n/(n + 1)$. Since the right part of the obtained inequality exceeds the left part by unity, there is only one integer m^* between them. The only exception is the case when both $(n + 1)p$ and $(n + 1)p - 1$ are integers. Then there are two very unlikely values $(n + 1)p$ and $(n + 1)p - 1$. If np is an integer, then $m^* = np$.

Example 2.13 The probability that the given electric bulb is faulty is equal to p. A bulb is fastened in a holder and the current is turned on. A faulty bulb burns out immediately and is replaced. Construct the order series (frequency table) for n bulbs and determine the mean value of the number of bulbs that must be tested.

Solution The order series for the random variable X of the number of bulbs is

$$X: \begin{array}{|c|c|c|c|c|c|c|} \hline 1 & 2 & 3 & \cdots & k & \cdots & n \\ \hline q & pq & p^2q & \cdots & p^{k-1}q & \cdots & p^{n-1} \\ \hline \end{array}, \qquad q = 1 - p$$

The mean value of the number of bulbs that must be tested is

$$\langle x \rangle = \sum_{k=1}^{n-1} k p^{k-1} q + n p^{n-1} = q \frac{d}{dp}\left(\frac{p-p^n}{1-p}\right) + n p^{n-1} = \frac{1-p^n}{1-p}$$

See also Example 2.3, Sec. 2.1.

Example 2.14 In each trial the probability of occurrence of the event of interest is p. What is the probability that this event will occur an even number of times during n trials?

Solution Let p_k be the probability that, as a result of k trials, the event will occur an even number of times. The following two hypotheses could be made prior to the kth trial:

1. The event occurred an even number of times in the $(k-1)$st trial.
2. The event occurred an odd number of times in the $(k-1)$st trial.

The probabilities of these hypotheses are p_{k-1} and $1 - p_{k-1}$, respectively. Then we have

$$p_k = p_{k-1}(1-p) + (1 - p_{k-1})p$$

so that

$$p_k = p + p_{k-1}(1 - 2p)$$

This expression can be writen as

$$p_k - \frac{1}{2} = (1 - 2p)\left(p_{k-1} - \frac{1}{2}\right), \qquad k = 1, 2, \ldots, n$$

Multiplying the left and the right parts of all the n such equalities, we obtain

$$\prod_{k=1}^{n} \left(p_k - \frac{1}{2}\right) = (1 - 2p)^n \prod_{k=1}^{n} \left(p_{k-1} - \frac{1}{2}\right)$$

Since

$$\prod_{k=1}^{n} \left(p_k - \frac{1}{2}\right) = \left(p_n - \frac{1}{2}\right) \prod_{k=1}^{n-1} \left(p_k - \frac{1}{2}\right)$$

and

$$\prod_{k=1}^{n} \left(p_k - \frac{1}{2}\right) = \left(p_0 - \frac{1}{2}\right) \prod_{k=2}^{n} \left(p_{k-1} - \frac{1}{2}\right)$$

$$= \left(p_0 - \frac{1}{2}\right) \prod_{k=1}^{n-1} \left(p_k - \frac{1}{2}\right)$$

we have

$$p_n - \frac{1}{2} = (1-2p)^n\left(p_0 - \frac{1}{2}\right)$$

With $p_0 = 1$, the sought probability is

$$p_n = \frac{1}{2}[1 + (1-2p)^n]$$

Example 2.15 A physical system contains a large number, n, of particles. The volume Γ of a phase space (this is defined as a six-dimensional space, whose coordinates are three coordinates of the particles locations and three coordinates of their velocities) is finite. Find the most likely number of particles in the ith element of this space. The volume of the ith element is γ_i. Consider that the total energy E of the phase space is constant and can be calculated as the sum of the energies of all the particles.

Solution The most likely number n_i of particles in the ith element of the phase space in question can be found as

$$n_i = n\frac{\gamma_i}{\Gamma},$$

where

$$n = \sum_{i=1}^{k} n_i$$

is the total number of particles. The total energy is

$$E = \sum_{i=1}^{n} n_i E_i$$

In order to determine the most likely number n_i of particles in the ith element of the phase space we use the method of the maximum likelihood. In accordance with this method, one introduces the *Lagrange function* (the function of the maximum likelihood), which depends not only on the actual parameters of the problem but on some additional parameters (*Lagrange multipliers*) as well. These additional parameters are defined in such a way that the Lagrange function assumes its maximum value. Typically, one takes the Lagrange function in the form of a natural logarithm of the quantity of interest, using the fact that $y = \ln x$ is a strictly increasing function. In the case in question we form the Lagrange function as follows:

$$L = \ln \frac{n!}{n_1! n_2! \cdots n_k!} \left(\frac{\gamma_1}{\Gamma}\right)^{n_1} \left(\frac{\gamma_2}{\Gamma}\right)^{n_2} \cdots \left(\frac{\gamma_k}{\Gamma}\right)^{n_k} + \alpha \sum_{i=1}^{k} n_i - \beta \sum_{i=1}^{k} n_i E_i$$

where α and β are the Lagrange multipliers. The expression (polynomial distribution)

$$\frac{n!}{n_1! n_2! \cdots m_k!} \left(\frac{\gamma_1}{\Gamma}\right)^{n_1} \left(\frac{\gamma_2}{\Gamma}\right)^{n_2} \cdots \left(\frac{\gamma_k}{\Gamma}\right)^{n_k}$$

defines the complete probability of a complex event that can be defined as follows: "As a result of n trials, n_1 particles will be found in the volume γ_1, n_2 particles will be found in the volume γ_2, etc." The probabilities of such "constituent" events are $p_1 = \gamma_1/\Gamma$, $p_2 = \gamma_2/\Gamma$, etc. Taking the partial derivatives of the Lagrange function L with respect to the Lagrange multipliers α and β, forming a sum of these derivatives, and equating this sum to zero, we obtain

$$-\frac{\partial \ln n_i!}{\partial n_i} + \ln\left(\frac{\gamma_i}{\Gamma}\right) + \alpha - \beta E_i = 0, \qquad i = 1, 2, \ldots, k$$

Since the total number n of particles is large, one can put

$$\frac{\partial n_i!}{\partial n_i} \simeq \ln(n+1)! - \ln n! = \ln(n+1) \simeq \ln n$$

Then the obtained equation can be written as

$$-\ln n_i + \ln\left(\frac{\gamma_i}{\Gamma}\right) + \alpha - \beta E_i = 0$$

This equation results in the following formula for the most likely number n_i of particles contained in the volume γ_i:

$$n_i = \frac{\gamma_i}{\Gamma} e^{\alpha - \beta E_i}$$

2.2.2. Poisson distribution: flow of events

A random variable X follows the *Poisson distribution* if the probabilities of its values can be calculated by the formula:

$$P_m = P\{X = m\} = \frac{a^m}{m!} e^{-a}, \qquad a > 0, \ m = 0, 1, 2, \ldots \qquad (2.11)$$

This distribution is named after the French mathematician Siméon D. Poisson (1781–1840). The Poisson distribution depends on just one dimensionless parameter a. The mean and the variance of this distribution are the same:

$$\langle x \rangle = a, \qquad D_x = a$$

The most likely value m of the Poisson distribution can be found from the condition

$$a - 1 \leq m \leq a$$

If a is not an integer, the most likely m value is equal to the nearest integer which is smaller than m. If $a < 1$, then P_m is the maximum for $m = 0$. If a is an integer, the maximum probability takes place for two m values: $m = \alpha - 1$ and $m = \alpha$.

The Poisson distribution can be obtained from the binomial distribution by putting $p \to 0$ and $n \to \infty$, given that $a = np = $ constant. The Poisson distribution is used to make approximations when one deals with a large number of independent trials, in each of which the event of interest is "rare," i.e., occurs with small probability. Examples are: earthquakes, floods, various accidents, telephone calls, winning in a lottery, passage of particles from a radioactive source through a Geiger counter, etc. The Poisson distribution can also be used in the case of a number of points falling in a given region of space (one-, two-, or three-dimensional) if the location of the points in space is random and their number is small.

In the case of a one-dimensional Poisson distribution, *flows of events* are said to take place. A flow of events, or *traffic*, is a sequence of homogeneous events occurring at random moments of time. The average number of events per unit time is the *intensity* of the flow. It can be either constant or time-dependent.

The flow of events is said to be *without aftereffects* if the probability of the number of events that can be found within the given time interval is independent of the number of events in any other non-overlapping interval. A flow of events is *ordinary* if the probability of occurrence of two or more events that can be found within an elementary time interval Δt is negligibly small compared to the probability of the occurrence of just one event. An ordinary flow of events without an after effect is a *Poisson flow*, or *Poisson traffic*. If certain events form such a flow, then the number X of events occurring within an arbitrary time interval $(t_0, t_0 + \tau)$ has a Poisson distribution, in which the parameter

$$a = \int_{t_0}^{t_0 + \tau} \lambda(t)\, dt$$

is the mean value of the number of points in this interval and $\lambda(t)$ is the intensity of the flow. If $\lambda = $ constant, the flow is said to be *steady-state* or *elementary* flow of the Poisson type. For an elementary flow, the number of events within any time interval of the duration ("length") τ has a Poisson distribution with the parameter $a = \lambda\tau$.

A *random field of points* is a collection of points randomly scattered over a plane (or in space). The *intensity* (or *density*) λ *of a field* is the mean number of points that fall within a unit area (volume). A field of points is of the *Poisson type* if it possesses the following properties:

1. The probability of the number of points falling into any region of a plane (space) is independent from the number of points falling into any other nonoverlapping region.
2. The probability of two or more points falling into the elementary region $\Delta x \, \Delta y$ is negligibly small compared to the probability of one point falling into the region (the property of *ordinarinesss*).

The number X of points in a Poisson field falling within any region S of a plane (space) has a Poisson distribution, in which a is the mean value of the number of points falling into the region S. If the intensity $\lambda(x, y)$ of the field is constant, then the field is *homogeneous* (a property similar to the stationary state of a flow). For a homogeneous field with an intensity λ, the parameter a can be calculated as $a = s\lambda$, where s is the area (volume) of the region S. If a field is inhomogeneous, then the a value can be calculated as an integral:

$$a = \iint\limits_{(s)} \lambda(x, y) \, dx \, dy$$

for a plane, and as an integral:

$$a = \iiint\limits_{(s)} \lambda(x, y, z) \, dx \, dy \, dz$$

for a space.

When dealing with problems associated with the Poisson distribution, it is convenient to use tabulated values of the functions

$$P(m, a) = \frac{a^m}{m!} e^{-a} \tag{2.12}$$

and

$$R(m, a) = \sum_{k=0}^{m} \frac{a^k}{k!} e^{-a} \tag{2.13}$$

encountered in practical applications. The function $R(m, a)$ is the probability that the random variable X, which follows the Poisson distribution, assumes the value a not exceeding m: $R(m, a) = P\{X \leq m\}$. The tables for the functions $P(m, a)$ and $R(m, a)$ are given in Appendix A.

Example 2.16 The probability of failure of a machine part is estimated to be $p = 10^{-4}$. If 1000 such parts are to be built, what is the probability that two of them will fail?

Solution In order to use Poisson's distribution (2.11), we calculate the parameter $a = np = 1000 \times 10^{-4} = 0.1$. The formula (2.11), with $m = 2$, yields

$$P_2 = \frac{0.1^2 e^{-0.1}}{2!} = 0.004\,524$$

If the more general formula, based on the binomial distribution (2.4), were used (this formula does not require that the probability p be small), then, with $n = 1000$, $m = 2$, $p = 10^{-4}$, and $q = 1 - 10^{-4} = 0.9999$, one would have

$$P_2 = C(1000, 2)(10^{-4})^2(0.9999)^{998} = 0.004\,520$$

This gives excellent agreement.

Example 2.17 What is the probability that in the previous example there will be not more than two failures?

Solution Using the formula (2.13), we obtain

$$R(2, 0.1) = \sum_{k=0}^{2} \frac{0.1^k e^{-0.1}}{k!} = e^{-0.1}\left(1 + 0.1 + \frac{0.01}{2}\right) = 0.9998$$

Example 2.18 It has been observed that the average number of cars making a left turn at a given intersection, while the green left-turn signal is on, is equal to six. The arrival of cars to the traffic light is a random event which follows Poisson's distribution. The designated left-hand turn lane is suggested to be effective 95 percent if the time. How many cars should this lane be able to accommodate?

Solution The formula (2.13) yields, for $a = 6$:

$$R(m, 6) = e^{-6} \sum_{k=0}^{m} \frac{6^k}{k!} = b(m)e^{-6}$$

The calculated values of the function $R(m, 6)$ are given in the Table 2.1.

As evident from Table 2.1, a lane should be able to accommodate not less than 10 cars in order to meet the requirement that six cars in average are served per cycle and be effective 95 percent of the time. The calculation can be simplified if the Appendix A data are used.

Example 2.19 The failure rate of the given type of electronic devices is $\lambda = 0.002$ 1/hour. What is the probability that three devices will fail during 1000 hours of operation?

Solution The expected number of hours the device will not operate within 1000 hours of the expected operation is $a = 0.002 \times 1000 = 2$. Then, assuming Poisson's distribution of failures, the probability that three devices will fail during 1000 hours is

$$P_3 = \frac{2^3 e^{-2}}{3!} = 0.1804$$

i.e., about 18 percent.

Example 2.20 The probability that m earthquakes of the magnitude M or higher will occur in a given seismic region during a given period of time t

TABLE 2.1 The Calculated Probability that the Number of Cars Arriving at the Intersection does not Exceed the Given Number (for the mean value a = 6)

k	0	1	2	3	4	5	6
$6^k/k!$	1	6	18	36	54	64.8	64.8
$R(m, 6)$	0.00248	0.01735	0.06197	0.15120	0.28506	0.44568	0.60630
k	7	8	9	10	11	12	>12
$6^k/k!$	55.543	41.543	27.668	16.609	9.049	4.544	3.539
$R(m, 6)$	0.74398	0.84695	0.91554	0.95670	0.97914	0.99040	0.9992

follows the Poisson distribution. The number of earthquakes per unit time can be evaluated as

$$\lambda = \alpha e^{-\beta M}$$

where α and β are the parameters characterizing the seismic conditions in the given region. What is the probability that there will be only one earthquake of the magnitude M or higher during the time t? What is the probability that there will be no earthquakes of the magnitude M or higher during this time? What is the probability that at least one earthquake of this magnitude will occur during the time t?

Solution The Poisson distribution for the earthquake in the given region is

$$P_m = \frac{a^m e^{-a}}{m!}$$

where $a = \lambda t$ is the mean number of earthquakes of the magnitude M or higher during the time t and λ is the mean number of earthquakes per unit time. Putting $m = 1$ in this formula and using the given formula for the λ value we obtain the following expression for the probability that one of the earthquakes, which are expected to occur during the time t, will be of the magnitude of M or higher:

$$P(t, M) = \alpha t e^{-\beta M} \exp(-\alpha t e^{-\beta M})$$

The probability that there will be no earthquakes of the magnitude M or higher during the time t can be obtained from the Poisson distribution by putting $m = 0$:

$$P(t, M) = e^{-\lambda t} = \exp(-\alpha t e^{-\beta M})$$

The probability that at least one earthquake of the magnitude M or higher will occur during the time t (the risk function) is

$$P_1(t, M) = 1 - \exp(-\alpha t e^{-\beta M})$$

Example 2.21 Using the conditions of the previous example, determine the duration t of time during which an earthquake of the magnitude M or higher will occur with the probability P.

Solution Solving the equation for the probability that at least one earthquake of the magnitude M or higher will occur during the time t (the risk function), obtained in the previous example, for the time t, we have

$$t = -\frac{e^{\beta M}}{\alpha} \ln(1 - P)$$

For small enough P values, $\ln(1 - P) \simeq -P$, and therefore

$$t \simeq \frac{e^{\beta M}}{\alpha} P$$

If, for instance, $\alpha = 500$, $\beta = 2$, and $M = 7$, then

$$t \simeq \frac{e^{14}}{5000} P = 2405 P$$

In earthquake engineering, the level $P = 0.05$ corresponds to the case of a moderate risk, when the damage from a would-be earthquake is estimated only from the economics standpoint. With such a risk level, the earthquake of the magnitude $M = 7$ is expected to occur once in 120 years. The level $P = 0.001$ corresponds to the case of a low allowable risk, when an earthquake of the given magnitude can be dangerous for people's lives. In this case we obtain $t = 2.4$ years. These data indicate that there is a definite incentive to undertake measures for improved seismic stability of structures located in seismic regions.

Example 2.22 How many lottery tickets should one buy so that the probability of winning would not be smaller than P?

Solution Let the total number of lottery tickets be N and the total number of winning tickets be M. Then the probability that the bought ticket is a winning one is M/N. The purchase of each particular ticket can be viewed as a separate trial with the probability $p = M/N$ of success in a series of n independent trials. If this probability is substantially smaller than the given probability P (which is typically the case), one has to buy a large number of tickets in order that the probability of buying a winning ticket would not be smaller than P. Therefore one can assume that the purchase of a winning ticket is a rare event, and therefore the number of winning tickets is distributed in accordance with Poisson's law:

$$P_m = \frac{a^m}{m!} e^{-a}$$

where

$$a = n \frac{M}{N}$$

The probability that at least one of the purchased tickets is a winning one is $1 - P_0 = 1 - e^{-a}$, so that the number n should be computed as the least integer for which

$$e^{-a} = 1 - P$$

Thus, the number of the lottery tickets one should buy is

$$n = -\frac{N}{M} \ln (1 - P)$$

If, for instance, $N = 100\,000$, $M = 500$, and $P = 0.5$, then one has to buy $n = 138.5 \simeq 139$ tickets to break even.

Example 2.23 A car is subjected to an inspection and maintenance. The number of faults detected during inspection has a Poisson distribution with the parameter a. If no faults are detected, the maintenance takes on average two hours. If one or two faults are detected, another half hour is needed to eliminate each defect. If more than two faults are found, it takes four hours on average to fix the car. Find the distribution of the average time T of maintenance and repair of the car, and its mean value $\langle t \rangle$.

Solution The distribution of the average time T is

$$T: \begin{array}{|c|c|c|c|} \hline 2 & 2.5 & 3 & 6 \\ \hline e^{-a} & ae^{-a} & \dfrac{1}{2}a^2e^{-a} & 1 - e^{-a}\left(1 + a + \dfrac{a^2}{2}\right) \\ \hline \end{array}$$

The mean value of this time can be computed as

$$\langle t \rangle = e^{-a}(2 + 2.5a + 1.5a^2) + 6\left[1 - e^{-a}\left(1 + a + \dfrac{a^2}{2}\right)\right]$$

$$= 6 - e^{-a}(4 + 3.5a + 1.5a^2)$$

Example 2.24 The stars in a certain cluster form a three-dimensional Poisson field of points with the density λ (the average number of stars per unit volume). An arbitrary star is fixed and the nearest neighbor, the next nearest neighbor, the third nearest neighbor, and so on, are considered. Find the distribution densities functions of the distance R_n from the fixed star to its nth nearest neighbor.

Solution The cumulative distribution function is

$$F_n(r) = 1 - \sum_{k=0}^{n-1} \dfrac{a^k}{k!} e^{-a}$$

where

$$a = \dfrac{4}{3}\pi r^3 \lambda, \qquad r > 0$$

The corresponding density function is

$$f_n(r) = \dfrac{dF_n(r)}{dr} = \dfrac{a^{n-1}}{(n-1)!} e^{-a} 4\pi \lambda r^2, \qquad r > 0$$

Example 2.25 The outer space particles hitting the surface of a satellite form a field with the density λ (particle per meter squared). The equipment, exposed to the "rain" of the outer space particles, occupies the area S on the surface of the satellite. This equipment fails if two particles hit the equipment at the surface of the satellite. The probability that the equipment fails as a result of being hit with one particle is p. What is the probability that the equipment fails?

Solution The mean number of particles hitting the surface S occupied by the equipment is $a = \lambda S$. We examine the probabilities of the following hypotheses:

$$H_1 = \{\text{one particles hits the equipment}\}$$

$$H_2 = \{\text{not less than two particles hit the equipment}\}$$

Using Poisson distribution, we find that

$$P(H_1) = \lambda S e^{-\lambda S}$$

$$P(H_2) = R_2 = 1 - P_0 - P_1 = \bar{R}(1, \lambda S) = 1 - e^{-\lambda S}(1 + \lambda S)$$

The conditional probabilities of the event A, "the equipment fails," are

$$P(A \mid H_1) = p, \qquad P(A \mid H_2) = 1$$

Then the sought probability that the equipment fails can be found as a complete probability as follows:

$$P(A) = \lambda S p e^{-\lambda S} + 1 - e^{-\lambda S}(1 + \lambda S)$$
$$= 1 - e^{-\lambda S}[1 + \lambda S(1 - p)]$$

Example 2.26 Observations of a river that is subject to flooding have been carried out over a hundred year period. The number of floods that have occurred are as follows: there were 10 years with no floods, 30 years with one flood, 28 years with two floods, 22 years with three floods, 7 years with four floods, 2 years with five floods, and one year with as many as six floods. These were no years with seven or more floods. The floods are rare events that follow the Poisson distribution. Find the probability of flooding for the next ten years.

Solution The total number of the observed floods is

$$n = 30 \times 1 + 28 \times 2 + \cdots + 1 \times 6 = 196,$$

so that the mean number of floods per year is

$$a = \frac{196}{100} = 1.96,$$

and $e^{-a} = 0.1409$. The probabilities that no floods, one flood, two, three, four, five or six floods will occur can be computed as

$$0.1409 \times \frac{1.96^0}{0!} = 0.1409, \qquad 0.1409 \times \frac{1.96^1}{1!} = 0.2761,$$

$$0.1409 \times \frac{1.96^2}{2!} = 0.2706, \qquad 0.1409 \times \frac{1.96^3}{3!} = 0.1768,$$

$$0.1409 \times \frac{1.96^4}{4!} = 0.0866, \qquad 0.1409 \times \frac{1.96^5}{5!} = 0.0340,$$

$$0.1409 \times \frac{1.96^6}{6!} = 0.0111.$$

The probability

$$R(7, 1.96) = 0.1409 + 0.2761 + \cdots + 0.0111 = 0.9961$$

is smaller then 1.0, inferring that there is a minimal probability 0.0039 that seven floods or more might occur. The number of years with no flood, one flood, two, three, four, five, or six floods during the next ten years can be found as

$0.1409 \times 10 = 1.409,$ $\quad 0.2761 \times 10 = 2.761,$ $\quad 0.2706 \times 10 = 2.706,$

$0.1768 \times 10 = 1.768,$ $\quad 0.0866 \times 10 = 0.866,$ $\quad 0.0340 \times 10 = 0.340,$

$0.0111 \times 10 = 0.111.$

Note that the total number of years is only 9.961 for the reason explained.

2.2.3. Geometric distribution

The probabilities of the possible values $x = 0, 1, 2, m, \ldots$ of a random variable X, which follows a *geometric distribution*, can be calculated as

$$P_m = pq^{m-1}, \quad m = 0, 1, 2, \ldots \quad (2.14)$$

For a sequence of values m the probabilities P_m form an infinitely decreasing geometric progression with a common ratio q. Geometric distributions are encountered when a number of independent trials are carried out to reach a desirable result. In each trial the result is achieved with the probability p. The random variable X is the number of "useless" trials (till the first trial, in which the event of interest occurs). The ordered series of the random variable X is

$$X: \begin{array}{|c|c|c|c|c|c|} \hline 0 & 1 & 2 & \cdots & m & \cdots \\ \hline p & qp & q^2p & \cdots & q^mp & \cdots \\ \hline \end{array}$$

The mean and the variance of the geometric distribution are as follows:

$$\langle x \rangle = \frac{q}{p}, \quad D_x = \frac{q}{p^2}$$

Sometimes a random variable $Y = X + 1$ is considered (the number of the performed trials till the desirable result is achieved, including the successive trial). The ordered series of the random variable Y is

$$Y: \begin{array}{|c|c|c|c|c|c|} \hline 1 & 2 & 3 & \cdots & m & \cdots \\ \hline p & qp & q^2p & \cdots & q^{m-1}p & \cdots \\ \hline \end{array}$$

The mean and the variance of the random variable Y are

$$\langle y \rangle = \frac{1}{p}, \quad D_y = \frac{q}{p^2}$$

Example 2.27 A number of attempts are made to start an engine. Each attempt is independent of the other attempts. The probability that the attempt is successful is equal to p. The duration of each attempt is τ. What is the distribution, the mean value, and the variance of the total time T needed to start the engine?

Solution The number X of trials (attempts) is a variable with a geometric distribution beginning with unity. The random variable $T = X\tau$ has the distribution

$$T: \begin{array}{|c|c|c|c|c|c|} \hline \tau & 2\tau & 3\tau & \cdots & m\tau & \cdots \\ \hline p & qp & q^2p & \cdots & q^{m-1}p & \cdots \\ \hline \end{array}$$

The mean value and the variance of this variable are

$$\langle t \rangle = \tau \langle x \rangle = \frac{\tau}{p}, \qquad D_T = \tau^2 D_x = \frac{\tau^2 q}{p^2}, \qquad q = 1 - p$$

Example 2.28 A number of independent trials are made, and in each of the trials an event A may occur with the probability p. The trials are terminated after the occurrence of this event. Construct the ordered series and the frequency polygon for the number X of trials. Obtain formulas for its mean value and variance.

Solution The random variable X has a geometric distribution beginning with unity. The ordered series of the variable X is

$$X: \begin{array}{|c|c|c|c|c|c|} \hline 1 & 2 & 3 & \cdots & m & \cdots \\ \hline p & qp & q^2p & \cdots & q^{m-1}p & \cdots \\ \hline \end{array}, \qquad q = 1 - p$$

The mean value and the variance of the random variable X are as follows:

$$\langle x \rangle = \sum_{m=1}^{\infty} m q^{m-1} p = p \sum_{m=1}^{\infty} m q^{m-1} = \frac{p}{(1-q)^2} = \frac{1}{p}, \qquad D_x = \frac{q}{p^2}$$

Example 2.29 The probability that an object will be detected during one cycle of a radar surveillance is equal to $p = 0.2$. Find the mean and the variance of the number X of the surveillance cycles that will not lead to a detection of the object and of the number Y of the surveillance cycles that one would have to conduct until the object is detected (including the one, at which the object is detected).

Solution The mean and the variance of the random variable X can be found as

$$\langle x \rangle = \frac{1-p}{p} = \frac{1-0.2}{0.2} = 4$$

$$D_x = \frac{1-p}{p^2} = \frac{0.8}{0.04} = 20$$

$$\sigma_x = \sqrt{D_x} = \sqrt{20} = 4.46$$

The mean of the random variable Y is $\langle y \rangle = \langle x \rangle + 1 = 4 + 1 = 5$ and the variance is the same as for the variable X.

2.2.4. Hypergeometric distribution

A random variable X with possible values $0, 1, \ldots, m, \ldots, a$ has a *hypergeometric distribution* with parameters n, a and b if the probability that this variable assumes the value m is

$$P_m = P\{X = m\} = \frac{C(a, m)C(b, n - m)}{C(a + b, n)}$$

$$= \frac{a!\,b!\,n!\,(a + b - n)!}{(a + b)!} \cdot \frac{1}{m!\,(a - m)!\,(n - m)!\,(b - n + m)!}$$

$$m = 0, 1, \ldots, a \quad (2.15)$$

The mean and the variance of the hypergeometric distribution are as follows:

$$\langle x \rangle = \frac{na}{a + b}$$

$$D_x = \frac{nab}{(a + b)^2} + n(n - 1)\left[\frac{a}{a + b}\frac{a - 1}{a + b - 1} - \left(\frac{a}{a + b}\right)^2\right]$$

The distribution has the maximum value when the variable $X = m$ is found between $(a + 1)(n + 1)/(a + b + 2) - 1$ and $(a + 1)(n + 1)/(a + b + 2)$. If the $a + b$ and n are large, the most likely m value can be computed as

$$m = \frac{an}{a + b}$$

Such a distribution occurs, for instance, when there is an urn which contains a white and b black balls, and n balls are drawn from it. The random variable X is the number of white balls drawn. The distribution of this variable is expressed by the formula (2.15). The hypergeometric distribution is used in quality control.

Hypergeometric distribution is a generalization of the binomial distribution. The binomial distribution considers two mutually exclusive outcomes for which the probabilities of occurrence are p and $q = 1 - p$. This distribution is applied to evaluate the probability that the number of occurrences in n trials is m. However, if one arbitrarily chooses n samples from $N = a + b$ and examines the probability of getting exactly m occurrences of the event of interest in these n samples, then the binomial distribution cannot be applied. The probability distribution of m is given by the hypergeometric distribution. When $a \to \infty$, $b \to \infty$, and $a/(a + b) = p$, the hypergeometric distribution approaches the binomial one with the parameters n and p.

Example 2.30 There are nine devices in a box. Five of them are sound and four are faulty. One takes at random four devices. Determine the mean and the variance of the number X of the sound devices among those that were taken out.

Solution From the formula (2.15), with $a = 5$ and $b = 4$, we find

$$P_0 = \frac{C(5, 0)C(4, 4)}{C(9, 4)} = \frac{1 \times 1}{\frac{9 \times 8 \times 7 \times 6}{1 \times 2 \times 3 \times 4}} = \frac{1}{126} = 0.008$$

$$P_1 = \frac{C(5, 1)C(4, 3)}{C(9, 4)} = \frac{5 \times 4}{126} = \frac{10}{53} = 0.159$$

$$P_2 = \frac{C(5, 2)C(4, 2)}{C(9, 4)} = \frac{10 \times 6}{126} = \frac{10}{21} = 0.476$$

$$P_3 = \frac{C(5, 3)C(4, 1)}{C(9, 4)} = 0.317$$

$$P_4 = \frac{C(5, 4)C(4, 0)}{C(9, 4)} = 0.040$$

The ordered series is

X:	0	1	2	3	4
	0.008	0.159	0.476	0.317	0.040

The mean value $\langle x \rangle$ can be found as

$$\langle x \rangle = 0 \times 0.008 + 1 \times 0.159 + 2 \times 0.476 + 3 \times 0.317 + 4 \times 0.040 = 2.222$$

or on the basis of the formula

$$\langle x \rangle = \frac{4 \times 5}{9} = 2.222$$

The variance can also be found on the basis of the ordered series, or using the formula

$$D_x = \frac{4 \times 5 \times 4}{(5 + 4)^2} + 4 \times 3\left(\frac{5}{9} \times \frac{4}{8} - \frac{25}{81}\right) = 0.617$$

2.3. Generating functions

The complete information about the probability distribution and its moments is contained in a single expression—the *generating function*. Application of generating functions enables one to obtain by formal differentiation the mean, the variance, and, if necessary, also the higher moments of a probability distribution. The probability generating function for a discrete probability distribution P_k is a polynomial in the variable t whose coefficients are the values P_k:

$$G(t) = \sum_k P_k t^k \tag{2.16}$$

For example, for a die with six faces, the generating function is

$$G(t) = \frac{1}{6}(t + t^2 + \cdots + t^6) = \frac{t}{6}\frac{1-t^6}{1-t}$$

The generating function of the binomial distribution (2.4):

$$P_k = C(n, k)p^k q^{n-k}, \qquad q = 1 - p \tag{2.17}$$

is

$$G(t) = \sum_{k=0}^{n} [C(n, k)p^k q^{n-k}]t^k = (q + pt)^n \tag{2.18}$$

Let us show what kind of information can be drawn from the series (2.16). From (2.16) we obtain

$$G(1) = \sum_k P_k = 1$$

$$G'(t) = \sum_k k P_k t^{k-1} \tag{2.19}$$

$$G'(1) = \sum_k k P_k = \langle x \rangle$$

If we multiply $G'(t)$ by t and then differentiate again, we obtain

$$[tG'(t)]' = tG''(t) + G'(t) = \sum_k k^2 P_k t^{k-1}$$

The second moment about the mean is

$$D_x = G''(1) + G'(1) - [G'(1)]^2 = \sum_k k^2 P_x - \langle x \rangle^2 \tag{2.20}$$

The mean and the variance of the binomial distribution (2.4) can be found as follows. The derivatives of the distribution (2.4) are

$$G'(t) = np(q + p')^{n-1}$$

$$G''(t) = n(n-1)p^2(q + p')^{n-2}$$

Then, bearing in mind that $p + q = 1$, we obtain

$$G(1) = 1, \qquad G'(1) = np, \qquad G''(1) = n(n-1)p^2$$

Hence, the mean and the variance are

$$\langle x \rangle = np, \qquad D_x = n(n-1)p^2 + np - (np)^2 = npq$$

The generating function has one more useful property. If one asks how the sum of the faces of two dice can arise, it can be seen that it is found as the product of the generating function with itself:

$$G(t)G(t) = [G(t)]^2 = (t + t^2 + \cdots + t^6)^2 = t^2 + 2t^3 + 3t^4 + \cdots + t^{12}$$

The coefficient of any power of t is, in effect, the number of ways that the sums of the original exponents can produce this power. Indeed, the sum 2 can occur in just one way $(1+1)$, and this is the coefficient of t^2; the sum 3 can occur in two ways $(1+2, 2+1)$, and this is the coefficient of t^3; the sum 4 can occur in three ways $(1+3, 2+2, 3+1)$, and this is the coefficient of t^4, and so on. The coefficient "counts" the number of ways the sum can arise. This is a general result. Since the exponent of t is the same as the value assigned to the outcome of the trial, the power series product of two generating functions as, for each power of t, a coefficient that "counts" how often that sum can occur. In general, if

$$G(t) = \sum_i a_i t^i, \qquad H(t) = \sum_j b_j t^j \qquad (2.21)$$

then

$$G(t)H(t) = \sum_k c_k t^k \qquad (2.22)$$

where

$$c_k = \sum_{i+j=k} a_i b_{k-i} \qquad (2.23)$$

The sequence $\{c_k\}$ is the convolution of the sequence $\{a_i\}$ with the sequence $\{b_j\}$.

Let us show how the concept of the generating function can be applied to the polynomial distribution (2.6). The probability $P_{n;\,n_1,\,n_2,\,\ldots,\,n_m}$ can be determined as the coefficient of $t_1^{n_1} t_2^{n_2} \cdots t_m^{n_m}$ in the following generating function:

$$G(t_1, t_2, \ldots, t_m) = (p_1 t_1 + p_2 t_2 + \cdots + p_m t_m)^n \qquad (2.24)$$

The generating function for $n + N$ independent trials can be found as the product of generating functions for n and N trials. The use of this property simplifies substantially the computation of the sought probabilities. For the same purpose one can use the corresponding substitution of the arguments in the generating function. If, for instance, one wishes to determine the probability that the event A_1 will occur in n trials by l times more than the event A_2, then $t_1 = t$, $t_2 = 1/t$, $t_j = 1$ ($j = 3, 4, \ldots, m$) should be put into the generating function. The sought probability is the coefficient of t^l in the expansion of the function

$$G(t) = \left(p_1 t + \frac{p_2}{t} + \sum_{j=3}^{m} p_j\right)^n \qquad (2.25)$$

If $p_k = 1/m$ ($k = 1, 2, \ldots, m$) and one has to determine the probability that the sum of the numbers of the occurring events is equal to r, then this probability can be found as the coefficient of t^r in the expansion of the function

$$G(t) = \frac{1}{m^n} t^n (1 + t + \cdots + t^{m-1})^n = \frac{t^n}{m^n} \left(\frac{1-t^m}{1-t}\right)^n \qquad (2.26)$$

When expanding the function $G(t)$ into a series, it is convenient to use for $(1-t)^{-n}$ the followig expansion:

$$(1-t)^{-n} = 1 + C(n, n-1)t + C(n+1, n-1)t^2$$
$$+ C(n+2, n-1)t^3 + \cdots \qquad (2.27)$$

Although in this book we address only generating functions for discrete random variables, application of generating functions is equally useful for evaluating moments of continuous random variables.

Example 2.31 Determine the probability of buying a "lucky" ticket with a six-digit number in which the sums of the first three digits is equal to the sum of the last three digits.

Solution Examine first the first three digits. One can consider that three trials ($n = 3$) are carried out, and the probability of occurrence of each of the digits is $p = 1/10$. The total number of events is $m = 10$, and the probabilities of these events are $p_k = 1/10$, $k = 0, 1, 2, \ldots, 9$. The generating function is

$$G_1(t_0, t_1, \ldots, t_9) = \frac{1}{10^3} \left(\sum_{k=0}^{9} t_k\right)^3$$

where the index k indicates that the number k appears as a result of the experiment. Let us put $t_k = t^k$. Then the coefficient of t^m in the function

$$G_1(t) = \frac{1}{10^3} \left(\sum_{k=0}^{9} t^k\right)^3 = \frac{1}{10^3} \left(\frac{1-t^{10}}{1-t}\right)^3$$

is equal to the probability that the sum of the first three digits in the ticket's number is equal to m. Similarly, the coefficient of t^{-m} in the function

$$G_2(t) = \frac{1}{10^3} \left(\frac{1-t^{-10}}{1-t^{-1}}\right)^3$$

is equal to the probability that the sum of the last three digits in the ticket's number is equal to m. Then the coefficient of t^0 in the function

$$G(t) = G_1(t)G_2(t) = \frac{1}{10^6 t^{27}} \left(\frac{1-t^{10}}{1-t} \right)^6$$

is equal to the sought probability that the sum of the first three digits is the same as the sum of the last three digits. Thus, we have

$$(1-t^{10})^6 = 1 - C(6,1)t^{10} + C(6,2)t^{30} - \cdots$$
$$(1-t)^{-6} = 1 + C(6,5)t + C(7,5)t^2 + \cdots$$

and therefore

$$p = \frac{1}{10^6} [C(32,5) - C(6,1)C(22,5) + C(6,2)C(12,5)] = 0.055\,25.$$

Example 2.32 Write the generating functions for the binomial and Poisson distributions.

Solution As has been discussed in Sec. 2.2.1, the random variable ξ which is characterized by the binomial distribution

$$P_\xi(m) = C(n,m) p^m q^{n-m}, \qquad q = 1-p, \; m = 0, 1, \ldots, n$$

can be represented as a sum of n independent variables ξ_m, $m = 1, 2, \ldots, n$:

$$\xi = \xi_1 + \xi_2 + \cdots + \xi_m$$

where

$$\xi_m = \begin{cases} 1, & \text{with probability } p \\ 0, & \text{with probability } q \end{cases}$$

The generating function $G_{\xi_m}(t)$ of each summand is

$$G_{\xi_m}(t) = pt + q$$

and the generating function of the variable ξ is

$$G_\xi(t) = (pt + q)^n$$

The succession of these generating functions, when $n \to \infty$, converges to the generating function of the Poisson distribution:

$$G(t) = \lim_{n \to \infty} G_\xi(t) = \lim_{n \to \infty} (pt + q)^n$$

$$= \lim_{n \to \infty} [1 + p(t-1)]^n = \lim_{n \to \infty} \left[1 - \frac{a(1-t)}{n} \right]^n$$

where $a = \lim_{n \to \infty} np$. It is known that for any a value the following limit takes place:

$$\lim_{n \to \infty} \left(1 - \frac{a}{n} \right)^n = e^{-a}$$

Hence, the generating function of Poisson's distribution is

$$G(t) = e^{-a(1-t)}$$

Further Reading (see Bibliography): 7, 20, 30, 32, 35, 38, 40, 44, 58, 59, 62, 84, 88.

Chapter 3

Continuous Random Variables

"Nothing is impossible. It is often merely for an excuse that we say things are impossible."
FRANCOIS DE LA ROCHEFOUCOULD

"Mathematical formulas have their own life, they are smarter than we, even smarter than their authors, and provide more than what has been put into them."
HEINRICH HERTZ

"Everything should be made as simple as possible, but not one bit simpler."
ALBERT EINSTEIN

3.1. Characteristics of Continuous Random Variables

A nondiscrete random variable is characterized by an uncountable set of possible values. The sets of the possible values of such a variable continuously fill a certain section along the abscissa axis. The *continuous random variable X* is defined if its *cumulative distribution function*

$$F(x) = P\{X < x\} \tag{3.1}$$

is known (Fig. 3.1). This function can be interpreted as the fraction of values in the population less than or equal to x. This function is continuous for any x, and so is its derivative

$$f(x) = \frac{dF(x)}{dx} \tag{3.2}$$

except, maybe, at some particular points. The function $f(x)$ is the *probability density function*. It can be interpreted as the fraction of

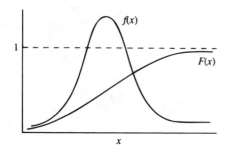

Figure 3.1 The cumulative distribution function $F(x)$ and the probability density function $f(x)$.

the population values that can be found in the interval dx. The probability of a particular individual value of a continuous random variable is equal to zero.

The probability that the random variable X can be found within the values $x = a$ and $x = b$ can be evaluated as

$$P\{a \leq X \leq b\} = F(b) - F(a) \tag{3.3}$$

As follows from (3.2),

$$F(x) = \int_{-\infty}^{x} f(x)\, dx \tag{3.4}$$

Since $F(\infty) = 1$, then

$$\int_{-\infty}^{\infty} f(x)\, dx = 1 \tag{3.5}$$

This formula expresses the *condition of normalization* for a continuous random variable.

The probability that a continuous random variable X falls within an interval from a to b can be calculated as

$$P\{a < X < b\} = \int_{a}^{b} f(x)\, dx = F(b) - F(a) \tag{3.6}$$

The *mean value* (mathematical expectation) and *variance* of a continuous random variable X are as follows:

$$\langle x \rangle = \int_{-\infty}^{\infty} x f(x)\, dx \tag{3.7}$$

$$D_x = \langle (x - \langle x \rangle)^2 \rangle = \int_{-\infty}^{\infty} (x - \langle x \rangle)^2 f(x)\, dx \tag{3.8}$$

The square root of variance is known as the *mean square deviation*, or *standard deviation*:

$$\sigma_x = \sqrt{D_x} \tag{3.9}$$

The standard deviation is often used to approximate the range of the possible values of a random variable. The *coefficient of variation*

$$\text{cov}_x = v_x = \frac{\sigma_x}{\langle x \rangle} \tag{3.10}$$

is used as a characteristic of the "degree of randomness" of a random variable. It is equal to zero for a deterministic variable and is equal to infinity for an "absolutely uncertain" variable.

The kth moment about the origin for a continuous random variable X is expressed as

$$m_k(x) = \langle x^k \rangle = \int_{-\infty}^{\infty} x^k f(x)\, dx \tag{3.11}$$

The kth central moment is

$$\mu_k(x) = \int_{-\infty}^{\infty} (x - \langle x \rangle)^k f(x)\, dx \tag{3.12}$$

The central moments can be expressed through the moments about the origin. The expression for the variance in terms of the second moment about the origin is of especial practical interest:

$$D_x = m_2(x) - \langle x \rangle^2 \tag{3.13}$$

Example 3.1 The probability density function of a random variable X is

$$f(x) = a \cos x, \qquad -\frac{\pi}{2} < x < \frac{\pi}{2}$$

Find the coefficient a, the probability distribution function $F(x)$, and the probability that the random variable X will be found between zero and $\pi/4$.

Solution Introducing the formula for the probability density function into the condition of normalization

$$\int_{-\infty}^{\infty} f(x)\, dx = \int_{-\pi/2}^{\pi/2} f(x)\, dx = 1$$

we obtain $a = 1/2$. The probability distribution function can be found as

$$F(x) = \begin{cases} 0, & x < -\pi/2 \\ \dfrac{1}{2}(1 + \sin x), & -\pi/2 < x < \pi/2 \\ 1, & x > \pi/2 \end{cases}$$

The probability that the random variable X will fall within the interval between zero and $\pi/4$ is

$$P\left(0 < X < \frac{\pi}{4}\right) = \int_0^{\pi/4} \frac{\cos x}{2} \, dx = \frac{\sqrt{2}}{4} = 0.353$$

Example 3.2 The probability of encountering a molecule of a gas inside a small volume dv is $a\,dv$. The distance of this molecule from the nearest adjacent one is a random variable. Find the probability distribution for this variable.

Solution The nearest adjacent molecule will be located at the distance between r and $r + dr$ from the given molecule if the following two conditions are fulfilled:

1. There should be a molecule inside a spheric layer $4\pi r^2 \, dr$.
2. The distance between the molecules should not be smaller than r.

The probability of the first event is $a(4\pi r^2 \, dr)$ and the probability of the second event is $1 - F(r)$. These two events are statistically independent and therefore the probability of the sought event is

$$f(r)\,dr = 4\pi a r^2 [1 - F(r)] \, dr$$

or, considering that $F'(r) = f(r)$,

$$\frac{1}{r^2} f'(r) - \frac{2}{r} f(r) + 4\pi a f(r) = 0$$

This equation has the following solution:

$$f(r) = Cr^2 \exp\left(-\frac{4}{3}\pi a r^3\right)$$

The constant C of integration can be found from the condition of normalization as follows:

$$C = 4\pi a$$

Thus, the probability density function $f(r)$ is

$$f(r) = 4\pi a r^2 \exp\left(-\frac{4}{3}\pi a r^3\right)$$

This formula indicates that the probability density of the distance r increases from zero to its maximum value when this distance increases from zero to $r = (2\pi a)^{-1/3}$ and then decreases to zero with a further increase in r.

3.2. Bayes Formula for Continuous Random Variables

If the probability of an event A depends on the value x assumed by a continuous random variable X whose probability density function is

$f(x)$, then the total probability of this event can be calculated on the basis of the following integral formula:

$$P(A) = \int_{-\infty}^{\infty} P(A/x) f(x) \, dx \tag{3.14}$$

Here

$$P(A/x) = P\{A/X = x\} \tag{3.15}$$

is the *conditional probability of the event A on the hypothesis* $\{X = x\}$, i.e., the probability that is calculated provided that the random variable X assumed the realization x.

If an event A occurs as a result of a certain experiment, and the probability of this event depends on the value assumed by the continuous random variable X, then the conditional probability density of the random variable X, where the occurrence of the event A is taken into account, is

$$f_A(x) = f(x) \frac{P(A/x)}{P(A)}$$

This relationship can be written, considering the formula (3.14), as follows:

$$f_A(x) = \frac{f(x) P(A/x)}{\int_{-\infty}^{\infty} P(A/x) f(x) \, dx} \tag{3.16}$$

This is the *Bayes formula* for a continuous random variable X.

Example 3.3 The probability of the event A is dependent on a random variable X. This variable uniformly distributed on the interval from 0 to 1:

$$f(x) = 1, \quad 0 < x < 1$$

The conditional probability of the event A for $X = x$ is

$$P(A \mid x) = x^2, \quad 0 < x < 1$$

Find the complete probability of the event A.

Solution Using the formula (3.14), we find

$$P(A) = \int_{-\infty}^{\infty} x^2 f(x) \, dx = \int_0^1 x^2 \, dx = \frac{1}{3}$$

Example 3.4 The weight X of an object is a continuous random variable with the probability density function $f(x)$. During the quality control of a batch of these objects, one rejects all those objects whose weights are outside the interval (x_1, x_2). A certain object was not rejected as a result of the quality control. Find the conditional probability density function of the weight X.

Solution Using the formula (3.16), we have

$$f_A(x) = \frac{f(x)P(A\,|\,x)}{P(A)}$$

As a result of the experiment, one found that the probability of the event A, "the object was not rejected," is

$$P(A) = \int_{x_1}^{x_2} f(x)\,dx$$

Since

$$P(A\,|\,x) = 0 \quad \text{for } X = x < x_1 \text{ and } X = x > x_2$$

and

$$P(A\,|\,x) = 1 \quad \text{for } X = x \in (x_1, x_2)$$

the sought probability density function is

$$f_A(x) = \frac{f(x)}{P(A)} = \frac{f(x)}{\int_{x_1}^{x_2} f(x)\,dx}$$

3.3. Frequently Encountered Continuous Distributions

3.3.1. Uniform distribution

A random variable X has a *uniform distribution* on the interval from a to b if its density on this interval is constant:

$$f(x) = \begin{cases} \dfrac{1}{b-a} & \text{for } x \in (a, b) \\ 0 & \text{for } x \notin (a, b) \end{cases} \qquad (3.17)$$

A uniform distribution is inherent in the measurement errors when using an instrument with large divisions and when the obtained value is rounded off to the nearest integer, or to the nearest smaller or a larger number. The error, when measuring the length of an object in centimeters with a ruler with centimeter divisions, has a uniform distribution on the interval $(-\frac{1}{2}, \frac{1}{2})$ if the obtained value is rounded off to the nearest integer or on the interval $(0, 1)$ if the obtained value is rounded down to the nearest smallest value. The error, in minutes, made using a watch with a jumping minute hand also has a uniform distribution on the interval $(0, 1)$. The rotation angle of a well-balanced wheel has a uniform distribution on the interval $(0, 2\pi)$ if it is set in motion and stops by friction. In Buffon's needle problem (Example 1.46, Sec. 1.5), the angle that defines the direction of the needle also has a uniform distribution, since the needle is thrown at random so that none of the values of this angle is more likely.

The cumulative distribution function of the uniform distribution is

$$F(x) = \frac{x-a}{b-a}$$

The mean value, variance, and coefficient of variation of this distribution are

$$\langle x \rangle = \frac{a+b}{2}, \quad D_x = \frac{(b-a)^2}{12}, \quad v_x = \frac{1}{\sqrt{3}}\frac{b-a}{b+a}$$

The uniform distribution has the largest entropy (uncertainty) of all the distributions on the given finite interval (a, b) (see Sec. 6.2).

Example 3.5 The subway trains are run with 2 min time intervals. A passenger arrived at the subway station at a random moment in time. What is the probability that the passenger will have to wait for a train not longer than $\frac{1}{2}$ min?

Solution The probability density function of the time T that the passenger will be waiting for a train is $f(t) = 1/2$, $0 < t < 2$. The probability that he or she will not wait for a train longer than one half of a minute is $P\{T < 1/2\} = 1/4$.

Example 3.6 A random variable X has a uniform distribution within the interval (a, b). Find the probability that as a result of an experiment it will deviate from its mean value more than three standard deviations

Solution The standard deviation of the variable X can be found as

$$\sigma_x = \frac{b-a}{2\sqrt{3}}$$

The distance of either of the limits a or b from the mean $\langle x \rangle = (a+b)/2$ is $(b-a)/2$. This distance is smaller than $3\sigma_x = \sqrt{3}(b-a)/2$. Hence,

$$P\{|X - \langle x \rangle| > 3\sigma_x\} = 0$$

3.3.2. Exponential distribution

A random variable X has an *exponential distribution* if its probability density function is (Fig. 3.2)

$$f(x) = \begin{cases} \lambda e^{-\lambda x}, & \text{for } x > 0 \\ 0, & \text{for } x < 0 \end{cases} \quad (3.18)$$

Here λ is the parameter of the distribution. An exponential distribution is especially significant in the theory of Markovian processes (see Chapter 11), in reliability engineering (see Chapter 10), and in

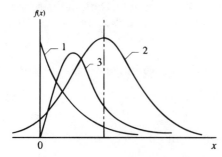

Figure 3.2 Different types of probability distribution: 1, exponential; 2, normal; 3, Rayleigh.

the queueing theory (see Chapter 11). If there is a simple Poisson flow with intensity λ on the time axis, then the time interval between two adjacent events has an exponential distribution with parameter λ.

The mean value, variance, standard deviation, and coefficient of variation for a random variable distributed in accordance with the exponential distribution are

$$\langle x \rangle = \frac{1}{\lambda}, \qquad D_x = \frac{1}{\lambda^2}, \qquad \sigma_x = \frac{1}{\lambda}, \qquad v_x = \frac{\sigma_x}{\langle x \rangle} = 1$$

The exponential distribution has the largest entropy (uncertainty) of all the distributions with the same mean (see Sec. 6.2).

Example 3.7 When a machine is operating, failures may occur at random moments of time. The time T of operation until the first failure occurs has an exponential distribution with parameter λ:

$$f(t) = \lambda e^{-\lambda t}, \qquad t > 0$$

When the failure occurs, the necessary corrections are made within a time t_0, after which the machine begins to operate again. Find the probability that the time interval Z between successive failures is larger than $2t_0$.

Solution The time interval Z between successive failures is related to the time T until the first failure as $Z = T + t_0$. Hence, the distribution of the random variable Z is

$$f(z) = \begin{cases} \lambda e^{-\lambda(z-t_0)}, & \text{for } z > t_0 \\ 0, & \text{for } z < t_0 \end{cases}$$

The sought probability can be found as

$$P\{Z > 2t_0\} = \int_{2t_0}^{\infty} f(z)\, dz = e^{-\lambda t_0}$$

Example 3.8 The time T between two malfunctions of a computer has an exponential distribution with parameter λ:

$$f(t) = \lambda e^{-\lambda t}, \qquad t > 0$$

A computer-aided solution to a certain problem requires that a computer runs failure-free for a time τ. If a malfunction occurs during the time τ, the solution of the problem must be restarted. A malfunction is detected only in a time τ after the solution has begun. We consider a random variable Z which is the time during which the problem will be solved. Find its distribution and the mean value (the average time needed to solve the problem), and the probability that no less than m problems will be solved during the time $t = k\tau$ $(k > m)$.

Solution The random variable is discrete and has an ordered series:

$$Z: \begin{array}{|c|c|c|c|c|} \hline \tau & 2\tau & \cdots & i\tau & \cdots \\ \hline p & pq & \cdots & pq(i-1) & \cdots \\ \hline \end{array}, \quad q = 1 - p$$

where $p = e^{-\lambda\tau}$ and $q = 1 - e^{-\lambda\tau}$. The mean value of the random variable Z is $\langle z \rangle = \tau/p = \tau e^{-\lambda\tau}$. Let us find the probability that no less than m problems will be solved during the time $t = k\tau$. We designate as $P_{m,k}$ the probability that exactly m problems will be solved during the time $t = k\tau$, i.e., the probability that exactly m out of the k time intervals τ will have no malfunctions. The probability that there will be no malfunctions during the time τ is $p = P(T > \tau) = e^{-\lambda\tau}$. According to the Bernoulli theorem on the repetition of trials,

$$P_{m,k} = C(k, m) p^m q^{k-m} = C(k, m) e^{-m\lambda\tau}(1 - e^{-\lambda\tau})^{k-m} \quad (3.19)$$

The probability that no less than m problems will be solved during the time $t = k\tau$ is

$$R_{m,k} = \sum_{i=m}^{k} P_{i,k} = \sum_{i=m}^{k} C(k, i) e^{-i\lambda\tau}(1 - e^{-\lambda\tau})^{k-i}$$

$$= 1 - \sum_{i=0}^{m-1} C(k, i) e^{-i\lambda\tau}(1 - e^{-\lambda\tau})^{k-i} \quad (3.20)$$

Example 3.9 Prove that the distribution of the intervals T between the successive events in a simple flow with intensity λ is exponential with parameter λ.

Solution We first find the distribution function $F(t)$ of the random variable T, which is the time between the successive events:

$$F(t) = P\{T < t\} = 1 - e^{-\lambda t}, \quad \text{for } t > 0 \quad (3.21)$$

This is the probability that at least one event occurs during the time t. Hence, the probability density function is

$$f(t) = \frac{dF(t)}{dt} = \lambda e^{-\lambda t}, \quad \text{for } t > 0 \quad (3.22)$$

Example 3.10 The probability density function of a random variable X is

$$f(x) = \frac{\lambda}{2} e^{-\lambda|x|}, \quad \lambda > 0$$

Find the mean value and the variance of this variable.

Solution The mean can be found as

$$\langle x \rangle = \int_{-\infty}^{\infty} x f(x) \, dx = \frac{\lambda}{2} \int_{-\infty}^{0} x e^{-\lambda x} \, dx + \frac{\lambda}{2} \int_{0}^{\infty} x e^{-\lambda x} \, dx$$

$$= \frac{\lambda}{2} \int_{\infty}^{0} y e^{-\lambda y} \, dy + \frac{\lambda}{2} \int_{0}^{\infty} x e^{-\lambda x} \, dx$$

$$= -\frac{\lambda}{2} \int_{0}^{\infty} y e^{-\lambda y} \, dy + \frac{\lambda}{2} \int_{0}^{\infty} x e^{-\lambda x} \, dx$$

Since

$$\int_{0}^{\infty} x^n e^{ax} \, dx = \frac{\Gamma(n+1)}{a^{n+1}}$$

then

$$\langle x \rangle = -\frac{\lambda}{2} \frac{\Gamma(2)}{\lambda^2} + \frac{\lambda}{2} \frac{\Gamma(2)}{\lambda^2} = 0$$

With $\langle x \rangle = 0$, we have

$$D_x = \int_{-\infty}^{\infty} x^2 f(x) \, dx = \frac{\lambda}{2} \int_{-\infty}^{0} x^2 e^{\lambda x} \, dx + \frac{\lambda}{2} \int_{0}^{\infty} x^2 e^{-\lambda x} \, dx$$

$$= \frac{\lambda}{2} \int_{0}^{\infty} y^2 e^{-\lambda y} \, dy + \frac{\lambda}{2} \int_{0}^{\infty} x^2 e^{-\lambda x} \, dx$$

Since

$$\int_{0}^{\infty} x^2 e^{-\lambda x} \, dx = \frac{\Gamma(3)}{\lambda^3}$$

we obtain

$$D_x = \frac{2}{\lambda^2}$$

3.3.3. Normal (Gaussian) distribution

This distribution was first introduced by De Moivre in 1733. However, it was Laplace in 1810 who implemented the normal distribution as a useful practical tool. The distribution is often called gaussian, after the famous German mathematician who showed (1809) that the distribution of the measurement errors must be normal. The probability density of this distribution is (see Fig. 3.2)

$$f(x) = \frac{1}{\sqrt{2\pi D_x}} \exp\left[-\frac{(x - \langle x \rangle)^2}{2 D_x}\right], \quad -\infty < x < \infty \quad (3.23)$$

The distribution is defined by two parameters: the mean value $\langle x \rangle$ and the variance D_x of the random variable X.

The normal distribution is "bell-shaped," i.e. symmetric with respect to the straight line $x = \langle x \rangle$. The maximum (most likely) value of the normal distribution takes place for $x = \langle x \rangle$ and is equal to

$$f_{max} = \frac{1}{\sqrt{2\pi D_x}} \tag{3.24}$$

The cumulative distribution function of the normal distribution can be found as

$$F(x) = \int_{-\infty}^{x} f(x)\, dx = \frac{1}{2} + \int_{0}^{x} f(x)\, dx = \frac{1}{2}[1 + \Phi(\alpha)] \tag{3.25}$$

where

$$\Phi(\alpha) = \frac{2}{\sqrt{\pi}} \int_{0}^{\alpha} e^{-t^2}\, dt, \qquad \alpha = \frac{x - \langle x \rangle}{\sqrt{2D_x}} \tag{3.26}$$

is the *Laplace function*, or *the probability integral*. We have already encountered this function in Example 2.4, Sec. 2.2.1. This function has the following major properties:

$$\Phi(0) = 0, \qquad \Phi(-\alpha) = -\Phi(\alpha), \qquad \Phi(\infty) = 1$$

The tabulated values of the Laplace function are given in Appendix B.

The probability that a normally distributed random variable X is found within the interval (a, b) can be computed as

$$P\{X \in (a, b)\} = \int_{a}^{b} f(x)\, dx = \frac{1}{\sqrt{2\pi D_x}} \int_{a}^{b} e^{(x - \langle x \rangle)^2/(2D_x)}\, dx$$

$$= \frac{1}{2}\left[\Phi\left(\frac{b - \langle x \rangle}{\sqrt{2D_x}}\right) - \Phi\left(\frac{a - \langle x \rangle}{\sqrt{2D_x}}\right)\right] \tag{3.27}$$

Normal distribution is not only the most widespread one used in practical problems, but is also the cornerstone of the probability theory. It arises when the variable X results from summation of a large number of independent (or weakly dependent) random variables comparable from the standpoint of their effect on the scattering of the sum. In other words, the distribution of the sum of a number of random variables converges, under very general conditions, to the normal distribution as the number of variables in the sum becomes large. This remarkable property of the normal distribution is known in the probability theory as the *central limit theorem*, which states that the sum of many small random effects is normally distributed. Examples are: displacements in Brownian motion, errors associated with repeated measurements, ordinates of irregular sea waves, drops of water from a leaky faucet that seep steadily into the ground, flow through porous media, etc. The central limit theorem does not require that the random variables entering the sum are described by the same distribution function or even that they be entirely independent. This is in part the reason why this theorem provides a simple and practical tool for approximating probabilities associated with the sum of collections of arbitrary distributed random variables.

The kth central moment of the normal distribution (see Sec. 3.1) can be found as

$$\mu_k = \frac{1}{\sqrt{2\pi D_x}} \int_{-\infty}^{\infty} (x - \langle x \rangle)^k \exp\left[-\frac{(x - \langle x \rangle)^2}{2D_x}\right] dx$$

$$= (k-1)D_x \frac{1}{\sqrt{2\pi D_x}} \int_{-\infty}^{\infty} (x - \langle x \rangle)^{k-2} \exp\left[-\frac{(x - \langle x \rangle)^2}{2D_x}\right] dx$$

Hence, the following formula is valid for the normal distribution:

$$\mu_k = (k-1)D_x \mu_{k-2}$$

If k is an odd number, all the central moments are equal to zero: the distribution is symmetric with respect to its mean value.

In many applications the *error function*

$$\text{erf } \alpha = \frac{1}{\sqrt{2\pi}} \int_{-\infty}^{\alpha} e^{-(t^2/2)} dt$$

is used instead of the Laplace function. The error function erf α is related to the Laplace function $\Phi(\alpha)$ as follows:

$$\text{erf } \alpha = \frac{1}{\sqrt{\pi}} \int_{-\infty}^{\alpha/\sqrt{(2)}} e^{-t^2} dt$$

$$= \frac{1}{\sqrt{\pi}} \int_{-\infty}^{0} e^{-t^2} dt + \frac{1}{\sqrt{\pi}} \int_{0}^{\alpha/\sqrt{(2)}} e^{-t^2} dt$$

$$= \frac{1}{2}\left[1 + \Phi\left(\frac{\alpha}{\sqrt{2}}\right)\right]$$

or

$$\Phi(\alpha) = 2 \text{ erf } (\alpha\sqrt{2}) - 1$$

While the function $\Phi(\alpha)$ changes from 0 to 1.0, the function erf α changes from $\frac{1}{2}$ to 1.0. The error function can also be defined as

$$\text{erf } \alpha = \frac{1}{\sqrt{2\pi}} \int_0^\alpha e^{-(t^2/2)} \, dt$$

In this case it is related to the Laplace function $\Phi(\alpha)$ as

$$\text{erf } \alpha = \frac{1}{2} \Phi\left(\frac{\pi}{\sqrt{2}}\right)$$

and changes from 0 to $\frac{1}{2}$. Note that in some books the opposite definitions of the functions Φ and erf are used.

In practical applications, the range of possible values of a normally distributed random variable is often assessed by using the *rule of three sigma*: the realizations of the random variable X that lie outside the region $(\langle x \rangle - 3\sigma_x, \langle x \rangle + 3\sigma_x)$ need not be considered. (see Example 3.6, Sec. 3.3.1). The probability of the event that the variable X falls within this region is

$$P = \int_{\langle x \rangle - 3\sigma_x}^{\langle x \rangle + 3\sigma_x} f(x) \, dx = \frac{1}{2}\left[\Phi\left(\frac{3}{\sqrt{2}}\right) - \Phi\left(-\frac{3}{\sqrt{2}}\right)\right]$$

$$= \Phi\left(\frac{3}{\sqrt{2}}\right) = 0.9973$$

This result implies that the probability that the random variable X deviates from the mean value $\langle x \rangle$ by more than $3\sigma_x$ is as low as 0.27 percent.

The normal law has the maximum entropy (uncertainty) among all the laws of probability distributions of continuous random variables for which only the expected (mean) value and the variance are known (see Sec. 6.2).

Example 3.11 Balls intended for the ball-bearings are sorted by passing them over two holes. If a ball does not pass through a hole of diameter d_1 but passes through a hole of diameter $d_2 > d_1$, it is accepted; otherwise it is rejected. The diameter D of a ball is a normally distributed random variable with the mean value $\langle d \rangle = (d_1 + d_2)/2$ and standard deviation $\sqrt{D_d} = (d_2 - d_1)/4$. Find the probability P that the ball is rejected.

Solution The interval (d_1, d_2) is symmetric with respect to the mean value $\langle d \rangle$. Setting $\sqrt{D_d} = (d_2 - d_1)/2$, we find that the probability that the ball is not rejected is

$$P\left\{|D - \langle d \rangle| < \frac{d_2 - d_1}{2}\right\} = \Phi\left(\frac{d_2 - d_1}{\sqrt{2D_d}}\right)$$

so that

$$P = 1 - \Phi\left(\frac{d_2 - d_1}{\sqrt{2D_d}}\right) = 1 - \Phi(2\sqrt{2}) = 0.0455$$

Example 3.12 The underkeel clearance M_s for a ship passing a shallow waterway in still water conditions is a normally distributed random variable:

$$f_s(m_s) = \frac{1}{\sqrt{2\pi D_s}} e^{-(m_s - \langle m_s \rangle)^2/(2D_s)}$$

Here $\langle m_s \rangle$ and D_s are the mean and the variance of the random variable M_s. The uncertainties in the evaluation of the M_s value are due to the fact that the ship's draft is affected by the errors in the prediction of this draft, ship's squat, water density, as well as by various uncertainties in the charted depth, sea bed relief, tide, siltation, etc. Determine the factor $\gamma_s = \langle m_s \rangle / \sqrt{2D_s}$ that would enable the ships navigator to safely pass the waterway, if the required probability of safe passing in the waterway is $P_s = 0.99999$.

Solution The cumulative probability distribution function of the random variable M_s is

$$F_s(m_s) = \int_{-\infty}^{m_s} f_s(m_s)\, ds = \frac{1}{2}\left[1 + \Phi\left(\frac{m_s - \langle m_s \rangle}{\sqrt{2D_s}}\right)\right]$$

where

$$\Phi(\alpha) = \frac{2}{\sqrt{\pi}} \int_0^\alpha e^{-t^2}\, dt$$

is the Laplace function (the probability integral). The probability of safe passing in the waterway can be computed as

$$P_s = 1 - F_s(0) = \frac{1}{2}[1 + \Phi(\gamma_s)]$$

where

$$\gamma_s = \frac{\langle m_s \rangle}{\sqrt{2D_s}}$$

is the factor of safety. The calculated values of this factor for different probabilities P_s are given in Table 3.1.

It is evident from this table that the factor of safety of $\gamma_s = 3.200$ corresponds to the probability $P_s = 1 - 10^{-5} = 0.99999$. The approach used to solve this problem is a typical one and is widely used in engineering. It will be discussed in greater detail in Chapter 10.

TABLE 3.1 Probability of Safe Passing a Waterway vs Factor of Safety

P_s	$1-10^{-3}$	$1-10^{-4}$	$1-10^{-5}$	$1-10^{-6}$	$1-10^{-7}$	1
γ_s	2.185	2.630	3.200	3.360	3.710	∞

3.3.4. Distributions associated with the normal law

3.3.4.1. Distribution of the absolute value of a normal random variable. If a random variable X follows the normal law of distribution with the mean $\langle x \rangle$ and the variance D_x, then its absolute value (module) $|X|$ has the following probability density function:

$$f(|x|) = \frac{1}{\sqrt{2\pi D_x}} [e^{-(|x|-\langle x \rangle)^2/(2D_x)} + e^{-(|x|+\langle x \rangle)^2/(2D_x)}].$$

In a special case, when $\langle x \rangle = 0$,

$$f(|x|) = \frac{2}{\sqrt{2\pi D_x}} e^{-|x|^2/(2D_x)}$$

The cumulative distribution function of the absolute value of a normal random variable is

$$F(|x|) = \frac{1}{2}\left[\Phi\left(\frac{|x|-\langle x \rangle}{\sqrt{2D_x}}\right) + \Phi\left(\frac{|x|+\langle x \rangle}{\sqrt{2D_x}}\right)\right]$$

The mean value of the module $|X|$ is

$$\langle |x| \rangle = \sqrt{\frac{2D_x}{\pi}}$$

Variance for $\langle x \rangle = 0$ is

$$D_{|x|} = \left(1 - \frac{2}{\pi}\right)D_x$$

3.3.4.2. Truncated normal distribution. If a random variable X is distributed in accordance with the normal law and its possible values fall within the interval (a, b), then its probability density function can be written as

$$f(x) = \frac{c}{\sqrt{2\pi D_x}} e^{-(x-\langle x \rangle)^2/(2D_x)}, \qquad a < x < b$$

where the factor c can be determined from the condition of normalization

$$\int_a^b f(x)\, dx = 1$$

as follows:

$$c = \frac{1}{\frac{1}{2}\left[\Phi\left(\frac{b - \langle x \rangle}{\sqrt{2D_x}}\right) - \Phi\left(\frac{a - \langle x \rangle}{\sqrt{2D_x}}\right)\right]}$$

If one assumes $a = \langle x \rangle - k\sqrt{D_x}$ and $b = \langle x \rangle + k\sqrt{D_x}$, then the factor c can be calculated as

$$c = \frac{1}{\Phi(k/\sqrt{2})}$$

For $k = 2$ we obtain $\Phi(\sqrt{2}) = 0.9543$ and $c = 1.004$. Hence, for $k = 2$ and larger, the c value is very close to unity.

3.3.4.3. Log-normal distribution. Let a random variable X be distributed in accordance with the normal law

$$f_1(x) = 1/\sqrt{2\pi D_x}\, e^{-(x - \langle x \rangle)^2/(2D_x)}$$

with the mean $\langle x \rangle$ and the variance D_x, and let this variable be related to a random variable Y as

$$X = \log Y, \quad 0 < y < \infty$$

Then the probability density function for the variable Y is

$$f_2(y) = \frac{M}{y} f_1(\log y)$$

or

$$f_2(y) = \frac{M}{y\sqrt{2\pi D_x}} e^{-(\log y - \log y_0)^2/(2D_x)}$$

Here $M = 0.430\,43$ is the coefficient of transfer of the natural logarithms to the decimal ones. The mean $\log y_0$ is equal to the mean $\langle x \rangle$.

The distribution $f_2(y)$ is known as the *log-normal distribution* (Fig. 3.3). This distribution is a flexible analytical model for reliability data. It is used to model variance failures due to chemical reac-

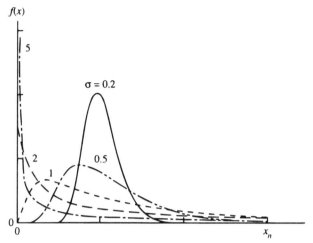

Figure 3.3 Lognormal distribution probability density functions.

tions, molecular diffusion, migration, insufficient strength, etc. Some types of brittle crack growth can also be described by log-normal distributions.

The cumulative probability distribution function for the random variable Y is

$$F(y) = \frac{1}{2}\left[1 + \Phi\left(\frac{\log y - \log y_0}{\sqrt{2D_x}}\right)\right]$$

The mean value and variance of the variable Y are as follows:

$$\langle y \rangle = y_0 \, e^{D_x/(2M^2)}$$

$$D_y = \langle y \rangle^2 \left[\left(\frac{\langle y \rangle}{y_0}\right)^2 - 1\right]$$

3.3.5. Rayleigh distribution

If X and Y are normally distributed random variables, then the variable

$$R = \sqrt{X^2 + Y^2}$$

is distributed in accordance with Rayleigh's law. Another way to introduce this law of distribution is as follows.

Let a random variable X, which obeys the normal law of distribution with variance D_x and zero mean, be changing periodically with a constant frequency but with a random amplitude R and random phase angle. Then the values of this amplitude obey the *Rayleigh law* of the probability distribution (see Fig. 3.2):

$$f(r) = \frac{r}{D_x} \exp\left(-\frac{r^2}{2D_x}\right), \quad R > 0 \qquad (3.28)$$

From this expression we find

$$F(r) = \int_0^r f(r)\, dr = 1 - \exp\left(-\frac{r^2}{2D_x}\right) \qquad (3.29)$$

The mean value and the variance of the random variable R are

$$\langle r \rangle = \int_0^\infty r f(r)\, dr = \sqrt{\frac{\pi}{2} D_x} \qquad (3.30)$$

$$D_r = \int_0^\infty (r - \langle r \rangle)^2 f(r)\, dr = \frac{4 + \pi}{2} D_x \qquad (3.31)$$

The coefficient of variation is

$$v_r = \frac{\sqrt{D_r}}{\langle r \rangle} = \sqrt{1 + \frac{4}{\pi}} \qquad (3.32)$$

The Rayleigh distribution was obtained by J. J. Thomson (Lord Rayleigh) in 1880 in connection with the investigation of the resultant intensity of a large number of independent sounds. Rayleigh's law is widely used in applied problems in which a random variable can assume positive values only (amplitudes, energies, etc). For instance, in oceanography and naval architecture this law is used to describe the short-term distribution of the heights of ocean waves, as well as various responses characterized by narrow-band spectra. From equation (3.29) we find:

$$D_x = -\frac{r^2}{2 \ln F^*(r)} \qquad (3.33)$$

where

$$F^*(r) = 1 - F(r)$$

is the probability that the random variable R exceeds a certain level r. The current scale of the intensity of ocean waves (see Sec. 15.2) is based on a wave height of the 1/3 significance; i.e., the probability $F^*(r) = 1/3$ that the given level r is exceeded is assumed to be a standard one. Then the formula (3.33) yields

$$D_x = 0.455 \left(\frac{h_{1/3}}{2} \right)^2$$

Example 3.13 Points are distributed randomly on a plane, in accordance with the Poisson law with a constant density λ. Prove that the distance R between the adjacent points follows the Rayleigh distribution and find the variance of this distance.

Solution We draw a circle of radius r about an arbitrary point in the field. For the distance R from this point to its nearest neighbor to be smaller than r, at least one other point must fall within the circle. The proper ties of the Poisson fields are such that the probability of this event is independent of whether there is a point in the center of the field or not. Therefore,

$$F(r) = 1 - e^{-\pi r^2 \lambda}, \quad r > 0$$

so that the probability density function is

$$f(r) = \frac{dF(r)}{dr} = 2\pi \lambda r e^{-\pi \lambda r^2}$$

This is Rayleigh distribution, in which $\lambda = 1/(4\langle r \rangle^2)$. The variance of this distribution is

$$D_r = \frac{4 - \pi}{4\pi\lambda}$$

Example 3.14 The coefficient of friction F is a random variable that has a Rayleigh distribution. Find the mode f_m of the random variable F, i.e., the abscissa of the maximum of its probability density function (the most likely F value), and the probability that the actual value of the coefficient of friction will not exceed the mode.

Solution By differentiating the probability density function

$$p(f) = \frac{f}{D_x} \exp\left(-\frac{f^2}{2D_x} \right)$$

with respect to f and equating the obtained expression to zero, we find

$$f_m = \sqrt{D_x}$$

The sought probability is

$$P\{F < f_m\} = \int_0^{f_m} p(f)\, df = \int_0^{f_m} \frac{f}{D_x} \exp\left(-\frac{f^2}{D_x} \right) df$$

$$= 1 - e^{-f_m^2/(2D_x)} = 1 - \frac{1}{\sqrt{e}} = 0.3935$$

Example 3.15 The coefficient F of friction in a frictional-contact axial clutch is distributed in accordance with the Rayleigh law:

$$p(f) = \frac{f}{f_m^2} e^{-f^2/(2f_m^2)}$$

where f_m is the most likely value ("mode") of this coefficient. The actuating force F_a in the clutch is related to the torque Q as follows:

$$F_a = C \frac{Q}{f}$$

where the factor C depends on whether a uniform wear or a uniform pressure condition takes place. Establish the magnitude of the actuating force in such a way that the probability that the actual torque Q will stay within the desired (required) limits $Q_1 < Q < Q_2$ is the largest.

Solution The torque Q treated as a random variable, which is a nonrandom function of the random variable f, is distributed as follows:

$$p_Q(Q) = \frac{Q}{Q_m} e^{-Q^2/(2Q_m)}$$

where

$$Q_m = \left(\frac{f_m}{C} F_a\right)^2$$

is the most probable value of the torque Q. The probability that the torque Q will be found between the values Q_1 and Q_2 is

$$P\{Q_1 < Q < Q_2\} = e^{-Q_1^2/(2Q_m)} - e^{-Q_2^2/(2Q_m)}$$

From this expression we find that the probability P will be the largest if

$$Q_m = \frac{Q_1^2 - Q_2^2}{4 \ln (Q_1/Q_2)}$$

or if the actuating force F_a assumes the following value:

$$F_a = \frac{C}{f_m} \sqrt{Q_m} = \frac{C}{2f_m} \frac{Q_1^2 - Q_2^2}{\ln (Q_1/Q_2)} = \frac{CQ_2}{2f_m} \sqrt{\frac{\eta^2 - 1}{\ln \eta}}$$

Here $\eta = Q_1/Q_2$ is the ratio of the extreme torque values. If the actuating force is established in accordance with this formula, then there is a reason to expect that the torque will remain within the design limits during the operation time of the clutch.

3.3.6. Rice distribution

If high-frequency oscillations superimpose the low-frequency oscillations of a random variable $X(t)$, then the distribution of the amplitudes of the maxima of the variable X might differ substantially from the Rayleigh law. In such a case the relative extrema

$$\eta_{max} = \frac{x_{max}}{\sqrt{D_x}}, \quad \eta_{min} = \frac{x_{min}}{\sqrt{D_x}}$$

of the random variable $X(t)$, distributed in accordance with the normal law, obey the following *Rice distribution* (see Fig. 3.2):

$$f(\eta) = \frac{1}{\sqrt{2\pi}(1+\sqrt{1-\varepsilon^2})}$$
$$\times \left[\varepsilon e^{-\eta^2/(2\varepsilon^2)} + \eta e^{-\eta^2/2}\sqrt{2(1-\varepsilon^2)} \int_{-\infty}^{\eta\sqrt{(1-\varepsilon^2)/(2\varepsilon^2)}} e^{-t^2} dt \right] \quad (3.34)$$

Here ε is the width of the energy spectrum of the random process $X(t)$. O. Rice was on the staff of AT&T Bell Laboratories in the mid-1940s, and in 1944 and 1945 carried out pioneering work in the field of the mathematical analysis of random noise. The Rice law is often employed in telecommunications engineering, oceanography, ocean engineering, and naval architecture. When $\varepsilon = 1$, the Rice law becomes normal law truncated at $\eta = 0$. When $\varepsilon = 0$, the Rice law reduces to the Rayleigh law. In practice, the Rayleigh law, because of its simplicity, is used up to $\varepsilon = 0.4$.

3.3.7. Weibull distribution

The Rayleigh law can be obtained also as a special case of the following *Weibull distribution* (Fig. 3.4):

$$f(x) = \frac{a}{b}\left(\frac{x}{b}\right)^{a-1} \exp\left[-\left(\frac{x-x_0}{b}\right)^a\right], \quad (3.35)$$

$$x \geq 0, \, a > 0, \, b > 0, \, x_0 > 0$$

where a is the *shape parameter*, b is the *scale parameter*, and x_0 is the *shift parameter* (the minimum possible value of the random variable X). This triparametric distribution was introduced in 1939 by a Swedish engineer, W. Weibull, on empirical ground in the statistical theory of the strength of materials.

The mean value and the variance of a random variable x obeying the Weibull distribution are

$$\langle x \rangle = b\Gamma\left(1+\frac{1}{a}\right) + x_0 \quad (3.36)$$

$$D_x = b^2\left[\Gamma\left(1+\frac{2}{a}\right) - \Gamma^2\left(1+\frac{1}{a}\right)\right] \quad (3.37)$$

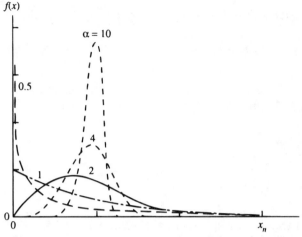

Figure 3.4 Weibull distribution probability density functions.

where

$$\Gamma(\alpha) = \int_0^\infty x^{\alpha-1} e^{-x} \, dx \tag{3.38}$$

is the *gamma function*. It is tabulated in Appendix C.

The cumulative distribution function of a variable X, which follows the Weibull distribution, is

$$F(x) = 1 - \exp\left[-\left(\frac{x - x_0}{b}\right)^a\right]$$

If $x_0 = 0$, the Weibull law with $a = 2$ becomes the Rayleigh law, with $a = 1$ it becomes the exponential law, and with $a = 3$ it results in a distribution that is close to the normal distribution.

The Weibull law is widely used in various engineering problems and, particularly, in reliability analyses and in the theory of fatigue of materials. The Weibull distribution is applicable also to the long-term distribution of ocean wave heights, wave periods, extreme winds, etc. It is a very flexible distribution with a wide variety of possible curve shapes. This law has proven to be a successful analytical model for many product failure mechanisms for all the major periods of the product use: infant mortality, normal (steady-state) operation and aging (see Chapter 10). It has been shown that the Weibull distribution can also be obtained as an extreme value distribution (see Chapter 9). In reliability engineering, it suggests its theoretical applicability when failure is due to a "weakest link" of many possible failure points. One expects that the Weibull distribution will apply when failure occurs at a defect site within a material and there are many such sites competing with each other to be the first to produce a failure.

3.3.8. χ^2-distribution. Gamma distribution. Erlang distribution

The χ^2 distribution is widely used in statistics and reliability engineering. It describes the distribution of the probabilities of a random variable

$$\chi^2 = \xi_1^2 + \xi_2^2 + \cdots + \xi_n^2$$

where $\xi_1, \ldots, \xi_n$ are independent random variables, distributed in accordance with the normal law with zero mean and unit standard deviation. The probability density function of the χ^2-*distribution* (or Pearson distribution) is expressed as

$$f(x) = \frac{1}{2^{n/2}\Gamma(n/2)} x^{n/2-1} e^{-x/2}, \qquad 0 < x < \infty \tag{3.39}$$

where n is the parameter of the distribution (the number of the "*degrees of freedom*"). The cumulative distribution function of the χ^2-distribution is

$$F(\chi) = 1 - P(\chi^2, n) \tag{3.40}$$

The values of the function

$$P(\chi^2, n) = \frac{2^{1-n/2}}{\Gamma(n/2)} \int_\chi^\infty x^{n-1} e^{-x^2/2} \, dx \tag{3.41}$$

are tabulated (see Appendix D). When $n = 2$, the χ^2 distribution becomes the Rayleigh distribution, and when $n = 1$ it becomes a distribution whose ordinates are twice as large as the ordinates of the positive branch of the normal distribution with zero mean.

The χ^2 distribution can be viewed as a special case of the *gamma distribution*, whose probability density is expressed as follows:

$$f(x) = \begin{cases} 0, & \text{if } x \leq 0 \\ \dfrac{b^a}{\Gamma(a)} x^{a-1} e^{-bx}, & \text{if } x > 0 \end{cases} \tag{3.42}$$

The parameters a and b are both positive, and $\Gamma(a)$ is the gamma function, given by the relationship (3.38). The parameters a and b of the gamma distribution are related to the mean, $\langle x \rangle$, and the variance, D_x, of the random variable X as

$$a = \frac{\langle x \rangle^2}{D_x}, \qquad b = \left(\frac{D_x}{\langle x \rangle}\right)^{-1/a}$$

When the a value is large (the coefficient of variation is small), the gamma distribution becomes a normal one. If one puts $a = k + 1$ and $b = (1/\lambda)^{1/a}$ in the gamma distribution, the following *Erlang distribution* can be obtained:

$$f_k(x) = \frac{\lambda(\lambda x)^k}{k!} e^{-\lambda x} = \lambda P(k, \lambda x), \qquad x > 0$$

where

$$P(m, a) = \frac{a^m}{m!} e^{-a}$$

Hence, the probability distribution function of a random variable X distributed in accordance with the Erlang law of the kth order can be expressed through a tabulated function of the Poisson distribution. The probability distribution function in the Erlang law can be found as

$$F_k(x) = \int_0^x f_k(x) \, dx = \lambda \int_0^x P(k, \lambda x) \, dx = 1 - R(k, \lambda x)$$
$$= \bar{R}(k, \lambda x)$$

Thus, the following relationships take place:

$$\frac{dR(k, a)}{da} = -P(k, a), \qquad R(k, a) = \int_a^\infty P(k, x) \, dx$$

The Erlang distribution is used in the theory of queues and in reliability engineering. By putting in the gamma distribution $a = n/2$ and $b = 1/2$, one obtains the χ^2 distribution. By putting $a = 1$, the exponential distribution can be obtained.

Example 3.16 The flow of the items manufactured on a conveyer is characterized by the parameter λ. All the items undergo quality control. The faulty ones are put in a special box. This box can accommodate not more than k items. The probability that the given item is faulty is p. Determine the distribution function for the random time T required to fill up the box with faulty items and the capacity k of the box, so that the probability that the box will not be overfilled during one shift is $P_b = 0.99$.

Solution The intensity of the flow of the faulty items is λp. The time T follows the Erlang distribution with parameters k and λp:

$$f_k(t) = \lambda p P(k, \lambda p t)$$

and therefore

$$\langle t \rangle = \frac{k+1}{\lambda p}, \qquad D_t = \frac{k+1}{(\lambda p)^2}$$

The number of faulty items arriving during the time t is distributed in accordance with the Poisson law with the parameter $\lambda p t$. Hence, the following condition must be fulfilled:

$$P_b = P(T < t) = F_k(t) = R(k, \lambda p t) = 0.99$$

If, for instance, $\lambda = 20$ (item/hour), $p = 0.1$, $t = 8.0$ hours, then the equation $R(k, 16) = 0.99$ yields $k = 26$.

3.3.9. χ-distribution

This is the distribution of the square root:

$$\sqrt{\chi^2} = \sqrt{\xi_1^2 + \xi_2^2 + \cdots + \xi_n^2}$$

where $\xi_1, \xi_2 \ldots, \xi_n$ are independent random variables distributed in accordance with the normal law with zero mean and unit standard deviation. The probability distribution function of the χ *distribution* is

$$f(x) = \begin{cases} 0, & \text{if } x \leq 0 \\ \dfrac{1}{2^{n/2-1}\Gamma(n/2)} x^{n-1} e^{-x^2/2}, & \text{if } x > 0 \end{cases} \qquad (3.43)$$

The moments of the random variable χ are expressed as follows:

$$m_{\chi^m} = \frac{1}{2^{n/2-1}\Gamma(n/2)} \int_0^\infty x^{n+m-1} e^{-x^2/2} \, dx = 2^{m/2} \frac{\Gamma[(n+m)/2]}{\Gamma(n/2)} \qquad (3.44)$$

3.3.10. Cauchy distribution

This probability distribution function of the *Cauchy distribution* is described by the formula

$$f(x) = \frac{1}{\pi(1+x^2)}, \qquad -\infty < x < \infty \qquad (3.45)$$

This formula can be obtained as the distribution of the ratio ξ_1/ξ_2 of two independent "standard" variables ξ_1 and ξ_2, having the same normal distribution with zero mean and unit standard deviation. The tangent tan α of a random variable α, uniformly distributed on a segment $(-\pi/2, \pi/2)$, also follows the Cauchy distribution.

The cumulative probability distribution function for the random variable X distributed in accordance with the Cauchy law is

$$F(x) = \int_{-\infty}^{x} f(x)\, dx = \frac{1}{\pi} \arctan x + \frac{1}{2} \qquad (3.46)$$

The mean value $\langle x \rangle$ and variance D_x do not exist for this distribution: the corresponding integrals diverge.

3.3.11. Beta distribution

The *beta distribution* was introduced by Karl Pearson in 1895. It is defined over the range (a, b) as

$$f(x) = C(x-a)^{\alpha}(b-x)^{\beta} \qquad (3.47)$$

where both the parameters α and β are greater than -1. If α and β are integers, the constant C can be found from the condition of normalization as follows:

$$C = \frac{(\alpha + \beta + 1)!}{\alpha!\,\beta!\,(b-a)^{\alpha+\beta+1}} \qquad (3.48)$$

The mean value and the variance of the beta distribution are

$$\langle x \rangle = a + \frac{\alpha + 1}{\alpha + \beta + 2}(b-a) \qquad (3.49)$$

$$D_x = \frac{(b-a)^2(\alpha+1)(\beta+1)}{(\alpha+\beta+2)^2(\alpha+\beta+3)} \qquad (3.50)$$

Solving the last two equations for α and β, we obtain

$$\alpha = \frac{X^2}{Y^2}(1-X) - (1+X), \qquad \beta = \frac{\alpha+1}{X} - (\alpha+2) \qquad (3.51)$$

where

$$X = \frac{\langle x \rangle - a}{b - a}, \qquad Y = \frac{\sqrt{D_x}}{b - a} \qquad (3.52)$$

The beta distribution is used in statistics and reliability engineering. If, for a certain random variable, the mean, variance, and the minimum and maximum values are available, then it is the beta distribution that results in the maximum entropy (uncertainty) (see Sec. 6.2).

3.3.12. Student distribution

Let the random variable X and a set of random variables Y_i, $i = 1, 2, \ldots, n$, be statistically independent normally distributed variables. Then the random variable

$$T = \frac{X\sqrt{n}}{\sqrt{Y_1^2 + \cdots + Y_n^2}}$$

(*Student statistics*, or T statistics) is distributed with the following probability density function:

$$S_n(t) = C_n\left(1 + \frac{t^2}{n}\right)^{-(n+1)/2} \tag{3.53}$$

where

$$C_n = \frac{1}{\sqrt{\pi n}} \frac{\Gamma[(n+1)/2]}{\Gamma(n/2)} \tag{3.54}$$

The Student distribution is widely used in mathematical statistics. It was introduced by W. S. Gosset who published his findings under pseudonym A. Student. If n is substituted by $n - 1$, the distribution can be also written as

$$S_n(t) = C_n\left(1 + \frac{t^2}{n-1}\right)^{-n/2}, \tag{3.55}$$

where

$$C_n = \frac{1}{\sqrt{\pi(n-1)}} \frac{\Gamma(n/2)}{\Gamma[(n-1)/2]} \tag{3.56}$$

The $k = n - 1$ value is referred to as the number of degrees of freedom of the distribution. One encounters this distribution when examining the confidence limits and confidence intervals for observed (measured) random variables. The tabulated values of the Student distribution are given in Appendix E.

Further Reading (*see Bibliography*): 7, 20, 30, 32, 38, 40, 44, 58, 59, 62, 84, 88

Chapter 4

Systems of Random Variables

> *"The infinite number of particles makes it impossible to apply a deterministic description and forces us to use statistical methods."*
> MAX BORN
> *Natural Philosophy of Cause and Chance*
>
> *"Probability is too important to be left to the mathematicians."*
> UNKNOWN ENGINEER
>
> *"The degree of understanding a phenomenon is inversely proportional to the number of variables used for its description."*
> UNKNOWN PHYSICIST

4.1. System of Two Random Variables

A system of two random variables (X, Y) can be geometrically interpreted as a *random point* with coordinates (X, Y) on the xy plane. It can be interpreted also as a random vector directed from the origin to the point (X, Y) whose components are random variables X and Y. The joint probability distribution of two random variables (X, Y) is the probability that both inequalities $X < x$ and $Y < y$ take place simultaneously, i.e.,

$$F(x, y) = P\{X < x, Y < y\} \qquad (4.1)$$

The function $F(x, y)$ of two variables x and y can be interpreted geometrically as the probability that a random point (X, Y) falls in a quadrant whose vortex is (x, y). The function $F(x, y)$ possesses the following properties:

(a) $F(-\infty, -\infty) = F(-\infty, y) = F(x, -\infty) = 0$
(b) $F(\infty, \infty) = 1$

(c) $F(x, \infty) = F_1(x)$, $F(\infty, y) = F_2(y)$

(d) $F(x, y)$ is a nondecreasing function of the arguments x and y

The joint probability density of two continuous random variables is the limit of the ratio of the probability of a random point falling in an element of a plane Δx, Δy, adjoining the point (x, y), to the area $\Delta x\, \Delta y$ of the element when its dimensions Δx, Δy tend to zero. The joint probability density function can be expressed in terms of the joint probability distribution function as

$$f(x, y) = \frac{\partial^2 F(x, y)}{\partial x\, \partial y} \tag{4.2}$$

The one-dimensional distributions can be found as

$$F_1(x) = \int_{-\infty}^{x} \int_{-\infty}^{\infty} f(x, y)\, dx\, dy$$

$$F_2(y) = \int_{-\infty}^{\infty} \int_{-\infty}^{y} f(x, y)\, dx\, dy$$

$$f_1(x) = \int_{-\infty}^{\infty} f(x, y)\, dy$$

$$f_2(y) = \int_{-\infty}^{\infty} f(x, y)\, dx$$

The joint probability density has the following properties

(a) $f(x, y) \geq 0$

(b) $\int_{-\infty}^{\infty} \int_{-\infty}^{\infty} f(x, y)\, dx\, dy = 1$

The conditional distribution of a random variable entering the system of two random variables is the distribution calculated under the condition that the other random variable has assumed a definite value. Then the rule of the multiplication of probability densities can be written as

$$f(x, y) = f_1(x) f_2(y/x)$$

or

$$f(x, y) = f_2(y) f_1(x/y)$$

Therefore

$$f_2(x, y) = \frac{f(x, y)}{f_1(x)}$$

and

$$f_1(x, y) = \frac{f(x, y)}{f_2(y)}$$

The random variables X, Y are *mutually statistically independent* if the conditional probability distribution of one of them is not affected by the value of the other:

$$f_1(x/y) = f_1(x), \qquad f_2(y/x) = f_2(y)$$

For mutually (statistically) independent random variables,

$$f(x, y) = f_1(x) f_2(y)$$

The moments of continuous random variables X and Y can be calculated by the formulas:

$$m_{ks} = \int_{-\infty}^{\infty} \int_{-\infty}^{\infty} x^k y^s f(x, y) \, dx \, dy \tag{4.3}$$

$$\mu_{ks} = \int_{-\infty}^{\infty} \int_{-\infty}^{\infty} (x - \langle x \rangle)^k (y - \langle y \rangle)^s f(x, y) \, dx \, dy \tag{4.4}$$

The *covariance* (bivariance) of two random variables X, Y is the second-order mixed central moment (correlation moment):

$$\mu_{11} = K_{xy} = m_{11} - \langle x \rangle \langle y \rangle = \int_{-\infty}^{\infty} \int_{-\infty}^{\infty} (x - \langle x \rangle)(y - \langle y \rangle) f(x, y) \, dx \, dy \tag{4.5}$$

For statistically independent random variables the covariance is zero. The inverse statement is correct for normally distributed random variables only. For random variables obeying other laws of distribution, statistical independence does not necessarily follow from the fact that the variables are uncorrelated.

When $X = Y$, the formula (4.5) yields

$$K_{xy} = D_x = \int_{-\infty}^{\infty} (x - \langle x \rangle)^2 f(x) \, dx \tag{4.6}$$

Hence, the covariance of random variables characterizes not only the statistical dependence of these variables but also their scattering with respect to their mean values as well.

The factors of correlation and scattering can affect the magnitude of the covariance in opposite directions. Therefore it is advisable to use, for characterizing the statistical correlation, a dimensionless *correlation coefficient* (normalized covariance)

$$r_{xy} = \frac{K_{xy}}{\sqrt{D_x D_y}} \qquad (4.7)$$

This coefficient is a measure of how closely the random variables are linearly related. If the random variables X and Y are linearly related, that is, $Y = aX + b$, where a and b are nonrandom parameters, then their correlation coefficient is $r_{xy} = \pm 1$. Here the "plus" sign or the "minus" sign corresponds to the sign of the coefficient a. For any two random variables X and Y, $|r_{xy}| \leq 1$.

The normal probability distribution of two random variables X and Y (normal distribution on a plane) is expressed as follows:

$$f(x, y) = \frac{1}{2\pi\sqrt{1 - r_{xy}^2}\sqrt{D_x D_y}} \exp\left\{-\frac{1}{2(1 - r_{xy}^2)}\right.$$
$$\left. \times \left[\frac{(x - \langle x \rangle)^2}{D_x} + \frac{(y - \langle y \rangle)^2}{D_y} - \frac{2r_{xy}(x - \langle x \rangle)(y - \langle y \rangle)}{\sqrt{D_x D_y}}\right]\right\} \qquad (4.8)$$

where r_{xy} is the correlation coefficient. The distribution (4.8) is often encountered in engineering practice. Examples are: errors due to two-dimensional measurements of a certain value, the coordinates of the landing point for a space apparatus, etc.

As has been mentioned, for normally distributed random variables, uncorrelation is equivalent to statistical independence of the variables. If the random variables X and Y are uncorrelated, then $r_{xy} = 0$, and the probability density function (4.8) is expressed as

$$f(x, y) = \frac{1}{2\pi\sqrt{D_x D_y}} \exp\left\{-\frac{1}{2}\left[\frac{(x - \langle x \rangle)^2}{D_x} + \frac{(y - \langle y \rangle)^2}{D_y}\right]\right\} \qquad (4.9)$$

If, in addition, $\langle x \rangle = \langle y \rangle = 0$, then the normal distribution of two random variables X and Y assumes its simplest *canonical form*:

$$f(x, y) = \frac{1}{2\pi\sqrt{D_x D_y}} \exp\left(-\frac{x^2}{2D_x} - \frac{y^2}{2D_y}\right) \qquad (4.10)$$

Example 4.1 The time-to-failure T_1 of an electronic device is distributed in accordance with an exponential law with parameter λ_1. The time-to-repair

T_2 of the device is also distributed in accordance with an exponential law with parameter $\lambda_2 = at_1$ that is proportional to the time-to-failure. Find the joint probability density function of the system of the random variables T_1 and T_2.

Solution The probability distribution of the time T_1 is

Then
$$f_1(t_1) = \lambda_1 e^{-\lambda_1 t_1}$$

$$f(t_1, t_2) = \lambda_1 e^{-\lambda_1 t_1} \lambda_2 e^{-\lambda_2 t_2} = \lambda_1 e^{-\lambda_1 t_1} at_1 e^{-at_1 t_2}$$

so that

$$f_2(t_2) = \int_{-\infty}^{\infty} f(t_1, t_2)\, dt_1 = \int_0^{\infty} \lambda_1 at_1 e^{-(\lambda_1 + at_2)t_1}\, dt_1$$

$$= \frac{\lambda_1 a}{(\lambda_1 + at_2)^2}$$

The conditional probability density functions can be found as

$$f_2(t_2|t_1) = \frac{f(t_1, t_2)}{f_1(t_1)} = at_1 e^{-at_1 t_2}$$

$$f_1(t_1|t_2) = \frac{f(t_1, t_2)}{f_2(t_2)} = (\lambda_1 + at_2)^2 t_1 e^{-(\lambda_1 + at_2)t_1}$$

Note that the conditional probability density function of the random variable T_1 for the fixed value $T_2 = t_2$ is the Erlang law of the second order with the parameter $\lambda + at_2$.

The mean value and the variance of the random variable T_1 are

$$\langle t_1 \rangle = \int_0^{\infty} \lambda_1 t_1 e^{-\lambda_1 t_1}\, dt_1 = \frac{1}{\lambda_1}, \qquad D_{t_1} = \frac{1}{\lambda_1^2}$$

The mean value and the variance of the random variable T_2 do not exist.

Example 4.2 A cylindrical machine part is characterized by its length L and diameter D. These are statistically independent random variables distributed in accordance with the normal law. The means and the standard deviations of these variables are: $\langle l \rangle = 100$ mm, $\sigma_l = 0.1$ mm, $\langle d \rangle = 10$ mm, $\sigma_d = 0.01$ mm. The machine part is accepted if its dimensions meet the conditions:

$$(100 - 0.1)_{mm} < L < (100 + 0.05)_{mm}, \quad (10 - 0.005)_{mm} < D < (10 + 0.007)_{mm}$$

Determine the percentage of the parts that will be rejected.

Solution The probability that the machine part will not be rejected can be calculated as

$$p = \left[\Phi\left(\frac{100 + 0.05 - 100}{0.1}\right) - \Phi\left(\frac{100 - 0.1 - 100}{0.1}\right)\right]$$
$$\times \left[\Phi\left(\frac{10 + 0.007 - 10}{0.01}\right) - \Phi\left(\frac{10 - 0.005 - 10}{0.01}\right)\right] \simeq 0.239$$

Hence, the percentage of the faulty parts is $q = 1 - p = 0.761$.

Example 4.3 The coordinates X, Y of the point of landing of a space apparatus are random variables with zero means and equal standard deviations $\sigma_x = \sigma_y = \sigma$. Find the radius R_0 of a circle in which the apparatus will land with the probability P.

Solution The sought radius can be found from the equation

$$P = P\{R < R_0\} = 1 - \exp\left(-\frac{R_0^2}{2\sigma^2}\right)$$

Solving this equation of R_0 we have

$$R_0 = \sigma\sqrt{-2\ln(1-P)}$$

If, for instance, $P = 0.95$ and $\sigma = 2$ km, then $R = 4.89$ km.

Example 4.4 A random point A represents an object on a circular radar screen of unit radius and has a uniform distribution within that circle. Find the joint probability density function $f(r, \phi)$ of the polar coordinates R, Φ of the point A and determine whether the random variables R and Φ are independent.

Solution Consider an elementary "rectangle" in polar coordinates. The "rectangle" corresponds to infinitesimal increments dr, $d\phi$ of the polar coordinates r, ϕ of the point inside the circle. Its area is $r\,dr\,d\phi$. Dividing this by the area of the circle (which is equal to π), we obtain the elementary probability as

$$f(r, \phi)\,dr\,d\phi = \frac{r\,dr\,d\phi}{\pi}$$

so that

$$f(r, \phi) = \frac{r}{\pi}, \qquad \text{for } 0 < r < 1,\ 0 < \phi < 2\pi$$

The probability density function can be found by integrating this formula in the whole range of the change of ϕ, that is,

$$f_1(r) = \int_0^{2\pi} \frac{r}{\pi}\,d\phi = 2r, \qquad 0 < r < 1$$

Similarly,

$$f_2(\phi) = \int_0^1 \frac{r}{\pi}\,dr = \frac{1}{2\pi}, \qquad 0 < \phi < 2\pi$$

The graph of the distribution $f_1(r)$ is a right triangle, while the distribution $f_2(\phi)$ is uniform over the region $(0, 2\pi)$. Multiplying $f_1(r)$ by $f_2(\phi)$ we obtain the joint probability density $f(r, \phi)$ of the system R, Φ, and, consequently, R and Φ are statistically independent. Note that, for the same case, the rectangular coordinates X, Y of a point in the interior of a circle are statistically dependent.

4.2. System of Many Random Variables

The joint distribution function of n random variables $X_1, X_2, \ldots, X_n$ is the probability that n inequalities of the type $X_i < x_i$ take place:

$$F(x_1, x_2, \ldots, x_n) = P\{X_1 < x_1, X_2 < x_2, \ldots, X_n < x_n\} \quad (4.11)$$

The joint probability density of n random variables is the nth mixed partial derivative of the probability distribution function:

$$f(x_1, x_2, \ldots, x_n) = \frac{\partial^n F(x_1 x_2, \ldots, x_n)}{\partial x_1 \partial x_2 \cdots \partial x_n} \quad (4.12)$$

The distribution function $F_i(x_i)$ for a particular variable X_i can be obtained from $F(x_1, x_2, \ldots, x_n)$ if all the arguments, except for x_i, are put equal to $+\infty$:

$$F_i(x_i) = F(\infty, \infty, \ldots, x_i, \infty, \ldots, \infty) \quad (4.13)$$

The probability density of a variable X_i entering the system $(X_1, X_2, \ldots, X_n)$ of random variables can be expressed in terms of the joint probability density by the formula

$$f_i(x_i) = \underbrace{\int_{-\infty}^{\infty} \cdots \int_{-\infty}^{\infty}}_{n-1} f(x_1, x_2, \ldots, x_n)\, dx_1\, dx_2 \cdots dx_{i-1}\, dx_{i+1} \cdots dx_n \quad (4.14)$$

If the random variables $X_1, X_2, \ldots, X_n$ are mutually independent (uncorrelated), then

$$f(x_1, x_2, \ldots, x_n) = f_1(x_1) f_2(x_2) \cdots f_n(x_n) \quad (4.15)$$

The *correlation matrix* of a system of n random variables $(X_1, X_2, \ldots, X_n)$ contains the pairwise covariances of all the variables:

$$\|K_{ij}\| = \begin{Vmatrix} K_{11} & K_{12} & \cdots & K_{1n} \\ K_{21} & K_{22} & \cdots & K_{2n} \\ \cdots & \cdots & \cdots & \cdots \\ K_{n1} & K_{n2} & \cdots & K_{nn} \end{Vmatrix} \quad (4.16)$$

The correlation matrix is symmetric with respect to its indices: $K_{ij} = K_{ji}$. The *normalized correlation matrix* is compiled from the pairwise correlation coefficients:

$$\|r_{ij}\| = \begin{Vmatrix} 1 & r_{12} & r_{13} & \cdots & r_{1n} \\ r_{21} & 1 & r_{23} & \cdots & r_{2n} \\ r_{31} & r_{32} & 1 & \cdots & r_{3n} \\ \cdots & \cdots & \cdots & \cdots & \cdots \\ r_{n1} & r_{n2} & r_{n3} & \cdots & 1 \end{Vmatrix} \qquad (4.17)$$

Example 4.5 A shell is fired at a point airborne target. The point where the shell explodes is normally distributed with the center of scattering at the target. The standard deviations are $\sigma_x = \sigma_y = \sigma_z = \sigma$. The target is destroyed if its distance from the point where the shell explodes does not exceed $r_0 = 2\sigma$. Find the probability p that one shell will be sufficient to destroy the target.

Solution For independent random variables X, Y, Z, the normal probability distribution in a three-dimensional space is

$$f(x, y, z) = \frac{1}{(2\pi)^{3/2} \sigma_x \sigma_y \sigma_z}$$

$$\times \exp\left\{ -\frac{1}{2} \left[\frac{(x - \langle x \rangle)^2}{\sigma_x^2} + \frac{(y - \langle y \rangle)^2}{\sigma_y^2} + \frac{(z - \langle z \rangle)^2}{\sigma_z^2} \right] \right\} \quad (4.18)$$

The probability that a random point (X, Y, Z) will be found within a domain E_k, bounded by an ellipsoid of equal probability density with semi-axes $a = k\sigma_x$, $b = k\sigma_y$, and $c = k\sigma_z$, is

$$P\{(X, Y, Z) \in E_k\} = \Phi\left(\frac{k}{\sqrt{2}}\right) - \sqrt{\frac{2}{\pi}} k e^{-k^2/2}$$

With $k = 2$, we obtain

$$p = P\{(X, Y, Z) \in E_2\} = \Phi\left(\frac{1}{\sqrt{2}}\right) - \sqrt{\frac{2}{\pi}} 2e^{-2} = 0.739$$

4.3. Complex Random Variables

In some problems it might be convenient to use complex random variables, such as

$$Z = X + iY \qquad (4.19)$$

where X and Y are real random variables and $i = \sqrt{-1}$ is the "imaginary unity." Treating the variable Z as a sum of the variables X and iY, one can use the following formula for its mean value:

$$\langle z \rangle = \langle x \rangle + i \langle y \rangle \qquad (4.20)$$

The variance of the random variable Z can be found as

$$D_z = \langle |\mathring{Z}|^2 \rangle = \langle \mathring{Z}\mathring{Z}^* \rangle = \langle \mathring{X}^2 + \mathring{Y}^2 \rangle = D_x + D_y \qquad (4.21)$$

where $\mathring{Z} = \mathring{X} + i\mathring{Y}$ is the centered random variable and $\mathring{Z}^* = \mathring{X} - i\mathring{Y}$ is the conjugate random variable.

The correlation moment of the two complex random variables $Z_1 = X_1 + iX_2$ and $Z_2 = Y_1 + iY_2$ can be evaluated as

$$K_{xy} = \langle (\mathring{X}_1 + i\mathring{X}_2)(\mathring{Y}_1 - i\mathring{Y}_2) \rangle$$
$$= \langle \mathring{X}\mathring{Y}^* \rangle = K_{x_1y_1} + K_{x_2y_2} + i(K_{x_2y_1} - K_{x_1y_2}) \qquad (4.22)$$

If $X = Y$, then

$$K_{xy} = \langle \mathring{X}\mathring{X}^* \rangle = X_1^2 + X_2^2 = D_x \qquad (4.23)$$

One can check that

$$K_{xy} = \langle \mathring{X}\mathring{Y}^* \rangle = \langle \mathring{X}^*\mathring{Y} \rangle = K_{yx}^* \qquad (4.24)$$

Further Reading (see Bibliography): 7, 20, 30, 32, 38, 40, 44, 58, 59, 62, 84, 88

Chapter

5

Functions of Random Variables

> "The practical value of mathematics is, in effect, a possibility to obtain, with its help, results simpler and faster."
> ANDREY KOLMOGOROV

> "All the general theories originate from examination of particular problems and do not make any sense, if they do not serve for classification of more specific problems and putting them in order."
> RICHARD COURANT

> "Be thankful for problems. If they were less difficult, someone with less ability might have your job."
> UNKNOWN ENGINEER

5.1. Characteristics of Functions of Random Variables

One of the most effective means of solving applied probability problems is to use numerical characteristics of random variables. These characteristics can often be found even without knowing the probability distributions. If, for instance, X is a continuous random variable with the probability density function $f(x)$ and $y = y(x)$ is a nonrandom relationship between the varibles x and y, then the mean value and the variance of the random variable Y can be evaluated as follows:

$$\langle y \rangle = \int_{-\infty}^{\infty} y(x) f(x) \, dx \tag{5.1}$$

$$D_y = \int_{-\infty}^{\infty} [y(x) - \langle y \rangle]^2 f(x) \, dx = \int_{-\infty}^{\infty} [y(x)]^2 f(x) \, dx - \langle y \rangle^2 \tag{5.2}$$

If X, Y is a system of continuous random variables with the joint probability density function $f(x, y)$ and $z = z(x, y)$ is a nonrandom function of these variables, then the mean value and the variance of the random variable Z is expressed as

$$\langle z \rangle = \int_{-\infty}^{\infty} \int_{-\infty}^{\infty} z(x, y) f(x, y) \, dx \, dy \tag{5.3}$$

$$D_z = \int_{-\infty}^{\infty} \int_{-\infty}^{\infty} [z(x, y) - \langle z \rangle]^2 f(x, y) \, dx \, dy$$

$$= \int_{-\infty}^{\infty} \int_{-\infty}^{\infty} [z(x, y)]^2 f(x, y) \, dx \, dy - \langle z \rangle^2 \tag{5.4}$$

For a system $X_1, X_2, \ldots, X_n$ of n continuous random variables with the probability density function $f(x_1, x_2, \ldots, x_n)$, the mean value and the variance of a random variable Y, which is related to the random variables $X_1, X_2, \ldots, X_n$ by the nonrandom function $y = y(x_1, x_2, \ldots, x_n)$, can be found by the formulas

$$\langle y \rangle = \underbrace{\int_{-\infty}^{\infty} \cdots \int_{-\infty}^{\infty}}_{n} y(x_1, x_2, \ldots, x_n) f(x_1, x_2, \ldots, x_n) \, dx_1 \, dx_2 \cdots dx_n \tag{5.5}$$

$$D_y = \underbrace{\int_{-\infty}^{\infty} \cdots \int_{-\infty}^{\infty}}_{n} [y(x_1, x_2, \ldots, x_n) - \langle y \rangle]^2$$

$$\times f(x_1, x_2, \ldots, x_n) \, dx_1 \, dx_2 \cdots dx_n$$

$$= \underbrace{\int_{-\infty}^{\infty} \cdots \int_{-\infty}^{\infty}}_{n} [y(x_1, x_2, \ldots, x_n)]^2$$

$$\times f(x_1, x_2, \ldots, x_n) \, dx_1 \, dx_2 \cdots dx_n - \langle y \rangle^2 \tag{5.6}$$

If X is a discrete random variable with an ordered series

$$X: \begin{array}{|c|c|c|c|} \hline x_1 & x_2 & \cdots & x_n \\ \hline p_1 & p_2 & \cdots & p_n \\ \hline \end{array}$$

and the variable Y is related to the variable X as a nonrandom functional relationship $y = y(x)$, then the mean value and the variance of the random variable Y can be computed as

$$\langle y \rangle = \sum_{i=1}^{n} y(x_i) p_i \tag{5.7}$$

and

$$D_y = \sum_{i=1}^{n} [y(x_i) - \langle y \rangle]^2 p_i = \sum_{i=1}^{n} [y(x_i)]^2 p_i - \langle y \rangle^2 \tag{5.8}$$

In some cases, one does not even need to know the distributions of the arguments to obtain the probabilistic characteristics of the functions. These characteristics can be obtained directly from the characteristics of the arguments. Here are the basic rules that can be used:

1. If c is a nonrandom variable, then

$$\langle c \rangle = c, \qquad D_c = 0$$

2. If c is a nonrandom variable and X is a random variable, then

$$\langle cx \rangle = c \langle x \rangle, \qquad D_{cx} = c^2 D_x$$

3. The mean value of a sum of random variables is equal to the sum of their mean values:

$$\langle x + y \rangle = \langle x \rangle + \langle y \rangle$$

In general,

$$\left\langle \sum_{i=1}^{n} x_i \right\rangle = \sum_{i=1}^{n} \langle x_i \rangle$$

4. The mean value of a linear function of several random variables

$$y = \sum_{i=1}^{n} a_i x_i + b$$

where a_i and b are nonrandom numbers, is equal to the same linear function of their mean values:

$$\langle y \rangle = \left\langle \sum_{i=1}^{n} a_i x_i + b \right\rangle = \sum_{i=1}^{n} a_i \langle x_i \rangle + b$$

5. The mean value of the product of two random variables X and Y is

$$\langle xy \rangle = \langle x \rangle \langle y \rangle + K_{xy}$$

6. The mean value of the product of two uncorrelated random variables X, Y is equal to the product of their mean values:

$$\langle xy \rangle = \langle x \rangle \langle y \rangle$$

7. If $X_1, X_2, \ldots, X_n$ are statistically independent random variables, then the mean value of their product is equal to the product of their mean values:

$$\left\langle \prod_{i=1}^{n} x_i \right\rangle = \prod_{i=1}^{n} \langle x_i \rangle$$

8. The variance of the sum of two random variables is

$$D_{x+y} = D_x + D_y + 2K_{xy}$$

9. The variance of the sum of several random variables is expressed by the formula

$$\text{var}\left(\sum_{i=1}^{n} x_i \right) = \sum_{i=1}^{n} D_{xi} + 2 \sum_{i<j} K_{x_i x_j}$$

10. The variance of the sum of two uncorrelated random variables X and Y is equal to the sum of their variances:

$$D_{x+y} = D_x + D_y$$

In general,

$$\text{var}\left(\sum_{i=1}^{n} X_i \right) = \sum_{i=1}^{n} D_{xi}$$

11. The variance of a linear function of several random variables, that is,

$$Y = \sum_{i=1}^{n} a_i X_i + b$$

where a_i and b are nonrandom variables, is given by the formula

$$D_y = \text{var}\left(\sum_{i=1}^{n} a_i x_i + b \right) = \sum_{i=1}^{n} a_i^2 D_{xi} + 2 \sum_{i<j} a_i a_j K_{x_i x_j}$$

When the variable $X_1, X_2, \ldots, X_n$ are uncorrelated, then

$$D_y = \text{var}\left(\sum_{i=1}^{n} a_i x_i + b \right) = \sum_{i=1}^{n} a_i^2 D_{xi}$$

12. When several uncorrelated random vectors are added, their covariances are added. For instance, if

$$X = X_1 + X_2, \quad Y = Y_1 + Y_2, \quad K_{x_i x_j} = K_{x_1 y_2} = K_{y_1 y_2} = K_{y_1 x_2} = 0$$

then

$$K_{xy} = K_{x_1 y_1} + K_{x_2 y_2}$$

5.2. Distributions of Functions of Random Variables: Convolution of Distributions

If X is a continuous random variable with a probability density function $f(x)$ and a random variable Y is in a functional (deterministic) relationship with the variable X, so that

$$Y = \phi(X)$$

then the probability density function of the random variable Y is expressed by the formula

$$g(y) = f(\psi(y)) |\psi'(y)|$$

where the function ψ is the inverse of ϕ.

This relationship can be generalized for the case of many random variables as follows:

$$f_2(y_1, y_2, \ldots, y_n) = f_1[\psi_1(y_1, y_2, \ldots, y_n), \quad \psi_2(y_1, y_2, \ldots, y_n), \ldots,$$
$$\psi_n(y_1, y_2, \ldots, y_n)] J \quad (5.9)$$

where

$$Y_i = \phi_i(X_1, X_2, \ldots, X_n), \quad X_i = \psi_i(Y_1, Y_2, \ldots, Y_n) \quad (5.10)$$

and

$$J = \begin{vmatrix} \dfrac{\partial \psi_1}{\partial y_1} & \dfrac{\partial \psi_1}{\partial y_2} & \cdots & \dfrac{\partial \psi_1}{\partial y_n} \\ \dfrac{\partial \psi_2}{\partial y_1} & \dfrac{\partial \psi_2}{\partial y_2} & \cdots & \dfrac{\partial \psi_2}{\partial y_n} \\ \cdots & \cdots & \cdots & \cdots \\ \dfrac{\partial \psi_n}{\partial y_1} & \dfrac{\partial \psi_n}{\partial y_2} & \cdots & \dfrac{\partial \psi_n}{\partial y_n} \end{vmatrix} \quad (5.11)$$

is known as the Jacobi determinant (*Jacobian* of the transformation).

For a function of several random variables, it is more convenient to seek first the cumulative distribution function rather than the probability density function. For instance, for a function of two arguments $Z = \phi(X, Y)$, the distribution function is

$$G(z) = \iint_{D(z)} f(x, y) \, dx \, dy \tag{5.12}$$

where $f(x, y)$ is the joint probability density of the variables X and Y, and $D(z)$ is the domain on the xy plane for which $\phi(x, y) < z$. The corresponding probability density function $g(z)$ can be determined by differentiation: $g(z) = G'(z)$.

Just as with discrete random variables, it is often desirable to find the distribution of the sums of continuous random variables. The probability density of the sum of two random variables $Z = X + Y$ is expressed by either of the formulas

$$g(z) = \int_{-\infty}^{\infty} f(x, z - x) \, dx, \qquad g(z) = \int_{-\infty}^{\infty} f(z - y, y) \, dy \tag{5.13}$$

where $f(x, y)$ is the joint probability density function of the variables X and Y. If the random variables X and Y are statistically independent, then

$$f(x, y) = f_1(x) f_2(y) \tag{5.14}$$

and

$$g(z) = \int_{-\infty}^{\infty} f_1(x) f_2(z - x) \, dx \tag{5.15}$$

or

$$g(z) = \int_{-\infty}^{\infty} f_1(z - y) f_2(y) \, dy \tag{5.16}$$

The distribution of the sum $g(z)$ is called the *composition* or the *convolution* of the distributions $f_1(x)$ and $f_2(y)$. The cumulative distribution function for the variable $Z = X + Y$, if the variables X and Y are statistically independent, can be calculated as

$$G_z(z) = \int_{-\infty}^{z} f_1(x) \, dx \int_{-\infty}^{z-x} f_2(y) \, dy = \int_{-\infty}^{z} f_1(x) F_x(z - x) \, dx$$

where

$$F_2(y) = \int_{-\infty}^{y} f_2(y) \, dy$$

is the cumulative distribution function of the variable Y.

If a random variable has a normal distribution and is subjected to a linear transformation, then the resulting random variable will also have a normal distribution. A convolution of two normal distribution functions $f_1(x)$ and $f_2(y)$ with characteristics $\langle x \rangle$, σ_x and $\langle y \rangle$, σ_y, respectively, results in a normal distribution with characteristics

$$\langle z \rangle = \langle x \rangle + \langle y \rangle, \qquad D_z = D_x + D_y$$

The addition of two normally distributed random variables X, Y with parameters $\langle x \rangle$, σ_x and $\langle y \rangle$, σ_y and the correlation coefficient r_{xy} results in a random variable Z, which also has a normal distribution with characteristics

$$\langle z \rangle = \langle x \rangle + \langle y \rangle, \qquad D_z = D_x + D_y + 2r_{xy}\sqrt{D_x D_y} \qquad (5.17)$$

Let Z be a linear function of several independent normally distributed random variables $X_1, X_2, \ldots, X_n$:

$$Z = \sum_{i=1}^{n} a_i X_i + b \qquad (5.18)$$

where a_i and b are nonrandom coefficients. The variable Z also has a normal distribution with parameters

$$\langle z \rangle = \sum_{i=1}^{n} a_i \langle x_i \rangle + b \qquad (5.19)$$

and

$$D_z = \sum_{i=1}^{n} a_i^2 D_{x_i} \qquad (5.20)$$

where $\langle x_i \rangle$ and D_{x_i} are the mean and variance of the random variable X_i, $i = 1, 2, \ldots, n$.

If the arguments $X_1, X_2, \ldots, X_n$ are correlated, then the distribution of a linear function Z of these arguments remains normal, and is characterized by the parameters

$$\langle z \rangle = \sum_{i=1}^{n} a_i \langle x_i \rangle + b \qquad (5.21)$$

$$D_z = \sum_{i=1}^{n} a_i^2 D_{x_i} + 2 \sum_{i<j} a_i a_j r_{x_i x_j} \sqrt{D_{x_i} D_{x_j}} \qquad (5.22)$$

where $r_{x_i x_j}$ is the correlation coefficient of the random variables X_i and X_j ($i = 1, 2, \ldots, n; i \neq j$).

A convolution of two normal distributions on a plane results in a normal distribution with parameters

$$\langle x \rangle = \langle x_1 \rangle + \langle x_2 \rangle, \qquad \langle y \rangle = \langle y_1 \rangle + \langle y_2 \rangle \tag{5.23}$$

$$D_x = D_{x_1} + D_{x_2}, \qquad D_y = D_{y_1} + D_{y_2} \tag{5.24}$$

and

$$K_{xy} = K_{x_1 y_1} + K_{x_2 y_2} \tag{5.25}$$

$$r_{xy} = \frac{r_{x_1 y_1}\sqrt{D_{x_1} D_{y_1}} + r_{x_2 y_2}\sqrt{D_{x_2} D_{y_2}}}{\sqrt{D_x D_y}} \tag{5.26}$$

Let us give formulas for the distribution functions for the difference $Z = X - Y$, the product $Z = XY$, and the ratio $Z = X/Y$ of two random variables X and Y. These formulas can be obtained by applying the relationship (5.9).

The distribution function of the random variable $Z = X - Y$ can be expressed by the same formulas (5.13) as for the sum of these variables, but with a different integration domain:

$$g(z) = \int_{-\infty}^{\infty} f(x, x - z)\, dx = \int_{-\infty}^{\infty} f(z + y, y)\, dy \tag{5.27}$$

For statistically independent variables X and Y, we have

$$g(z) = \int_{-\infty}^{\infty} f_1(x) f_2(x - z)\, dx = \int_{-\infty}^{\infty} f_1(z + y) f_2(y)\, dy \tag{5.28}$$

The distribution function of the random variable $Z = XY$ is

$$g(z) = \int_{-\infty}^{\infty} \frac{f(X, Z/X)\, dx}{|x|} = \int_{-\infty}^{\infty} \frac{f(z/y, y)\, dy}{|y|}, \tag{5.29}$$

For statistically independent variables X and Y,

$$g(z) = \int_{-\infty}^{\infty} \frac{f_1(x) f_2(z/x)\, dx}{|x|} = \int_{-\infty}^{\infty} \frac{f_1(z/y) f_2(y)\, dy}{|y|} \tag{5.30}$$

Finally, the distribution function of the random variable $Z = X/Y$ is

$$g(z) = \int_{-\infty}^{\infty} |y| f(zy, y)\, dy \tag{5.31}$$

or, if the variables X and Y are statistically independent,

$$g(z) = \int_{-\infty}^{\infty} |y| f_1(zy) f_2(y) \, dy \qquad (5.32)$$

Example 5.1 A continuous random variable X is distributed in the interval $(0, 1)$ and has a probability density function $f(x) = 2x$. Find the mean value and variance of the function $Y = X^2$.

Solution The mean and variance of the square of the random variable X are as follows:

$$\langle y \rangle = m_2(x) = \int_0^1 x^2(2x) \, dx = \frac{1}{2}$$

$$D_y = m_2(y) - \langle y \rangle^2 = \int_0^1 (x^2)^2 (2x) \, dx - \left(\frac{1}{2}\right)^2 = \frac{1}{12}$$

Example 5.2 A positive random variable X is exponentially distributed and is characterized by the probability density $f(x) = \lambda e^{-\lambda x}$ for $x > 0$. Find the mean value, variance, and the coefficient of variation of the random variable $Y = e^{-X}$.

Solution The mean value, variance, and the coefficient of variation of the random variable Y are as follows:

$$\langle y \rangle = \int_0^\infty e^{-x}(\lambda e^{-\lambda x}) \, dx = \frac{\lambda}{\lambda + 1}$$

$$D_y = m_2(y) - \langle y \rangle^2 = \int_0^\infty e^{-2x}(\lambda e^{-\lambda x}) \, dx - \left(\frac{\lambda}{\lambda + 1}\right)^2 = \frac{\lambda}{(\lambda + 1)(\lambda + 2)}$$

$$v_y = \sqrt{\frac{\lambda + 1}{\lambda(\lambda + 2)}}$$

Example 5.3 An analytical balance is used to weigh an object. The acutal (unknown) mass of the object is a. Because of the inevitable errors, the result of each weighing is a random variable which follows a normal distribution with parameters a and σ. To reduce the errors, the object is weighed n times and the arithmetic mean of the results of n weighings is taken as an approximate value of the mass of the object:

$$Y(n) = \frac{1}{n} \sum_{i=1}^n X_i$$

Find the mean value and the standard deviation of the random variable $Y(n)$. How many times must the object be weighed to reduce the mean square error of evaluating the mass by a factor of ten?

Solution The mean value of the random variable $Y(n)$ is

$$\langle y(n) \rangle = \frac{1}{n} \sum_{i=1}^n \langle x_i \rangle$$

Since the object is weighed every time under the same conditions, then $\langle x_i \rangle = a$ for any i, and therefore

$$\langle y(n) \rangle = \frac{1}{n} \sum_{i=1}^{n} a = a$$

Assuming that the errors in each weighing are independent, the variance of the random variable $Y(n)$ is as follows:

$$D_{y(n)} = \frac{1}{n^2} \sum_{i=1}^{n} D_{x_i} = \frac{1}{n^2} \sum_{i=1}^{n} \sigma^2 = \frac{\sigma^2}{n}$$

The number of weighings resulting in a tenfold reduction of the mean square error can be found from the condition

$$\sqrt{D_{y(n)}} = \frac{\sigma}{\sqrt{n}} = \frac{\sigma}{10}$$

Hence, $n = 100$. The object must be weighed a hundred times to reduce the mean square error of evaluating the mass of the object by a factor of ten.

Example 5.4 Each of two points X_1 and X_2 may independently occupy random position within the interval (0, 1) of the axis x. The probability density is constant for both variables. Find the mean value of the distance R between the points and the mean of the square of this distance.

Solution The mean value of the distance $R = |X_2 - X_1|$ is $\langle r \rangle = \langle |x_2 - x_1| \rangle$. One can treat the system X_1, X_2 of random variables as a random point on the x_1, x_2 plane, distributed with constant density $(x_1, x_2) = 1$ within a square whose side is equal to unity. In the domain $D_1: X_1 > X_2$, $|X_2 - X_1| = X_1 - X_2$. In the domain $D_2: X_2 > X_1$, $|X_2 - X_1| = X_2 - X_1$. Thus, the mean value $\langle r \rangle$ of the distance R can be computed as

$$\langle r \rangle = \iint\limits_{(D_1)} (x_1 - x_2) \, dx_1 \, dx_2 + \iint\limits_{(D_2)} (x_2 - x_1) \, dx_1 \, dx_2$$

$$= \int_0^1 dx_1 \int_0^{x_1} (x_1 - x_2) \, dx_2 + \int_0^1 dx_2 \int_0^{x_2} (x_2 - x_1) \, dx_1 = \frac{1}{3}$$

The mean of the random variable R^2 is

$$\langle r^2 \rangle = \langle |x_2 - x_1|^2 \rangle = m_2(x_2) + m_2(x_1) - 2\langle x_1 \rangle^2 = \frac{1}{6}$$

Example 5.5 Statistically independent random variables X and Y have probability densities $f_1(x)$ and $f_2(y)$. Find the mean value and variance of the modulus of their difference $Z = |X - Y|$.

Solution We have

$$\langle z \rangle = \int_{-\infty}^{\infty} \int_{-\infty}^{\infty} |x - y| \, f_1(x) f_2(y) \, dx \, dy$$

A straight line $y = x$ divides the xy plane into two domains, I and II. In domain I, $x > y$ and $|x - y| = x - y$. In domain II, $y < x$ and $|x - y| = y - x$. Hence,

$$\langle z \rangle = \iint_{(I)} (x - y) f_1(x) f_2(y) \, dx \, dy + \iint_{(II)} (y - x) f_1(x) f_2(y) \, dx \, dy$$

$$= \int_{-\infty}^{\infty} x f_1(x) \, dx \int_{-\infty}^{x} f_2(y) \, dy - \int_{-\infty}^{\infty} y f_2(y) \, dy \int_{y}^{\infty} f_1(x) \, dx$$

$$+ \int_{-\infty}^{\infty} y f_2(y) \, dy \int_{-\infty}^{y} f_1(x) \, dx - \int_{-\infty}^{\infty} x f_1(x) \, dx \int_{x}^{\infty} f_2(y) \, dy$$

$$= \int_{-\infty}^{\infty} x f_1(x) F_2(x) \, dx - \int_{-\infty}^{\infty} y f_2(y) [1 - F_1(y)] \, dy$$

$$+ \int_{-\infty}^{\infty} y f_2(y) F_1(y) \, dy - \int_{-\infty}^{\infty} x f_1(x) [1 - F_2(x)] \, dx$$

where

$$F_1(x) = \int_{-\infty}^{\infty} f_1(x) \, dx, \qquad F_2(y) = \int_{-\infty}^{\infty} f_2(y) \, dy$$

are the cumulative probability distribution functions. In the above expression, combining the first integral with the fourth one and the second integral with the third one, we obtain

$$\langle z \rangle = \int_{-\infty}^{\infty} [2x f_1(x) F_2(x) - x f_1(x)] \, dx + \int_{-\infty}^{\infty} [2y f_2(y) F_1(y) - y f_2(y)] \, dy$$

$$= 2 \int_{-\infty}^{\infty} x f_1(x) F_2(x) \, dx + 2 \int_{-\infty}^{\infty} y f_2(y) F_1(y) \, dy - \langle x \rangle - \langle y \rangle$$

Since X and Y are statistically independent, then

$$m_2(z) = \langle |x - y|^2 \rangle = \langle (x - y)^2 \rangle = \langle x^2 \rangle + \langle y \rangle^2 - 2\langle x \rangle \langle y \rangle$$
$$= m_2(x) + m_2(y) - 2\langle x \rangle \langle y \rangle = D_x + D_y + (\langle x \rangle - \langle y \rangle)^2$$

Then we have

$$D_z = m_2(z) - \langle z \rangle^2 = D_x + D_y + (\langle x \rangle - \langle y \rangle)^2 - \langle z \rangle^2$$

Example 5.6 Assume that n messages are being sent over a communication channel. The lengths of the messages are random, independent, and have the same mean value m and variance D. Find the mean value and variance of the total time T during which all of the n messages will be transmitted. Using the three-sigma rule (see Sec. 3.3.3), assess the maximum practically possible time T_{max} during which the messages can be transmitted.

Solution The total time T is a sum of the durations of all n messages:

$$T = \sum_{i=1}^{n} T_i$$

Applying the addition rule for the expectations and variances, we obtain

$$\langle t \rangle = \left\langle \sum_{i=1}^{n} t_i \right\rangle = \sum_{i=1}^{n} \langle t_i \rangle = nm$$

and

$$D_t = \sum_{i=1}^{n} D_{t_i} = nD$$

Using the three-sigma rule, we have $T_{\max} = nm + 3\sqrt{nD}$.

Example 5.7 A device consists of n units, each of which may fail independently of the others. The time of the failure-free operation of the ith unit has an exponential distribution with parameter λ_i:

$$f_i(t) = \lambda_i e^{-\lambda_i t}, \quad t > 0$$

A unit that fails is immediately replaced by a new one and is sent for repair. The repair of the ith unit lasts for a random time. This time is also distributed in accordance with the exponential law:

$$\phi_i(t) = \mu_i e^{-\mu_i t}, \quad t > 0$$

The device operates during a time τ. Find the mean value and variance of the number of units that will be replaced, and the mean value of the total time T that will be taken for the repairs of the units that failed.

Solution The random number X_i of units of the ith type that fail during the time τ is distributed in accordance with the Poisson law. Therefore its mean value is $\langle x_i \rangle = \lambda_i \tau$ and the variance is $D_{x_i} = \lambda_i \tau$. The total number X of units that failed during the time τ is $X = \sum_{i=1}^{n} X_i$. Then the mean value of the total number of failed units is

$$\langle x \rangle = \sum_{i=1}^{n} \langle x_i \rangle = \tau \sum_{i=1}^{n} \lambda_i$$

For mutually independent variables X_i, the variance of the total number of failed units is

$$D_x = \sum_{i=1}^{n} D_{x_i} = \tau \sum_{i=1}^{n} \lambda_i$$

The total time T_i needed to repair all the units of the type i that failed during the time τ can be found as the sum

$$T_i = T_i^{(1)} + T_i^{(2)} + \cdots + T_i^{(X_i)} = \sum_{k=1}^{X_i} T_i^{(k)}$$

where $T_i^{(k)}$ is a random time needed to have k units of the ith type repaired. This time has an exponential distribution with parameter μ_i.

The mean value of the random variable T_i can be found on the basis of the formula for calculating the complete mean. Assuming that the random variable X_i assumes a certain value m, the mean value of the time T_i can be evaluated as

$$\langle t_i/m \rangle = \sum_{k=1}^{m} \langle t_i^{(k)} \rangle = \sum_{k=1}^{m} \frac{1}{\mu_i} = \frac{m}{\mu_i}$$

Multiplying this conditional expectation by the probability P_m that the random variable X_i assumes the value m and summing up the products, we obtain the complete (absolute) expectation of the variable T_i:

$$\langle t_i \rangle = \sum_{m=1}^{\infty} P_m \frac{m}{\mu} = \frac{1}{\mu_i} \sum_{m=1}^{\infty} m P_m = \frac{1}{\mu_i} \langle x_i \rangle = \frac{\lambda_i}{\mu_i} \tau$$

Using the rule for adding the means (expectations), we have

$$\langle t \rangle = \tau \sum_{i=1}^{n} \frac{\lambda_i}{\mu_i}$$

The same result could be obtained also on the basis of the following simplified considerations. The average number of failures of the units of the ith type during the time τ is $\lambda_i \tau$. The average repair time for one unit of that type is $1/\mu_i$ and the average time for repairing all the units of type i that failed during the time τ is $\lambda_i \tau/\mu_i$. Then the average time that is required to repair all the units of all the types is

$$\langle t \rangle = \tau \sum_{i=1}^{n} \frac{\lambda_i}{\mu_i}$$

Example 5.8 The conditions of the previous problem are changed so that each unit that fails is sent to a repair shop and the equipment is stopped for the time needed to repair the failed parts. Clearly, when the equipment does not operate, no units can fail. Find the mean value of the number of stops of the equipment during the time τ and the mean value of the portion of the time τ during which the equipment will be idle (which is the mean time spent on repairs).

Solution Let the number of stops during the time τ be X. Using a simplified approach, we represent the operating process of the equipment (device), infinite and continuous in time, as a sequence of "cycles," each of which consists of a period of operation of the system and a period of repair. Hence, the duration of each cycle is the sum of two random variables: T_o, the duration of operation, and T_r, the time to repair. The average duration $\langle t_o \rangle$ of the operation of the equipment can be calculated as the mean time between two consecutive failures in the flow of failures of intensity

$$\lambda = \sum_{i=1}^{n} \lambda_i$$

This average time is

$$\langle t_o \rangle = \frac{1}{\lambda} = \frac{1}{\sum_{i=1}^{n} \lambda_i}$$

The mean time of repairs $\langle t_r \rangle$ can be determined on the basis of the complete expectation formula on the hypothesis H_i, "a unit of type i is being repaired," $i = 1, 2 \ldots, n$. The probability of each hypothesis is proportional to the parameter λ_i:

$$P(H_i) = \frac{\lambda_i}{\sum_{i=1}^{n} \lambda_i} = \frac{\lambda_i}{\lambda}$$

The conditional expectation of the time taken by the repair on this hypothesis is $1/\mu_i$. Hence,

$$\langle t_r \rangle = \sum_{i=1}^{n} \frac{\lambda_i}{\lambda \mu_i} = \frac{1}{\lambda} \sum_{i=1}^{n} \frac{\lambda_i}{\mu_i}$$

The average time of a cycle is

$$\langle t_c \rangle = \langle t_o \rangle + \langle t_r \rangle = \frac{1}{\lambda} + \frac{1}{\lambda} \sum_{i=1}^{n} \frac{\lambda_i}{\mu_i} = \frac{1}{\lambda} \left(1 + \sum_{i=1}^{n} \frac{\lambda_i}{\mu_i} \right)$$

Let us now represent the sequence of stages of the device as a sequence of random points on the t axis, partitioned by intervals of the duration $\langle t_r \rangle$. The average number of stops during the time τ is equal to the average number of random points on the interval of the duration τ. Therefore the mean value $\langle x \rangle$ of the number of stops during the time τ can be found as

$$\langle x \rangle = \frac{\tau}{\langle t_r \rangle} = \frac{\lambda \tau}{1 + \sum_{i=1}^{n} (\lambda_i/\mu_i)}$$

In order to find the mean value of the portion of the time τ during which the equipment will be idle (will be under repair), one can use the fact that during each cycle the device will be idle for the average time of

$$\langle t_r \rangle = \frac{1}{\lambda} \sum_{i=1}^{n} \frac{\lambda_i}{\mu_i}$$

Hence, the average idle time can be computed as

$$\langle x \rangle \langle t_r \rangle = \frac{\lambda \tau}{1 + \sum_{i=1}^{n} (\lambda_i/\mu_i)} \frac{1}{\lambda} \sum_{i=1}^{n} \frac{\lambda_i}{\mu_i} = \tau \frac{\sum_{i=1}^{n} (\lambda_i/\mu_i)}{1 + \sum_{i=1}^{n} (\lambda_i/\mu_i)}$$

The availability index of the equipment can be evaluated as follows:

$$K_a = \frac{\langle t_o \rangle}{\langle t_o \rangle + \langle t_r \rangle} = \frac{1}{1 + (\sum_{i=1}^{n} \lambda_i/\mu_i) \sum_{i=1}^{n} (\lambda_i/\mu_i)}$$

The problem of the evaluation of the reliability of a repairable item will be revisited in Sec. 10.6 and, on the basis of the methods of the theory of markovian processes, in Sec. 11.3.7.

Example 5.9 A hole may occur in a cubical fuel tank with equal probability in each of the six walls and at any point on each wall. All the tank's dimensions are the same and are equal to unity. If the hole appears, all the fuel above the hole leaks out of the tank. When sound, the tank is kept three-quarters full. Find the average amount of fuel that will remain in the tank after a hole appears.

Solution Let the height of the hole be X and the amount of the remaining fluid Y. Since the base area of the tank is equal to unity, we have

$$Y = \begin{cases} X, & \text{for } X < 0.75 \\ 0.75, & \text{for } 0.75 < X < 1 \end{cases}$$

If the hole occurs higher than 0.75 of the distance from the bottom of the tank ($X > 0.75$), then no fuel will leak out and the same amount of fuel, $Y = 0.75$, will remain in the tank. The probability of such an event is equal to the fraction of the surface area of the tank located above the 0.75 level:

$$P\{Y = 0.75\} = P\{X > 0.75\} = \frac{1}{6} + \frac{4}{6} \times \frac{1}{4} = \frac{1}{3}.$$

If the hole appears in the bottom of the tank ($X = 0$), then all the fuel will run out. The probability of this event is equal to the fraction of the bottom area of the tank:

$$P\{Y = 0\} = P\{X = 0\} = \frac{1}{6}$$

If the hole appears in one of the walls at the distance $X < 0.75$ from the bottom, then the amount $Y = X$ of fuel will remain in the tank. The probability density of the amount of fuel in the interval (0, 0.75) is constant and is equal to $(1 - 1/3 - 1/6)/0.75 = 2/3$. The average amount of fuel remaining in the tank is

$$\langle y \rangle = 0.75 \times \frac{1}{3} + 0 \times \frac{1}{6} + \int_0^{0.75} x\left(\frac{2}{3}\right) dx = 0.44$$

Example 5.10 A point a is fixed within the interval (0, 1) and a random point X is uniformly distributed in that interval. Find the coefficient of correlation between the random variable X and the distance R from the point a to X (the distance R is always assumed to be positive). Find the value of a for which the variables X and R are uncorrelated.

Solution The correlation moment K_{xr} is expressed by the formula

$$K_{xr} = \langle xr \rangle - \langle x \rangle \langle r \rangle$$

The means $\langle xr \rangle$ and $\langle r \rangle$ can be calculated as

$$\langle xr \rangle = \langle x|a - x|\rangle = \int_0^1 x|a - x| f(x)\, dx = \int_0^1 x|a - x|\, dx$$

$$= \int_0^a x(a - x)\, dx - \int_a^1 x(a - x)\, dx = \frac{a^3}{3} - \frac{a}{2} + \frac{1}{3} \quad \langle x \rangle = \frac{1}{2}$$

$$\langle r \rangle = \int_0^1 |a - x|\, dx = \int_0^a (a - x)\, dx - \int_0^1 (a - x)\, dx = a^2 - a + \frac{1}{2}$$

Hence, the correlation moment is

$$K_{xr} = \frac{a^3}{3} = \frac{a}{2} + \frac{1}{3} - \frac{1}{2}\left(a^2 - a + \frac{1}{2}\right) = \frac{a^3}{3} - \frac{a^2}{2} + \frac{1}{12}$$

Then we find

$$D_x = \frac{1}{12}, \quad \sigma_x = \frac{1}{2\sqrt{3}}, \quad D_r = m_2(r) - \langle r \rangle^2$$

and

$$m_2(r) = \int_0^1 (a-x)^2 \, dx = a^2 - a + \frac{1}{3}$$

$$D_r = 2a^3 - a^4 - a^2 + \frac{1}{12}$$

$$\sigma_r = \sqrt{D_r}$$

The coefficient of correlation is expressed as follows:

$$r_{xr} = \frac{a^3/3 - a^2/2 + 1/12}{(1/2\sqrt{3})2a^3 - a^4 - a^2 + 1/12}$$

The equation $a^3/3 - a^2/2 + 1/12 = 0$ has only one root within the interval $(0, 1)$: $a = 1/2$. Therefore, the random variables X and R will be uncorrelated only for $a = 1/2$.

Example 5.11 State troopers are stationed at random along the highway. The random number of traffic cops per unit highway length has a Poisson distribution with parameter λ. There is a linear relationship between the probability of being stopped and the car's speed v:

$$p(v) = kv \ (0 \le v \le v_{max}), \quad k = \frac{1}{v_{max}}$$

Forgetting the fines and "points," find a rational speed v_r at which the car will cover the whole route s in a minimum average time.

Solution The average time to cover the distance s is

$$t = \frac{s}{v} + \lambda s p(v) t_0 = \frac{s}{v} + \lambda s k v t_0$$

Where t_0 is the average time the car is delayed by a cop. If the minimum of this function lies within the interval $(0, v_{max})$, then it can be found from the equation

$$\frac{\partial t}{\partial v} = -\frac{s}{v^2} + \lambda s k t_0 = 0$$

so that

$$v = v_r = \sqrt{\frac{v_{max}}{\lambda t_0}} = \sqrt{\frac{1}{\lambda k t_0}}$$

This formula is valid for $v_r < v_{max}$, i.e., for $v_{max} > 1/(\lambda t_0)$. If, for instance, $v_{max} = 100$ miles/hour, $\lambda = 1/20$ miles, and $t_0 = 20$ min, then the rational speed is

$$v_r = \sqrt{\frac{100}{1/20 \times 1/3}} \approx 77.5 \text{ miles/hour}$$

If $v_{max} < 1/(\lambda t_0)$, then the minimum of the function $t = s/v + \lambda s p(v) t_0$ lies outside the interval $(0, v_{max})$, and the speed $v_r = v_{max}$ is the most advantageous. For instance, if, for the above data, the time the car is delayed by a cop decreases to 12 min, then the rational speed is $v_r = v_{max} = 100$ miles/hour.

Example 5.12 A random variable X (phase angle) is uniformly distributed on the interval $(-\pi/2, \pi/2)$. Find the distribution of the random variable $Y = \sin X$.

Solution The sought probability density function can be determined by the formula

$$g(y) = f(\psi(y)) |\psi'(y)|$$

It is convenient to arrange the solution of the problem as two columns, writing the designations of the functions in a general case on the left and those of the specific functions, corresponding to the example in question, on the right:

$f(x)$	$1/\pi$ for $x \in (-\pi/2, \pi/2)$
$y = \phi(x)$	$y = \sin x$
$x = \psi(y)$	$x = \arcsin y$
$\psi'(y)$	$1/\sqrt{1-y^2}$
$g(y) = f(\psi(y))\psi'(y)$	$g(y) = 1/(\pi\sqrt{1-y^2})$ for $y \in (-1, 1)$

Example 5.13 A mass is subjected to harmonic oscillations

$$x = a \sin \omega t$$

where a is the amplitude of the oscillations, ω is their frequency, and x is the coordinate of the mass at the moment of time t. Determine the probability density function for an event "at the arbitrary moment of time t the mass will be found at the distance x from the equilibrium position."

Solution One can assume that the probability of the event, "the mass will be found at an arbitrary moment of time within the interval $(x, x + dx)$," is directly proportional to the length dx of this interval and is inversely proportional to the mass's speed, that is,

$$dP(x \leq X \leq x + dx) = c\frac{dx}{\dot{x}} = c\,dt$$

where c is the factor of proportionality. Since $dP = f(x)\,dx$, we obtain

$$f(x) = c\frac{dt}{dx} = \frac{c}{a\omega \cos \omega t} = \frac{c}{a\omega\sqrt{1 - \sin^2 \omega t}}$$

$$= \frac{c}{a\omega\sqrt{1 - (x/a)^2}} = \frac{c}{\omega\sqrt{a^2 - x^2}}$$

The c value can be determined from the condition of normalization

$$\int_{-\infty}^{\infty} f(x)\,dx = \int_{-a}^{a} f(x)\,dx = 1$$

which yields $c = \omega/\pi$. Thus, the sought probability density function is

$$f(x) = \frac{1}{\pi\sqrt{a^2 - x^2}}$$

The probability that the mass will be found within the interval (x_1, x_2) at an arbitrary moment of time is expressed by the formula:

$$P(x_1 \leq X \leq x_2) = \int_{x_1}^{x_2} f(x)\,dx = \frac{1}{\pi}\left(\arcsin\frac{x_2}{a} - \arcsin\frac{x_1}{a}\right)$$

Example 5.14 A wheel fixed at its center (say, a "wheel of fortune") is rotated. As a result of friction, a fixed point A on the rim of the wheel stops at a certain distance H from the horizontal line that passes through the center of the wheel. Find the distribution of the distance H and the distribution of its absolute value.

Solution The relationship between the distance H and the angle θ at which the rotation terminates is

$$H = r\sin\theta$$

The angle θ is a random variable uniformly distributed within the interval $(0, 2\pi)$. The solution of the problem will evidently not change if the random variable θ is assumed to be uniformly distributed within the interval $(-\pi/2, \pi/2)$. The distribution density of the random variable H is

$$g(h) = \frac{1}{\pi r\sqrt{1 - (h/r)^2}}, \quad \text{for } -r < h < r$$

The distribution density of its absolute value $d = |H|$ is

$$g_1(d) = \frac{2}{\pi r\sqrt{1 - (d/r)^2}}, \quad \text{for } 0 < d < r$$

Example 5.15 The natural logarithm of a random variable X has a normal distribution with a center of scattering m and a standard deviation σ. Find the probability density of the variable X.

Solution From $Y = \ln X$, we obtain $X = e^Y$, and therefore
$$f(y) = \frac{1}{\sigma\sqrt{2\pi}} \exp\left[-\frac{(y-m)^2}{2\sigma^2}\right]$$
Then we find
$$g(x) = \frac{1}{\sigma\sqrt{2\pi}\,x} \exp\left[-\frac{(\ln x - m)^2}{2\sigma^2}\right]\frac{1}{x} = \frac{1}{\sigma\sqrt{2\pi}\,x} \exp\left[-\frac{(\ln x - m)^2}{2\sigma^2}\right], \quad x > 0$$

This distribution is a log-normal distribution (see Sec. 3.3.4.3).

Example 5.16 Form a convolution of two experimental distributions with parameter λ, i.e., find the distribution of the sum of two independent random variables X_1 and X_2 that have the following probability density functions:
$$f_1(x_1) = \lambda e^{-\lambda x_1}, \quad f_2(x_2) = \lambda e^{-\lambda x_2}, \quad x_1 > 0, x_2 > 0$$

Solution Using the general formula (5.15) for the convolution of distributions, we have
$$g(z) = \int_{-\infty}^{\infty} f_1(x_1) f_2(z - x_1)\, dx_1$$
$$= \int_0^z f_1(x_1) f_2(z - x_1)\, dx_1 = \lambda^2 \int_0^z e^{-\lambda x_1} e^{-\lambda(z - x_1)}\, dx_1$$
$$= \lambda^2 z e^{-\lambda z} \quad (z > 0)$$

This distribution is known as *Erlang's distribution of order 2*. It is named in honor of A. K. Erlang, a Danish mathematician engaged for many years by the Copenhagen Telephone Company.

Physically, the obtained distribution originates as follows. Assume that there is an elementary flow of events with intensity λ on the t axis, and only every second point (event) is retained in this flow, while the intermediate points are deleted. Then the intervals between adjacent events in the "rarefied" flow has an Erlang order 2 distribution. The Erlang distribution can also be obtained from the gamma distribution (see Sec. 3.3.8).

Example 5.17 Form a convolution of two exponential distributions with different parameters λ_1 and λ_2:
$$f_1(x_1) = \lambda_1 e^{-\lambda_1 x_1}, \quad f_2(x_2) = \lambda_2 e^{-\lambda_2 x_2}, \quad x_1 > 0, x_2 > 0$$

Solution In accordance with the general formula (5.15) for the convolution of distributions
$$g(x) = \int_{-\infty}^{\infty} f_1(x_1) f_2(x - x_1)\, dx_1$$
we obtain the distribution of the sum X of two variables X_1 and X_2 as follows:
$$g(x) = \int_0^x \lambda_1 e^{-\lambda_1 x_1} \lambda_2 e^{-\lambda_2(x - x_1)}\, dx_1 = \frac{\lambda_1 \lambda_2}{\lambda_2 - \lambda_1}(e^{-\lambda_1 x} - e^{-\lambda_2 x}), \quad x > 0$$

This distribution is known as *Erlang's generalized distribution of order 2*.

Example 5.18 Form a convolution of n exponential distributions with parameter λ, i.e., the distribution of the sum of n independent random variables $X_1, X_2, \ldots, X_n$, which are exponentially distributed with parameter λ.

Solution One could solve the problem by consequently finding the convolutions of two, three, etc., distributions, but it is simpler to proceed from the elementary flow, retaining every nth point in it and deleting the intermediate points, as shown in Fig. 5.1. One can evaluate the probability density $f_n(x)$ of the random variable

$$X = \sum_{i=1}^{n} X_i$$

where X_i is a random variable with an exponential distribution, by finding first the element of probability $f_n(x)\,dx$. This is the probability that the random variable X is found within the elementary interval $(x, x + dx)$. For X to fall within this interval, it is necessary that exactly $n-1$ events are found within the interval x and one event within the interval dx. The probability of occurrence of such an event is $f_n(x)\,dx = P_{n-1}\lambda\,dx$, where P_{n-1} is the probability that $n-1$ events are found within the interval x. Since the number of events of the elementary flow falling on the interval x has a Poisson distribution with parameter $a = \lambda x$, then

$$f_n(x)\,dx = \frac{(\lambda x)^{n-1}}{(n-1)!} \lambda e^{-\lambda x}\,dx$$

so that

$$f_n(x) = \frac{\lambda(\lambda x)^{n-1}}{(n-1)!} e^{-\lambda x}, \qquad x > 0$$

This distribution is known as *Erlang's distribution of order n* (see also Sec. 3.3.8).

Example 5.19 The hydrodynamic impact pressure during ship slamming (impact of the ship's bottom on the incoming wave) in rough seas can be assumed to be proportional to the impact velocity squared: $p = Kv^2$. Assuming that the impact velocity is a random variable that obeys the Rayleigh law

$$f_v(v) = \frac{v}{v_0^2} e^{-v^2/(2v_0^2)}$$

where v_0 is the most likely v value, determine the probability density function of the impact pressure.

Solution Using the general formula (5.9), we obtain

$$f_p(p) = [f_v(v)]_v = \sqrt{p/K}\,\frac{dv}{dp} = \frac{1}{Kv_0} e^{-p/(Kv_0)}, \qquad p \geq 0$$

Thus, the random pressure p has an exponential distribution.

First point   ninth point

$x_1\ x_2 \quad xi \qquad xn \to x$

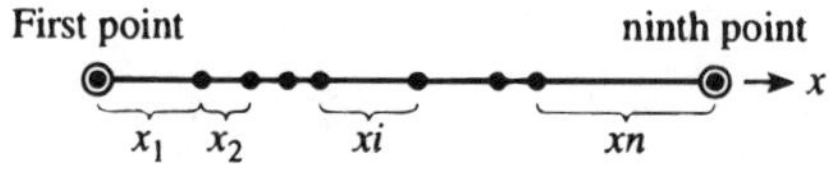

Figure 5.1 Elementary flow.

Example 5.20 An ocean-going ship, like many other dynamic systems characterized by low damping, behaves, when sailing in irregular seas, as a narrow-band filter which selectively enhances those components of the random "input" process of wave excitation that have frequencies close to the ship's own natural frequencies and suppresses all the other components. This results in a concurrent two-frequency bending of the ship's hull: low-frequency bending as a solid body in its motion in heave-and-pitch (these motions have, typically, very close frequencies) and high-frequency bending as an elastic nonprismatic beam (wave-excited hull vibration). Such a "high-frequency" bending has, in effect, the lowest (fundamental) frequency of the natural elastic vibrations of the hull. The higher modes of elastic vibrations usually need not be accounted for, as they are characterized by relatively high damping and contribute little to the magnitude of the total bending moment and the induced dynamic stresses. The low-frequency bending moment ("wave moment") amidship M_1 is proportional to the wave height, while the high-frequency moment ("vibration moment") M_2, if due to the "nonstraightwallness" of the ship's forebody ("whipping" type of elastic vibration), was found to be proportional to the wave height squared:

$$M_1 = C_1 r_0, \qquad M_2 = C_2 r_0^2$$

Here r_0 is half the wave height (wave "radius") and the coefficients C_1 and C_2 depend on the hull shape and dimensions, ship's speed, etc. Assuming that the wave radii r_0 are distributed in accordance with the Rayleigh law, determine the probability density functions for the bending moments M_1 and M_2, the probability distribution of the total moment $M = M_1 + M_2$, and the probability that the total bending moment M will exceed a certain level m^*.

Solution The probability density functions $f_1(m_1)$ and $f_2(m_2)$ for the random bending moments M_1 and M_2 can be found using the Rayleigh law:

$$f(r_0) = \frac{r_0}{D} e^{-r_0^2/(2D)}$$

for the wave radii r_0 as follows:

$$f_1(m_1) = f[r_0(m_1)] \frac{dr_0}{dM_1} = \frac{m_1}{D_1} \exp\left(-\frac{m_1^2}{2D_1}\right)$$

$$f_2(m_2) = f[r_0(m_2)] \frac{dr_0}{dM_2} = \frac{1}{\sqrt{D_2}} \exp\left(-\frac{m_2}{\sqrt{D_2}}\right)$$

where $D_1 = C_1^2 D$ and $D_2 = (2C_2 D)^2$. Thus, while the wave bending moments are distributed, like the wave radii, according to the Rayleigh law, the vibration moments caused by the "whipping" type of the hull vibration follow an exponential law of distribution. Note that, in accordance with the formula (3.28), the D value is related to the variance D_0 of the wave radii as

$$D = \frac{2}{4 + \pi} D_0$$

The distribution of the total bending moment $M = M_1 + M_2$ can be found, using the convolution formulas (5.15) or (5.16), as follows:

$$f_m(m) = \int_0^m f_1(m_1) f_2(m - m_1)\, dm_1$$

$$= \frac{\sqrt{D_2}}{D_1} \exp\left(-\frac{m}{\sqrt{D_2}}\right) \int_0^{m/\sqrt{D_2}} z \exp\left(z - \frac{D_2}{2D_1} z^2\right) dz$$

$$= \frac{1}{\sqrt{D_2}} \exp\left(-\frac{m}{\sqrt{D_2}}\right) - \frac{1}{\sqrt{D_2}} \exp\left(\frac{m^2}{2D_1}\right)$$

$$+ \frac{1}{D_2}\sqrt{\frac{\pi D_1}{2}} \exp\left(\frac{D_1}{2D_2} - \frac{m}{\sqrt{D_2}}\right)\left[\Phi\left(\frac{m}{\sqrt{2D_1}} - \sqrt{\frac{D_1}{2D_2}}\right) + \Phi\left(\sqrt{\frac{D_1}{2D_2}}\right)\right]$$

where the function $\Phi(\alpha)$ is the probability integral expressed by formula (3.22).

The probability that the total bending moment M exceeds the level m^* is

$$P = \int_{m^*}^\infty f_m(m)\, dm = 1 - \int_0^{m^*} f_1(m_1)\, dm_1 \int_0^{m^*-m_1} f_2(m_2)\, dm_2$$

$$= \exp\left(-\frac{m^*}{\sqrt{D_2}}\right) + \sqrt{\frac{\pi D_1}{2D_2}} \exp\left(\frac{D_1}{2D_2} - \frac{m^*}{\sqrt{D_2}}\right)$$

$$\times \left[\Phi\left(\frac{m^*}{\sqrt{2D_1}} - \sqrt{\frac{D_1}{2D_2}}\right) + \Phi\left(\sqrt{\frac{D_1}{2D_2}}\right)\right]$$

Typically, the variance D_2 of the vibration moment is substantially smaller than the variance D_1 of the wave moment. With this in mind, the obtained formula for the probability P can be simplified, using the asymptotic expansion for the error function:

$$\Phi(\beta) \approx 1 - \frac{e^{-\beta^2}}{\beta\sqrt{\pi}} \left[1 - \frac{1}{2\beta^2} + \frac{1.3}{(2\beta^2)^3} - \frac{1.3.5}{(2\beta^2)^4} + \cdots\right]$$

(which is valid for large β values), as follows:

$$P \simeq \sqrt{\frac{\pi D_1}{2D_2}} \exp\left(\frac{D_1}{2D_2} - \frac{m^*}{\sqrt{D_2}}\right)\left[1 - \Phi\left(\sqrt{\frac{D_1}{2D_2}} - \frac{m^*}{\sqrt{D_1}}\right)\right]$$

In the case $D_2 = 0$, when only the wave moment exists, one obtains

$$P = \exp\left[-\frac{(m^*)^2}{2D_1}\right]$$

Comparing the last two formulas, we conclude that the factor

$$\eta = \sqrt{\frac{\pi D_1}{2D_2}} e^{\beta^2}[1 - \Phi(\beta)], \qquad \beta = \sqrt{\frac{D_1}{2D_2}} - \frac{m^*}{\sqrt{2D_1}}$$

reflects the effect of the "whipping" type of wave-excited hull vibration on the level of the total bending moment due to the action of the irregular sea waves.

The obtained formula for the factor η can be further simplified, using the asymptotic expansion for the error function, as follows:

$$\eta \simeq \frac{1}{1 - m^*(\sqrt{D_2/D_1})} = \frac{1}{1 - v_2/v_1^2} \approx 1 + \frac{v_2}{v_1^2}$$

where

$$v_1 = \frac{\sqrt{D_1}}{m^*}, \quad v_2 = \frac{\sqrt{D_2}}{m^*}$$

are coefficients of variation for the bending moments with respect to the level m^*.

Example 5.21 The roof of a shelter can be idealized as a simply supported elongated plate whose edges are fixed and cannot get closer when the roof (plate) is subjected to uniform lateral loading. This loading can be due, for instance, to a heavy layer of snow, and results, in addition to bending stresses, also in the in-plane ("membrane") stresses, thereby making the load-deflection relationship nonlinear. Assuming that the lateral loading produced by the layer of snow is a random variable that follows the Rayleigh distribution, determine the probability that the maximum total stress σ in the roof will exceed the (allowable) level σ^*.

Solution The deflection function $w(x)$ of the roof (plate) can be sought in the form

$$w(x) = f \cos \frac{\pi x}{a}$$

where f is the maximum deflection and a is the width of the roof. The origin of the coordinate x is in the center of the roof. The total stress σ in the roof is due to the bending stress

$$\sigma_b = -\frac{6D}{h^2} \frac{\partial^2 w}{\partial x^2} = \frac{\pi^2 E}{2(1-v^2)} \frac{hf}{a^2} \cos \frac{\pi x}{a}$$

and the in-plane ("membrane") stress

$$\sigma_0 = \frac{E}{2a} \int_{-a/2}^{a/2} \left(\frac{\partial w}{\partial x}\right)^2 dx = \frac{\pi^2 E}{4} \frac{f^2}{a^2}$$

In these formulas, h is the roof's (plate's) thickness, $D = Eh^3/[12(1-v^2)]$ is its flexural rigidity, E is Young's modulus of the material, and v is Poisson's ratio.

The maximum total stress occurs in the center of the roof ($x = 0$) and can be evaluated as

$$\sigma = \frac{\pi^2 E}{4} \bar{f}\left(\bar{f} + \frac{2}{1-v^2} \frac{h}{a}\right)$$

where $\bar{f} = f/a$ is the deflection-to-width ratio. As evident from this formula, the stress σ will exceed a certain level σ^* if the maximum deflection f exceeds the level f^*. This level can be determined from the equation

$$(\bar{f}^*)^2 + \frac{2}{1-v^2}\frac{h}{a}\bar{f}^* - \frac{4\sigma^*}{\pi^2 E} = 0, \qquad \bar{f}^* = \frac{f^*}{a}$$

This biquadratic equation has the following solution:

$$\bar{f}^* = \eta_f (\bar{f}^*)^0$$

where

$$(\bar{f}^*)^0 = \frac{2(1-v^2)}{\pi^2}\frac{a}{h}\frac{\sigma^*}{E}$$

is the deflection-to-width ratio in the case of small deflections (i.e., in the linear case, when the in-plane stress need not be accounted for), and the factor

$$\eta_f = \frac{\sqrt{1+2\bar{\alpha}}-1}{\bar{\alpha}}, \qquad \bar{\alpha} = \frac{2(1-v^2)^2}{\pi^2}\frac{a^2}{h^2}\frac{\sigma^*}{E} = (1-v^2)\frac{a}{h}(\bar{f}^*)^0$$

reflects the effect of the nonlinearity (caused by large deflections and, as a consequence of that, by elevated membrane stresses) on the maximum deflection.

The differential equation of bending is

$$D\frac{\partial^4 w}{\partial x^4} - \sigma_0 h \frac{\partial^2 w}{\partial x^2} = q$$

where q is the lateral random load per unit roof (plate) area. This equation is nonlinear, since the "membrane" stress σ_0 is deflection-dependent. The equation of bending can be solved using Galerkin's method. In accordance with this method, the sought solution

$$w(x) = f \cos\frac{\pi x}{a}$$

should be introduced into the governing equation; then the obtained expression should be multiplied by $\cos \pi x/a$ and integrated over the roof's (plate's) width a. This results in the following cubic equation for the deflection-to-width ratio $\bar{f}$:

$$\bar{f}^3 + \frac{h^2}{3(1-v^2)a^2}\bar{f} = \frac{16a}{\pi^5 Eh}q$$

If the load q exceeds a certain level q^*, the deflection-to-width ratio $\bar{f}$ will exceed the level $\bar{f}^*$, which, as follows from the obtained equation, is related to the threshold q^* value as follows:

$$q^* = \frac{\pi^5 E h}{16a} \bar{f}^* \left[(\bar{f}^*)^2 + \frac{h^2}{3(1-v^2)a^2} \right]$$

$$= \frac{\pi^3}{24} \frac{h^2}{a^2} \sigma^* \eta_f \left[1 + \frac{12(1-v^2)^3}{\pi^4} \frac{a^4}{h^4} \frac{(\sigma^*)^2}{E^2} \eta_f^2 \right].$$

If the load q is a random variable distributed according to the Rayleigh law. The probability that this load exceeds a certain level q^* is

$$P = \exp \left[-\frac{(q^*)^2}{2q_0^2} \right]$$

where q_0 is the most likely value of the load q. Hence, the probability that the maximum stress in the roof exceeds the level σ^* is

$$P = \exp \left\{ -\left(\frac{\pi^3}{96} \frac{h^2}{a^2} \frac{\sigma^*}{q_0} \eta_f \right)^2 \left[1 + \frac{12(1-v^2)^3}{\pi^4} \frac{a^4}{h^4} \frac{(\sigma^*)^2}{E^2} \eta_f^2 \right]^2 \right\}$$

For small deflections, the second term in the brackets is small in comparison with unity, and so this formula can be simplified to

$$P = \exp \left[-\left(\frac{\pi^3}{96} \frac{h^2}{a^2} \frac{\sigma^*}{q_0} \right)^2 \right]$$

The obtained formulas can be used to establish the required thickness of the roof for the given probability P that the actual maximum stress exceeds its allowable level σ^*. In the case of small deflections, this results in the following simple formula for the required thickness h:

$$h = \frac{4a}{\pi} \sqrt{\frac{6}{\pi} \frac{q_0}{\sigma^*}} \sqrt{-\ln P}$$

This formula indicates that, if one is interested in bringing the thickness of the roof structure down, it is the width of the roof that plays the most important role. Then follows the ratio of the most likely load to the allowable stress level, and, finally, the probability that this level is exceeded.

Example 5.22 Two groups of similar machine parts, some of which are faulty, are mixed up. The number of parts in these groups are n_1 and n_2, respectively. The numbers X and Y of faulty parts in each group have binomial distributions:

$$P(X = m) = C(n_1, m) p^m q^{n_1 - m}$$

$$P(Y = m) = C(n_2, m) p^m q^{n_2 - m}$$

Find the distribution of the total number Z of faulty parts: $Z = X + Y$.

Solution The sought distribution is

$$P(Z=z) = \sum_{x=0}^{z} C(n_1, x)p^x q^{n_1-x} C(n_2, z-x) p^{z-x} q^{n_2-z+x}$$

$$= p^z q^{n_1+n_2-z} \sum_{x=0}^{z} C(n_1, x) C(n_2, z-x)$$

$$= C(n_1+n_2, z) p^z q^{n_1 n_2 - z}, \qquad z = 0, 1, 2, \ldots, n_1 + n_2$$

Example 5.23 The random variables X and Y are statistically independent and have normal distributions with zero means and unit standard deviations. Determine the probability density function of their ratio $Z = X/Y$.

Solution The sought probability density function is

$$f(z) = \int_{-\infty}^{\infty} |y| \frac{1}{2\pi} e^{-(z^2+1)y^2/2} \, dy$$

$$= \frac{1}{\pi} \int_{0}^{\infty} y e^{-(z^2+1)y^2/2} \, dy = \frac{1}{\pi(z^2+1)} \int_{0}^{\infty} e^{-t} \, dt$$

$$= \frac{1}{\pi} \frac{1}{z^2+1}$$

This probability density function is known as the Cauchy distribution (see Sec. 3.3.10).

Example 5.24 A solder joint in a flip-chip design of an electronic package of an integrated circuit device experiences a shear strain because of the thermal expansion (contraction) mismatch of the chip and the substrate materials. This strain can be assessed by the following simplified formula:

$$\gamma = l \frac{\Delta \alpha \, \Delta t}{h} = l \frac{\varepsilon}{h}$$

Here $\varepsilon = \Delta \alpha \, \Delta t$ is the thermal mismatch strain, $\Delta \alpha$ is the difference in the coefficients of thermal expansion of the soldered components (the chip and the substrate), Δt is the change in temperature, h is the solder bump's height (stand-off), and l is its distance from the center of the chip. Let the strain ε and the height h be normally distributed random variables with the probability density functions

$$f_\varepsilon(\varepsilon) = \frac{1}{\sqrt{2\pi D_\varepsilon}} e^{-(\varepsilon - \langle \varepsilon \rangle)^2/(2D_\varepsilon)}, \qquad f_h(h) = \frac{1}{\sqrt{2\pi D_h}} e^{-(h - \langle h \rangle)^2/(2D_h)}$$

Find the distribution of the shear strain γ responsible for the long-term reliability of the joint.

Solution The distribution of the shear strain γ can be found as

$$f_\gamma(\gamma) = \frac{1}{l} f_z(z)$$

where $Z = \varepsilon/h$ is the ratio of the random variables ε and h. Using the formula (5.32) for the probability density function of the ratio of two random variables, we obtain:

$$f_z(z) = \int_{-\infty}^{\infty} |h| f_\varepsilon(zh) f_h(h) \, dh$$

$$= \frac{1}{2\pi\sqrt{D_\varepsilon D_h}} \int_{-\infty}^{\infty} |h| \exp\left[-\frac{(zh - \langle\varepsilon\rangle)^2}{2D_\varepsilon} - \frac{(h - \langle h\rangle)^2}{2D_h}\right] dh$$

$$= \frac{1}{2\pi\sqrt{D_\varepsilon D_h}} \int_0^{\infty} h \exp\left[-\left(\frac{z^2}{2D_\varepsilon} + \frac{1}{2D_h}\right)h^2 \right.$$

$$\left. + \left(\frac{z\langle\varepsilon\rangle}{D_\varepsilon} + \frac{\langle h\rangle}{D_h}\right)h - \left(\frac{\langle\varepsilon\rangle^2}{2D_\varepsilon} + \frac{\langle h\rangle^2}{2D_h}\right)\right] dh$$

$$= \frac{1}{2\pi\sqrt{D_\varepsilon D_h}} \int_0^{\infty} h e^{-ah^2 + bh - c} \, dh$$

where the following notation is used:

$$a = \frac{z^2 D_h + D_\varepsilon}{2D_\varepsilon D_h}, \quad b = \frac{\langle\varepsilon\rangle z D_h + \langle h\rangle D_\varepsilon}{D_\varepsilon D_h}, \quad c = \frac{\langle\varepsilon\rangle^2 D_h + \langle h\rangle^2 D_\varepsilon}{2D_\varepsilon D_h}$$

The integral in the obtained formula can be evaluated as

$$\int_0^{\infty} h e^{-ah^2 + bh - c} \, dh = \frac{1}{a} e^{\beta^2 - c} \int_0^{\infty} \xi e^{-(\xi - \beta)^2} \, d\xi$$

$$= \frac{1}{a} e^{\beta^2 - c} \int_{-\beta}^{\infty} (t + \beta) e^{-t^2} \, dt$$

$$= \frac{1}{a} e^{\beta^2 - c} \left(\frac{1}{2} e^{-\beta^2} + \beta \int_{-\beta}^{\infty} e^{-t^2} \, dt\right)$$

$$= \frac{1}{a} e^{-c} \left[\frac{1}{2} + \beta e^{\beta^2} \frac{\sqrt{\pi}}{2} \left(\frac{2}{\sqrt{\pi}} \int_0^{\beta} e^{-t^2} \, dt + 1\right)\right]$$

$$= \frac{1}{2a} e^{-c} \{1 + \sqrt{\pi} \beta e^{\beta^2} [1 + \Phi(\beta)]\}$$

where the parameter β is expressed as

$$\beta = \frac{b}{2\sqrt{a}} = \frac{\langle\varepsilon\rangle z D_h + \langle h\rangle D_\varepsilon}{\sqrt{2D_\varepsilon D_h(z^2 D_h + D_\varepsilon)}}$$

The variance D_h of the solder joint height is typically much smaller than the variance D_ε of the thermal mismatch strain, and therefore the formula for the β value can be simplified:

$$\beta = \frac{\langle h\rangle}{\sqrt{2D_h}}$$

This value is rather large since the mean value of the solder joint height is substantially larger than its standard deviation. Then the following approximate formula can be used for the evaluation of the function $\Phi(\beta)$:

$$\Phi(\beta) \simeq 1 - \frac{1}{\beta\sqrt{\pi}} e^{-\beta^2}$$

and the expression for the probability density function of the shear strain γ can be written as

$$f_\gamma(\gamma) = \frac{1}{\pi l \sqrt{D_\varepsilon D_h}} \frac{\sqrt{\pi}}{2a} \beta e^{-c+\beta^2}$$

$$= \frac{1}{\sqrt{2\pi}} \frac{1}{l} \frac{\langle\varepsilon\rangle z D_h + \langle h\rangle D_\varepsilon}{(z^2 D_h + D_\varepsilon)^{3/2}} \exp\left[-\frac{(\langle\varepsilon\rangle - \langle h\rangle z)^2}{2(z^2 D_h + D_\varepsilon)}\right]$$

$$= \frac{1}{\sqrt{2\pi}} \frac{\langle\varepsilon\rangle(D_h/l^2)\gamma + (\langle h\rangle/l)D_\varepsilon}{[(D_h/l^2)\gamma^2 + D_\varepsilon]^{3/2}} \exp\left\{-\frac{[\langle\varepsilon\rangle - (\langle h\rangle/l)\gamma]^2}{2[(D_h/l^2)\gamma^2 + D_\varepsilon]}\right\}$$

If only the randomness of the thermal expansion mismatch were considered, then, putting $D_h = 0$ and $\langle h\rangle = h$, we would obtain

$$f_\gamma(\gamma) = \frac{1}{\sqrt{2\pi}} \frac{\langle h\rangle}{l\sqrt{D_\varepsilon}} \exp\left\{-\frac{[\langle\varepsilon\rangle - (h/l)\gamma]^2}{2D_\varepsilon}\right\}$$

$$= \frac{1}{\sqrt{2\pi D_\gamma}} \exp\left[-\frac{(\gamma - \langle\gamma\rangle)^2}{2D_\gamma}\right]$$

where

$$\langle\gamma\rangle = \frac{l}{h}\langle\varepsilon\rangle$$

is the mean value of the shearing (angular) strain and

$$D_\gamma = D_\varepsilon \frac{l^2}{h^2}$$

is its variance. Note that the above distribution for the shear strain γ could have been obtained directly from the probability density function $f_\varepsilon(\varepsilon)$ and the relationship $\varepsilon = (h/l)\gamma$. Indeed,

$$f_\gamma(\gamma) = f_\varepsilon[\varepsilon(\gamma)]\frac{d\varepsilon}{d\gamma} = \frac{h/l}{\sqrt{2\pi D_\varepsilon}} \exp\left\{-\frac{[(h/l)\gamma - (h/l)\langle\gamma\rangle]^2}{2(h/l)^2 D_\gamma}\right\}$$

$$= \frac{1}{\sqrt{2\pi D_\gamma}} e^{-(\gamma - \langle\gamma\rangle)^2/(2D_\gamma)}$$

5.3. Limit Theorems of the Probability Theory

Probability theory limit theorems form two groups: the law of large numbers and the central limit theorem.

5.3.1. The law of large numbers

The *law of large numbers* has several forms, each establishing the stability of the average for a large number of observations:

1. *Bernoulli's theorem.* As the number n of independent trials, in each of which an event A occurs with the probability p, tends to infinity, the frequency p_n^* of this event converges in probability to the probability p of the event. This statement can be written as follows:

$$\lim_{n \to \infty} P\{|p_n^* - p| < \varepsilon\} = 1$$

where ε is an arbitrary small positive number.

2. *Poisson's theorem.* As the number n of independent trials, in each of which an event A occurs with probabilities $p_1, p_2, \ldots, p_n$, tends to infinity, the frequency p_n^* of this event converges in probability to the average probability of the event:

$$\lim_{n \to \infty} P\left\{\left|p_n^* - \frac{1}{n}\sum_{i=1}^n p_i\right| < \varepsilon\right\} = 1$$

3. *Chebyshev's theorem.* As the number n of independent trials, in each of which a random variable X with the expectation $\langle x \rangle$ assumes a value X_i, tends to infinity, the arithmetic mean of these values converges in probability to the expectation of the random variable X (the law of large numbers):

$$\lim_{n \to \infty} P\left(\left|\frac{1}{n}\sum_{i=1}^n X_i - \langle x \rangle\right| < \varepsilon\right) = 1$$

4. *Markov's theorem.* If $X_1, X_2, \ldots, X_n$ are independent random variables with expectations $\langle x_1 \rangle, \langle x_2 \rangle, \ldots, \langle x_n \rangle$ and variances $D_{x_1}, D_{x_2}, \ldots, D_{x_n}$, all the variances being bounded from above by the same number L, that is, $D_{x_i} < L$ ($i = 1, 2, \ldots, n$), then, as n trends to infinity, the arithmetic mean of the observed values of the random variables converges in probability to the arithmetic mean of their expectations (the law of large numbers for varying experimental conditions);

$$\lim_{n \to \infty} P\left\{\left|\frac{1}{n}\sum_{i=1}^n X_i - \frac{1}{n}\sum_{i=1}^n \langle x_i \rangle\right| < \varepsilon\right\} = 1$$

When the speed of convergence in probability of various averages to constant values is to be estimated, the following *Chebyshev's inequality* is used:

$$P\{|X - \langle x \rangle| \geq \alpha\} \leq \frac{D_x}{\alpha^2}$$

where $\alpha > 0$ and $\langle x \rangle$ and D_x are the mean value and variance of the random variable X.

5.3.2. The central limit theorem

The *central limit theorem* has also several forms:

1. *Laplace's theorem.* If n independent trials are made, in each of which an event A occurs with probability p, then, as $n \to \infty$, the distribution of a random variable X, the number of times the event occurs, tends to the normal distribution with parameters $m = np$ and $\sigma = \sqrt{npq}$ ($q = 1 - p$). We can go on from this to calculate the probability that the random variable X falls on any interval (α, β): for a sufficiently large n,

$$P\{X \in (\alpha, \beta)\} \simeq \frac{1}{2}\left[\Phi\left(\frac{\beta - np}{\sqrt{2npq}}\right) - \Phi\left(\frac{\alpha - np}{\sqrt{2npq}}\right)\right]$$

2. *The central limit theorem.* If $X_1, X_2, \ldots, X_n$ are similarly distributed independent random variables with mean value $\langle x \rangle$ and mean square deviation $\sigma_x = \sqrt{D_x}$, then, for a sufficiently large n, their sum is

$$Y = \sum_{i=1}^{n} X_i$$

and as an approximately normal distribution with parameters

$$\langle y \rangle = n\langle x \rangle, \qquad D_y = nD_x$$

This means that even though the distribution of the X value in the entire population ("universe") can be of any shape, the distribution of the Y value tends to be close to the normal distribution. The frequency distribution of the average Y values more closely approaches the normal curve the larger the sample size n and the more nearly normal the distribution of the variable X.

3. *Lyapunov's theorem.* If $X_1, X_2, \ldots, X_n$ are independent random variables with mean values $\langle x_1 \rangle, \langle x_2 \rangle, \ldots, \langle x_n \rangle$ and variances D_{x1},

$D_{x_2}, \ldots, D_{x_n}$ and the restriction

$$\lim_{n \to \infty} \frac{\sum_{i=1}^n \langle |x_i|^3 \rangle}{(\sum_{i=1}^n D_{x_i})^{3/2}} = 0 \tag{5.33}$$

is satisfied, then, for a sufficiently large n, the random variable

$$Y = \sum_{i=0}^n X_i$$

has an approximately normal distribution with parameters

$$\langle y \rangle = \sum_{i=1}^n \langle x_i \rangle, \qquad D_y = \sum_{i=1}^n D_{x_i}$$

The meaning of the restriction (5.33) is that the random variables should be comparable from the viewpoint of their effect on the scattering of the sum.

Further Reading (see Bibliography): 7, 20, 30, 32, 38, 40, 44, 58, 59, 62, 84, 88.

Chapter 6

Entropy and Information

> "If a man will begin with certainties he will end with doubts; but if he will be content to begin with doubts, he shall end in certainties."
>
> FRANCIS BACON

> "It is common sense to take a method and try it. If it fails, admit it frankly, and try another. But above all, try something."
>
> FRANKLIN D. ROOSEVELT

> "Say not 'I have found the truth,' but rather, 'I have found a truth'."
>
> KAHLIL GIBRON, *The Prophet*

6.1. Uncertainty of an Experiment

Every random phenomenon is associated with some uncertainty. Therefore, when planning activities aimed at analyzing random phenomena or when designing engineering systems intended for applications that cannot be foreseen with sufficient certainty, one has to account for the inevitable uncertainties.

In order to come up with a suitable approach for a quantitative assessment of the uncertainties in random phenomena, let us assume that, when participating in some mutual effort, one carries out an experiment and that it is necessary to inform the other participants about the outcome of this experiment, in order for them to be able to carry out their part of the effort. Transmission of information is conducted by means of signals. In order to transmit information, one has to develop a certain sequence of signals, to "code information." The binary code, based on just two numbers, "zero" and "one," is the simplest code, which enables one to code information in terms of a series of "zero" and "ones." The minimum number of characters required for the most certain and the most economical way of coding information can be assumed as a quantitative measure of the uncertainty of information. The unit of uncertainty is one binary digit, known as a *bit*.

To implement the most practically feasible way of coding, characterized by the minimum expectancy of the number of bits that are necessary for the transmission of information about the outcome of an experiment, one can break down all the outcomes of the experiment into two groups in such a way that the total probability of outcomes in each group is as close to 1/2 as possible. One of the groups is labeled with the digit "zero" ("no"), and the other with the digit "one" ("yes"). Then each of these two groups is divided into two subgroups, so that the total probability of the outcomes within each subgroup is as close to 1/4 as possible. So one continues breaking down the subgroups until a subgroup containing a single outcome of the experiment results. Since other subgroups might contain more than a single outcome, the process is carried on until all the subgroups contain just one outcome. Then more likely outcomes will contain less bits than less likely ones.

Example 6.1 Code the information about the outcomes $X = \{x_i\}$, $i = 1, 2, \ldots,$ 8, if all the outcomes have equal probabilities. Show that the obtained code is optimal.

Solution When coding the information, all the outcomes are put down in the decreasing order of their probabilities and are broken down into two groups in such a way that the sum of the probabilities in each group would be, to an extent possible, close to 1/2. All the outcomes of the "upper" group are labeled with the character "0" and the outcomes of the "lower" group are labeled with the character "1." Then each of these groups is broken, in a similar fashion, into subgroups with possibly equal total probabilities, and the "upper" subgroups are labeled with the character "0" and the "lower" with the character "1." This process is terminated when only one item of information remains in each subgroup. The process of coding is shown in the Table 6.1.

The mean number of characters in a single outcome is

$$\langle n \rangle = \sum_{i=1}^{n} n_i p(x_i) = 3 \times \frac{1}{8} + 3 \times \frac{1}{8} + \cdots + 3 \frac{1}{8} = 3 \text{ bits}$$

The *entropy* (uncertainty) of the set of information being coded can be calculated as

$$H(X) = \log N = \log 8 = 3 \text{ bits}$$

Since the mean length of the code "word" $\langle n \rangle$ is equal to the entropy $H(X)$, the code is an optimal one.

Example 6.2 Let eight possible outcomes of an experiment might occur with probabilities 1/2, 1/4, 1/8, 1/16, 1/64, 1/64, 1/64, 1/64. Code the information

TABLE 6.1 Coding of Information

Item of information	$p(x_i)$	Breaking down			Code "word"
x_1	1/8	0	0	0	000
x_2	1/8	0	0	1	001
x_3	1/8	0	1	0	010
x_4	1/8	0	1	1	011
x_5	1/8	1	0	0	100
x_6	1/8	1	0	1	101
x_7	1/8	1	1	0	110
x_8	1/8	1	1	1	111

about these outcomes in terms of bits. What is the uncertainty in this experiment?

Solution The coding can be conveniently carried out in the form of a table (Table 6.2). This table indicates that the information about the most likely outcome, having a 1/2 probability, can be transmitted by single character "0," the information about the second, less likely, outcome is transmitted by using two characters "1" and "0," and so on.

The uncertainty in the information about the conducted experiment is measured as the mathematical expectancy (mean) of the number of characters required to transmit this information. The possible number of characters containing information about the outcomes of the experiment are 1, 2, 3, 4 and 6, with the probabilities 1/2, 1/4, 1/8, 1/16 and $4 \times 1/64 = 1/16$, respectively. Therefore the mean value of the number of characters in the information about the outcome of the experiment is

$$H = 1 \times \frac{1}{2} + 2 \times \frac{1}{4} + 3 \times \frac{1}{8} + 4 \times \frac{1}{16} + 6 \times \frac{1}{16} = 2 \text{ bits}$$

TABLE 6.2 Coding of Information

i	P_i	Digits					
		1	2	3	4	5	6
1	1/2	0					
2	1/4	1	0				
3	1/8	1	1	0			
4	1/16	1	1	1	0		
5	1/64	1	1	1	1	0	0
6	1/64	1	1	1	1	0	1
7	1/64	1	1	1	1	1	0
8	1/64	1	1	1	1	1	1

Example 6.3 An experiment has n possible outcomes. The probabilities of the different outcomes in this experiment can be written as 2^{-m}, where m is an integer. What is the expectancy of the number of characters in the information about the outcome of the experiment?

Solution Based on the approach taken in the previous example one can evaluate the sought expectancy as

$$H = \sum m 2^{-m} \tag{6.1}$$

where the summation is carried out for all possible outcomes of the experiment. If some of the outcomes have equal probabilities, the corresponding terms in the sum are equal as well.

6.2. Entropy

If the probability of an outcome i of an experiment is $p_i = 2^{-m}$, where m is an integer, the number of digits in the information about this outcome is calculated as

$$m = -\log_2 p_i, \quad i = 1, 2, \ldots, n$$

and the total expectancy of the number of digits in the information about the outcome is

$$H = -\sum_{i=1}^{n} p_i \log_2 p_i \tag{6.2}$$

This value is known as *entropy*. It characterizes the degree of uncertainty in the outcome of an experiment with a discrete and finite number of outcomes. The term "entropy" was coined by Clausius in 1850 in connection with the second law of thermodynamics and means "transformation" (in thermodynamics, transformation of work into heat).

The concept of entropy does not have to be necessarily applied to an outcome of an experiment, but can be applied to evaluate the state of uncertainty of any system. In addition, the basis of the logarithm in the formula for the entropy does not have to be necessarily equal to 2, i.e., entropy can be computed as

$$H = -\sum_{i=1}^{n} p_i \log_a p_i \tag{6.3}$$

It is noteworthy, however, that when $a \neq 2$, the entropy evaluated for $a = 2$ should be multiplied by $\log_a 2$ to account for the completeness of the group of the mutually exclusive equipossible events.

The scale of uncertainty was first provided by C. Shannon (1948) at Bell Laboratories. He proposed that the entropy expressed by the function H should satisfy the following requirements:

1. Entropy should be a function of all the component probabilities.
2. If all the constituent probabilities are equally likely ($p_i = 1/n$), entropy should increase with an increase in the number of the constituent probabilities.
3. The entropy of several independent outcomes should be calculated as the sum of their entropies.
4. The entropy should be independent of how the problem is broken down.

One can easily check that the function (6.3) meets all these requirements.

For a continuous random variable X, characterized by the probability density function $f(x)$, the entropy can be expressed, by analogy with formula (6.3), as follows:

$$H = -\int_{-\infty}^{\infty} f(x) \log_a f(x) \, dx \qquad (6.4)$$

For a system $X_1, X_2, \ldots, X_n$ of random variables, this formula can be generalized as follows:

$$H = -\int_{-\infty}^{\infty} \cdots \int_{-\infty}^{\infty} f(x_1, x_2, \ldots, x_n)$$
$$\times \log_a f(x_1, x_2, \ldots, x_n) \, dx_1 \, dx_2 \cdots dx_n \qquad (6.5)$$

A similar function

$$H = -\int_a^b f(x) \ln f(x) \, dx$$

was introduced in 1872 by the Austrian physicist L. Boltzmann in the kinetic theory of gases and is widely used in statistical mechanics (physics). The entropies of Shannon and Boltzmann are, in fact, examples of the same idea. In thermodynamics entropy is thought of as a quantitative measure of the disorder (molecular randomness) in a physical system. The second law of thermodynamics states that in an adiabatic (closed) system entropy (disorder) must increase.

The entropy is always positive, since the logarithms of probabilities are always negative. The entropy of an experiment with a single possible outcome is equal to zero. The entropy of an experiment with equipossible outcomes is the maximum one and increases with an increase in the number of outcomes. Among all the laws of probability distributions for a continuous random variable $X(t)$ with a given mean and variance, the normal law has the maximum entropy. In the case of zero mean, the entropy of the normal law can be calculated as

$$H = -\frac{1}{\sqrt{2\pi D}} \int_{-\infty}^{\infty} e^{-x^2/(2D)} \left(-\frac{1}{2} \log_2 2\pi D - \frac{x^2}{2D} \log_2 e \right) dx$$
$$= \log_2 \sqrt{2\pi e D} \qquad (6.6)$$

The entropies of the uniform and the exponential distributions are

$$H = -\int_a^b \frac{1}{b-a} \log_2 \frac{1}{b-a} dx = \log_2 (b-a) \qquad (6.7)$$

and

$$H = -\int_0^{\infty} A e^{-Ax} (\log_2 \lambda - \lambda x \log_2 e) dx = \log_2 \frac{e}{\lambda} \qquad (6.8)$$

respectively. The uniform distribution has the largest entropy of all the distributions on the given finite interval (a, b). Additional constraints introduce more information, which reduces the "uncertainty," "lack of knowledge," and thereby alters the probability distribution of the outcomes. The exponential distribution has the largest entropy of all the distributions with the same mean. This distribution is suggested if no information except the expected value is available. If only the expected (mean) value and the variance are known, then, as has been mentioned earlier, the normal distribution is produced. If the mean, the variance, and the minimum and the maximum values are available, then it is the beta distribution that results in the maximum uncertainty, hence, in the maximum entropy. Finally, knowing the mean occurrence rate between the independent events produces the Poisson distribution. The expression for the entropy $H(X)$ of some continuous laws of distribution is shown in Table 6.3.

Example 6.4 Two shots have been made at the first target and three shots at the second target. The probabilities of hitting the targets in a single shot are 1/2 and 1/3, respectively. Determine the entropy of the number of hittings of each target.

Solution The number of hittings of a target follows a binomial law:

$$P(X = m) = C(n, m) p^m (1-p)^{n-m}$$

The order series is

$$X_1 : \begin{array}{|c|c|c|} \hline 0 & 1 & 2 \\ \hline 1/4 & 1/2 & 1/4 \\ \hline \end{array}$$

for the first target ($n = 2, p = 1/2$) and

TABLE 6.3 Entropy of Some Continuous Laws of Distribution

Distribution	Probability density function	Entropy $H(X) = -\int_{-\infty}^{\infty} f(x) \log f(x)\, dx$
Uniform	$f(x) = \begin{cases} \dfrac{1}{b-a}, & a < x < b \\ 0, & x < a,\ x > b \end{cases}$	$H(X) = \log(b-a) = \log(2\sqrt{3}\,\sigma)$
Exponential	$f(x) = \lambda e^{-\lambda x},\quad x > 0$	$H(X) = \log \dfrac{e}{\lambda} = \log(\sigma e)$ $\sigma =$ standard deviation
Normal	$f(x) = \dfrac{1}{\sqrt{2\pi}\,\sigma} e^{-(x-\langle x \rangle)^2/(2\sigma^2)}$	$H(X) = \log(\sqrt{2\pi e}\,\sigma)$
Poisson	$P_m = \dfrac{a^m}{m!} e^{-a}$	$H(X) = a \log \dfrac{e}{a} + \sum_{m=1}^{\infty} \dfrac{a^m}{m!} e^{-a} \log(m!)$
Rayleigh	$f(x) = \dfrac{x}{\sigma^2} e^{-x^2/(2\sigma^2)},\quad x > 0$	$H(X) = \left(\dfrac{C}{2} + 1\right)\log e$ $C = 0.5772$ (Euler's number)
Cauchy	$f(x) = \dfrac{1}{\pi} \dfrac{1}{1+x^2}$	$H(X) = \log 4\pi$

$$X_2: \begin{array}{|c|c|c|c|} \hline 0 & 1 & 2 & 3 \\ \hline 1/27 & 2/9 & 4/9 & 8/27 \\ \hline \end{array}$$

for the second target ($n = 3$, $p = 1/3$). The corresponding entropies are

$$H_1 = -\frac{1}{4}\log\frac{1}{4} - \frac{1}{2}\log\frac{1}{2} - \frac{1}{4}\log\frac{1}{4} = 0.452 \text{ bits}$$

and

$$H_2 = -\frac{1}{27}\log\frac{1}{27} - \frac{2}{9}\log\frac{2}{9} - \frac{4}{9}\log\frac{4}{9} - \frac{8}{27}\log\frac{8}{27} = 0.511 \text{ bits}$$

Hence, the outcome of shooting at the first target is more certain.

Example 6.5 Show that if the probability p is a small number, the entropy $H = -p \log p$ will also be small.

Solution Let, for instance, $p = 1/2^n$. In this case, $\log p = -n$ and $H = -p \log p = n/2^n$. The fraction $n/2^n$ when the number n is large will be small, since the denominator of this fraction increases with an increase in n much more rapidly than the numerator. Hence, the entropy H decreases without any limit when the probability p tends to zero.

The behavior of the fraction $n/2^n$ has to do with the famous legend about the Iranian inventor of the chess game. This inventor asked, as a reward, for as many grains of wheat as one would get if one grain was put on the first box of the chess board, two on the second, four on the third, and so on. The Shah thought at the beginning that it was a rather modest reward. In reality, however, the number $2^{64} - 1$ exceeds all the available wheat on earth.

Example 6.6 Based on long-term weather observations, it has been established that for a certain geographical area the probability that it will be raining on 4 July is equal to 0.4 and that it will be raining on 25 December is 0.65. It has also been established that the probability that it will be snowing on 25 December is 0.15. Hence, the probabilities that there will be no precipitations on 4 July and 25 December are 0.6 and 0.2, respectively. One is interested in whether there will be precipitations on these days and what kind of precipitations can be expected. For which of these days can a more certain weather forecast be made?

Solution The entropies of the weather conditions on 4 July and 25 December can be calculated as follows:

$$H_1 = -0.4 \log 0.4 - 0.6 \log 0.6 = 0.97 \text{ bits}$$
$$H_2 = -0.65 \log 0.65 - 0.15 \log 0.15 - 0.2 \log 0.2 = 1.28 \text{ bits}$$

The weather on 4 July is more certain. The obtained results are dependent, however, on how one interprets the word "weather." If no distinction is made between the two types of precipitations, then the second entropy should be computed as

$$H_2 = -0.8 \log 0.8 - 0.2 \log 0.2 = 0.72$$

and the conclusion will be different.

6.3. Information

For the complete removal of uncertainty of a random variable one has to carry out an experiment to determine the actual value of the variable. Quite often, however, the random variable X of interest does not lend itself to a direct observation. In these cases one has to observe other random variables and on this basis make a judgement on the variable X. If such an approach is possible then the observed random variable is said to contain *information* regarding the variable X. Obviously, only a (random) variable that depends, in one way or another, on the variable X of interest can contain information about this variable. Therefore information can be viewed as one of the forms of the manifestation of the correlation between random variables. It is natural to accept the decrease in the uncertainty in the evaluation of the random variable X, obtained as a result of the observation of a random variable Y, as a suitable measure of the amount of information about the variable X. Such a decrease can be evaluated as the difference between the entropy of the variable X and its mean conditional entropy with respect to the variable Y. Based on this reasoning, the amount of information $I_y(x)$ contained in the random variable Y, about the random variable X, can be evaluated by the formula

$$I_y(x) = H(x) - H_y(x) \geq 0 \qquad (6.9)$$

The amount of information cannot be negative. It is equal to zero if the variables X and Y are independent. If these variables are dependent, each of them contains a certain amount of information about the other. It is intuitively clear and can be proven that the amount of information that the variable Y contains about the variable X is equal to the amount of information that the variable X contains about Y, i.e., that

$$I_y(x) = I_x(y) \qquad (6.10)$$

Example 6.7 One has 12 coins of the same value. One of these coins is a false one and is therefore different in weight. What is the minimum number of weighings on a lever scale that would enable one to detect the false coin and to find out whether it is heavier or lighter than the rest of the coins?

Solution Each of the 12 coins can turn out to be false and be either heavier or lighter than the rest of the 11 coins. The total amount of outcomes is 24. The entropy of such an experiment is

$$H_1 = \log_2 24 = 4.58$$

Every weighing has three possible outcomes. These, assuming that their probabilities are the same, results in the following entropy:

$$H_2 = \log_2 3 = 1.58$$

The minimum number of weighings cannot be smaller than

$$n = \frac{H_1}{H_2} = \frac{4.58}{1.58} = 2.90$$

i.e., cannot be smaller than 3. Let us examine this problem using just common sense. In order that the total number of weighings be minimum, each weighing should produce the maximum amount of information. At the first weighing let one put i coins on each scale. This test can have three outcomes: (1) the scales are in equilibrium; (2) the right scale outweighs; (3) the left scale outweighs. In the case of the first outcome, the false coin can be found among the remaining $12 - 2i$ coins, and therefore the probability of this outcome is

$$P_1 = \frac{12 - 2i}{12}$$

In the case of the second or the third outcomes, the false coin is among the coins that were weighed. The probabilities of these outcomes are

$$P_2 = P_3 = \frac{i}{12}$$

In order that the weighing produces the maximum information, the distribution of the probabilities of the outcomes must have the maximum entropy. This will take place if all the outcomes have the same probability. This yields

$$\frac{12 - 2i}{12} = \frac{i}{12}$$

so that $i = 4$. Thus, four coins should be put on each scale at the first weighing. We do not examine here the most feasible weighing strategies. These can be established by calculating the entropies of different cases.

Example 6.8 If the 26 letters of the English alphabet were equally probable and perfectly random, how many guesses would be necessary to identify a letter?

Solution Using the pattern from the previous example, we have

$$n = \frac{\ln 26}{\ln 2} = 4.7$$

Let us solve the problem using a different approach. The following strategy can be used. Instead of asking about a single letter, one could divide the 26 letters in half by asking: is the given letter in the first or in the second half of the alphabet? The answer will divide the letters in the alphabet in half. The process is then continued, dividing the remaining possible numbers into two equal groups. Since one encounters two choices n times, the total number $N = 26$ of letters can be written as $N = 2^n$. Therefore, $n = \ln N/\ln 2 = \ln 26/\ln 2 = 4.7$. Note that with actual frequencies p_i of the letters in the contemporary English alphabet, the number n can be found as

$$n = -\frac{1}{\ln 2} \sum p_i \ln p_i = 4.14$$

Example 6.9 The probabilities of appearance of the symbols of the Morse alphabet in the given information are: 0.51 for a dot, 0.31 for a dash, 0.12 for the space between the symbols, and 0.06 for the space between the words. Determine the mean value of the amount of information in a piece of information that contains 500 independent symbols of information.

Solution The sought mean is

$$I = 500(-0.51 \times \log_2 0.51 - 0.31 \times \log_2 0.31 - 0.12 \times \log_2 0.12$$
$$- 0.06 \times \log_2 0.06) = 815 \text{ bits}$$

Example 6.10 There are 100 devices in a batch. 5 percent of these devices are faulty. Five devices are taken at random for inspection. Three of them turned out to be faulty. Determine the amount of information that is contained in this outcome.

Solution The random number X of faulty devices follows the hypergeometric law of the probability distribution:

$$P_n(k) = P_n(X = k) = \frac{C(M, k)C(N - M, n - k)}{C(N, n)}$$

The random variable X can assume the following values: $x_1 = 0$, $x_2 = 1$, $x_3 = 2$, $x_4 = 3$, $x_5 = 4$, $x_6 = 5$. In the case in question, $N = 100$, $n = 5$, $M = 100 \times 0.05 = 5$, and $k = 3$. Hence, the probability $P_5(3)$ that there will be three faulty devices in a randomly selected sample is

$$P_5(3) = \frac{C(5, 3)C(100 - 5, 5 - 3)}{C(100, 5)} = \frac{\frac{5!}{3!2!} \frac{95!}{2!93!}}{\frac{100!}{5!95!}}$$

$$= \frac{95 \times 47}{99 \times 98 \times 97 \times 8} = 0.000\,953$$

The sought amount of information is

$$I(x_4) = -\log P_5(3) = 10.7 \text{ bits}$$

Example 6.11 Find the maximum value of the entropy $H(X)$ of the alphabet $\{x_i\}$ with the finite set of symbols x_i, $i = 1, 2, \ldots, n$.

Solution The expression for the entropy can be written as

$$H = -\sum_{i=1}^{n} p_i \log p_i$$

where $p_i = p(x_i)$. Using the method of Lagrange multipliers (see Example 12, Section 2.2.1) we seek the p_i values that result in the extremum of the function

$$\Phi = -\sum_{i=1}^{n} p_i \log p_i + \lambda \sum_{i=1}^{n} p_i$$

Here λ is the Lagrange multiplier. Differentiating this expression with respect to p_i, $i = 1, 2, \ldots, n$, and equating the obtained derivatives to zero, we obtain the following system of equations:

$$\log p_i + \log e + \lambda = 0, \qquad i = 1, 2, \ldots, n$$

or

$$\log p_i = -\lambda - \log e = \text{constant} \qquad i = 1, 2, \ldots, n$$

The condition of normalization of probabilities requires that

$$\sum_{i=1}^{n} p_i = 1$$

Then we obtain

$$p_1 = p_2 = \cdots = p_n = \frac{1}{n}$$

and the maximum entropy is

$$H(X) = H_m(X) = \log n$$

6.4. Telecommunication Systems

Let $x(t)$ be a realization of a continuous "input" information process $X(t)$ and $y(t)$ be a realization of a continuous "output" process $Y(t)$. Let these processes be characterized by the probability density function $f_1(x)$, $f_2(y)$, and $f(x, y)$. Let $f_1(x|y)$ be the conditional probability of the realization $x = x(t)$ for the known realization $y(t)$ and $f_2(y|x)$ be the conditional probability of the realization $y(t)$ for the given $x(t)$. Then the mutual information $I(x, y)$ between a certain value x of the "input" process and the value y of the "output" process, the mean value $I(X, y)$ of the conditional information, and the total mean mutual information $I(X, Y)$ can be expressed by the formulas:

$$I(x, y) = \log \frac{f(x, y)}{f_1(x) f_2(y)} = \log \frac{f_1(x|y)}{f_1(x)} = \log \frac{f_2(y|x)}{f_2(y)} = I(y, x) \quad (6.11)$$

$$I(X, y) = \int_X I(x, y) f_1(x|y) \, dx \quad (6.12)$$

$$I(X, Y) = \int_Y f_2(y) I(X, y) \, dy = \int_Y \int_X f(x, y) I(x, y) \, dx \, dy = I(Y, X) \quad (6.13)$$

The mean mutual information $I(X, Y)$ can also be found as

$$I(X, Y) = \int_Y \int_X f(x, y) \log \frac{f_1(x|y)}{f_1(x)} \, dx \, dy$$

$$= -\int_X f_1(x) \log f_1(x) \, dx + \int_X \int_Y f(x, y) \log f_1(x|y) \, dx \, dy \quad (6.14)$$

The entropies of the information contained in the processes $Y(t)$ and $X(t)$ can be evaluated by the formulas:

$$H(Y) = -\int_Y f_2(y) \log f_2(y) \, dy$$

$$= -\int_{-\infty}^{\infty} f_2(y) \log f_2(y) \, dy \qquad (6.15)$$

$$H(X|Y) = -\int_X \int_Y f(x, y) \log f_1(x|y) \, dx \, dy$$

$$= -\int_{-\infty}^{\infty} \int_{-\infty}^{\infty} f(x, y) \log f_1(x|y) \, dx \, dy \qquad (6.16)$$

and

$$H(Y|X) = -\int_X \int_Y f(x, y) \log f_2(y|x) \, dx \, dy$$

$$= -\int_{-\infty}^{\infty} \int_{-\infty}^{\infty} f(x, y) \log f_2(y|x) \, dx \, dy \qquad (6.17)$$

The entropy of the product of the variables (processes) X and Y can be evaluated as

$$H(XY) = H(X) + H(Y|X) = H(Y) + H(X|Y) \qquad (6.18)$$

If these processes are statistically independent, then

$$H(XY) = H(X) + H(Y) \qquad (6.19)$$

The total mean mutual information $I(X, Y)$ contained in the processes $X(f)$ and $Y(f)$ can be determined by the formula

$$I(X, Y) = H(X) - H(X|Y) = H(Y) - H(Y|X) \qquad (6.20)$$

Example 6.12 A process

$$Y(t) = X(t) + n(t)$$

acting on a receiving device consists of the useful signal $X(t)$ and a noise $n(t)$. Both are independent normal random processes with zero means and variances $D_s = \sigma_s^2$ and $D_n = \sigma_n^2$, respectively. Determine the amount of information $I(x, y)$ contained in the process $Y(t)$ about the signal $X(t)$ and the total mean mutual information $I(X, Y)$.

Solution From the distribution

$$f_1(x) = \frac{1}{\sqrt{2\pi D_x}} e^{-x^2/(2D_x)}$$

of the useful signal and the conditional probability density function

$$f_2(y|x) = \frac{1}{\sqrt{2\pi D_n}} e^{-(y-x)^2/(2D_n)} = \frac{1}{\sqrt{2\pi D_n}} e^{-n^2/(2D_n)} = f(n)$$

for the process accepted by the receiving unit, considering the statistical independence of the processes $X(t)$ and $n(t)$, we obtain

$$f_2(y) = \frac{1}{\sqrt{2\pi D_y}} e^{-y^2/(2D_y)}$$

Here

$$D_y = D_x + D_n$$

is the variance of the process $Y(t)$. Using the formula (6.11) we find the following expression for the information received by the device:

$$I(x, y) = \log \frac{f_2(y|x)}{f_2(y)} = \frac{1}{\ln 2} \ln \frac{(1/\sqrt{2\pi D_n})e^{-n^2/(2D_n)}}{(1/\sqrt{2\pi D_y})e^{-y^2/(2D_y)}}$$

$$= \frac{1}{\ln 2} \ln \sqrt{\frac{D_x + D_n}{D_n}} + \frac{1}{\ln 2} \left[-\frac{n^2}{2D_n} + \frac{y^2}{2(D_x + D_n)} \right]$$

Formula (6.13) enables one to obtain the expression for the total mean mutual information about the variables X and Y:

$$I(X, Y) = \int_X \int_Y f(x, y) I(x, y) \, dx \, dy = \langle |I(x, y)| \rangle$$

$$= \log \sqrt{1 + \frac{D_x}{D_n}} + \frac{1}{\ln 2} \left[\frac{\langle y^2 \rangle}{2(D_x + D_n)} - \frac{\langle n^2 \rangle}{2D_n} \right]$$

Since $\langle y^2 \rangle = D_x + D_n$ and $\langle n^2 \rangle = D_n$, we have

$$I(X, Y) = \log \sqrt{1 + \frac{D_x}{D_n}} = \frac{1}{2} \log \left(1 + \frac{D_x}{D_n} \right)$$

Example 6.13 The mean power of the transmitted signals is D_x. Determine the maximum entropy of these signals.

Solution If one seeks the maximum value of the integral

$$I = \int_a^b F[x, f(x)] \, dx \quad (6.21)$$

under additional conditions (constraints)

$$\left. \begin{array}{l} \int_a^b \phi_1[x, f(x)] \, dx = k_1 = \text{constant} \\[4pt] \int_a^b \phi_2[x, f(x)] \, dx = k_2 = \text{constant} \\[2pt] \dotfill \\[2pt] \int_a^b \phi_n[x, f(x)] \, dx = k_n = \text{constant} \end{array} \right\} \quad (6.22)$$

then the function $f(x)$, which results in the maximum value of the integral I, can be found using the method of Lagrange multipliers. In accordance with this method, the function $f(x)$ can be found from the equation

$$\frac{\partial F}{\partial f} + \lambda_1 \frac{\partial \phi_1}{\partial f} + \lambda_2 \frac{\partial \phi_2}{\partial f} + \cdots + \lambda_n \frac{\partial \phi_n}{\partial f} = 0 \tag{6.23}$$

where $\lambda_1, \lambda_2, \ldots, \lambda_n$ are Lagrange multipliers. These can be determined by introducing the function $f(x)$, found from Eq. (6.23), into the conditions (6.22). In the case in question, one seeks a function $f(x)$, for which the entropy

$$H(X) = -\int_{-\infty}^{\infty} f(x) \ln f(x) \, dx$$

achieves its maximum value, while the following conditions are fulfilled:

$$\int_{-\infty}^{\infty} x^2 f(x) \, dx = \sigma^2, \qquad \int_{-\infty}^{\infty} f(x) \, dx = 1 \tag{6.24}$$

The first condition is the definition of the variance and the second one is the condition of normalization. The functions F, ϕ_1, and ϕ_2, are as follows:

$$F[x, f(x)] = -f(x) \ln f(x)$$

$$\phi_1[x, f(x)] = x^2 f(x)$$

$$\phi_2[x, f(x)] = f(x)$$

Hence,

$$\frac{\partial F}{\partial f} = -[1 + \ln f(x)], \qquad \frac{\partial \phi_1}{\partial f} = x^2, \qquad \frac{\partial \phi_2}{\partial f} = 1$$

Introducing these partial derivatives into the equation (6.23), we find

$$-1 - \ln f(x) + \lambda_1 x^2 + \lambda_2 = 0$$

and therefore

$$f(x) = e^{\lambda_2 - 1} e^{\lambda_1 x^2} \tag{6.25}$$

Introducing this solution into the second formula in (6.24), we obtain

$$\int_{-\infty}^{\infty} f(x) \, dx = \int_{-\infty}^{\infty} e^{\lambda_2 - 1} e^{\lambda_1 x^2} \, dx = 2 e^{\lambda_2 - 1} \int_{0}^{\infty} e^{\lambda_1 x^2} \, dx = 1$$

The λ_1 value must be negative, otherwise the obtained integral will diverge. Since

$$\int_0^\infty e^{-a^2 x^2}\, dx = \frac{\sqrt{\pi}}{2a}$$

we have

$$e^{\lambda_2 - 1} = \sqrt{-\frac{\lambda_1}{\pi}}$$

and Eq. (6.25) yields

$$f(x) = \sqrt{-\frac{\lambda_1}{\pi}}\, e^{\lambda_1 x^2} \tag{6.26}$$

Introducing this equation into the first formula in (6.24), we obtain the following formula for the variance $D_x = \sigma^2$:

$$\sigma^2 = \int_{-\infty}^\infty x^2 f(x)\, dx = 2\sqrt{-\frac{\lambda_1}{\pi}} \int_0^\infty x^2 e^{\lambda_1 x^2}\, dx$$

$$= 2\sqrt{-\frac{\lambda_1}{\pi}}\, \frac{1}{4}\sqrt{\frac{\pi}{(-\lambda_1)^3}} = -\frac{1}{2\lambda_1}$$

Hence,

$$\lambda_1 = -\frac{1}{2\sigma^2}$$

and the formula (6.26) yields

$$f(x) = \frac{1}{\sqrt{2\pi}\,\sigma}\, e^{-x^2/(2\sigma^2)}$$

Thus, when the signals are limited in their mean power, the signals with the normal distribution possess the maximum entropy.

The entropy of a normally distributed random variable is

$$H(X) = -\int_{-\infty}^\infty f(x) \ln f(x)\, dx = -\int_{-\infty}^\infty f(x) \ln\left[\frac{1}{\sqrt{2\pi}\,\sigma}\, e^{-x^2/(2\sigma^2)}\right] dx$$

$$= \int_{-\infty}^\infty f(x) \ln(\sqrt{2\pi}\,\sigma)\, dx + \int_{-\infty}^\infty f(x)\, \frac{x^2}{2\sigma^2}\, dx$$

Considering (6.24), we have

$$H(X) = H_m(X) = \ln(\sqrt{2\pi}\,\sigma) + \frac{\sigma^2}{2\sigma^2} = \ln(\sqrt{2\pi}\,\sigma) + \frac{1}{2}$$

Since $1/2 = \ln \sqrt{e}$, we can write the obtained formula as follows (see also Table 6.3):

$$H(X) = \ln (\sqrt{2\pi e}\, \sigma) = \frac{1}{2} \ln (2\pi e \sigma^2)$$

Further Reading (see Bibliography): 1, 4, 9, 19, 26, 33, 39, 43, 61, 73, 86, 101, 105, 117, 129, 131.

Chapter 7

Random Processes: Correlation Theory

> *"Give me a fruitful error any time, full of seeds, bursting with its own corrections. You can keep your sterile truth for yourself."*
> — VILFREDO PARETO

> *"I will quote our famous mathematician P. L. Chebyshev. He said: 'In ancient times the mathematical problems were set by the Gods, as, for example, the problem of the doubling of a cube, in connection with the alteration in the dimensions of the Delphic altar. Then, in a later period, the problems were set by the demigods: Newton, Euler, Lagrange. Today we have a third period, when problems are set by everyday needs.' I think that these words are completely true, with just one correction: in my opinion, only the third period has ever existed."*
> — DIMITRY GRAVE
> *Encyclopedia of Mathematics,*
> *St Petersburg, Russia, 1912*

> *"Mathematicians are like French: when you talk to them, they translate your words into their language, and at once you cease to understand them."*
> — JOHANN GOETHE

7.1. Random Functions and Processes

A function $X(t)$ is a *random function* if its value is a random variable for any given argument t. If this argument is time, the random function is a *random (stochastic) process*. The specific form that a random function (process) assumes as a result of an experiment is a *realization* of the random function.

The concept of a random function is a generalization of the concept of a random variable. Since we can consider a random variable X to be a function of an elementary event ω: $X = \phi(\omega)$, $\omega \in \Omega$, where Ω is the sample space, it follows that the random function $X(t)$ can be represented as

$$X(t) = \phi(t, \omega), \qquad \omega \in \Omega, t \in T$$

where t is a nonrandom argument and T is the domain of the function $X(t)$. Examples of random functions are: air temperature $T(h)$ at a given point at a given moment of time for the given altitude h above the ground, ordinate $X(t)$ of an irregular sea wave, time-dependent supply voltage $V(t)$, etc.

A random process $X(t)$ can be characterized by a collection of realizations $x_1(t)$, $x_2(t)$, ..., $x_n(t)$ which can be obtained as a result of actual trials (experiments). For a fixed moment t of time the random process is an ordinary random variable (a *section* of a random process). Thus, a random process can be viewed either as a set of realizations or as a set of random variables depending on the parameter t, i.e., a set of sections. If one considers several sections as a series of points t_1, t_2, ..., t_m rather than one section of a random process, then an m-dimensional random vector can be obtained which describes the random process with a certain degree of approximation.

Stochastic processes are classified according to a number of criteria. A process taking place in a certain system whose transformations from one state to another occur at the predetermined moments of time is a *process with discrete time* or a *random sequence*. Examples are: a machine or device inspected at given moments of time, a computer that changes its state at discrete moments of time, a dynamic system subjected to periodic impacts, etc. If the transitions from one state to another occur in the system at random moments of time in a continuous manner, the random processes in such a system are *processes with continuous time*. Examples are: functioning of a device that fails and becomes reconditioned at random moments of time, change in the supply voltage of a computer, stresses in a ship hull in irregular waves, etc. If the number of possible states in a system is finite or countable, the random processes in this system are *processes with discrete states*. An example of such a system is a device consisting of two units for which the possible states of the system are s_1, "both units are sound," s_2, "the first unit is faulty and the second unit is sound," s_3, "the first unit is sound and the second unit is faulty," and s_4, "both units are faulty." If the set of possible states of the system is uncountable, the random processes occurring in this system are *processes with continuous states*. Examples are: the supply voltage of a computer, the process of landing of an aircraft in a predetermined spot, the height of an irregular ocean wave, etc. Thus, random processes can be classified as processes with discrete states and discrete time, processes with discrete states and continuous time, processes with continuous states and discrete time, and processes with continuous states and continuous time.

7.2. Characteristics of Random Processes

A *univariate distribution* of a random process $X(t)$ is the distribution of its section $X(t)$ for any moment of time t. If the random process $X(t)$ is continuous, this distribution is the probability density function $f(x, t)$ of the section $X(t)$. If the random process $X(t)$ is discrete, the univariate distribution of this process is an ordered series, i.e., the probability $P(x_i, t)$ that the random variable $X(t)$ assumes a value x_i at a moment of time t. A *bivariate distribution* of a random process $X(t)$ is a joint distribution of two of its sections $X(t_1)$ and $X(t_2)$ for any moments t_1 and t_2 of time. This is a function of four arguments: x_1, x_2, t_1, t_2. Thus, a random process is characterized by a one-dimensional probability density function $f(x, t)$ of a section $X(t)$ and by a two-dimensional probability density function $f(x_1, x_2, t_1, t_2)$ of the system of its two sections $X(t_1)$ and $X(t_2)$.

The *mean value* $\langle x(t) \rangle$ of a random process $X(t)$ is a nonrandom function which is calculated as the mean value of the section of the random process at an arbitrary moment of time t:

$$\langle x(t) \rangle = \int_{-\infty}^{\infty} x(t) f(x, t) \, dx \qquad (7.1)$$

The *variance* $D_x(t)$ of a random process $X(t)$ is a nonrandom function which is calculated as the variance of the section of the random process at an arbitrary moment of time t:

$$D_x(t) = \int_{-\infty}^{\infty} [x(t) - \langle x(t) \rangle]^2 f(x, t) \, dx \qquad (7.2)$$

The *correlation function* (or *autocorrelation function*) $K_x(t_1, t_2)$ of a random process $X(t)$ is a nonrandom function which is calculated as the covariance (see Sec. 4.1) of the sections $X(t_1)$ and $X(t_2)$ for an arbitrary pair of moments t_1 and t_2 of time:

$$K_x(t_1, t_2) = \int_{-\infty}^{\infty} \int_{-\infty}^{\infty} [x_1 - \langle x(t_1) \rangle][x_2 - \langle x(t_2) \rangle] f(x_1, x_2, t_1, t_2) \, dx_1 \, dx_2$$

$$(7.3)$$

For $t_1 = t_2$ the correlation function turns into a variance of the random process. In Fig. 7.1 realizations of two random processes are shown. The processes $X(t)$ and $Y(t)$ have approximately the same mean values and variances but different correlation functions: the correlation between two sections in Fig. 7.1a is small, while in Fig. 7.1b it is significant.

The main properties of a correlation function are as follows:

1. $K_x(t_1, t_2) = K_x(t_2, t_1)$, i.e., the function $K_x(t_1, t_2)$ does not change if t_1 is replaced by t_2.
2. $|K_x(t_1, t_2)| \le \sqrt{D_x(t_1)D_x(t_2)}$, i.e., the absolute value of the correlation function never exceeds the product of the standard deviations of the corresponding sections.

The correlation function $K_{xy}(t_1, t_2)$ for two random processes $X(t)$ and $Y(t)$ can be written in a way similar to (7.3):

$$K_{xy}(t_1, t_2) = \int_{-\infty}^{\infty} \int_{-\infty}^{\infty} [x - \langle x(t_1) \rangle][y - \langle y(t_2) \rangle] f(x, y, t_1, t_2) \, dx \, dy$$

(7.4)

This is a *bivariate correlation function*.

If the multidimensional (multivariate) laws of the distribution of random variables are known, formulas of the type (7.4) enable one to obtain a complete information about the system of random processes. The use of such law often leads, however, to rather cumbersome derivations. On the other hand, complete information about a system of

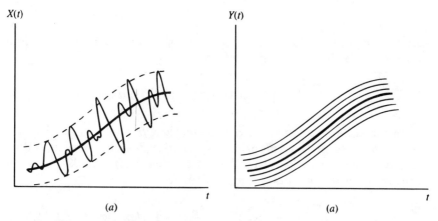

Figure 7.1 The processes $X(t)$ and $Y(t)$ have the same means and variances, but different correlation functions: the correlation between two sections in (a) is small, while in (b) it is significant.

random processes is not always needed. In practical applications, it is often sufficient to be able to evaluate only the first two moments of the random processes—the mean and the correlation function. The correlation function of two centered random processes $\overset{\circ}{X}(t)$ and $\overset{\circ}{Y}(t)$ for two moments of time t and t' can be found as

$$K_{xy}(t, t') = \langle \overset{\circ}{X}(t)\overset{\circ}{Y}(t') \rangle$$

For complex random processes $\overset{\circ}{X}(t)$ and $\overset{\circ}{Y}(t)$, the correlation function can be calculated as

$$K_{xy}(t, t') = \langle \overset{\circ}{X}(t)\overset{\circ}{Y}{}^*(t') \rangle$$

where $\overset{\circ}{Y}{}^*(t)$ is the conjugated, with respect to the process $\overset{\circ}{Y}(t)$, complex random function. It can be shown that the correlation function $K_{xy}(t, t')$ of complex random processes $X(t)$ and $Y(t)$ possesses the following property:

$$K_{xy}(t, t') = K_{yx}^*(t', t)$$

where $K_{yx}^*(t', t)$ is the conjugated, with respect to $K_{yx}(t', t)$, complex correlation function. A theory of random processes based on the use of the means, variances, and correlation functions of random processes is known as the *correlation theory of random processes*.

Example 7.1 The random process $X(t)$ is described by the equation

$$X(t) = Ae^{-\alpha t}$$

where A is a random variable with the known mean $\langle a \rangle$ and variance D_a. Determine the correlation function of this process and the coefficient of correlation.

Solution The mean $\langle x \rangle$ and the variance D_x of the process $X(t)$ can be found as

$$\langle x \rangle = \langle Ae^{-\alpha t} \rangle = \langle a \rangle e^{-\alpha t}$$
$$D_x = \langle (Ae^{-\alpha t} - \langle x \rangle)^2 \rangle = e^{-2\alpha t}\langle (A - \langle a \rangle)^2 \rangle = e^{-2\alpha t}D_a$$

The correlation function of the random process $X(t)$ is

$$K_x(t, t') = \langle (Ae^{-\alpha t} - \langle a \rangle e^{-\alpha t})(Ae^{-\alpha t'} - \langle a \rangle e^{-\alpha t'}) \rangle$$
$$= e^{-\alpha t}e^{-\alpha t'}\langle (A - \langle a \rangle)^2 \rangle = e^{-\alpha(t+t')}D_a$$

The standard deviations for the process $X(t)$ at the moments of time t and t' are as follows:

$$\sigma_x(t) = \sqrt{D_x(t)} = \sigma_a e^{-\alpha t}, \qquad \sigma_x(t') = \sqrt{D_x(t')} = \sigma_a e^{-\alpha t'}$$

The coefficient of correlation (the normalized correlation function) is

$$r_x(t, t') = \frac{K_x(t, t')}{\sigma_x(t)\sigma_x(t')} = \frac{D_a e^{-\alpha(t+t')}}{\sigma_a^2 e^{-\alpha t}e^{-\alpha t'}} = 1$$

Example 7.2 Find the correlation function of the random process

$$X(t) = A \sin \omega t + B \cos \omega t$$

where A and B are correlated random variables with zero means. Determine this function for the case where these variables are uncorrelated and have the same variance.

Solution By definition, the sought correlation function can be calculated as

$$K_x(t, t') = \langle X(t)X(t') \rangle = \langle (A \sin \omega t + B \cos \omega t)(A \sin \omega t' + B \cos \omega t') \rangle$$

$$= \langle A^2 \sin \omega t \sin \omega t' + AB(\sin \omega t \cos \omega t' + \cos \omega t \sin \omega t')$$

$$+ B^2 \cos \omega t \cos \omega t' \rangle$$

$$= \langle A^2 \rangle \sin \omega t \sin \omega t' + \langle AB \rangle \sin \omega (t + t') + \langle B^2 \rangle \cos \omega t \cos \omega t'$$

$$= \sigma_a^2 \sin \omega t \sin \omega t' + K_{ab} \sin \omega (t + t') + K_b^2 \cos \omega t \cos \omega t'$$

In a special case, when $K_{ab} = 0$ (the random variables A and B are uncorrelated) and $\sigma_a^2 = \sigma_b^2 = \sigma^2$ (the variables A and B have the same variance), we have

$$K_x(t, t') = \sigma^2 \cos \omega (t - t')$$

Example 7.3 The process $X(t)$ of random vibrations can be described as

$$X(t) = A \sin \omega t$$

where ω is a non-random parameter (frequency) and A is a random variable (amplitude) with the mean value $\langle a \rangle$ and variance D_a. Determine the correlation function and the variance of the time derivatives $\dot{X}(t)$ and $\ddot{X}(t)$, i.e., of the velocity and acceleration of the process $X(t)$.

Solution The correlation function of the process $X(t)$ is

$$K_x(t, t') = \langle [(A - \langle a \rangle) \sin \omega t][(A - \langle a \rangle) \sin \omega t'] \rangle$$

$$= D_a \sin \omega t \sin \omega t'$$

The mean value of the process $\dot{X}(t)$ is

$$\langle \dot{x} \rangle = \frac{d}{dt} \langle x \rangle = 0$$

and the correlation function of this process can be calculated as

$$K_{\dot{x}}(t, t') = \langle [\dot{X}(t) - \langle \dot{x} \rangle][\dot{X}(t') - \langle \dot{x} \rangle] \rangle$$

$$= \left\langle \frac{d\overset{\circ}{X}}{dt} \frac{d\overset{\circ}{X}}{dt'} \right\rangle = \frac{\partial^2}{\partial t \, \partial t'} \langle \overset{\circ}{X}(t)\overset{\circ}{X}(t') \rangle = \frac{\partial^2 K_x(t, t')}{\partial t \, \partial t'}$$

$$= D_a \omega^2 \cos \omega t \cos \omega t'$$

The variance of the random process $\dot{X}(t)$ is

$$D_{\dot{x}}(t) = D_a \omega^2 \cos^2 \omega t$$

Similarly, one can obtain the following correlation function for the process $\ddot{X}(t)$:

$$K_{\ddot{x}}(t,\, t') = \frac{\partial^4 K_x(t,\, t')}{\partial t^2 \, \partial t'^2} = D_a \omega^4 \sin \omega t \sin \omega t' = \omega^4 K_x(t,\, t')$$

Example 7.4 The mean value of the random process $X(t)$ is zero and its correlation function is $K(t,\, t')$. Determine the correlation function of the process

$$Y(t) = a(t)X(t) + b(t)\dot{X}(t)$$

where $a(t)$ and $b(t)$ are nonrandom functions of time. One encounters processes of the type $Y(t)$ when dealing with vibration systems with small masses, so that the inertial forces need not be accounted for.

Solution The correlation function of the process $Y(t)$ can be calculated as

$$\begin{aligned}K_y(t,\, t') = \mathring{Y}(t)\mathring{Y}(t')\rangle &= a(t)a(t')\langle X(t)X(t')\rangle \\&+ a(t)b(t')\langle X(t)\dot{X}(t')\rangle + a(t')b(t)\langle \dot{X}(t)X(t')\rangle \\&+ b(t)b(t')\langle \dot{X}(t)\dot{X}(t')\rangle\end{aligned}$$

where $\mathring{Y}(t)$ and $\mathring{Y}(t')$ are centered random processes. Since

$$\langle X(t)\dot{X}(t')\rangle = \frac{\partial}{\partial t'} \langle X(t)X(t')\rangle = \frac{\partial}{\partial t'} K_x(t,\, t')$$

we obtain

$$\begin{aligned}K_y(t,\, t') = a(t)a(t')K_x(t,\, t') &+ a(t)b(t') \frac{\partial K_x(t,\, t')}{\partial t'} \\&+ a(t')b(t) \frac{\partial K_x(t,\, t')}{\partial t} + b(t)b(t') \frac{\partial^2 K_x(t,\, t')}{\partial t \, \partial t'}\end{aligned}$$

Example 7.5 The random vibration processes

$$X = A \cos \omega_1 t, \qquad Y = B \cos \omega_2 t$$

have random amplitudes A and B, with the given characteristics $\langle a \rangle$, $\langle b \rangle$, D_a, D_b, and K_{ab}. Determine the bivariate correation function K_{xy}.

Solution Using the formula (7.4) for the correlation function of two random processes, we obtain

$$K_{xy}(t,\, t') = \langle (A - \langle a \rangle) \cos \omega_1 t (B - \langle b \rangle) \cos \omega_2 t' \rangle = K_{ab} \cos \omega_1 t \cos \omega_2 t'$$

For $t = t'$ this formula yields

$$K_{xy}(t,\, t) = K_{ab} \cos \omega_1 t \cos \omega_2 t$$

Example 7.6 A simply supported beam of a circular cross section is loaded at its left end ($z = 0$) by two concentrated moments, M_1 and M_2, which are located in the vertical and the horizontal planes, respectively. The moments M_1 and M_2 are statistically dependent random variables with the means $\langle m_1 \rangle$ and $\langle m_2 \rangle$, variances D_1 and D_2, and the correlation moment K_{12}. Find the mean and the variance of the maximum bending stress σ_{max} for an arbitrary cross section of the beam. The length of the beam is l.

Solution The bending moments acting in an arbitrary cross section z of the beam are

$$M_x = M_1\left(1 - \frac{z}{l}\right), \quad M_y = M_2\left(1 - \frac{z}{l}\right)$$

Here x is the horizontal transverse axis, y is the vertical axis, and the axis z is oriented along the beam. The maximum normal (bending) stress in the cross section z is

$$\sigma_{max} = \frac{r}{I}(M_x + M_y) = \frac{4}{\pi r^3}(M_1 + M_2)\left(1 - \frac{z}{l}\right)$$

where r is the radius of the beam's cross section and $I = (\pi/4)r^4$ is the moment of inertia of the cross section. The mean and the correlation function of the normal stress can be found as follows:

$$\langle \sigma \rangle = \frac{4}{\pi r^3}\left(1 - \frac{z}{l}\right)(\langle m_1 \rangle + \langle m_2 \rangle)$$

$$K_\sigma(z, z') = \left(\frac{4}{\pi r^3}\right)^2 \left\langle (M_1 + M_2)^2 \left(1 - \frac{z}{l}\right)\left(1 - \frac{z'}{l}\right) \right\rangle$$

$$= \left(\frac{4}{\pi r^3}\right)^2 \left(1 - \frac{z}{l}\right)\left(1 - \frac{z'}{l}\right)[\langle\langle m_1\rangle^2\rangle + 2\langle\langle m_1\rangle\langle m_2\rangle\rangle + \langle\langle m_2\rangle^2\rangle]$$

$$= \left(\frac{4}{\pi r^3}\right)^2 \left(1 - \frac{z}{l}\right)\left(1 - \frac{z'}{l}\right)(D_1 + 2K_{12} + D_2)$$

The variance of the stress σ_{max} is

$$D_\sigma = \left(\frac{4}{\pi r^3}\right)^2 \left(1 - \frac{z}{l}\right)^2 (D_1 + 2K_{12} + D_2)$$

The mean and the variance are the largest at the origin ($z = 0$) where the moments M_1 and M_2 are applied

$$\langle \sigma \rangle_{max} = \frac{4}{\pi r^3}(\langle m_1 \rangle + \langle m_2 \rangle)$$

$$D_{\sigma, max} = \left(\frac{4}{\pi r^3}\right)^2 (D_1 + 2K_{12} + D_2)$$

Example 7.7 Using the results of the previous example, find the required diameter of the beam, assuming that the maximum stress σ follows the normal law of the probability distribution. The allowable probability that a permissible stress level σ^* is exceeded is equal to P.

Solution The probability that a normally distributed random variable σ exceeds a certain level σ^* can be found as

$$P(\sigma > \sigma^*) = \frac{1}{\sqrt{2\pi D_\sigma}} \int_{\sigma^*}^{\infty} \exp\left[-\frac{(\sigma - \langle\sigma\rangle)^2}{2D_\sigma}\right] d\sigma$$

$$= \frac{1}{2}\left[1 - \Phi\left(\frac{\sigma^* - \langle\sigma\rangle}{\sqrt{2D_\sigma}}\right)\right] = \frac{1}{2}[1 - \Phi(\alpha)]$$

where $\Phi(\alpha)$ is the Laplace function. Using the given probability P and the tables for the function $\Phi(\alpha)$ from Appendix B, one can find the allowable stress level σ^* as

$$\sigma^* = \langle\sigma\rangle + \alpha\sqrt{2D_\sigma}$$

The mean value $\langle\sigma\rangle$ and the variance D_σ at the origin, where the induced stress is the highest, are

$$\langle\sigma\rangle = \frac{4}{\pi r^3}(\langle m_1\rangle + \langle m_2\rangle), \quad D_\sigma = \left(\frac{4}{\pi r^3}\right)^2 (D_1 + 2K_{12} + D_2)$$

The required diameter of the beam is

$$d = 2r = 2\left[\frac{4}{\pi}\frac{\langle m_1\rangle + \langle m_2\rangle + \alpha\sqrt{2(D_1 + 2K_{12} + D_2)}}{\sigma^*}\right]^{1/3}$$

In this formula, the allowable stress level σ^* is a function of the parameter $\alpha = (\sigma^* - \langle\sigma\rangle)/\sqrt{2D_\sigma}$ and, hence, of the allowable probability P.

7.3. Stationary Random Processes

A random process is a *stationary* one if its mean value is constant:

$$\langle x(t)\rangle = \langle x\rangle = \int_{-\infty}^{\infty} xf(x)\, dx = \text{constant} \quad (7.5)$$

and the correlation function depends only on the difference $\tau = t_2 - t_1$ of the arguments:

$$K_x(t_1, t_2) = K_x(\tau) = \int_{-\infty}^{\infty}\int_{-\infty}^{\infty} (x_1 - \langle x\rangle)(x_2 - \langle x\rangle) f(x_1, x_2, \tau)\, dx_1\, dx_2$$

$$(7.6)$$

In this case the variance is also constant and is equal to

$$D_x(t) = D_x = K_x(0) = \int_{-\infty}^{\infty} (x - \langle x \rangle)^2 f(x)\, dx \qquad (7.7)$$

The correlation function of a stationary random process possesses the following major properties:

1. $\lim_{\tau \to \infty} |K_x(\tau)| = 0$, i.e., the correlation function tends to zero for a sufficiently large difference τ in the moments of time t_1 and t_2.
2. $K_x(\tau) = K_x(-\tau)$, i.e., the correlation function is an even function: it is symmetric with respect to the axis $\tau = 0$.
3. $|K_x(\tau)| \le K_x(0) = D_x$, i.e., the correlation function has its maximum value at $\tau = 0$ and this value is equal to the variance of the random process.

A typical correlation function of a stationary random process is shown in Fig. 7.2.

Many stationary random processes possess also the important property of *ergodicity*. This property means that the characteristics of the process evaluated on the basis of the above formulas, i.e., by means of *ensemble averaging*, can be obtained also by means of *time averaging* of a single realization of a sufficiently large duration T. In accordance with the property of ergodicity, the mean and the corre-

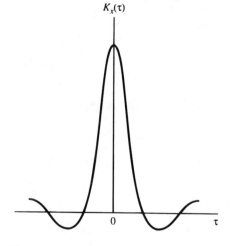

Figure 7.2 Typical correlation function for a stationary random process.

lation function can be calculated as follows:

$$\langle x \rangle \approx \frac{1}{T} \int_0^T X(t)\, dt \tag{7.8}$$

and

$$K_x(\tau) \approx \frac{1}{T} \int_0^T [X(t) - \langle x \rangle][X(t + \tau) - \langle x \rangle]\, dt$$

$$= \frac{1}{T} \int_0^T \mathring{X}(t)\mathring{X}(t + \tau)\, dt \tag{7.9}$$

Then the variance can be computed as

$$D_x = K_x(0) = \frac{1}{T} \int_0^T [X(t) - \langle x \rangle]^2\, dt \tag{7.10}$$

An ensemble can be defined as a collection of records representing the variation of a random variable as a function of time. An ergodic process is "homogeneous" in the time domain, and each realization of such a process is a "typical representative" of the process. This useful property results in the fact that one can, when it is needed, carry out time averaging, instead of ensemble averaging, and vice versa. In an ergodic process, the statistical properties of an ensemble do not change with a spatial (time) shift.

The sufficient condition for ergodicity of a stationary process with respect to the mean value is tending of the correlation function $K_x(\tau)$ of the process to zero as $\tau \to \infty$:

$$\lim_{\tau \to \infty} K_x(\tau) = 0$$

As far as the variance is concerned, it is the process $Y(t) = X^2(t)$ that should possess a similar property:

$$\lim_{\tau \to 0} K_y(\tau) = 0$$

For a random function to be ergodic with respect to its correlation function, it is necessary that the function $Z(t, \tau) = X(t)X(t + \tau)$ possesses the property

$$\lim_{\tau \to 0} K_z(\tau) = 0$$

Note that if a random process is ergodic with respect to its mean value, it does not necessarily mean that it is ergodic also with respect to its correlation function.

Let us find out at what condition the relationship (7.8) holds, i.e., at what condition the process $X(t)$ can be considered an ergodic one. The integral in the expression (7.8) has different values for different realizations and is, in the general case, a random variable unless the time T is very large:

$$E_T = \frac{1}{T} \int_0^T X(t)\, dt$$

The mean and variance of this variable can be calculated, in accordance with the general rule, as follows:

$$\langle E_T \rangle = \left\langle \int_0^T \frac{1}{T} X(t)\, dt \right\rangle = \frac{1}{T} \int_0^T \langle x \rangle\, dt = \langle x \rangle$$

$$D_{E_T} = \langle (E_T - \langle x \rangle)^2 \rangle = \left\langle \left[\frac{1}{T} \int_0^T (X - \langle x \rangle)\, dt \right]^2 \right\rangle$$

$$= \frac{1}{T^2} \int_0^T \int_0^T K_x(t - t')\, dt\, dt'$$

The obtained equation for the variance indicates that the expression (7.8) holds only for large enough T values, making the variance D_{E_T} very small. This means that the relationship

$$\langle x \rangle = \lim_{T \to \infty} \frac{1}{T} \int_0^T X(t)\, dt$$

is fulfilled only in the case when the condition

$$L = \lim_{T \to \infty} D_{E_T} = \lim_{T \to \infty} \frac{1}{T^2} \int_0^T \int_0^T K_x(t - t')\, dt\, dt' = 0$$

is fulfilled. The formula for the L value can be written by introducing a new variable $\tau = t - t'$ (which changes from $-T$ to T when the moments t and t' of time change from zero to T), as follows:

$$L = \frac{1}{T^2} \int_0^T \int_{-t'}^{T-t'} K_x(\tau)\, d\tau\, dt' = \frac{1}{T^2} \int_{-T}^{T} \left[\int_{|\tau|}^{T} K_x(\tau)\, dt' \right] d\tau$$

$$= \frac{1}{T} \int_{-T}^{T} \left(1 - \frac{|\tau|}{T}\right) K_x(\tau)\, d\tau = \frac{2}{T} \int_0^T \left(1 - \frac{\tau}{T}\right) K_x(\tau)\, d\tau$$

Thus, a random process $X(t)$ possesses the property of ergodicity if its correlation function satisfies the condition

$$\lim_{T \to \infty} \frac{1}{T} \int_0^T \left(1 - \frac{\tau}{T}\right) K_x(\tau) \, d\tau = 0 \tag{7.11}$$

This condition is both necessary and sufficient.

An important characteristic of a stationary random process is its mean ("*effective*") *period*. This can be calculated as

$$\tau_e = 2\pi \sqrt{\frac{D_x}{D_{\dot{x}}}} \tag{7.12}$$

where D_x and $D_{\dot{x}}$ are variances of the processes $X(t)$ and $\dot{X}(t)$, respectively. The term "effective period" is due to the fact that, for periodic ("regular") processes

$$x(t) = A \cos \omega t, \qquad \dot{x}(t) = -A\omega \sin \omega t$$

with the frequency $\omega = 2\pi/\tau$, the "variances" D_x and $D_{\dot{x}}$ are

$$D_x = \int_0^\tau x^2(t) \, dt = \frac{A^2}{2}, \qquad D_{\dot{x}} = \int_0^\tau \dot{x}^2(t) \, dt = \frac{\omega^2 A^2}{2} \tag{7.13}$$

The "effective" period in this case is simply the actual period of the process:

$$\tau_e = 2\pi \sqrt{\frac{D_x}{D_{\dot{x}}}} = \tau$$

The parameter

$$\omega_e = \frac{2\pi}{\tau_e} = \sqrt{\frac{D_{\dot{x}}}{D_x}} \tag{7.14}$$

is the mean ("effective") frequency of a stationary process $X(t)$.

Example 7.8 A stationary random process $X(t)$ has a correlation function

$$K_x(\tau) = D_x e^{-\alpha|\tau|}$$

Is this process an ergodic one?

Solution In order for a stationary random process $X(t)$ to be ergodic, it is necessary and sufficient that its correlation function meets the requirement (7.11). This requirement is fulfilled. Indeed,

$$\lim_{T \to \infty} \frac{1}{T} \int_0^T \left(1 - \frac{\tau}{T}\right) D_x e^{-\alpha \tau} \, d\tau$$

$$= \lim_{T \to \infty} \frac{1}{T} \left[-\frac{1}{\alpha}(e^{-\alpha T} - 1) - \frac{D_x}{T}\left(-\frac{T}{\alpha} e^{-\alpha T} - \frac{1}{\alpha^2} e^{-\alpha T} + \frac{1}{\alpha^2}\right) \right] = 0$$

Example 7.9 Considering that a deterministic function $\phi(t)$ can be viewed as a special case of the random function $X(t) = \phi(t)$, find its mean value $\langle t \rangle$, variance $D_x(t)$, and the correlation function $K_x(t_1, t_2)$. Is the "random" function $X(t)$ a stationary one?

Solution The characteristics of the function $\phi(t)$ are $\langle t \rangle = \phi(t)$, $D_x(t) = 0$, and $K_x(t_1, t_2) = 0$. The random function $X(t)$ is nonstationary since its mean value changes with time.

Example 7.10 The random process $X(t)$ is described by the equation

$$X(t) = \sum_{i=1}^{\infty} (A_i \cos \omega_i t + B_i \sin \omega_i t)$$

where A_i and B_i are random statistically independent variables with zero means and equal variances $D_{A_i} = D_{B_i} = D_i$. One encounters such processes, particularly, in the theory of irregular sea waves. Is the process $X(t)$ a stationary one? Find the variance of this process.

Solution The correlation function of the process $X(t)$ is

$$K_x(t, t') = \left\langle \sum_{i=1}^{\infty} (A_i \cos \omega_i t + B_i \sin \omega_i t) \sum_{i=1}^{\infty} (A_k \cos \omega_k t' + B_k \sin \omega_k t') \right\rangle$$

$$= \sum_{i=1}^{\infty} D_i \cos \omega_i (t - t')$$

The obtained expression for the correlation function $K_x(t, t')$ depends on the difference $\tau = t - t'$ only. Hence, the random process $X(t)$ is a stationary one. Its variance is

$$D_x = \sum_{i=1}^{\infty} D_i$$

Example 7.11 Determine the mean value and the correlation function of the random process $Y(t) = \dot{X}(t)$ if the correlation function of the process $X(t)$ is

$$K_x(\tau) = D_x e^{-\alpha |\tau|}$$

Solution The mean value of the random variable $Y(t) = \dot{X}(t)$ can be found as

$$\langle y(t) \rangle = \langle \dot{x}(t) \rangle = \frac{d}{dt} \langle x(t) \rangle$$

i.e., the mean value of the derivative of a variable $X(t)$ can be evaluated as a derivative of its mean value.

Let us obtain the relationship between the correlation functions $K_y(t, t')$ and $K_x(t, t')$:

$$K_y(t, t') = \langle [Y(t) - \langle y(t) \rangle][Y(t') - \langle y(t') \rangle] \rangle = \left\langle \frac{d\mathring{X}}{dt} \frac{d\mathring{X}}{dt'} \right\rangle$$

$$= \frac{\partial^2}{\partial t \, \partial t'} [\mathring{X}(t)\mathring{X}(t')] = \frac{\partial^2 K_x(t, t')}{\partial t \, \partial t'}$$

For stationary processes, when the correlation function depends on the time difference $\tau = t' - t$ only, we have

$$K_y(\tau) = -\frac{d^2 K_x(\tau)}{d\tau^2}$$

Then the sought correlation function can be found as

$$K_y(\tau) = -\frac{d^2 K_x(\tau)}{d\tau^2} = -\frac{d}{d\tau}\left(-D_x \alpha e^{-\alpha|\tau|} \frac{d|\tau|}{d\tau}\right)$$

$$= D_x \alpha \frac{d}{d\tau}(e^{-\alpha|\tau|} \operatorname{sign} \tau)$$

$$= D_x \alpha \left[-\alpha e^{-\alpha|\tau|}(\operatorname{sign} \tau)^2 + e^{-\alpha|\tau|} \frac{d}{d\tau} \operatorname{sign} \tau \right]$$

$$= D_x \alpha[2\delta(\tau) - \alpha(\operatorname{sign} \tau)^2]e^{-\alpha|\tau|}$$

The functions $\operatorname{sign} \tau$ and $\delta(\tau)$, used in this derivation, possess the following major properties:

(1) $\operatorname{sign} \tau = \dfrac{d|\tau|}{d\tau} = 1,$ for $\tau > 0$

(2) $\operatorname{sign} \tau = 0,$ for $\tau = 0$

(3) $\operatorname{sign} \tau = -1,$ for $\tau < 0$

(4) $|\tau| = \tau \operatorname{sign} \tau$

(5) $\tau = |\tau| \operatorname{sign} \tau$

(6) $f(|\tau|) = f(\tau) \operatorname{sign} \tau$

(7) $(\operatorname{sign} \tau)^2 = 1,$ for $\tau \neq 0$

(8) $(\operatorname{sign} \tau)^2 = 0,$ for $\tau = 0$

(9) $\operatorname{sign} \tau = 2\mathbf{1}(\tau) - 1$

(10) $\mathbf{1}(\tau) = \displaystyle\int_{-\infty}^{\tau} d(\operatorname{sign} \tau)$

and

(1) $\delta(t) = 0,$ for $t \neq 0$

(2) $\delta(0) = \infty,$ for $t = 0$

(3) $\int_{-\infty}^{\infty} \delta(t)\, dt = 1$

(4) $\int_{-\infty}^{\infty} f(t-a)\delta(t)\, dt = f(a)$

(5) $1(\tau) = \int_{-\infty}^{\tau} \delta(\tau)\, d\tau$

(6) $\delta(\tau) = \dfrac{1}{2}\dfrac{d}{d\tau} \text{sign } \tau = \dfrac{1}{2}\dfrac{d^2|\tau|}{d\tau^2} = \dfrac{d1(\tau)}{d\tau}$

where $1(\tau)$ is the Heaviside unit step function (see Example 7.14).

The function $\delta(t)$ is known as the *Dirac delta function*. It was introduced to quantum mechanics by an English physicist P. A. Dirac and is often employed in applied problems as a useful means of describing highly localized phenomena, such as instantaneous impulses, concentrated loads, point charges, etc.

For $\tau \neq 0$ we find

$$K_y(\tau) = D_x \alpha^2 e^{-\alpha|\tau|}$$

Example 7.12 From the correlation function for the process $\dot{X}(t)$, found in the previous example, determine the correlation function for the process $\ddot{X}(t)$.

Solution Using the same reasoning as before, we find

$$K_y(t, t') = \frac{\partial^4 K_x(t, t')}{\partial t^2\, \partial t'^2}$$

Then, with $\tau = t' - t$, we have

$$K_z(\tau) = \frac{d^4 K_x(\tau)}{d\tau^4} = \frac{d^2 K_y(\tau)}{d\tau^2} = D_x \alpha \frac{d^2}{d\tau^2}\{[2\delta(\tau) - \alpha(\text{sign }\tau)^2]e^{-\alpha|\tau|}\}$$

For $\tau \neq 0$ we obtain

$$K_2(\tau) = D_x \alpha^4 e^{-\alpha|\tau|}$$

Example 7.13 The random process $X(t)$ is described by the series

$$X(t) = \sum_{j=1}^{\infty} A_j e^{i\omega_j t}$$

where $A_j = \bar{A}_j + i\bar{\bar{A}}_j$ are the complex amplitudes and $i = \sqrt{-1}$ is the "imaginary unity." The random variables $\bar{A}_j$ and $\bar{\bar{A}}_j$ have zero means and the same variances D_j. Determine the correlation function of the process $X(t)$.

Solution The sought correlation function can be found as

$$K_x(t, t') = \left\langle \sum_{j=1}^{\infty} A_j e^{i\omega_j t} \sum_{k=1}^{\infty} A_k^* e^{-i\omega_k t'} \right\rangle$$

where $A_k^* = \bar{A}_k - i\bar{\bar{A}}_k$ is a conjugate complex amplitude. After carrying out the calculations we finally obtain

$$K_x(t, t') = \sum_{j=1}^{\infty} \langle A_j A_j^* \rangle e^{i(t-t')\omega_j}$$

where

$$\langle A_j A_j^* \rangle = D_{A_j} = 2D_j$$

Example 7.14 Consider an elementary flow of events with intensity λ (Fig. 7.3) on the t axis (points on the t axis) and a related function $X(t)$, which is the number of events occurring during the time $(0, t)$. At the moment an event occurs, the random function $X(t)$ makes a jump, thereby increasing its value by unity. Such a random function is known as a *Poisson process*. Find the univariate distribution of the Poisson process and its characteristics $\langle x(t) \rangle$, $D_x(t)$, $K_x(t_1, t_2)$, $r_x(t_1, t_2)$, as well as the characteristics of the random process $Z(t) = X(t) - \lambda t$.

Solution The distribution of the random process, the section of $X(t)$, is a Poisson distribution with the parameter $a = \lambda t$. Therefore, the probability that the random variable $X(t)$ will assume the value m can be found by the formula

$$P_m = (\lambda t)^m \frac{e^{-\lambda t}}{m!}, \quad m = 0, 1, 2, \ldots$$

The mean value and variance of the random function $X(t)$ are

$$\langle x(t) \rangle = D_x(t) = \lambda t$$

In order to find the correlation function $K_x(t_1, t_2)$, we assume that $t_2 > t$, consider the time interval $(0, t_2)$, as shown in Fig. 7.3, and partition the interval $(0, t_2)$ into intervals $(0, t_1)$ and (t_1, t_2). The number of events over the entire interval $(0, t_2)$ is equal to the sum of the numbers of events on the intervals $(0, t)$ and (t_1, t_2):

$$X(t_2) = X(t_1) + Y(t_2 - t_1)$$

where $Y(t_2 - t_1)$ is the number of events which occur on the interval (t_1, t_2). Since the flow is stationary, the random process $Y(t_2 - t_1)$ has the same distribution as the process $X(t_1)$. In addition, in accordance with the properties of the Poisson flow of events, the random variables $X(t_1)$ and

Figure 7.3 Elementary flow of events.

$Y(t_2 - t_1)$ are statistically uncorrelated. Therefore, the correlation function $K_x(t_1, t_2)$ can be found as

$$K_x(t_1, t_2) = \langle \mathring{X}(t_1)\mathring{X}(t_2) \rangle = \langle \mathring{X}(t_1)[\mathring{X}(t_1) + \mathring{Y}(t_2 - t_1)] \rangle$$
$$= \langle [\mathring{X}(t_1)]^2 \rangle = D_x(t_1) = \lambda t_1$$

Similarly, for $t > t_2$, we obtain $K_x(t_1, t_2) = \lambda t_2$. Thus,

$$K_x(t_1, t_2) = \min\{\lambda t_1, \lambda t_2\} = \lambda \min\{t_1, t_2\}$$

where $\min\{t_1, t_2\}$ is the smaller one of the two values t_1 and t_2 (if $t_1 = t_2$, one can take either t_1 or t_2 as the minimum value).

Using the *Heaviside unit step function* (a unit jump) $1(x)$, one can write the expression for the correlation function as

$$K_x(t_1, t_2) = \lambda t_1 1(t_2 - t_1) + \lambda t_2 1(t_1 - t_2)$$

The function $1(\tau)$ is defined as follows:

$$1(\tau) = \begin{cases} 1, & \text{for } \tau > 0 \\ \tfrac{1}{2}, & \text{for } \tau = 0 \\ 0, & \text{for } \tau < 0 \end{cases}$$

(Some properties of this function have been indicated in Example 7.11). For the correlation coefficient $r_x(t_1, t_2)$ we have

$$r_x(t_1, t_2) = \frac{K_x(t_1, t_2)}{\sqrt{D_x(t_1)D_x(t_2)}} = \sqrt{\frac{t_1}{t_2}}\, 1(t_2 - t_1) + \sqrt{\frac{t_2}{t_1}}\, 1(t_1 - t_2)$$

A Poisson process is a process with independent increments since its increment at any interval is the number of events occurring on the interval; for an elementary flow the numbers of events falling on non-overlapping intervals are independent.

The process $Z(t) = X(t) - \lambda t$ results from a nonhomogeneous linear transformation of the random process $X(t)$. Consequently, $\langle z(t) \rangle = \langle x(t) \rangle - \lambda t = 0$ and $D_z(t) = D_x(t) = \lambda t$. Since a homogeneous linear transformation does not change the correlation function of the process $X(t)$, then

$$K_z(t_1, t_2) = K_x(t_1, t_2) = \lambda \min(t_1, t_2)$$

7.4. Crossing the Given Level

If the realization of a random process $X(t)$ crosses (upwards) a straight line ("level") that is parallel to the t axis and is located at the distance x^* from it, then the process $X(t)$ is said to "*cross*" the level x^*. The number of level crossings during the given time is a discrete random variable. If such crossings are infrequent, then their number can be considered to have a Poisson distribution.

In order to determine the number of upward crossings of the level x^* by the random process $X(t)$, one should know the two-dimensional probability density function $f(x, \dot{x})$ for the process $X(t)$ and its derivative $\dot{X}(t)$. The expression

$$f(x^*, \dot{x})\, dx\, d\dot{x} = f(x^*, \dot{x})\dot{x}\, d\dot{x}\, dt$$

determines the probability that the values of the function $X(t)$ and $\dot{X}(t)$ are found within the intervals $(x^*, x^* + dx)$ with $(\dot{x}, \dot{x} + d\dot{x})$, respectively. It is clear that upward crossings of the level x^* take place for all positive values of the derivative $\dot{x}$, i.e., for $0 < \dot{x}(t) < \infty$. Therefore the complete probability of crossing the level x^* during the time dt can be computed as

$$P_1 = dt \int_0^\infty f(x^*, \dot{x})\dot{x}\, d\dot{x}$$

Since the duration dt is small, one can expect that upward crossing of the level x^* during the time dt does not take place more than once. With such an assumption, one can view the probability P_1 as the probability of a single upward crossing of the level x^* by the random process $X(t)$ during the time dt, i.e., within the time interval $(t, t + dt)$.

Let P_0 be the probability that there will be no upward crossing of the level x^* for the time dt. Then the mean number of upward crossings within the time dt can be found as

$$1 \times P_1 + 0 \times P_0 = P_1$$

i.e., this number is the same as the probability P_1. Thus, the mean number of upward crossings per unit time can be computed by the formula

$$N_{x^*} = \int_0^\infty f(x^*, \dot{x})\dot{x}\, d\dot{x}$$

If the ordinates $X(t)$ and the velocities $\dot{X}(t)$ are statistically independent random processes, the joint probability density function $f(x, \dot{x})$ of these processes can be represented as

$$f(x, \dot{x}) = f_1(x) f_2(\dot{x})$$

Then the obtained formula for the number of crossings yields

$$N_{x^*} = f_1(x^*) \int_0^\infty f_2(\dot{x})\dot{x}\, d\dot{x}$$

This formula indicates that, for a stationary process with a statistically independent derivative, the mean value of the upward crossings of the given level is proportional to the probability density on this level. This result can be used, particularly, for the experimental evaluation of the probability density functions for such processes, simply by counting the number of upward crossings on different levels.

If the process $X(t)$ and its time derivative $\dot{X}(t)$ are normally distributed, then the number of crossings of the level x^* can be calculated as

$$N_{x^*} = \frac{1}{\tau_e} \exp\left[-\frac{(x^* - \langle x \rangle)^2}{2D_x}\right] \qquad (7.15)$$

7.5. Duhamel Integral: Transformation of Random Processes by Linear Dynamic Systems

The response $h(t)$ of a dynamic system to a unit pulse excitation is called an *impulse response* of the system, the *weighting function*, or the *Green function* of this system. The use of the impulse response function is an effective means of evaluating the response of the system to an arbitrary excitation, whether deterministic or random, stationary or nonstationary. The system's response at any moment of time t to an elementary impulse ds, applied at the preceding moment of time ξ, can be evaluated as

$$dy = h(t - \xi)\, ds(\xi)$$

Treating a continuous excitation $x(t)$ as a sequence of successive elementary impulses $ds(\xi) = x(\xi)\, d\xi$, we obtain

$$y(t) = \int_0^t x(\xi) h(t - \xi)\, d\xi \qquad (7.16)$$

This formula is called the *Duhamel integral*. This integral can be written also as follows:

$$y(t) = \int_0^t x(t - \tau) h(\tau)\, d\tau = \int_{-\infty}^t x(\tau) h(t - \tau)\, d\tau = \int_0^\infty x(t - \tau) h(\tau)\, d\tau \qquad (7.17)$$

Let us show how the Duhamel integral can be used to find the response $Y(t)$ of a linear dynamic system to a stationary random excitation $X(t)$. The mean value of the "output" process $Y(t)$ is

$$\langle y \rangle = \frac{1}{T} \int_0^T y(t)\, dt = \frac{1}{T} \int_0^T \int_0^\infty x(t - \tau) h(\tau)\, d\tau\, dt = \int_0^\infty h(\tau) \langle x(t - \tau) \rangle\, d\tau$$

where

$$\langle x \rangle = \frac{1}{T} \int_0^T x(t - \tau) \, dt \simeq \frac{1}{T} \int_0^T x(t) \, dt$$

is the mean value of the "input" process $X(t)$. Thus, the mean values of the output and the input processes are related as

$$\langle y \rangle = \langle x \rangle \int_0^\infty h(\tau) \, d\tau$$

For the correlation function $K_y(\tau)$ of the "output" process $Y(t)$ we have

$$K_y(\tau) = \frac{1}{T} \int_0^T y(t) y(t + \tau) \, dt$$

$$= \frac{1}{T} \int_0^T dt \int_0^\infty x(t_1 - \tau_1) h(\tau_1) \, d\tau_1 \int_0^\infty x(t_2 - \tau_2) h(\tau_2) \, d\tau_2$$

$$= \int_0^\infty \int_0^\infty h(\tau_1) h(\tau_2) \, d\tau_1 \, d\tau_2 \left[\frac{1}{T} \int_0^T x(t_1 - \tau_1) x(t_2 - \tau_2) \, dt \right]$$

$$= \int_0^\infty \int_0^\infty h(\tau_1) h(\tau_2) K_x(\tau + \tau_1 - \tau_2) \, d\tau_1 \, d\tau_2, \qquad \tau = t_2 - t_1 \quad (7.18)$$

where

$$K_x(\tau + \tau_1 - \tau_2) = \frac{1}{T} \int_0^T x(t_1 - \tau_1) x(t_2 - \tau_2) \, dt$$

$$= K_x[(t_2 - \tau_2) - (t_1 - \tau_1)] \quad (7.19)$$

The variance of the system's response is

$$D_y = K_y(0) = \int_0^\infty \int_0^\infty h(\tau_1) h(\tau_2) K_x(\tau_1 - \tau_2) \, d\tau_1 \, d\tau_2 \quad (7.20)$$

Transformation of random processes by linear systems employed in telecommunication engineering is often referred to as *linear filtering*. This will be addressed in greater detail in Sec. 11.3.8.

Example 7.15 Because of the nonuniform thrust and random vibration of the ship propeller, the shaft system and the power plant of a sea-going ship experience random disturbance ΔQ of the torque Q. Assuming that this torque is proportional to the angular velocity ω (revolutions per minute) of the propeller, determine the correlation function $K_{\Delta\omega}$ and the variance $D_{\Delta\omega}$ of the change $\Delta\omega$ in the angular velocity of the propeller for the given (measured) correlation function of the "input" random process $\Delta Q(t)$.

Solution The differential equation of the disturbed rotation of the propeller can be written as

$$\frac{d}{dt}(I\,\Delta\omega) + R\,\Delta\omega = \Delta Q$$

where I is the effective moment of inertia of the rotating parts of the shaft system and R is the damping coefficient. In the analysis that follows we assume that the moment of inertia I is time independent (in effect, this moment will exhibit some time dependence, because of the periodic change in the added hydrodynamic moment which is affected by the change in the angular velocity of the propeller); the equation of the propeller motion in terms of the change $\Delta\omega$ in its angular velocity can be written as follows:

$$\frac{d\,\Delta\omega}{dt} + r\,\Delta\omega = \frac{\Delta Q}{I}, \qquad r = \frac{R}{I}$$

Using the Duhamel integral in the form (7.16) or (7.17) and considering the initial value $\Delta\omega_0$ of the velocity change, one can write the solution to this equation as follows:

$$\Delta\omega = \Delta\omega_0\,e^{-rt} + \frac{1}{I}\int_0^t e^{-r(t-\xi)}\,\Delta Q(\xi)\,d\xi$$

Here $\Delta\omega_0$ is the initial value of the velocity change and

$$h(t-\xi) = \frac{1}{I}e^{-r(t-\xi)}$$

is the impulse response of the system. The mean value of the random process $\Delta\omega(t)$ can be calculated as follows:

$$\langle\Delta\omega\rangle = \langle\Delta\omega_0\,e^{-rt}\rangle + \frac{1}{I}\int_0^t \langle e^{-r(t-\xi)}\,\Delta Q(\xi)\rangle\,d\xi$$

$$= \langle\Delta\omega_0\rangle e^{-rt} + \frac{1}{I}\int_0^t e^{-r(t-\xi)}\langle\Delta Q\rangle\,d\xi$$

where $\langle\Delta\omega_0\rangle$ is the mean value of the random variable $\Delta\omega_0$ and $\langle\Delta Q\rangle$ is the mean value of the torque $\Delta Q(t)$. The correlation function of the random process $\Delta\omega(t)$ is

$$K_{\Delta\omega}(t,t') = \left\langle\left[\Delta\mathring{\omega}_0\,e^{-rt} + \frac{1}{I}\int_0^t e^{-r(t-\xi)}\,\Delta\mathring{Q}(\xi)\,d\xi\right]\right.$$

$$\left. \times \left[\Delta\mathring{\omega}_0\,e^{-rt'} + \frac{1}{I}\int_0^{t'} e^{-r(t'-\xi')}\,\Delta\mathring{Q}(\xi')\,d\xi'\right]\right\rangle$$

where $\Delta\mathring{\omega}_0(t)$ and $\Delta\mathring{Q}(t)$ are centered random variables of the change in the initial angular velocity and the change in the torque. Since the random processes $\Delta Q(t)$ and $\Delta\omega(t)$ are statistically independent, the correlation function $K_{\Delta\omega}$ of the velocity change $\Delta\omega$ can be found as follows:

$$K_{\Delta\omega}(t, t') = D_0 e^{-r(t+t')} + \frac{1}{I^2} \int_0^t \int_0^{t'} e^{-r(t-\xi)} e^{-r(t'-\xi')} K_{\Delta Q}(\xi, \xi')\, d\xi\, d\xi'$$

Here D_0 is the variance of the change $\Delta\omega_0$ in the initial angular velocity ω_0 and $K_{\Delta Q}$ is the correlation function of the disturbance $\Delta Q(t)$ in the torque. In the case when the mean value of the torque disturbance is zero ($\langle \Delta Q \rangle = 0$) and its variance $D_{\Delta Q}$ is constant, the mean value $\langle \Delta\omega_0 \rangle$ is zero as well, and the correlation function of the velocity change can be found as

$$K_{\Delta\omega}(t, t') = \frac{D_{\Delta Q}}{r^2 I^2}(e^{-rt} - 1)(e^{-rt'} - 1)$$

Putting $t = t'$ we obtain the following formula for the variance:

$$D_{\Delta\omega}(t) = \frac{D_{\Delta Q}}{r^2 I^2}(e^{-rt} - 1)^2 = \frac{D_{\Delta Q}}{R^2}(e^{-rt} - 1)^2$$

When $t \to \infty$, the variance of the disturbance in the shaft revolutions per minute becomes

$$D_{\Delta\omega} = \frac{D_{\Delta Q}}{r^2 I^2} = \frac{D_{\Delta Q}}{R^2}$$

This formula indicates that the variance of the random change in the angular velocity of the shaft (propeller) rotations is strongly affected by the system's damping. If the mechanical and hydrodynamic damping is significant, this variance is small, no matter how large the variance of the torque disturbance might be.

Example 7.16 The output process $Y(t)$ in a one-degree-of-freedom dynamic system is related to the input process $X(t)$ by the equation

$$\frac{d^2 Y}{dt^2} + 2r \frac{dY}{dt} + \omega_0^2 Y = X(t)$$

The mean value $\langle x(t) \rangle$ and the correlation function $K_x(t, t')$ of the process $X(t)$ are known. Determine the mean and the correlation function of the output process $Y(t)$.

Solution Using the Duhamel integral in the form (7.16) or (7.17) and considering the initial conditions Y_0 and $\dot{Y}_0$ for the response function $Y(t)$, one can write the solution to the given differential equation of motion as follows:

$$Y(t) = e^{-rt}\left[\frac{\dot{Y}_0 + rY_0}{\omega} \sin \omega t + Y_0 \cos \omega t\right]$$

$$+ \frac{1}{\omega} \int_0^t e^{-r(t-\xi)} \sin \omega(t-\xi) X(\xi)\, d\xi, \qquad \omega = \sqrt{\omega_0^2 - r^2}$$

In the case where the initial conditions Y_0, $\dot{Y}_0$, and the excitation $X(t)$ are statistically independent, one can write

$$\langle y \rangle = \langle y_0 \rangle f_1(t) + \langle \dot{y}_0 \rangle f_2(t) + \int_0^t f(t, \xi) \langle x(t) \rangle \, d\xi$$

where the following notation is used:

$$f_1(t) = e^{-rt}\left(\cos \omega t + \frac{r}{\omega} \sin \omega t\right)$$

$$f_2(t) = e^{-rt} \frac{1}{\omega} \sin \omega t$$

$$f(t, \xi) = \frac{1}{\omega} e^{-r(t-\xi)} \sin \omega(t - \xi)$$

The correlation function of the process $Y(t)$ can be obtained as follows:

$$K_y(t, t') = f_1(t)f_1(t')D_{y_0} + f_2(t)f_2(t')D_{\dot{y}_0} + \int_0^t \int_0^{t'} f(t, \xi)f(t', \xi')K_x(\xi, \xi') \, d\xi \, d\xi'$$

The variance of the process $Y(t)$ can be calculated, by putting $t = t'$, as

$$D_y(t) = f_1^2(t)D_{y_0} + f_2^2(t)D_{\dot{y}_0} + \int_0^t \int_0^{t'} f(t, \xi)f(t', \xi')K_x(\xi, \xi') \, d\xi \, d\xi'$$

The obtained expressions enable the non-steady-state response of a one-degree-of-freedom dynamic system to a nonstationary random excitation to be evaluated.

Example 7.17 A light elastic vertical mast carries a concentrated heavy mass M at its upper end. The mass is subjected to a horizontal suddenly applied constant random force F. The mean value of this force is $\langle f \rangle$ and the variance is D_f. The differential equation of the motion of the mass mounted on the mast is

$$\ddot{Y}(t) + 2r\dot{Y}(t) + \omega_0^2 Y(t) = \frac{F(t)}{M}$$

In this equation, $Y(t)$ is the horizontal displacement of the mass, ω_0 is the frequency of its free undamped vibrations, r is the damping coefficient, and the excitation force $F(t)$ is equal to zero for $t < 0$ and has a constant value F for $t \geq 0$. The initial conditions for the mass displacement and the velocity are zero: $Y(0) = 0$, $\dot{Y}(0) = 0$. Using the Duhamel integral, determine the mean and the variance of the lateral displacement $Y(t)$ of the mast. The mass of the mast itself is small in comparison with the mass M and need not be considered.

Solution To solve the problem one can use the solution obtained in the previous example. Putting

$$f(t, \xi) = \frac{F}{M\omega} 1(t) e^{-r(t-\xi)} \sin \omega(t - \xi), \qquad \omega = \sqrt{\omega_0^2 - r^2}$$

where $1(t)$ is the Heaviside unit step function (see Example 7.14, Sec. 7.3), we obtain the following expressions for the mean value and the correlation function of the lateral displacement $Y(t)$ of the mast:

$$\langle y \rangle = \frac{\langle f \rangle}{M\omega} \int_0^t e^{-r(t-\xi)} \sin \omega(t-\xi) \, d\xi$$

$$= \frac{\langle f \rangle}{M\omega} \frac{1}{\omega^2 + r^2} [\omega(1 - e^{-rt} \cos \omega t) - re^{-rt} \sin \omega t]$$

$$K_y(t, t') = \frac{D_f}{M^2 \omega^2} \int_0^t \int_0^{t'} e^{-r(t-\xi)} e^{-r(t'-\xi')} \sin \omega(t-\xi) \sin \omega(t'-\xi') \, d\xi \, d\xi'$$

$$= \frac{D_f}{M^2 \omega^2} \frac{1}{(\omega^2 + r^2)^2} [-re^{-rt} \sin \omega t + \omega(1 - e^{-rt} \cos \omega t)]$$

$$\times [-re^{-rt'} \sin \omega t' + \omega(1 - e^{-rt'} \cos \omega t')]$$

The variance of the lateral displacement can be found as

$$D_y = \frac{D_f}{M^2 \omega^2} \frac{1}{(\omega^2 + r^2)^2} [-re^{-rt} + \omega(1 - e^{-rt} \cos \omega t)]^2$$

$$= \frac{D_f}{M^2 \omega^2 \omega_0^4} [-re^{-rt} + \omega(1 - e^{-rt} \cos \omega t)]^2$$

At the initial moment of time $(t = 0)$,

$$D_g = \frac{D_f r^2}{M^2 \omega^2 \omega_0^4}$$

The steady-state variance $(t \to \infty)$ of the random displacement can be obtained as follows:

$$D_y = \frac{D_f}{M^2} \frac{1}{\omega_0^4}$$

This formula indicates that the steady-state variance of the mass displacement is strongly affected by the frequency ω_0 of the free undamped vibrations of the system.

Example 7.18 (Svetlitsky, 1976) A concentrated mass M is fastened to the free end of a horizontal elastic element of a measuring device. The element can be idealized as a cantilever beam, whose clamped end is attached to a base. The base is subjected to a random vertical excitation. This can be modeled as a stationary random process $Y_0(t)$ with zero mean and the correlation function

$$K_{y_0}(\tau) = D_{y_0} e^{-\alpha|\tau|}$$

Before the device is turned on, the mass M cannot move with respect to the device's base. The mass M is released (becomes activated) at the initial moment of time when the device is turned on. Determine the variance of the force acting on the mass at the given moment of time t.

Solution The equation of motion of the mass M can be written as

$$\ddot{Y}(t) + 2r[\dot{Y}(t) - \dot{Y}_0(t)] + \omega_0^2[Y(t) - Y_0(t)] = 0,$$

where $Y_0(t)$ is the (absolute) displacement of the device base, $Y(t)$ is the (absolute) displacement of the mass M, r is the coefficient of damping,

$$\omega_0 = \sqrt{\frac{M}{K}}$$

is the frequency of free undamped vibrations of the mass M, and K is the spring constant of the elastic element. Writing the equation of motion as

$$\ddot{Y}(t) + 2r[\dot{Y}(t) - \dot{Y}_0(t)] + \omega_0^2[Y(t) - Y_0(t)] = 0,$$

or

$$\ddot{Y}(t) + 2r\dot{Y}(t) + \omega_0^2 Y(t) = 2r\dot{Y}_0(t) + \omega_0^2 Y_0(t)$$

and using the Duhamel integral (7.16) to solve the given equation we obtain the mean value and the correlation function of the displacement $Y(t)$ for zero initial conditions (the conditions $Y_0(t)$ and $\dot{Y}_0(t)$ play here the role of the excitation force) as follows:

$$\langle y(t) \rangle = \frac{\omega_0^2}{\omega} \int_0^t h(t - \xi) \langle y_0 \rangle \, d\xi + \frac{2r}{\omega} \int_0^t h(t - \xi) \langle \dot{y}_0 \rangle \, d\xi$$

$$K_y(t, t') = \frac{\omega_0^4}{\omega^2} \int_0^t \int_0^{t'} h(t - \xi) h(t' - \xi') K_{y_0}(\xi, \xi') \, d\xi \, d\xi'$$

$$+ \frac{2\omega_0^2 r}{\omega^2} \int_0^t \int_0^{t'} h(t - \xi) h(t' - \xi') \frac{\partial K_{y_0}(\xi, \xi')}{\partial \xi'} \, d\xi \, d\xi'$$

$$+ \frac{2\omega_0^2 r}{\omega^2} \int_0^t \int_0^{t'} h(t - \xi) h(t' - \xi') \frac{\partial K_{y_0}(\xi, \xi')}{\partial \xi} \, d\xi \, d\xi'$$

$$+ \frac{4r^2}{\omega^2} \int_0^t \int_0^{t'} h(t - \xi) h(t' - \xi') \frac{\partial^2 K_{y_0}(\xi, \xi')}{\partial \xi \, d\xi'} \, d\xi \, d\xi'$$

In the case in question, the impulse response function $h(t - \xi)$ of the system is

$$h(t - \xi) = e^{-r(t - \xi)} \sin \omega(t - \xi)$$

and $\omega = \sqrt{\omega_0^2 - r^2}$ is the frequency of free damped vibrations.

In accordance with the conditions of the problem, the correlation function of the input process $Y_0(t)$ is

$$K_{y_0}(\xi, \xi') = K_{y_0}(\tau) = D_{y_0} e^{-\alpha|\tau|}$$

where $\tau = \xi - \xi'$. Then, using the results obtained in Example 7.11, Sec. 7.3, we have

$$\frac{\partial K_{y_0}}{\partial \xi} = D_{y_0} \frac{\partial}{\partial \xi} e^{-\alpha|\tau|} = D_{y_0} \frac{\partial}{\partial |\tau|} e^{-\alpha|\tau|} \frac{d|\tau|}{d\tau} \frac{d\tau}{d\xi} = -D_{y_0} \alpha e^{-\alpha|\tau|} \text{sign } \tau$$

$$\frac{\partial K_{y_0}}{\partial \xi'} = D_{y_0} \frac{\partial}{\partial \xi'} e^{-\alpha|\tau|} = D_{y_0} \alpha e^{-\alpha|\tau|} \text{sign } \tau$$

$$\frac{\partial^2 K_{y_0}}{\partial \xi \, \partial \xi'} = D_{y_0} \alpha \frac{d}{d\tau} (e^{-\alpha|\tau|} \text{sign } \tau) = D_{y_0} \alpha \left[\frac{d \text{ sign } \tau}{d\tau} - \alpha(\text{sign } \tau)^2 \right] e^{-\alpha|\tau|}$$

$$= D_{y_0} \alpha [2\delta(\tau) - \alpha] e^{-\alpha|\tau|}$$

where the properties of the "sign" function sign τ and the Dirac delta function $\delta(\tau)$ are indicated in Example 7.11, Sec. 7.3.

The expression for the correlation function of the process $Y(t)$ can then be obtained as follows:

$$K_y(t, t') = \frac{D_{y_0}}{\omega^2} \int_0^t \int_0^{t'} h(t - \xi) h(t' - \xi') e^{-\alpha|\xi - \xi'|}$$

$$\times [\omega_0^4 + 8r^2\alpha\delta(\xi - \xi') - 4r^2\alpha^2] \, d\xi \, d\xi'$$

$$= \frac{D_{y_0}(\omega_0^4 - 4r^2\alpha^2)}{\omega^2} \int_0^t \int_0^{t'} h(t - \xi) h(t' - \xi') e^{-\alpha|\xi - \xi'|} \, d\xi \, d\xi'$$

$$+ \frac{8D_{y_0} r^2\alpha}{\omega^2} \int_0^{t'} h(t' - \xi') h(t - \xi') \, d\xi'$$

The variance of the process $Y(t)$ can be obtained by putting $t = t'$ in this formula:

$$D_y(t) = \frac{D_{y_0}(\omega_0^2 - 4r^2\alpha^2)}{\omega^2} \int_0^t \int_0^{t'} h(t - \xi) h(t - \xi') e^{-\alpha|\xi - \xi'|} \, d\xi \, d\xi'$$

$$+ \frac{8D_{y_0} r^2\alpha}{\omega^2} \int_0^t h^2(t - \xi) \, d\xi$$

The elastic force acting on the mass M is due to the difference in the displacements of the device base and the displacement of the mass itself and can be represented as

$$F = K(y_0 - y)$$

where K is the spring constant of the elastic element of the device. In order to evaluate the standard deviation of this force, one has to determine first its correlation function

$$K_f(t, t') = K^2 \langle [y_0(t) - y(t)][y_0(t') - y(t')] \rangle$$
$$= K^2 [K_{y_0}(t, t') + K_y(t, t') - \langle y_0(t) y(t') \rangle - \langle y_0(t') y(t) \rangle]$$

The mean $\langle y_0(t)y(t')\rangle$ entering this formula can be found as

$$M_1 = \langle y_0(t)y(t')\rangle$$

$$= \langle y_0(t)\frac{1}{\omega}\int_0^{t'} h(t'-\xi')[\omega_0^2 y_0(\xi') + 2r\dot{y}_0(\xi')]\,d\xi'\rangle$$

$$= \frac{1}{\omega}\left[\omega_0^2\int_0^{t'} h(t'-\xi')\langle y_0(t)y_0(\xi')\rangle\,d\xi' + 2r\int_0^{t'} h(t'-\xi')\langle y_0(t)\dot{y}_0(\xi')\rangle\,d\xi\right]$$

The mean values entering the integrands in this expression can be evaluated as follows:

$$\langle y_0(t)y_0(\xi')\rangle = K_{y_0}(t,\xi') = D_{y_0}e^{-\alpha|t-\xi'|}$$

$$\left\langle y_0(t)\frac{\partial y_0}{\partial \xi_1}\right\rangle = \frac{\partial}{\partial \xi_1}\langle y_0(t)y_0(\xi')\rangle$$

$$= \frac{\partial}{\partial \xi'}D_{y_0}e^{-\alpha|t-\xi'|} = \alpha D_{y_0}e^{-\alpha|t-\xi'|}\,\text{sign}\,(t-\xi')$$

Then we have

$$M_1 = \langle y_0(t)y(t')\rangle = \frac{D_{y_0}}{\omega}\int_0^{t'} h(t'-\xi')[\omega_0^2 + 2r\alpha\,\text{sign}\,(t-\xi')]e^{-\alpha|t-\xi'|}\,d\xi'$$

Similarly, we obtain

$$M_2 = \langle y_0(t')y(t)\rangle = \frac{D_{y_0}}{\omega}\int_0^t h(t-\xi)[\omega_0^2 + 2r\alpha\,\text{sign}\,(t'-\xi)]e^{-\alpha|t'-\xi|}\,d\xi$$

The variance of the force F can be calculated as follows:

$$D_f(t) = K^2[D_{y_0} + D_y(t)] - 2K_f(t,t')$$

$$= K^2\left[D_{y_0} + D_y(t) - 2\frac{D_{y_0}}{\omega}\int_0^t h(t-\xi)[\omega_0^2 + 2r\alpha\,\text{sign}\,(t-\xi)]e^{-\alpha(t-\xi)}\,d\xi\right]$$

$$= K^2\left[D_{y_0} + D_y(t) - \frac{2D_{y_0}}{\omega}(\omega_0^2 + 2r\alpha)\int_0^t h(t-\xi)e^{-\alpha|t-\xi|}\,d\xi\right]$$

Further Reading (see Bibliography): 13, 22, 28, 30, 34, 53, 66, 72, 85, 86, 94, 97, 101, 125, 135, 144, 145.

Chapter 8

Random Processes: Spectral Theory

> "Khintchin and Kolmogorov, two most outstanding Russian specialists in the Probability Theory, worked for a long time in the same field as I. For over twenty years we stepped on each other heels...."
>
> NORBERT WIENER
> *I Am a Mathematician*

> "There are truths which are like new lands: the best way to them becomes known only after trying many other ways."
>
> DENI DIDREAU
> *French Philosopher*

> "We should be grateful to God for creating the world such that all that is simple is true, and all that is complicated is false."
>
> GREGORY SKOVORODA
> *Ukrainian Philosopher*

8.1. Theory of Spectra

The theory of spectra is widely used in engineering and applied science, and is not necessarily associated with application of a probabilistic approach. In this chapter we will first briefly describe the general principles of this theory and will then address various probabilistic problems in which the spectral theory of random processes can be successfully used.

8.1.1 Fourier series

With few exceptions, any periodic function $x(t)$ of period T can be expanded in a trigonometric *Fourier series:*

$$x(t) = \sum_{k=1}^{\infty} (a_k \cos \omega_k t + b_k \sin \omega_k t) \tag{8.1}$$

The *Fourier coefficients* a_k and b_k are expressed as

$$a_k = \frac{2}{T} \int_{T/2}^{T/2} x(t) \cos \omega_k t \, dt, \quad b_k = \frac{2}{T} \int_{T/2}^{T/2} x(t) \sin \omega_k t \, dt, \quad k = 0, 1, 2, \ldots \tag{8.2}$$

where
$$\omega_k = \frac{2\pi}{T} k, \quad k = 0, 1, 2, \ldots \tag{8.3}$$

The set of the Fourier coefficients provides the *frequency spectrum* of the function $x(t)$. If $x(t)$ is an even function, i.e., if $x(t) = x(-t)$, then $b_k \equiv 0$. If $x(t)$ is an odd function, i.e., if $x(t) = -x(t)$, then $a_k \equiv 0$. Formally, one can also expand a nonperiodic function given on a finite interval into a Fourier series. The fourier series for such a function coincides with the Fourier series for a function periodically extended over the entire t axis.

The function $x(t)$ is an even one with respect to the index k, i.e., it does not change if the numbers k are replaced by the numbers $-k$. Therefore the expansion (8.1) can be rewritten as follows:

$$x(t) = \frac{1}{2} \sum_{k=-\infty}^{\infty} (a_k \cos \omega_k t + b_k \sin \omega_k t) \tag{8.4}$$

Then, using Euler's formulas

$$\cos \theta = \frac{e^{i\theta} + e^{-i\theta}}{2} \tag{8.5}$$

$$\sin \theta = \frac{e^{i\theta} - e^{-i\theta}}{2i} \tag{8.6}$$

one can write the series (8.1) and (8.4) in the complex form:

$$x(t) = \sum_{k=-\infty}^{\infty} C_k e^{i\omega_k t} \tag{8.7}$$

where the complex Fourier coefficients are expressed as follows:

$$C_k = \frac{1}{2}(a_k - ib_k) = \frac{1}{T} \int_{T/2}^{T/2} x(t) e^{-i\omega_k t} \, dt \tag{8.8}$$

The quantities $A_k = 2C_k = a_k - ib_k$ are known as *complex amplitudes* and the set of the amplitudes A_k is known as the *complex frequency spectrum*. The sets of modules

$$|A_k| = \sqrt{a_k^2 + b_k^2} \qquad (8.9)$$

and the arguments

$$\varepsilon_k = \arg A_k = \arctan \frac{b_k}{a_k} \qquad (8.10)$$

of complex amplitudes are known as the *amplitude frequency spectrum* and the *phase frequency spectrum* of the periodic function $x(t)$, respectively. The spectra $|A_k|$ and ε_k can be represented graphically as vertical lines perpendicular to the frequency axis ω_k.

8.1.2. Fourier integral

The Fourier series can be generalized for the case of a nonperiodic function given on an infinite interval. This can be done by treating this function as a periodic one with an infinitely large period. From Eqs (8.7) and (8.8) we obtain

$$x(t) = \frac{1}{T} \sum_{k=-\infty}^{\infty} e^{i\omega_k t} \int_{T/2}^{T/2} x(t) e^{-i\omega_k t} \, dt \qquad (8.11)$$

Let us find the limit of this expression when the time T approaches infinity. Since the quantity $2\pi/T$ is, in effect, the frequency interval between two adjacent harmonics, one can use the following substitutions: $2\pi/T \to d\omega$, $\omega_k \to \omega$. This yields

$$x(t) = \frac{1}{2\pi} \int_{-\infty}^{\infty} e^{i\omega t} \, d\omega \int_{-\infty}^{\infty} x(t) e^{-i\omega t} \, dt$$

$$= \frac{1}{2\pi} \int_{-\infty}^{\infty} d\omega \int_{-\infty}^{\infty} x(\xi) e^{i\omega(t-\xi)} \, d\xi \qquad (8.12)$$

The obtained formula is known as the *Fourier integral formula*. The integral with respect to the variable ξ in the right part of this formula is the *Fourier integral*.

The formula (8.12) can be written in real form as follows:

$$x(t) = \frac{1}{\pi} \int_0^{\infty} d\omega \int_{-\infty}^{\infty} x(\xi) \cos \omega(t-\xi) \, d\xi \qquad (8.13)$$

8.1.3. Fourier transforms: spectrum of a function

The formula (8.12) can be rewritten as follows:

$$x(t) = \frac{1}{2}\int_{-\infty}^{\infty} G_x(\omega)e^{i\omega t}\, d\omega$$
$$G_x(\omega) = \frac{1}{\pi}\int_{-\infty}^{\infty} x(t)e^{-i\omega t}\, dt \tag{8.14}$$

These equations are known as the *Fourier transforms*, and the function $G_x(\omega)$ known as the *complex spectrum*, *complex spectral density*, or *complex spectral characteristic* of the function $x(t)$. The function $G_x(\omega)$ plays in the theory of the Fourier integral the same role as the complex amplitudes play in the theory of the Fourier series.

The module

$$S_x(\omega) = |G_x(\omega)| \tag{8.15}$$

of the complex spectrum is called the *spectrum* or *spectral density* of the function $x(t)$. Unlike the complex amplitude spectrum, the function $S_x(\omega)$ is continuous, not discrete. The quantity

$$\varepsilon_x(\omega) = \arg G_x(\omega) \tag{8.16}$$

is called the *phase spectrum* of the function $x(t)$.

The complex spectrum $G_x(\omega)$ can be written in algebraic form as follows:

$$G_x(\omega) = S_1(\omega) - iS_2(\omega) \tag{8.17}$$

where the functions $S_1(\omega)$ and $S_2(\omega)$ are expressed as

$$S_1(\omega) = \frac{1}{\pi}\int_{-\infty}^{\infty} x(t)\cos\omega t\, dt$$
$$S_2(\omega) = \frac{1}{\pi}\int_{-\infty}^{\infty} x(t)\sin\omega t\, dt \tag{8.18}$$

If $x(t)$ is an even function, then $S_2(\omega) = 0$ and

$$G_x(\omega) = S_1(\omega) = S_x(\omega) = \frac{2}{\pi}\int_{0}^{\infty} x(t)\cos\omega t\, dt \tag{8.19}$$

In such a case the first formula in (8.14) yields

$$x(t) = \frac{1}{2}\int_{-\infty}^{\infty} G_x(\omega) \cos \omega t \, d\omega = \int_0^{\infty} S_x(\omega) \cos \omega t \, dt \qquad (8.20)$$

The formulas (8.19) and (8.20) are known as the *Fourier cosine transforms*. If $x(t)$ is an odd function, then the formulas (8.14) result in the following relationship:

$$x(t) = \int_0^{\infty} S_x(\omega) \sin \omega t \, d\omega$$

$$S_x(\omega) = \frac{2}{\pi}\int_0^{\infty} x(t) \sin \omega t \, dt \qquad (8.21)$$

which are known as the *Fourier sine transforms*. If the function $x(t)$ is given on an interval $(0, \infty)$, the formula (8.20) extends it over the entire t axis in an "even way," while the first formula in (8.21) extends the function $x(t)$ over the t axis in an "odd way." For $t > 0$, both formulas lead to the same result. The complex spectrum $G_x(\omega)$ tends to zero when the frequency ω tends to either $+\infty$ or $-\infty$.

In telecommunications engineering, one often encounters, when transmitting various signals, time functions having finite frequency spectra, i.e., spectra that do not contain frequencies beyond a certain range. Such functions possess an important property: they are completely determined by a finite number of ordinates, i.e., by a finite number of their values within a finite frequency interval. In other words, if one knows a finite number of values of a time function with a finite frequency spectrum, this function can be restored in a continuous fashion and in a unique way. In applications to telecommunications systems this means that any function $f(t)$ containing frequencies from zero to λ_c can be transmitted, with any required accuracy, by means of a train of discrete numbers that follow each other with a period of $1/(2\lambda_c)$ seconds.

Indeed, a function $f(t)$ with a finite spectrum $G(\omega)$ defined within the frequency range $(-\omega_c, \omega_c)$ can be expressed through this spectrum by the formula

$$f(t) = \frac{1}{2\pi}\int_{-\omega_c}^{\omega_c} G(\omega) e^{i\omega t} \, d\omega$$

The spectrum $G(\omega) = 0$ for $\omega > \omega_c$ and $\omega < -\omega_c$. The function $G(\omega)$ can be written in the form of a Fourier series as follows:

$$G(\omega) = \sum_{k=-\infty}^{\infty} C_k e^{i\pi k(\omega/\omega_c)}$$

where the coefficients C_k are expressed as

$$C_k = \frac{1}{2\omega_c} \int_{-\omega_c}^{\omega_c} G(\omega) e^{-i\pi k(\omega/\omega_c)} \, d\omega$$

and the value $2\omega_c$ plays the role of "the period with respect to the frequency." Introducing the formula for the spectrum $G(\omega)$ into the formula for the function $f(t)$, we obtain:

$$f(t) = \frac{1}{2\pi} \int_{-\omega_c}^{\omega_c} \left(\sum_{k=-\infty}^{\infty} C_k e^{-i\pi k(\omega/\omega_c)} \right) e^{i\omega t} \, d\omega$$

$$= \frac{1}{2\pi} \sum_{k=-\infty}^{\infty} C_k \int_{-\omega_c}^{\omega_c} e^{i\omega[t + k(\pi/\omega_c)]} \, d\omega$$

$$= \frac{1}{\pi} \sum_{k=-\infty}^{\infty} C_k \frac{\sin \omega_c[t + k(\pi/\omega_c)]}{t + k(\pi/\omega_c)}$$

Since

$$f\left(-k\frac{\pi}{\omega_c}\right) = \frac{1}{2\pi} \int_{-\omega_c}^{\omega_c} G(\omega) e^{-i\pi k(\omega/\omega_c)} \, d\omega,$$

we conclude that the coefficients C_k can be expressed as follows:

$$C_k = \frac{\pi}{\omega_c} f\left(-k\frac{\pi}{\omega_c}\right) = (\Delta t)\omega(-k\,\Delta t),$$

where the time increment $\Delta t = \pi/\omega_c$. Substituting this formula into the obtained expression for the function $f(t)$ and changing the signs in front of the index k (since the summation is carried out for all the k values, from $-\infty$ to $+\infty$), we obtain:

$$f(t) = \sum_{k=-\infty}^{\infty} f(k\,\Delta t) \frac{\sin \omega_c(t - k\,\Delta t)}{\omega_c(t - k\,\Delta t)}$$

Thus, a function $f(t)$ with a finite frequency spectrum can be effectively approximated by a series whose coefficients are values of this function determined for the arguments taken with an increment

$$\Delta t = \frac{\pi}{\omega_c} = \frac{1}{2\lambda_c}$$

8.1.4. Parseval's formula

Applying the first formula in (8.14) to a process $x_1(t)$, we have

$$x_1(t) = \frac{1}{2} \int_{-\infty}^{\infty} G_{x_1}(\omega) e^{i\omega t} \, d\omega$$

Multiplying this formula by $x_2(t)$ and integrating the obtained equation over the entire t axis, we have

$$\int_{-\infty}^{\infty} x_1(t)x_2(t)\, dt = \frac{1}{2}\int_{-\infty}^{\infty} x_2(t)\, dt \int_{-\infty}^{\infty} G_{x_1}(\omega)e^{i\omega t}\, d\omega$$

$$= \frac{1}{2}\int_{-\infty}^{\infty} G_{x_1}(\omega)\, d\omega \int_{-\infty}^{\infty} x_2(t)e^{i\omega t}\, dt$$

$$= \frac{\pi}{2}\int_{-\infty}^{\infty} G_{x_1}(\omega)G_{x_2}(-\omega)\, d\omega \qquad (8.22)$$

This formula expresses *Parseval's theorem*. This theorem enables one to determine the integral of the product of two functions from their spectra.

When $x_1(t) = x_2(t) = x(t)$, the relationship (8.22) results in the following *Parseval's formula*:

$$\int_{-\infty}^{\infty} x^2(t)\, dt = \frac{\pi}{2}\int_{-\infty}^{\infty} S_x^2(\omega)\, d\omega \qquad (8.23)$$

The integral in the left-hand part of this formula is the power of the process $x(t)$. Hence, the quantity $S_x^2(\omega)$ is proportional, for the given frequency ω, to the power per unit frequency range. Since, in its turn, the power of a periodic process $x(t)$ is proportional to the amplitude of this process squared, the spectrum of a function can be used as a characteristic of the distribution of the amplitudes of a periodic process over the frequency axis. This explains the physical meaning of the spectrum of a function.

8.1.5. Solution of ordinary differential equations using the Fourier integral

After differentiating the first formula in (8.14) n times with respect to the argument t, we obtain

$$x^{(n)}(t) = \frac{1}{2}\int_{-\infty}^{\infty} (i\omega)^n G_x(\omega)e^{i\omega t}\, d\omega \qquad (8.24)$$

On the other hand, the first formula in (8.14), if applied directly to the function $x^{(n)}(t)$, yields

$$x^{(n)}(t) = \frac{1}{2}\int_{-\infty}^{\infty} G_{x^{(n)}}(\omega)e^{i\omega t}\, d\omega \qquad (8.25)$$

Comparing the formulas (8.24) and (8.25), we obtained the following relationship:

$$G_{x^{(n)}}(\omega) = (i\omega)^n G_x(\omega) \qquad (8.26)$$

This formula expresses the *theorem of the spectrum of a derivative*: the spectrum of this function by the factor $(i\omega)^n$.

Let the behavior of a linear dynamic system be described by the following ordinary differential equation:

$$a_n y^{(n)} + a_{n-1} y^{(n-1)} + \cdots + a_1 y' + a_0 y = x(t) \qquad (8.27)$$

Here $x(t)$ is the input process and $y(t)$ is the output process (system's response). After applying the Fourier transform to both parts of this equation and using the formula (8.26), we obtain:

$$G_y(\omega) = G_x(\omega)\Phi(i\omega) \qquad (8.28)$$

where the function

$$\Phi(i\omega) = \frac{1}{a_n(i\omega)^n + a_{n-1}(i\omega)^{n-1} + \cdots + a_1(i\omega) + a_0} \qquad (8.29)$$

is known as the *complex frequency characteristic, complex transfer coefficient*, or the *mechanical impedance* of the dynamic system. The formula

$$y(t) = \frac{1}{2}\int_{-\infty}^{\infty} G_x(\omega)\Phi(i\omega)e^{i\omega t}\, d\omega \qquad (8.30)$$

provides the solution to Eq. (8.27).

The solution to Eq. (8.27) can also be obtained on the basis of the *Laplace transform*. This approach, closely related to the Fourier transform, underlies the calculus of operations. It leads, for steady-state conditions, to the following formula for the ratio of the "output" and the "input" *images* of the dynamic system (8.27):

$$\Phi(p) = \frac{Y(p)}{X(p)} = \frac{1}{a_n p^n + a_{n-1}p^{n-1} + \cdots + ap + a_0} \qquad (8.31)$$

The function $\Phi(p)$ is known in Laplace transform theory as the *transfer function*. Since for $p = i\omega$ the formula (8.31) results in the formula (8.29) for the complex frequency characteristic, the term "transfer function" is often used with respect to the frequency characteristic as well. The equivalency of the formulas (8.29) and (8.31) is due to the fact that the Laplace transform is, in effect, a generalization of the Fourier transform for functions that have diverging Fourier integrals. Examples of such functions are the simple functions $x = $ constant and $x = t^2$. The "close relationship" between the Fourier and the Laplace transforms enables one to use the existing tables for images of different functions to evaluate spectra of these functions.

The formula (8.29), when applied to harmonic "input" $x(t)$ and "output" $y(t)$ processes,

$$x(t) = A_i e^{i\omega t}, \qquad y(t) = A_0 e^{i\omega t}$$

yields
$$\Phi(i\omega) = \frac{y(t)}{x(t)} = \frac{A_0}{A_i} \qquad (8.32)$$

Thus, the complex frequency characteristic of a dynamic system can be obtained as the ratio of the complex displacements of the complex amplitudes of the "output" and the "input" harmonic processes. By analogy with the *static impedance* ("compliance," in the case of a mechanical system), which is defined as the ratio of the static response ("displacement") to the applied excitation ("force"), the complex frequency characteristic can be called *dynamic impedance* of the system. The modulus

$$a(\omega) = |\Phi(i\omega)| \qquad (8.33)$$

of the complex frequency characteristic is known as the *frequency response function of the amplitudes*, or the *gain factor*. The argument

$$\theta(\omega) = \arg \Phi(i\omega) \qquad (8.34)$$

of the complex frequency characteristic is known as the *frequency response function for the phase angles*, or the *phase factor*.

As an example, examine a dynamic response of a single-degree-of-freedom system to a harmonic excitation. The motions of this system are described by the equation

$$\ddot{y}(t) + 2r\dot{y}(t) + \omega_0^2 y(t) = x(t) \qquad (8.35)$$

After putting $n = 2$ in the formula (8.29), we have

$$\Phi(i\omega) = \frac{1}{a_2(i\omega)^2 + a_1(i\omega) + a_0} \qquad (8.36)$$

Comparing the formula (8.35) with (8.27), we obtain $a_2 = 1$, $a_1 = 2r$, $a_0 = \omega_0^2$. Therefore, the complex frequency characteristic is

$$\Phi(i\omega) = \frac{1}{\omega_0^2 - \omega^2 + 2ir\omega} \qquad (8.37)$$

and the solution to Eq. (8.35) can be written as

$$y(t) = \frac{1}{2}\int_{-\infty}^{\infty} \frac{G_x(\omega)e^{i\omega t}}{\omega_0^2 - \omega^2 + 2ir\omega}\,d\omega \qquad (8.38)$$

The frequency response function (transfer function) for the amplitudes is

$$a(\omega) = |\Phi(i\omega)| = \sqrt{\Phi(i\omega)\Phi(-i\omega)} = \frac{1}{\sqrt{(\omega_0^2 - \omega^2)^2 + (2r\omega)^2}} \qquad (8.39)$$

In the case $\omega \to 0$, corresponding to static conditions, this formula yields

$$a_{st} = \frac{1}{\omega_0^2}$$

The expression for the dynamic factor of the system can be written as follows:

$$\alpha(\omega) = \frac{a(\omega)}{a_{st}} = \frac{1}{\sqrt{[1 - (\omega^2/\omega_0^2)]^2 + (2r\omega/\omega_0^2)^2}} \qquad (8.40)$$

The factor $a(\omega)$ is plotted in Fig. 8.1 as a function of the ratio ω/ω_0 of the frequency ω of the excitation to the frequency ω_0 of the system's free vibrations.

8.1.6. The complex frequency characteristic as a spectrum of the impulse response of a dynamic system

The unit pulse excitation can be approximated by the Dirac delta function (see Example 7.11, Sec. 7.3). The complex spectrum of this function, as predicted by the second formula in (8.14), is

$$G_\delta(\omega) = \frac{1}{\pi}\int_{-\infty}^{\infty} \delta(t)e^{-i\omega t}\,dt = \frac{1}{\pi} \qquad (8.41)$$

i.e., is frequency independent. Using $G_\delta(\omega)$ as an "input" spectrum and applying the formula (8.30), one obtains the following formula for the impulse response function:

$$h(t) = \frac{1}{2\pi}\int_{-\infty}^{\infty} \Phi(i\omega)e^{i\omega t}\,d\omega \qquad (8.42)$$

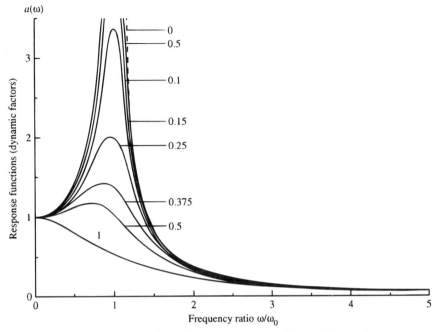

Figure 8.1 Response functions for an harmonic oscillator (dynamic factors).

On the other hand, applying the first formula in (8.14) to the function $h(t)$, we have

$$h(t) = \frac{1}{2} \int_{-\infty}^{\infty} G_h(\omega) e^{i\omega t} \, d\omega \qquad (8.43)$$

where $G_h(\omega)$ is the spectrum of the impulse response function $h(t)$. Comparing the formulas (8.42) and (8.43), we conclude that the complex frequency characteristic $\Phi(i\omega)$ of a dynamic system is related to the spectrum $G_h(\omega)$ of its impulse response function $h(t)$ as follows:

$$\Phi(i\omega) = \pi G_h(\omega) \qquad (8.44)$$

Thus, instead of evaluating, experimentally or theoretically, the complex frequency characteristic $\Phi(i\omega)$ for different frequencies, one can evaluate the impulse response function for a single impulse excitation, compute the spectrum of this function on the basis of the second formula in (8.14), and then evaluate the frequency characteristic by the formula (8.44).

For instance, the frequency response function for a sea-going ship, irrespective of whether her motion characteristics or wave-bending moments are of interest, can be determined in a model basin by a series of model experiments in regular waves over the range of wave frequencies. However, if the set of regular wave tests could be replaced by a single test experiment that uses a wave disturbance having energy distribution over all the wave frequencies, then a considerable amount of test time would be saved. Such a wave disturbance is called the transient wave. A transient wave can be produced by generating a wave train at a far distance from the test location (the ship model) in such a manner that the frequency of this wave train decreases linearly with time from the highest to the lowest frequency desired. The fast-moving (low-frequency) waves would then catch up with the slower (high-frequency) ones and coalesce at the test point, to produce a very large wave which may be thought as a unit impulse. The frequency response function that relates wave heights measured at two points separated by a distance in the direction of travel can be derived based on an equation of the (8.28) type as follows:

$$S_y(\omega) = S_x(\omega) G_h(\omega)$$

Here $S_x(\omega)$ and $S_y(\omega)$ are the Fourier transforms (spectra) of the input process $X(t)$ and the output process $Y(t)$, respectively, and $G_h(\omega)$ is the frequency response function (spectrum of the impulse response function).

8.1.7. The duration of a process and the width of its spectrum

Taking the modules of the left-hand and right-hand parts of the relationship (8.44), we have

$$a(\omega) = \pi S_h(\omega) \tag{8.45}$$

Here $a(\omega)$ is the frequency response function for the amplitudes (transfer function) and $S_h(\omega)$ is the spectrum of the impulse response of the system. From Eq. (8.40) we obtain

$$a(\omega) = \alpha(\omega) a_{\text{st}} = \frac{\alpha(\omega)}{\omega_0^2} \tag{8.46}$$

Introducing Eq. (8.46) into Eq. (8.45), we obtain the following relationship:

$$S_h(\omega) = \frac{\alpha(\omega)}{\pi \omega_0^2} \tag{8.47}$$

Hence, the spectrum of the impulse response function coincides, up to a constant factor, with the dynamic factor of the system.

The following parameter can be used as a suitable characteristic of the width of a function $f(\omega)$ of the frequency ω $(0 < \omega < \infty)$:

$$\varepsilon_f = \sqrt{1 - \frac{m_2^2}{m_0 m_4}} \tag{8.48}$$

where

$$m_k = \int_0^\infty f(\omega)\omega^k \, d\omega, \qquad k = 0, 1, 2, \ldots \tag{8.49}$$

are the moments of the function $f(\omega)$. As follows from the formula (8.48), the parameter ε_f changes from $\varepsilon_f = 0$ to $\varepsilon_f = 1$, when the function $f(\omega)$ changes from a delta function (sharp peak) to a constant value.

As far as the duration of the process $h(t)$ is concerned, this duration can be characterized, for instance, by the time required for the ordinate of this process to reduce, because of the damping in the system, by the given factor. As evident from Fig. 8.1, the decrease in damping, leading to an increase in the duration of the response process $h(t)$, results in an increase in the width ε_α of the curve $\alpha(\omega)$ and, as follows from the formula (8.47), in an increase in the width of the spectrum $S_h(\omega)$. Thus, the shorter the process, the wider its frequency spectrum. In the extreme case of an ideally narrow ("delta-like") process, the spectrum becomes, as evident from the formula (8.41), constant, i.e., "infinitely wide," and is independent of the frequency. Conversely, for a "delta-like" spectrum $S_x(\omega) = \delta(\omega)$, the process $x(t)$, as one can easily obtain from the formula (8.20) and the third property of the delta function (see Example 7.11, Sec. 7.3), lasts for an infinitely long time. Therefore it is impossible to localize an external excitation in time and, at the same time, to enhance the selectivity of this excitation to different frequencies: the more intense in time the external excitation, the larger is the number of the induced harmonics with different frequencies. This explains, in particular, the poor convergence, or even divergence, of the series in the method of principal coordinates (see, for instance, Sec. 13.1.1), when this method is used for the evaluation of the response of an elastic system to a short-term (shock) excitation.

Note that the inability to localize the external excitation in time and, at the same time, to enhance the selectivity of this excitation to a narrow frequency range is one of the manifestations of a more general *principle of uncertainty*, which plays an important role in modern physics. In quantum mechanics the energy of a particle is associated with the frequency of the wave function. The wave function ψ of a particle with a certain energy E is proportional to $e^{-iEt/\hbar}$, where $\hbar = 1.05 \times 10^{-27}$ dn cm s is Planck's constant. If the particle is observed during a short time Δt, then, as follows from the previous discussion, one can determine the frequency of its wave function, i.e., the $E/\hbar$ value, but only with a very low accuracy. The product $\Delta E \, \Delta t$ is on the order of $\hbar$. Similarly, a particle with a certain momentum $p = mV_x$ in the direction of the x axis has a wave function that is proportional to $e^{ipx/\hbar}$. If the particle is known to be located in the interval between x and $x + \Delta x$, then the wave function has a non-zero value only on the interval. Expanding the wave function into the Fourier integral, one can conclude that the momentum of the particle is known only with a certain accuracy Δp: the terms containing the factor $e^{ipx/\hbar}$ (with $p_1 < p < p_1 + \Delta p$) in the Fourier integral will be large. Hence, the product $\Delta p \, \Delta x$ is on the order of $\hbar$.

8.2. Wiener–Khinchin's Formulas

The theory of spectra can be applied to the stationary random processes and be effectively used for evaluating the responses of dynamic systems to random excitations. This can be done by applying the theory of the Fourier transforms to the correlation function $K_x(\tau)$ of the random process $X(t)$:

$$K_x(\tau) = \frac{1}{2} \int_{-\infty}^{\infty} G_x(\omega) e^{i\omega\tau} \, d\omega$$

$$G_x(\omega) = \frac{1}{\pi} \int_{-\infty}^{\infty} K_x(\tau) e^{-i\omega\tau} \, d\tau$$

(8.50)

These formulas are known as *Wiener–Khinchin's formulas*. They were obtained independently by the American mathematician N. Wiener and the Russian mathematician A. J. Khinchin in the early 1940s.

Since $K_x(\tau)$ is an even function, in the real domain the formulas (8.50) become the Fourier cosine transforms:

$$K_x(\tau) = \int_0^{\infty} S_x(\omega) \cos \omega\tau \, d\omega$$

$$S_x(\omega) = \frac{2}{\pi} \int_0^{\infty} K_x(\tau) \cos \omega\tau \, d\tau$$

(8.51)

The variance of a stationary random process can be obtained as

$$D_x = K_x(0) = \int_0^{\infty} S_x(\omega) \, d\omega \quad (8.52)$$

From this formula we find

$$S_x(\omega) = \frac{dD_x(\omega)}{d\omega} \tag{8.53}$$

Hence, the spectrum of the correlation function of a random process determines the density of the distribution of the variances over the frequencies of the continuous spectrum. The variance of an elementary harmonic determines the power of this harmonic and therefore the function $S_x(\omega)$ characterizes the distribution of the power of the process $X(t)$ over the range of frequencies. For this reason this function is often called the *power spectral density*, or the *power spectrum*.

The major incentive for the substitution of the correlation function $K_x(\tau)$ of a random process $X(t)$ by its power spectrum $S_x(\omega)$ is due to the fact that the use of the function $S_x(\omega)$ enables one to simplify considerably the procedure of the transformation of stationary random processes by linear dynamic systems. The theory of random processes based on the Wiener–Khinchin's formulas (8.50) is known as the *spectral theory of random processes*. It should be pointed out that the application of the Fourier transformation to the random process $X(t)$ directly is not advisable, since the resulting spectrum is also a random process. However, such a "current" spectrum can be successfully used for obtaining the spectrum $S_x(\omega)$ without the preliminary evaluation of the correlation function.

Examine a stationary random process $X(t)$ with zero mean and introduce a "truncated" realization $x_T(t)$ of this process:

$$x_T(t) = \begin{cases} x(t), & \text{for } -T \leq t \leq T \\ 0, & \text{for } -T > t > T \end{cases}$$

The current spectrum of such a realization is

$$G_T(\omega) = \frac{1}{\pi} \int_{-\infty}^{\infty} x_T(t) e^{-i\omega t} \, dt$$

The power spectrum of the process $X(t)$ can be evaluated as

$$\begin{aligned} S_x(\omega) &= \frac{1}{\pi} \int_{-\infty}^{\infty} K_x(\tau) e^{-i\omega\tau} \, d\tau \\ &= \frac{1}{\pi} \int_{-\infty}^{\infty} \left[\frac{1}{2T} \int_{-T}^{T} x_T(t) x_T(t+\tau) \, dt \right] e^{-i\omega\tau} \\ &= \frac{1}{2\pi T} \int_{-\infty}^{\infty} e^{-i\omega\tau} \, d\tau \int_{-\infty}^{\infty} x_T(t) x_T(t+\tau) \, dt \end{aligned}$$

$$= \frac{1}{2\pi T} \int_{-\infty}^{\infty} x_T(t)e^{i\omega t}\, dt \int_{-\infty}^{\infty} x_T(t+\tau)e^{-i\omega(t+\tau)}\, d\tau$$

$$= \frac{\pi}{2T} G_T(-\omega)G_T(\omega) = \frac{\pi}{2T} |G_T(\omega)|^2 \qquad (8.54)$$

The obtained formula indicates that the power spectrum $S_x(\omega)$ of a random process $X(t)$ (i.e., in effect, the spectrum of the correlation function of this process) can be obtained from the "current" spectrum $G_T(\omega)$ of the process $X(t)$. The relationship (8.54) is the more accurate, the longer is the time T.

Various observable characteristics of a random process can be obtained from the initial moments of the spectrum:

$$m_k = \int_0^{\infty} S_x(\omega)\omega^k\, d\omega, \qquad k = 0, 1, 2, \ldots$$

Particularly,
$$m_0 = \int_0^{\infty} S_x(\omega)\, d\omega$$

is the mean square value of the random process,

$$T_{-1} = 2\pi \frac{m_{-1}}{m_0}$$

is the average period of the process energy,

$$T_1 = 2\pi \frac{m_0}{m_1}$$

is the average mean period of the process,

$$T_2 = 2\pi \sqrt{\frac{m_0}{m_2}}$$

is the average zero-crossing period,

$$T_4 = 2\pi \sqrt{\frac{m_2}{m_4}}$$

is the average crest-to-crest period,

$$\gamma = \frac{m_3}{m_0^{3/2}}$$

is the skewness of the spectrum,

$$\varepsilon = \sqrt{1 - \frac{m_2^2}{m_0 m_4}}$$

is its width (broadness), and the ratio

$$\beta = \frac{m_4}{m_2}$$

characterizes the spectrum flatness.

8.3. "White Noise," Wiener's Process, and Narrow-Band Processes

"*White noise*" is an important special case of a stationary random process. It is a process with a constant, frequency-independent, power spectrum. The name "white noise" appeared by analogy with the white light which has, in the visible portion, a uniform continuous spectrum. The correlation function of a white noise $n(t)$ is

$$K_x(\tau) = \langle n(t)n(t+\tau)\rangle = \pi N_0 \, \delta(\tau) \tag{8.55}$$

where N_0 is the spectrum of the process and $\delta(\tau)$ is the delta function. Since the delta function is one of the multipliers in the expression for the correlation function of the random process, the "white noise" process is often referred to as a *delta-correlated process*. The "white noise" is an "absolutely random process," since its sections, no matter how close they are, are uncorrelated.

A real stationary random process can be approximated as a "white noise" if it has a rapidly decreasing correlation function. Examine, for instance, a process $X(t)$ with the correlation function

$$K_x(\tau) = D_x e^{-\alpha|\tau|} \tag{8.56}$$

where α is a large parameter. The spectrum of this process is

$$S_x(\omega) = \frac{2}{\pi} \int_0^\infty K_x(\tau) \cos \omega\tau \, d\tau = \frac{2}{\pi} \frac{\alpha D_x}{\alpha^2 + \omega^2} \tag{8.57}$$

If $\omega \ll \alpha$, i.e., if the period of oscillations $\tau_0 = 2\pi/\omega$ exceeds substantially the correlation time $\tau_k = 1/\alpha$, then the spectral density can be expressed as

$$S_x(\omega) \simeq \frac{2}{\pi} \frac{D_x}{\alpha} = N_0 \tag{8.58}$$

and can be considered constant. In practice, it is sufficient to make sure that the power spectrum of the given process is constant, or next to constant, within the frequency response function of the dynamic system. Many actual dynamic systems characterized by low damping satisfy this condition. These systems, when subjected to forced oscillations, behave like narrow-band filters, which selectively enhance the input harmonics with frequencies close to the fundamental frequency of free vibrations of the system and suppress all the other harmonics. The response functions of such systems have a narrow, strongly pronounced peak in the vicinity of the system's own frequency. Within this peak, even a rapidly changing spectral density function can be considered constant.

The major reason for the idealization of a real random process by a "white noise" is due to the fact that, when a narrow-band system is subjected to the excitation of the "white noise" type, the system's response can be satisfactorily approximated by a quasiharmonic process with slow-changing amplitude and phase angle. Such a response can be treated as a Markovian process, and a well-developed and powerful mathematical "equipment" of the theory of Markovian processes can be effectively applied.

Examine now a random process $\zeta(t)$ whose behavior is described by the equation

$$\dot{\zeta}(t) = n(t)$$

where $n(t)$ is a "white noise." The ordinate $\zeta(t)$ of this process can be found as

$$\zeta(t_0 + T) = \zeta(t_0) + \int_{t_0}^{t_0+T} n(t) \, dt$$

Here t_0 is the initial and T is the final moments of observation. The change in this ordinate is

$$\Delta\zeta(T) = \zeta(t_0 + T) - \zeta(t_0) = \int_{t_0}^{t_0+T} n(t) \, dt$$

Since the mean value of the "white noise" is zero, the mean value of the process $\Delta\zeta(T)$ is zero as well. The variance of this process can be evaluated as

$$D_{\Delta\zeta} = \langle [\Delta\zeta(T)]^2 \rangle = \left\langle \int_{t_0}^{t_0+T} n(t_1) \, dt_1 \int_{t_0}^{t_0+T} n(t_2) \, dt_2 \right\rangle$$

$$= \int_{t_0}^{t_0+T} \int_{t_0}^{t_0+T} \langle n(t_1) n(t_2) \rangle \, dt_1 \, dt_2$$

$$= \pi N_0 \int_{t_0}^{t_0+T} \int_{t_0}^{t_0+T} \delta(t_1 - t_2) \, dt_1 \, dt_2 = \pi N_0 T$$

Random Processes: Spectral Theory 213

The "white noise" has a normal probability density function, and so does the process $\Delta\zeta(T)$:

$$p(\Delta\zeta, T) = \frac{1}{\pi\sqrt{2N_0 T}} \exp\left[-\frac{(\Delta\zeta)^2}{2\pi N_0 T}\right]$$

If one assumes, for the sake of simplicity, that $t_0 = 0$, $\zeta(t_0) = 0$, and $T = t$, then

$$\zeta(t) = \int_0^t n(t) \, dt$$

and the probability density function is expressed as follows:

$$p(\zeta, t) = \frac{1}{\pi\sqrt{2N_0 t}} e^{-\zeta^2/(2\pi N_0 t)}$$

The normal Markovian process $\zeta(t)$ is known as *Wiener's process*.

Another extreme case of a stationary random process (opposite to a "white noise") is a *narrow-band process*. This is a process in which a significant portion of its spectral density is confined to a narrow frequency band, whose width is small in comparison with the central frequency of the band. The amplitudes R of a narrow-band, normally distributed random process $X(t)$ follow the Rayleigh distribution (see Sec. 3.3.5):

$$f(r) = \frac{r}{D_x} \exp\left(-\frac{r^2}{2D_x}\right), \qquad 0 < r < \infty$$

where $D_x = r_0^2$ is the parameter of the distribution and r_0 is the most likely value of the random variable R. The cumulative probability distribution function and the probability of exceeding the level x^* by the process $R(t)$ are

$$F(r) = \int_0^t f(r) \, dr = 1 - e^{-r^2/(2D_x)}$$

and

$$P(r) = 1 - F(r^*) = e^{-r_*^2/(2D_x)}$$

respectively. Denoting by $r_{1/3}$ the lower limit of the highest one-third of a sample r exhibiting Rayleigh distribution, we find

$$P(R \geq r_{1/3}) = \int_{r_{1/3}}^{\infty} f(r) \, dr = \frac{1}{3}$$

This formula leads to the relationship:

$$r_{1/3} = \sqrt{2D_x \ln 3} = 1.482\sqrt{D_x}$$

The average of the highest one-third value, $\langle x_{1/3} \rangle$, of the process, is called the *significant value*. This can be obtained by taking the moment of its probability density function about the origin. It can be shown that

$$\langle x_{1/3} \rangle = \{\sqrt{\ln 3} + 3\sqrt{\pi}[1 - \Phi(\sqrt{2 \ln 3})]\}\sqrt{2D_x} \approx 2.00\sqrt{D_x}$$

where
$$\Phi(\xi) = \frac{1}{2}\left[1 + \text{erf}\left(\frac{\xi}{\sqrt{2}}\right)\right]$$

is the Laplace function and erf $(\xi/\sqrt{2})$ is the error function. Similarly, the average of the highest nth of a random process from Rayleigh distribution is given by the formula

$$\langle x_{1/n} \rangle = \{\sqrt{\ln n} + n\sqrt{\pi}[1 - \Phi(\sqrt{2 \ln n})]\}\sqrt{2D_x}$$

When n becomes large,

$$\langle x_{1/n} \rangle = \sqrt{2D_x \ln n}$$

The obtained relationships are frequently used in statistical evaluations. We would like to note that in the case of a non-narrow-band spectrum ($\varepsilon \neq 0$), when the Rice law should be applied instead of the Rayleigh law, the area under the spectrum curve should be multiplied by a factor of $1 - 1/(2\varepsilon^2)$. Thus, wherever D_x (which is equal to the mean square value m_0) occurs in the formulas for the narrow-band case, it should be replaced by the factor $m_0/[1 - 1/(2\varepsilon^2)]$ for the non-narrow-band case.

8.4. Spectral Theory of the Transformation of Stationary Random Process by Linear Dynamic Systems

Let us establish the relationship between the power spectra for the "input" and the "output" stationary random processes. The response spectrum can be evaluated as

$$S_y(\omega) = \frac{1}{\pi}\int_{-\infty}^{\infty} K_y(\tau)e^{-i\omega\tau}\, d\tau = \frac{2}{\pi}\int_0^{\infty} K_y(\tau)e^{-i\omega\tau}\, d\tau$$

$$= \frac{2}{\pi}\int_0^{\infty}\int_0^{\infty}\int_0^{\infty} e^{-i\omega\tau}h(\tau_1)h(\tau_2)K_x(\tau + \tau_1 - \tau_2)\, d\tau_1\, d\tau_2\, d\tau$$

$$= \frac{2}{\pi}\int_0^{\infty}\int_0^{\infty}\int_0^{\infty} e^{-i\omega(\tau+\tau_1-\tau_2)}e^{i\omega\tau_1}e^{-\omega\tau_2}h(\tau_1)h(\tau_2)K_x$$

$$\times (\tau + \tau_1 - \tau_2)\, d\tau_1\, d\tau_2\, d\tau$$

$$= \frac{2}{\pi} \int_0^\infty h(\tau_1) e^{i\omega\tau_1}\, d\tau_1 \int_0^\infty h(\tau_2) e^{-i\omega\tau_2}\, d\tau_2$$

$$\times \int_0^\infty K_x(\tau + \tau_1 - \tau_2) e^{-i\omega(\tau + \tau_1 - \tau_2)}\, d\tau$$

$$= S_x(\omega) \int_0^\infty h(t) e^{i\omega t}\, dt \int_0^\infty h(t) e^{-i\omega t}\, dt \qquad (8.59)$$

where

$$S_x(\omega) = \frac{2}{\pi} \int_0^\infty K_x(\tau + \tau_1 - \tau_2) e^{-i\omega(\tau + \tau_1 - \tau_2)}\, d\tau$$

$$= \frac{2}{\pi} \int_0^\infty K_x(\tau) e^{-i\omega\tau}\, d\tau \qquad (8.60)$$

is the spectrum of the "input" process. The spectrum $G_h(\omega)$ of the impulse response function can be obtained from the second formula in (8.14) as follows:

$$G_h(\omega) = \frac{1}{\pi} \int_{-\infty}^\infty h(t) e^{-i\omega t}\, dt = \frac{1}{\pi} \int_0^\infty h(t) e^{-i\omega t}\, dt$$

Introducing this formula into the relationship (8.44), we obtain

$$\Phi(i\omega) = \int_0^\infty h(t) e^{-i\omega t}\, dt \qquad (8.61)$$

Hence,

$$\Phi(-i\omega) = \int_0^\infty h(t) e^{i\omega t}\, dt \qquad (8.62)$$

The relationship (8.59) can be written, with consideration of the formulas (8.61) and (8.62), as follows:

$$S_y(\omega) = S_x(\omega)\Phi(i\omega)\Phi(-i\omega) = S_x(\omega)|\Phi(i\omega)|^2 = S_x(\omega)a^2(\omega) \qquad (8.63)$$

Thus, the relationship between the spectral densities $S_x(\omega)$ and $S_y(\omega)$ of the "input" process (response) $X(t)$ and the "output" process $Y(t)$ is substantially simpler that the relationship (7.18) between the correlation functions of these processes.

Example 8.1 A stationary random process $X(t)$ has the following correlation function (Fig. 7.2):

$$K_x(\tau) = D_x e^{-\alpha|\tau|} \cos \beta\tau$$

Determine the spectral density of this process.

Solution Using the cosine Fourier transform in accordance with the second formula in (8.51), we have

$$S_x(\omega) = \frac{2D_x}{\pi} \int_0^\infty e^{-\alpha|\tau|} \cos \beta\tau \cos \omega\tau \, d\tau = \frac{2D_x}{\pi} \alpha \frac{\omega^2 + \alpha^2 + \beta^2}{(\omega^2 - \beta^2 - \alpha^2)^2 + 4\alpha^2\omega^2}$$

The maximum of the function $S_x(\omega)$ takes place for the frequency

$$\omega = \sqrt{\sqrt{(\alpha^2 + \beta^2)}[2\beta - \sqrt{(\alpha^2 + \beta^2)}]}$$

As can be obtained from this equation, when α is small, the spectral function $S_x(\omega)$ has a strongly pronounced peak at $\omega = \beta$. However, for sufficiently large α, the spectral density can be assumed, in an approximate analysis, to be frequency independent:

$$S_x(\omega) = \frac{2D_x}{\pi\alpha} = N_0 = \text{constant}$$

In this case the process $X(t)$ can be treated as a "white noise" within a wide range of frequencies.

Example 8.2 A stationary random process $X(t)$ has a constant power spectrum $S_x(\omega) = N_0$. Determine the correlation function and the variance of this process.

Solution The correlation function $K_x(\tau)$ can be determined, in accordance with the first formula in (8.50), as follows:

$$K_x(\tau) = \frac{1}{2} \int_{-\infty}^{\infty} N_0 e^{i\omega\tau} \, d\omega = \frac{N_0}{2} \int_{-\infty}^{\infty} e^{i\omega\tau} \, d\omega = \frac{N_0}{2} \delta(\tau)$$

Since the correlation function is proportional to the delta function, this process is a "white noise." Its variance is infinite.

Example 8.3 Determine the correlation function and the power spectrum of a random telegraphic signal—a random stationary process that assumes, with an equal probability, the values $+1$ and -1, while the change in these values can occur only at the moments of time separated by the time interval T_0 (Fig. 8.2a).

Solution Let us examine two moments of time, t_1 and t_2, and find the mean of the product

$$X(t_1)X(t_2) = X(t_1)X(t_1 + \tau)$$

where $\tau = t_2 - t_1$. If $\tau > T_0$, the moments t_1 and t_2 belong to different time intervals T_0, and the product can be equal to either $+1$ or -1 with the same probability, so that the mean is equal to zero. If, however, $\tau < T_0$, then two cases are possible: case A, when the times t_1 and t_2 belong to the same time interval T_0 and, hence, the above product is equal to 1, and case B, when these times belong to different time intervals T_0 and the product of interest can be either $+1$ or -1 with the same probability. Therefore the mean of the product $X(t_1)X(t_2)$ is equal, for $\tau < T_0$, to the probability $P(A)$ of the event A, "both sections t_1 and t_2 are found within the same interval T_0." It is clear that such an event takes place when the section t_1

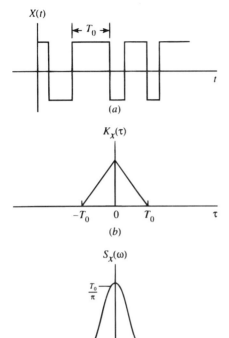

Figure 8.2 (a) Random telegraphic signal, (b) its correlation function and (c) its spectrum.

is separated from the beginning of the time interval T_0 by the time not exceeding $T_0 - |\tau|$. The probability of this event is $1 - |\tau|/T_0$. Thus, the correlation function of the process in question is (Fig. 8.2)

$$K_x(\tau) = \begin{cases} 1 - \dfrac{|\tau|}{T_0}, & \text{for } |\tau| \le T_0 \\ 0, & \text{for } |\tau| > T_0 \end{cases}$$

The variance of the signal is $D_x = K_x(0) = 1$. The spectral density of the process $X(t)$ is (Fig. 8.2c)

$$S_x(\omega) = \frac{2}{\pi} \int_0^\infty \left(1 - \frac{\tau}{T_0}\right) \cos \omega\tau \, d\tau = \frac{2}{\pi} \int_0^{T_0} \left(1 - \frac{\tau}{T_0}\right) \cos \omega\tau \, d\tau$$

$$= \frac{2}{\pi} \int_0^{T_0} \cos \omega\tau \, d\tau - \frac{2}{\pi T_0} \int_0^{T_0} \tau \cos \omega\tau \, d\tau = \frac{2}{\pi\omega} \sin \omega T_0$$

$$- \frac{2}{\pi T_0} \left(\frac{T_0}{\omega} \sin \omega T_0 + \frac{1}{\omega^2} \cos \omega T_0 - \frac{1}{\omega^2}\right)$$

$$= \frac{2(1 - \cos \omega T_0)}{\pi T_0 \omega^2} = \frac{T_0}{\pi} \left(\frac{\sin (\omega T_0/2)}{\omega T_0/2}\right)^2$$

Example 8.4 Determine the bivariate spectral density for two stationary correlated random processes $X(t)$ and $Y(t)$.

Solution By analogy with the bivariate correlation function, we use the Fourier transform to evaluate the bivariate spectral density:

$$S_{xy}(\omega) = \frac{1}{\pi} \int_{-\infty}^{\infty} K_{xy}(\tau) e^{-i\omega\tau} \, d\tau = \frac{1}{\pi} \int_{0}^{\infty} [K_{xy}(\tau) e^{-i\omega\tau} + K_{xy}(-\tau) e^{i\omega\tau}] \, d\tau$$

In the general case, the bivariate correlation function is not symmetric with respect to its arguments, that is, $K_{xy}(t, t') \neq K_{xy}(t', t)$ or, for stationary processes, $K_{xy}(\tau) \neq K_{xy}(-\tau)$. However, when both the arguments and the indices are interchanged, one can always write $K_{xy}(t, t') = K_{yx}(t', t)$ or, for stationary processes, $K_{xy}(\tau) = K_{yx}(-\tau)$. Therefore, to ensure that the sought bivariate spectral density is a real, not an imaginary, function of the real variable ω, we examine the bivariate spectral density, which corresponds to the bivariate correlation function $K_{yx}(\tau)$:

$$S_{yx}(\omega) = \frac{1}{\pi} \int_{0}^{\infty} [K_{yx}(\tau) e^{-i\omega\tau} + K_{yx}(-\tau) e^{i\omega\tau}] \, d\tau$$

Summing up the two spectra $S_{xy}(\omega)$ and $S_{yx}(\omega)$ and considering that $K_{xy}(\tau) = K_{yx}(-\tau)$, we obtain

$$S_{xy} + S_{yx} = \frac{1}{\pi} \int_{0}^{\infty} [K_{xy}(\tau) + K_{yx}(\tau)](e^{i\omega\tau} + e^{-i\omega\tau}) \, d\tau$$

$$= \frac{2}{\pi} \int_{0}^{\infty} [K_{xy}(\tau) + K_{yx}(\tau)] \cos \omega\tau \, d\tau$$

Example 8.5 A stationary random process $X(t)$ has a zero mean and a correlation function

$$K_x(\tau) = D_x e^{-\alpha|\tau|}.$$

Determine the bivariate correlation function and the bivariate spectral density for this process and its time derivative $\dot{X}(t)$.

Solution The bivariate correlation function can be found as

$$K_{x\dot{x}}(t, t') = \left\langle \dot{X}(t) \frac{\partial}{\partial t'} \dot{X}(t') \right\rangle = \frac{\partial}{\partial t'} \langle \dot{X}(t) \dot{X}(t') \rangle$$

$$= \frac{\partial}{\partial t'} K_x(\tau) = -D_x \alpha e^{-\alpha|\tau|} \frac{d|\tau|}{d\tau} = \frac{dK_x(\tau)}{d\tau} = -D_x \alpha e^{-\alpha|\tau|} \operatorname{sign} \tau$$

where the function sign $\tau = \tau/|\tau|$ is equal to 1 for $\tau > 0$, to -1 for $\tau < 0$, and to zero for $\tau = 0$ (see Example 7.11, Sec. 7.3). The bivariate spectral density can be found, considering the solution to the previous example, as follows:

$$S_{x\dot{x}}(\omega) = -D_x \alpha \int_{0}^{\infty} (e^{-\alpha\tau} e^{-i\omega\tau} + e^{-\alpha\tau} e^{i\omega\tau}) \, d\tau = -\frac{D_x \alpha(2i\omega)}{\alpha^2 + \omega^2}$$

where $i = \sqrt{-1}$ is the "imaginary unity."

8.5. Response of a Dynamic System to Dual Inputs

In practice it is often necessary to be able to evaluate the response of a dynamic system to more than one random input. For instance, the vertical motions of a ship's forebody, which are responsible for the severity of the slamming conditions, are due to both heave and pitch motions. The random motions of a car are due to the excitations coming from both its front and rear wheels. The random response of a lightwave coupler in a fiber optic transmission line is affected by two light beams.

Let the response of a linear dynamic system be due to two input processes $X_1(t)$ and $X_2(t)$. This response can be written, using the Duhamel integral, as follows:

$$y(t) = \int_{-\infty}^{\infty} x_1(t - \xi) h_1(\xi) \, d\xi + \int_{-\infty}^{\infty} x_2(t - \xi) h_2(\xi) \, d\xi$$

The correlation function of the response is

$$K_y(\tau) = \lim_{T \to \infty} \left[\frac{1}{T} \int_0^T y(t) y(t + \tau) \, dt \right]$$

$$= \lim_{T \to \infty} \left\{ \frac{1}{T} \int_0^T \left[\int_{-\infty}^{\infty} x_1(t - \xi) h_1(\xi) \, d\xi + \int_{-\infty}^{\infty} x_2(t - \xi) h_2(\xi) \, d\xi \right] \right.$$

$$\left. \times \left[\int_{-\infty}^{\infty} x_1(t + \tau - \xi') h_1(\xi') \, d\xi' + \int_{-\infty}^{\infty} x_2(t + \tau - \xi') h_2(\xi') \, d\xi' \right] dt \right\}$$

$$= \lim_{T \to \infty} \left\{ \frac{1}{T} \int_0^T \left[\int_{-\infty}^{\infty} x_1(t - \xi) h_1(\xi) \, d\xi \int_{-\infty}^{\infty} x_1(t + \tau - \xi') h_1(\xi') \, d\xi' \right] dt \right.$$

$$+ \frac{1}{T} \int_0^T \left[\int_{-\infty}^{\infty} x_2(t - \xi) h_2(\xi) \, d\xi \int_{-\infty}^{\infty} x_1(t + \tau - \xi') h_1(\xi') \, d\xi' \right] dt$$

$$+ \frac{1}{T} \int_0^T \left[\int_{-\infty}^{\infty} x_1(t - \xi) h_1(\xi) \, d\xi \int_{-\infty}^{\infty} x_2(t + \tau - \xi') h_2(\xi') \, d\xi' \right] dt$$

$$\left. + \frac{1}{T} \int_0^T \left[\int_{-\infty}^{\infty} x_2(t - \xi) h_2(\xi) \, d\xi \int_{-\infty}^{\infty} x_2(t + \tau - \xi') h_2(\xi') \, d\xi' \right] dt \right\}$$

$$= \int_{-\infty}^{\infty} \int_{-\infty}^{\infty} [K_{11}(\xi + \tau - \xi') h_1(\xi) h_1(\xi')$$

$$+ K_{21}(\xi + \tau - \xi') h_2(\xi) h_1(\xi')$$

$$+ K_{12}(\xi + \tau - \xi') h_1(\xi) h_2(\xi')$$

$$+ K_{22}(\xi + \tau - \xi') h_2(\xi) h_2(\xi')] \, d\xi \, d\xi'$$

Here $K_{11}(\tau)$ is the correlation function of the random process $X_1(t)$, $K_{22}(\tau)$ is the correlation function of the random process $X_2(t)$, $K_{12}(\tau)$ is the bivariate correlation function of the random processes $X_1(t)$ and $X_2(t)$, and $K_{21}(\tau)$ is the bivariate correlation function of the processes $X_2(t)$ and $X_1(t)$. Then, the spectral density function $S_y(\omega)$ of the system's response can be obtained as

$$G_y(\omega) = \frac{1}{\pi} \int_{-\infty}^{\infty} K_y(\tau) e^{-i\omega\tau} d\tau$$
$$= S_{11}(\omega) |\Phi_1(i\omega)|^2 + S_{21}(\omega) \Phi_2^*(i\omega) \Phi_1(i\omega)$$
$$+ S_{12}(\omega) \Phi_1^*(i\omega) + S_{22}(\omega) |\Phi_2(i\omega)|^2 \quad (8.64)$$

where $S_{11}(\omega)$ is the spectral density of the process $X_1(t)$, $S_{22}(\omega)$ is the spectral density of the process $X_2(t)$, $S_{12}(\omega)$ is the bivariate spectral density of the processes $X_1(t)$ and $X_2(t)$, $S_{21}(\omega)$ is the bivariate spectral density of the processes $X_2(t)$ and $X_1(t)$, $\Phi_1(i\omega)$ is the complex frequency characteristic (frequency response function) of the process $X_1(t)$, $\Phi_1^*(i\omega) = \Phi_1(-i\omega)$ is the conjugate complex frequency characteristic of this process, $\Phi_2(i\omega)$ is the complex frequency characteristic of the process $X_2(t)$, and $\Phi_2^*(i\omega) = \Phi_2(-i\omega)$ is the conjugate complex frequency characteristic of the process $X_2(t)$.

Let us show how the first term on the right-hand part of the formula (8.64) can be obtained:

$$\frac{1}{\pi} \int_{-\infty}^{\infty} \left[\int_{-\infty}^{\infty} \int_{-\infty}^{\infty} K_{11}(\xi + \tau - \xi') h_1(\xi) h_1(\xi') e^{-i\omega\tau} d\xi d\xi' \right] d\tau$$
$$= \frac{1}{\pi} \int_{-\infty}^{\infty} K_{11}(\xi + \tau - \xi') e^{-i\omega(\xi+\tau-\xi')} d\tau$$
$$\times \int_{-\infty}^{\infty} h_1(\xi) e^{i\omega\xi} d\xi \int_{-\infty}^{\infty} h_1(\xi') e^{-i\omega\xi'} d\xi'$$
$$= S_{11}(\omega) \Phi_1^*(i\omega) \Phi_1(i\omega) = S_{11}(\omega) |\Phi_1(i\omega)|^2$$

where
$$\Phi_1(i\omega) = \int_{-\infty}^{\infty} h_1(\xi) e^{-i\omega\xi} d\xi$$

is the complex frequency characteristic of the random process $X_1(t)$ and

$$\Phi_1^*(i\omega) = \Phi_1(-i\omega) = \int_{-\infty}^{\infty} h_1(\xi) e^{i\omega\xi} d\xi$$

is the conjugate complex frequency characteristic of this process.

If the input processes $X_1(t)$ and $X_2(t)$ are uncorrelated, then $S_{21}(\omega) = S_{12}(\omega) = 0$ and the formula (8.64) yields

$$G_y(\omega) = S_{11}(\omega)|\Phi_1(i\omega)|^2 + S_{22}(\omega)|\Phi_2(i\omega)|^2 \tag{8.65}$$

Thus, if the system's inputs are statistically independent random processes, the spectral density of the output process can be obtained as a sum of the spectral densities of the system's responses to each of the input processes.

8.6. Information Capacity of a Communication Channel

The information capacity of a communication channel can be computed as

$$C = \lim_{T \to \infty} \max_{f_1(x)} \left[\frac{1}{T} I(X, Y) \right] \tag{8.66}$$

Here the maximum $I(X, Y)$ value of the transmitted information (see Chapter 6) can be achieved by choosing an optimal distribution $f_1(x)$ of the input signal $x(t)$ for a sufficiently large time interval T.

In the case of a communication channel with a finite frequency band, the information capacity can be determined by the *Shannon formula*:

$$C = F \log \left(1 + \frac{\sigma_s^2}{N_0 F} \right) \tag{8.67}$$

Here F is the frequency band of the channel, N_0 is the constant spectral frequency of the normal stationary noise $n(t)$ within the frequency band $0 < \omega < F$, and $\sigma_s^2 = \langle x^2(t) \rangle = $ constant is the finite mean power of the signal $X(t)$. This signal can be treated as a stationary normal random process with a uniform spectral density within this band. The process $Y(t)$ acting on the receiving device consists of the useful signal $x(t)$ and the "additive" noise $n(t)$:

$$Y(t) = X(t) + n(t) \tag{8.68}$$

The processes $X(t)$ and $n(t)$ are assumed to be statistically independent. If $\sigma_s^2/(N_0 F) \gg 1$ (the ratio of the mean power of the signal to the power of the noise is large), then the formula for the information capacity can be simplified as

$$C \simeq F \log \frac{\sigma_s^2}{N_0 F} \tag{8.69}$$

If $\sigma_s^2/(N_0 F) \ll 1$ (the ratio of the mean power of the signal to the power of the noise is small), then

$$C \simeq 1.443 \frac{\sigma_s^2}{N_0} \tag{8.70}$$

The formula (8.67) can also be written as follows:

$$C = \int_{\omega_1}^{\omega_2} \log\left[1 + \frac{S(\omega)}{N(\omega)}\right] d\omega \tag{8.71}$$

where $S(\omega)$ is the spectral density of the normal random process $X(t)$ and $N(\omega)$ is the spectral density of the additive normal noise $n(t)$.

Example 8.6 The signal has a power P_s and the normal noise has a spectral density $N(\omega)$. Determine the spectral density of the signal that would result in the fastest transmission of the information.

Solution From the formula (8.71) we have

$$C = \int_{\omega_1}^{\omega_2} \ln\left[1 + \frac{S(\omega)}{N(\omega)}\right] d\omega$$

where $S(\omega)$ is the sought spectral density of the signal. We seek the maximum value of this integral under the condition

$$P_s = \int_{\omega_1}^{\omega_2} S(\omega)\, d\omega$$

Using the method of Lagrange multipliers and applying the formula (6.23) we have

$$\frac{\partial F}{\partial S} + \lambda \frac{\partial \phi}{\partial S} = \frac{\partial}{\partial S} \ln\left(1 + \frac{S}{N}\right) + \lambda \frac{\partial S}{\partial S} = 0$$

Since

$$\frac{\partial}{\partial S} \ln\left(1 + \frac{S}{N}\right) = \frac{1}{1 + (S/N)} \frac{1}{N}$$

then

$$\frac{1}{S(\omega) + N(\omega)} = -\lambda$$

Hence,

$$S(\omega) = -\frac{1}{\lambda} - N(\omega)$$

We conclude that the power spectrum $S(\omega)$ must be such that, after being added to the spectrum of the noise $N(\omega)$, it would ensure that this sum is constant and frequency independent. The quantity λ should be chosen in such a way that the total power of the useful signal would be equal to the given power P_s.

8.7. Characteristic Functions

In Sec. 2.3 we described a generating function of a random variable. Such a function turns out to be very useful not only for evaluating moments of the random variables but for other applications of the probability theory as well. Another useful function, which contains the information about the random variable and its probability distribution, is the *characteristic function*—the Fourier transform of the probability density function:

$$E(\omega) = \int_{-\infty}^{\infty} e^{i\omega x} f(x) \, dx \tag{8.72}$$

The inverse Fourier transform is

$$f(x) = \frac{1}{2\pi} \int_{-\infty}^{\infty} e^{-i\omega x} E(\omega) \, d\omega \tag{8.73}$$

For a discrete random variable,

$$E(\omega) = \sum_{k=1}^{n} p_k e^{i\omega x_k} \tag{8.74}$$

The inverse relationship gives the probability density function as a sum of delta functions.

A principal use of the characteristic function is to calculate the moments of the random variable. Differentiating the function (8.72) or (8.74) with respect to $i\omega$ and setting $\omega = 0$, we obtain

$$\left. \frac{dE(\omega)}{d(i\omega)} \right|_{\omega=0} = \int_{-\infty}^{\infty} x f(x) \, dx = \langle x \rangle \tag{8.75}$$

Continuing this process, one can find for the kth moment

$$m_k = \langle x^k \rangle = \left. \frac{d^k E(\omega)}{d(i\omega)^k} \right|_{\omega=0} \tag{8.76}$$

The Taylor series for the characteristic function $E(\omega)$ in the neighborhood of the origin can be obtained from the function (8.76) as follows:

$$E(\omega) = \sum_{k=1}^{\infty} \frac{(i\omega)^k}{k!} \langle x^k \rangle \tag{8.77}$$

Thus, an easy way to find the moments of a random variable is to expand its characteristic function about $\omega = 0$. This series must be terminated by a remainder term just before this first infinite moment, if any such exists. The method for evaluating moments of a random variable by using characteristics functions can be generalized to the case of many random variables.

Example 8.7 Determine the characteristic function of the random variable X with the probability density function

$$f(x) = \frac{1}{2} e^{-|x|}, \quad -\infty < x < \infty$$

Evaluate the mean and variance of the variable X.

Solution Using the formula (8.72) we have

$$E(\omega) = \int_{-\infty}^{\infty} e^{i\omega x} f(x) \, dx = \frac{1}{2} \int_{-\infty}^{\infty} e^{i\omega x - |x|} \, dx$$

$$= \frac{1}{2} \int_{0}^{\infty} e^{(i\omega - 1)x} \, dx + \frac{1}{2} \int_{-\infty}^{0} e^{(i\omega + 1)x} \, dx$$

$$= \frac{1}{2}\left(\frac{1}{1 - i\omega} - \frac{1}{1 + i\omega}\right) = \frac{1}{1 + \omega^2}$$

The mean and variance can be calculated as

$$\langle x \rangle = \frac{1}{i} E'(0) = 0, \quad D_x = \frac{1}{i^2} E''(0) = 2$$

Example 8.8 Determine the moments with respect to the origin for a random variable X with the characteristic function

$$E(\omega) = \frac{1}{1 + \omega^2}$$

Find the mean value and the variance.

Solution The formula (8.76) yields

$$m_k = \frac{1}{i^k} \frac{d^k E(\omega)}{d\omega^k}\bigg|_{\omega = 0}$$

The derivatives entering this formula can be found as coefficients of $u^k/k!$ in the expansion of the function $E(\omega)$ in the Macloren series:

$$E(\omega) = \sum_{k=1}^{\infty} \frac{d^k E(\omega)}{d\omega^k}\bigg|_{\omega=0} \frac{\omega^k}{k!}$$

On the other hand, the function $E(\omega)$ can be found, for $|\omega| < 1$, as a sum of a geometric progression:

$$E(\omega) = \frac{1}{1 - (i\omega)^2} = \sum_{m=0}^{\infty} (i\omega)^{2m} = \sum_{m=0}^{\infty} i^{2m}(2m)! \frac{\omega^{2m}}{(2m)!}$$

Thus, the Macloren series contains only even powers of ω. Therefore we conclude that

$$\left.\frac{d^k E(\omega)}{d\omega^k}\right|_{\omega=0} = \begin{cases} i^k k!, & \text{for even } k \\ 0, & \text{for odd } k \end{cases}$$

and the initial moments are

$$m_k = \begin{cases} k!, & \text{for even } k \\ 0, & \text{for odd } k \end{cases}$$

The mean value m_1 is zero and the variance is $m_2 = 2! = 2$.

Further Reading (see Bibliography): 13, 21, 22, 28, 30, 34, 53, 66, 72, 85, 86, 89, 94, 96, 97, 101, 102, 125.

Chapter 9

Extreme Value Distributions

" If you do not raise your eyes, you will think that you are at the highest point."
ANTONIO PORCHIA

"Avoid extremes."
ALEXANDER POPE

"In the long run we are all dead"
JOHN MAYNARD KEYNES

9.1. Extreme Value Statistics

In the analysis and design of engineering systems it is often desirable to predict the magnitude of the system's response in extreme operation conditions. The knowledge of the largest response (extreme value) that the system might encounter during its lifetime is needed to be able to assess the likelihood of the possible failure under the action of a single load of a very high level. The interest in the distribution of the extremes goes back to the works of N. Bernulli in the early eighteenth century. The modern mathematical theory of extreme value distributions was developed by R. von Mises (1923, 1936), M. Fréchet (1927), R. A. Fisher and L. H. C. Tippett (1928), B. V. Gnedenko (1943), E. J. Gumbel (1954, 1958), and others.

The *extreme value* is defined, in general, as the largest value expected to occur in a certain (limited) number of observations (sample size) or during a certain (finite) period of time. The number of observations or the time period should have to be established beforehand, prior to applying an extreme value evaluation technique. In addition, one should consider, before this technique is applied, the response of the system to the random excitation at every cycle of loading (excitation) and determine the "regular," "nonextreme," probability density function $f(x)$ and the corresponding cumulative distribution function $F(x)$. These are "initial," "basic," "generating," "parent" functions, which characterize the system's response to the random excitation $X(t)$.

The extreme response Y_n to the nth random loading is also a random variable. Its distributions, $g(y_n)$ and $G(y_n)$, are related to, but are different from, the "basic" distributions $f(x)$ and $F(x)$. The extreme value distributions can be obtained from the initial distributions in a straightforward way, using *order statistics*. This results in the following formulas:

$$g(y_n) = n\{f(x)[F(x)]^{n-1}\}_{x=y_n} \tag{9.1}$$

$$G(y_n) = \{[F(x)]^n\}_{x=y_n} \tag{9.2}$$

These equations enable one to obtain, for sufficiently large n numbers, various statistical characteristics of the extreme values Y_n of the random process $X(t)$: the most probable extreme value, i.e., the value that is most likely to occur in n observations, the probability of exceeding a certain level y^*, etc.

Rationale underlying the formulas (9.1) and (9.2) becomes clear from the following reasoning. Let us consider, instead of a random process $X(t)$, a stationary random succession

$$y_0 = y(0), \quad y_1 = y(\tau_e), \ldots, y_i = y(i\tau_e), \ldots, y_n(n\tau_e) = y(\Delta t)$$

where $\tau_e = \Delta t/n$ is the duration of each time segment representing a term of the succession, Δt is the duration of the whole succession, and n is the total number of terms in the succession. The probability $P_i(y^*)$ that the random variable Y_i, $i = 0, 1, 2, \ldots, n$, assumes the value y^* at the point y_i, while it is smaller than the y^* value at all the other points, can be evaluated by the integral

$$P_i(y^*) = \underbrace{\int_{-\infty}^{y^*}\int_{-\infty}^{y^*}\cdots\int_{-\infty}^{y^*}}_{n} f(n)(y_0, y_1, \ldots, y_{i-1}, y^*, y_{i+1}, \ldots, y_n)\, dy_0\, dy_1 \cdots dy_{i-1}\, dy_{i+1} \cdots dy_n$$

Here $f_n(y_0, y_1, \ldots, y_n)$ is the multidimensional probability density distribution function for the set of random variables Y_i. In the case when these variables are statistically independent, the function f_n can be written as

$$f_n(y_0, y_1, \ldots, y_n) = f_{y_0}(y_0) f_{y_1}(y_1) \cdots f_{y_n}(y_n)$$

For a stationary succession,

$$f_n(y_0, y_1, \ldots, y_n) = f_y(y_0) f_y(y_1) \cdots f_y(y_n)$$

so that

$$P_i(y^*) = f_y(y^*) \int_{-\infty}^{y^*} f_y(y_0)\, dy_0 \int_{-\infty}^{y^*} f_y(y_1)\, dy_1 \cdots \int_{-\infty}^{y^*} f_y(y_n)\, dy_n$$

$$= f_y(y^*) \left[\int_{-\infty}^{y^*} f_y(y)\, dy \right]^n$$

The total probability that, for an arbitrary point of the succession, the random variable Y_n is equal to y^* and is smaller than y^* for all other points can be found as

$$f_{y^*}(y^*) = C \sum_{i=0}^{n} P_i(y^*) = C f_y(y^*)[F_y(y^*)]^n$$

where

$$F_y(y^*) = \int_{-\infty}^{y^*} f_y(y)\, dy$$

is the cumulative distribution function for the random variable Y^*. The condition of normalization

$$\int_{-\infty}^{\infty} f_{y^*}(y^*)\, dy^* = 1$$

yields $C = n + 1$, so that

$$f_{y^*}(y^*) = (n+1) f_y(y^*)[F_y(y^*)]^n, \qquad n = 0, 1, 2, \ldots$$

The formula (9.1) can be obtained from this expression by shifting the origin for counting the i values from zero to unity and putting $y^* = y_n$. The formula (9.2) follows from the formula (9.1).

In a simplistic reasoning, the formula (9.2) can be based on a rule of multiplication probabilities. Indeed, let n loadings take place during a certain period of time t. Then the distribution of the absolute maximum of the process $X(t)$ of loading should coincide with the distribution of the maximum value of the random variable X in n series of observations, with n observations in each series. Then, using the rule of the multiplication of probabilities, one can conclude that the cumulative probability density function for the absolute maximum in n loadings should be expressed as

$$G(x) = [F(x)]^n$$

i.e. by the formula (9.2).

9.2. Exceeding the Given Level by Normal Processes

In Sec. 7.4 we obtained a formula for the mean number of upward crossings of the given level by a random process. In this section we will derive a formula for the probability that a normal process exceeds the given level. This formula will then be used to obtain an EVD for a stationary normal process.

The probability that the ordinate $X(t)$ of a stationary random process and its time derivative $\dot{X}(t)$ will be found within the intervals $(x, x + dx)$ and $(\dot{x}, \dot{x} + d\dot{x})$, respectively, can be written as

$$f(x, \dot{x})\, dx\, d\dot{x} = f(x, \dot{x})\dot{x}\, d\dot{x}\, dt \simeq \Delta t f(x, \dot{x})\dot{x}\, d\dot{x}$$

Here $f(x, \dot{x})$ is the two-dimensional probability density function for the processes $X(t)$ and $\dot{X}(t)$. In order for a certain level x^* to be exceeded, the process $X(t)$ has to reach this level and, in addition to that, the instantaneous value of the process $\dot{X}(t)$ must be positive, i.e., the velocity $\dot{X}(t)$ must be directed upwards. Since the level x^* will be exceeded for all the positive $\dot{X}(t)$ values, the total probability of exceeding this level within the time interval $(t, t + \Delta t)$ can be computed as follows:

$$P(X > x^*) = \Delta t \int_0^\infty f(x^*, \dot{x})\dot{x}\, d\dot{x} \qquad (9.3)$$

If the time interval Δt and the probability P are small, this formula can be used to evaluate the probability that the level x^* is exceeded just once during the time Δt. It should be pointed out that the mathematical requirement for the time Δt to be small (which is necessary to be confident that the level x^* will be exceeded by the random process $X(t)$ within the time interval Δt only once) by no means signifies that this time should be small from a physical point of view. When the events $X(t) > x^*$ are rare, the time Δt can be rather durable and, for vibration processes, can exceed many times the period of oscillations.

For statistically independent processes $X(t)$ and $\dot{X}(t)$, the two-dimensional probability density function $f(x, \dot{x})$ can be written as

$$f(x, \dot{x}) = f_x(x) f_{\dot{x}}(\dot{x})$$

Then the formula (9.3) yields

$$P = \Delta t f_x(x) \int_0^\infty f_{\dot{x}}(\dot{x})\dot{x}\, d\dot{x} \qquad (9.4)$$

If the distributions $f_x(x)$ and $f_{\dot{x}}(\dot{x})$ are normal ones with zero means, so that

$$f_x(x) = \frac{1}{\sqrt{2\pi D_x}} e^{-x^2/(2D_x)}, \quad f_{\dot{x}}(\dot{x}) = \frac{1}{\sqrt{2\pi D_{\dot{x}}}} e^{-\dot{x}^2/(2D_{\dot{x}})} \quad (9.5)$$

then the probability that the process $X(t)$ will exceed the level x^* can be computed as

$$\begin{aligned} P &= \frac{\Delta t}{2\pi\sqrt{D_x D_{\dot{x}}}} e^{-(x^*)^2/(2D_x)} \int_0^\infty e^{-\dot{x}^2/(2D_{\dot{x}})} \, d\dot{x} \\ &= \frac{\Delta t}{\pi\sqrt{2D_x}} e^{-(x^*)^2/(2D_x)} \int_0^\infty e^{-\eta^2} \, d\eta \end{aligned} \quad (9.6)$$

Since the Laplace function

$$\Phi(\alpha) = \frac{2}{\sqrt{\pi}} \int_0^\alpha e^{-\eta^2} \, d\eta$$

is equal to unity for $\alpha \to \infty$, the formula (9.6) can be written in the following simple way:

$$P = \lambda \, \Delta t \quad (9.7)$$

where

$$\lambda = \frac{1}{2\pi} \sqrt{\frac{D_{\dot{x}}}{D_x}} e^{-(x^*)^2/(2D_x)} \quad (9.8)$$

is the number of upward crossings the level x^* per unit time.

If the random processes $X(t)$ and $\dot{X}(t)$ have narrow spectra, these processes can be characterized by their effective period, given by the formula (7.12):

$$\tau_e = \frac{2\pi}{\omega_e} = 2\pi \sqrt{\frac{D_x}{D_{\dot{x}}}} \quad (9.9)$$

With the expression (9.9), the formula (9.8) for the number of upward crossings the level x^* can be written as

$$\lambda = \frac{1}{\tau_e} e^{-(x^*)^2/(2D_x)} \quad (9.10)$$

This formula enables one to evaluate the number of times a certain level x^* is exceeded per unit time by the random process $X(t)$.

9.3. Extreme Value Distribution for Stationary Normal Processes

If the stationary process $X(t)$ is *homogeneous* (i.e., that the probability P that the given level x^* is exceeded depends only on the duration of the time interval Δt and is independent of the initial moment of time), and *ordinary* (i.e., none of the events $X > x^*$ occurs simultaneously with another event), and all the events $X > x^*$ are statistically independent, then the "flow" of such events can be considered to be a Poisson process (see Sec. 11.1). The probability that the level x^* is exceeded (crossed upwards) by the process $X(t)$ can be found as

$$P_n(\Delta t) = \frac{(\lambda \, \Delta t)^n}{n!} e^{-\lambda \Delta t} \qquad (9.11)$$

It is assumed that the events $X > x^*$ are rare and the mean time Δt between these events is distributed exponentially:

$$f_t(\Delta t) = \lambda e^{-\lambda \Delta t} \qquad (9.12)$$

The corresponding cumulative distribution function is

$$F_t(\Delta t) = \int_0^{\Delta t} f_t(\Delta t) d(\Delta t) = 1 - e^{-\lambda \Delta t} \qquad (9.13)$$

When the level x^* increases, the time interval Δt between the two adjacent upward crossings of this level increases as well. In order to keep the "flow" of the events $X > x^*$ finite, when the level x^* tends to infinity, one must impose a requirement that the product $\lambda \, \Delta t$ be always finite. This requirement can be mathematically formulated as follows:

$$P = \lambda(x^*) \, \Delta t = e^{-\chi} \qquad (9.14)$$

where χ is a random parameter. In order to establish the law of distribution for this parameter, one can proceed from the Poisson distribution (9.11). The probability that level x^* will not be crossed during the time Δt can be obtained from (9.11) by putting $n = 0$ in this formula:

$$P_0 = P_0(\Delta t) = e^{-\lambda \Delta t} \qquad (9.15)$$

This probability, considering (9.14), can be written as

$$P_0 = P_0(\chi) = \exp(-e^{-\chi}), \qquad -\infty \leq \chi \leq \infty \qquad (9.16)$$

This double exponential formula defines the probability distribution function for the random variable χ. This variable can assume any value from $-\infty$ to $+\infty$. The corresponding probability density function can be obtained by differentiation:

$$f_\chi(\chi) = \frac{dP_0(\chi)}{d\chi} = \exp(-\chi - e^{-\chi}) \qquad (9.17)$$

This "*double exponential*" *probability distribution function* is the major asymptotic extreme value distribution. This distribution is applicable when the extremes of the process $X(t)$ are not restricted in the positive direction and, in addition to that, the "right tails" of the ordinates of this process exponentially decrease. Maximum values of snow or wind loading and long-term distribution of ocean wave heights are examples of such a process.

The distribution (9.17) can be generalized as follows:

$$f_\chi(\chi) = \frac{1}{\beta} \exp\left[-\frac{\alpha - \chi}{\beta} - \exp\left(\frac{\alpha - \chi}{\beta}\right)\right],$$

$$-\infty < \chi < \infty, \; -\infty < \alpha < \infty, \; \beta > 0$$

The corresponding probability distribution function is

$$F_\chi(\chi) = \exp\left[-e^{(\alpha - \chi)/\beta}\right]$$

The parameters α and β are related to the mean $\langle\chi\rangle$ and the variance D_χ of the random variable χ as follows:

$$\langle\chi\rangle = \alpha + 0.5776\beta, \qquad D_\chi = 1.645\beta^2$$

Solving Eq. (9.10) for the x^* value, we obtain:

$$x^* = \sqrt{-2D_x \ln(\lambda\tau_e)} \qquad (9.18)$$

or, considering (9.14),

$$x^* = \sqrt{2D_x(\chi + \ln n)} \qquad (9.19)$$

where

$$n = \frac{\Delta t}{\tau_e} \qquad (9.20)$$

is the number of oscillations during the time interval Δt between two adjacent upward crossings of the level x^*. Treating the x^* value as a deterministic function of the random variable χ (see Chapter 5) and considering the distribution law (9.17), we obtain the following formula for the probability density function of the random variable X^*:

$$f_{x^*}(x^*) = f_\chi[\chi(x^*)] \frac{d\chi}{dx^*} = \frac{x^*}{D_x} \exp\left[-\chi(x^*) - e^{-\chi(x^*)}\right]$$

$$= \frac{x^*}{D_x} n \exp\left[-\frac{(x^*)^2}{2D_x} - ne^{-(x^*)^2/(2D_x)}\right] \qquad (9.21)$$

The corresponding cumulative probability distribution function is

$$F_{x^*}(x^*) = \int_0^{x^*} f^*(x^*)\, dx^* = \exp\left[-ne^{-x^{*2}/(2D_x)}\right] - e^{-n} \qquad (9.22)$$

For large enough n values, the second term in this formula is small and can be omitted:

$$F_{x^*}(x^*) = \exp\left[-ne^{-x^{*2}/(2D_x)}\right] \qquad (9.23)$$

The formulas (9.21) and (9.22) define the *extreme value distribution* for stationary normal random processes. Unlike the formulas (9.1) and (9.2), based on the order statistics, the formulas (9.21) and (9.22) provide the exact solution to the extreme value distribution (EVD) problem.

Dimensionless probability density curves $f_{x^*}(x^*)\sqrt{D_x}$ are plotted in Fig. 9.1 for four n values ($n = 50, 100, 200, 1000$). As one can see from these plots, the probability density curves represent asymmetric, skew-normal distributions. They grow narrower and shift to the right, towards larger x^* values, with an increase in the number n of oscillations (observations). Such a behavior is due to the fact that the absolute maxima x^* increase with an increase in the time interval Δt, while the scattering of the maxima of different responses must decrease due to the reduction in the entropy of the distribution.

The most likely value x_0 of the distribution (9.21) can be determined from the equation

$$\frac{df_{x^*}(x^*)}{dx^*} = 0$$

This equation yields

$$n = \left(1 - \frac{1}{\xi_0^2}\right) e^{\xi_0^2/2} \qquad (9.24)$$

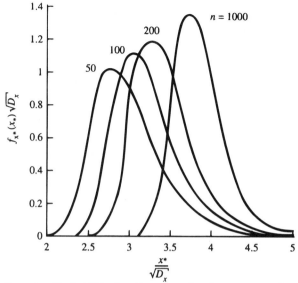

Figure 9.1 Probability density curves for the extreme value distribution (EVD).

where

$$\zeta_0 = \frac{x_0}{\sqrt{D_x}} \tag{9.25}$$

For sufficiently large n values (or large ζ_0 values), the formulas (9.24) and (9.25) yield

$$x_0 = \sqrt{2D_x \ln n} \tag{9.26}$$

When ζ_0 is close to 3 ($n \approx 80$), the error from using the formula (9.24), based on the order statistics, instead of the formula (9.26), based on the double exponential extreme value distribution function, is only 1.3 percent. When ζ_0 is about 4 ($n \approx 2800$), this error is as small as 0.4 percent.

As follows from the formula (9.22), the probability that the process $X^*(t)$ exceeds in its turn, a certain level x^{**} can be found as

$$P(X^* > x^{**}) = P(x^{**}) = 1 - F_{x^*}(x^{**})$$
$$= 1 - \exp\left[-n e^{-(x^{**})^2/(2D_x)}\right] + e^{-n} \tag{9.27}$$

For large n numbers,

$$P(x^{**}) \simeq 1 - \exp\left[-n e^{-(x^{**})^2/(2D_x)}\right] \tag{9.28}$$

Hence, for the given probability $P = P(x^{**})$ of exceeding the level x^{**}, this the level x^{**} can be defined as follows:

$$x^{**} = \sqrt{2D_x\{\ln n - \ln[-\ln(1-P)]\}} \qquad (9.29)$$

Substituting Eq. (9.24) into Eq. (9.21), one can obtain the following expression for the envelope of the probability density curves for different n numbers:

$$f_{x0}(x_0) = \frac{1}{\sqrt{D_x}} \exp\left(-\frac{1}{\sqrt{e}}\right)\left(x_0 - \frac{1}{x_0}\right) \qquad (9.30)$$

The corresponding cumulative distribution function, defined for the points $x^* = x_0$, is

$$F_{x^*}(x_0) = \exp\left[-\left(1 - \frac{1}{\xi_0^2}\right)\right] - \exp\left[-\left(1 - \frac{1}{\xi_0^2}\right)e^{\xi_0^2/2}\right] \qquad (9.31)$$

The functions (9.30) and (9.31) are plotted in Fig. 9.2. When ξ_0 tends to infinity, the function $F_{x^*}(x_0)$ tends to $1/e = 0.3679$. Hence, because of the skewed shape of the EVD, one may expect to find, for large enough dimensionless most-probable values ξ_0 (say, $\xi_0 > 6$), that about 37 percent of the measured values are smaller than the most-probable value. The degree of skewness can be considered to be practically constant if the ξ_0 value exceeds 6. The skewness is smaller for lower ξ_0 values. When ξ_0 is approximately equal to 3, one may expect

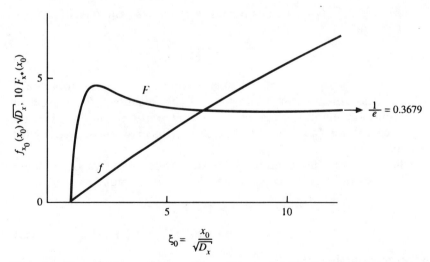

Figure 9.2 Dimensionless envelope values of the probability density $f_{x_0}(x_0)$ for the extreme value distribution (EVD) and the cumulative distribution function $Fx_0(x_0)$ determined for the points $x^* = x_0$.

to find about 42 percent of the measured absolute maxima to be smaller than the theoretically predicted most probable value.

Example 9.1 A ship in a seaway encounters a "packet" of n intensive successive waves. Assuming that the amplitude $R(t)$ of the bending moment amidships, when the ship encounters a single wave ($n = 1$), is distributed in accordance with the Rayleigh law, determine the most likely extreme value of the amplitude of the bending moment for a large n number.

Solution From the "initial" distribution

$$f(r) = \frac{r}{D_x} e^{-r^2/(2D_x)}$$

using the formulas (9.1) and (9.2), we obtain

$$g(y_n) = n\zeta_n^2 e^{-\zeta_n^2/2}(1 - e^{-\zeta_n^2/2})^{n-1} \qquad (9.32)$$

where

$$\zeta_n = \frac{y_n}{\sqrt{D_x}}$$

and D_x is the variance of the ordinates $X(t)$ of the bending moment. The condition $g'(y_n) = 0$ yields

$$\zeta_n^2(ne^{-\zeta_n^2/2} - 1) - (e^{-\zeta_n^2/2} - 1) = 0 \qquad (9.33)$$

In the case of large n, the second term in this equation is substantially smaller than the first one, and need not be accounted for. Then the most likely value y_n^* can be found as follows:

$$y_n^* = \zeta^{**}\sqrt{D_x} = \sqrt{2D_x \ln n} \qquad (9.34)$$

As evident from this formula, the most likely extreme value y_n^* of the amplitude of the bending moment is a function of the number n of oscillations (loadings) and is by a factor of $\sqrt{2 \ln n}$ larger than the most likely value $\sqrt{D_x}$ in the case of a single loading. The obtained formula indicates also that when the number n is large, its further increase has a small effect on the most likely extreme value of the bending moment.

Example 9.2 Determine the first five moments of the EVD (9.21) about the origin using the moment generation function method (see Sec. 2.3).

Solution The kth moment of the extreme value distribution (9.21) about the origin is as follows:

$$m_k = \int_0^\infty f_{x^*}(x^*) x^{*k} \, dx^*, \qquad k = 0, 1, 2, \ldots \qquad (9.35)$$

Because it is difficult to obtain the analytical expression for the m_k values directly, one can use the relationship (9.19) solved for the parameter χ

$$\chi = \frac{(x^*)^2}{2D_x} - \ln n$$

as a new variable χ of integration. Since, as one can see from (9.21),

$$f_{x*}(x^*)\, dx^* = f_\chi(\chi)\, d\chi$$

the formula (9.35) yields:

$$m_k = (2D_x \ln n)^{k/2} \int_{-\ln n}^{\infty} \left(1 + \frac{\chi}{\ln n}\right)^{k/2} f_\chi(\chi)\, d\chi \tag{9.36}$$

For sufficiently large n values (say, $n > 50$), the lower limit in this integral can be put equal to $-\infty$ (since the values of the function $f_\chi(\chi)$ become for $\chi < -\ln n$, negligibly small). Then the binomial in the parentheses in the integrand can be substituted by the series

$$\left(1 + \frac{\chi}{\ln n}\right)^{k/2} = \sum_{i=0}^{\infty} \frac{(k/2)(k/2-1)\cdots(k/2-i+1)}{i!(\ln n)^i} \chi^i$$

$$= 1 + \frac{k/2}{1!\ln n}\chi + \frac{(k/2)(k/2-1)}{2!(\ln n)^2}\chi^2$$

$$+ \frac{(k/2)(k/2-1)(k/2-2)}{3!(\ln n)^3}\chi^3 + \cdots$$

and the formula (9.36) can be written as

$$m_k = (2D_x \ln n)^{k/2} \sum_{i=0}^{\infty} \frac{(k/2)(k/2-1)\cdots(k/2-i+1)}{i!(\ln n)^i} m_i^0$$

where

$$m_i^0 = \int_{-\infty}^{\infty} \chi^i f_\chi(\chi)\, d\chi$$

are the moments of the distribution $f_\chi(\chi)$.

The moments m_i^0 can be found by using the moment generating function method (see Sec. 2.3). The moment generating function for the distribution $f_\chi(\chi)$ is

$$G_\chi(t) = \int_{-\infty}^{\infty} e^{\chi t} f_\chi(\chi)\, d\chi = \int_{-\infty}^{\infty} e^{-(1-t)\chi - e^{-\chi}}\, d\chi$$

$$= \int_0^{\infty} x^{-t} e^{-x}\, dx = \Gamma(1-t)$$

where

$$\Gamma(\alpha) = \int_0^{\infty} e^{-t} t^{\alpha-1}\, dt$$

is the gamma function.

In accordance with the moment generating function method, the ith moment m_i^0 about the origin can be obtained as the kth derivative of the moment generating function for $t = 0$. The first four derivatives of the function $\Gamma(1-t)$, calculated for $t = 0$, are

$$m_0^0 = \Gamma(1) = 1$$

$$m_1^0 = \Gamma'(1) = C = 0.5772$$

$$m_2^0 = \Gamma''(1) = C^2 + \frac{\pi^2}{6} = 1.9781$$

$$m_3^0 = \Gamma'''(1) = C^3 + C\frac{\pi^2}{6} = 5.4449$$

$$m_4^0 = \Gamma^{IV}(1) = 5C^4 + 3C^2\pi^2 + \frac{3\pi^4}{20} = 23.5615$$

where $C = 0.5772$ is Euler's constant. Then the first five moments of the EVD of the random process $X(t)$ are about the origin are

$$m_1 \simeq \sqrt{2D_x \ln n}\left[1 + \frac{m_1^0}{2 \ln n} - \frac{m_2^0}{8(\ln n)^2} + \frac{m_3^0}{16(\ln n)^3} - \frac{5m_4^0}{128(\ln n)^4}\right]$$

$$m_2 \simeq 2D_x \ln n\left(1 + \frac{m_1^0}{\ln n}\right)$$

$$m_3 \simeq (2D_x \ln n)^{3/2}\left[1 + \frac{3m_1^0}{2 \ln n} + \frac{3m_2^0}{8(\ln n)^2} - \frac{m_3^0}{16(\ln n)^3} + \frac{3m_4^0}{64(\ln n)^4}\right]$$

$$m_4 \simeq (2D_x \ln n)^2\left[1 + \frac{2m_1^0}{\ln n} + \frac{m_2^0}{(\ln n)^2}\right]$$

$$m_5 \simeq (2D_x \ln n)^{5/2}\left[1 + \frac{5m_1}{2 \ln n} + \frac{15m_2}{8(\ln n)^2} + \frac{5m_3}{16(\ln n)^3} - \frac{5m_4}{128(\ln n)^4}\right]$$

As evident from these expressions, one can use, for large enough n values, the approximate formula

$$m_k \simeq (2D_x \ln n)^{k/2}$$

to evaluate the low-order moments.

Example 9.3 Determine the mean and the variance of the EVD (9.21).

Solution The mean value of the distribution (9.21) is given by the formula for the mean value m_1 in the solution to the previous example:

$$\langle x^* \rangle = m_1 = \sqrt{2D_x \ln n}\left[1 + \frac{0.2886}{\ln n} - \frac{0.2473}{(\ln n)^2} + \frac{0.3403}{(\ln n)^3} - \frac{0.9204}{(\ln n)^4}\right]$$

The variance of the random variable X^* is

$$D_{x^*} = m_2 - m_1^2$$

$$= \frac{D_x}{2 \ln n} D_x\left\{1 - \frac{m_3^0 - m_1^0 m_2^0}{D_x(2 \ln n)} + \frac{5m_4^0 - (m_2^0)^2 - 4m_1^0 m_3^0}{D_x(4 \ln n)^2} + \frac{5m_1^0 m_4^0 + 2m_2^0 m_3^0}{D_x[4(2 \ln n)^3]}\right\}$$

where

$$D_\chi = m_2^0 - (m_1^0)^2 = 1.6449$$

is the variance of the random variable χ. Using the values m_i^0 from the solution to the previous example, we obtain

$$D_{x^*} = \frac{D_x}{2 \ln n} D_\chi \left[1 - \frac{1.3080}{\ln n} + \frac{3.8498}{(\ln n)^2} + \frac{1.7011}{(\ln n)^3} \right]$$

As evident from this formula, the variance D_{x^*} tends to zero when the n value tends to infinity.

Example 9.4 Determine the level x^{**} corresponding to the probability $P = 1/3$ of exceeding this level by the extreme value X^*. Find the relationship between the level x^{**} and the most likely value x_0 of the extreme value distribution (9.21). Determine the relationship between the most likely value x_0 and the significant value $x_{1/3}$ of the Rayleigh distribution.

Solution The "significant" level x^{**} corresponding to the probability $P = 1/3$ can be determined from the formula (9.29) as follows:

$$x^{**} = x_{1/3} = \sqrt{2D_x(\ln n + 0.9027)}$$

The most likley value x_0 of the distribution (9.21) can be found from Eq. (9.26). It is related to the significant value $x_{1/3}$ as

$$x_0 = \frac{x_{1/3}}{\sqrt{1 + 0.9027(\ln n)^{-1}}}$$

and to the significant value

$$x_{1/3}^R = \sqrt{-2D_x \ln P} = 1.4823\sqrt{D_x}$$

for the Rayleigh distribution as

$$x_0 = 0.954 x_{1/3}^R \sqrt{\ln n}$$

For large n values (say, $n > 80$) one can conservatively assume

$$x_0 = x_{1/3}^R \sqrt{\ln n}$$

Various characteristics of the EVD (9.21) are plotted in Fig. 9.3 as functions of the log n.

Example 9.5 Evaluate the sensitivity of the EVD (9.21) to the variations in the variance and the number of oscillations.

Solution The distribution (9.21) is a two-parametric one, and depends on the variance D_x and the "sample size" n. The complete differential $dF_x(x^*)$ of the cumulative probability distribution function $F_{x^*}(x^*)$ can be found as

$$dF_{x^*}(x^*) = \frac{\partial F_{x^*}}{\partial D_x} dD_x + \frac{\partial F_{x^*}}{\partial n} dn$$

$$= -\frac{x^*}{2D_x} f_{x^*}(x^*) \, dD_x - \left[\frac{D_x}{nx^*} f_{x^*}(x^*) - e^{-n} \right] dn$$

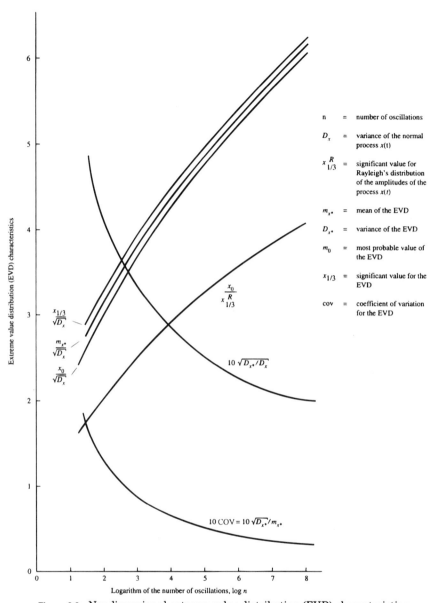

Figure 9.3 Nondimensional extreme value distribution (EVD) characteristics.

For large n values the term e^{-n} may be neglected, so that

$$dF_{x^*}(x^*) \simeq -f_{x^*}(x^*) \frac{x^*}{2D_x} \left[dD_x + \frac{2}{n} \frac{D_x^2}{(x^*)^2} dn \right]$$

For large n values (say, $n > 80$), the ratio $(x^*)^2/D_x$, if substituted by the ratio $\langle x^* \rangle^2/D_x$, is, as one can see from the solution obtained in the

Example 9.2, on the order of $2 \ln n$. Then the coefficient in front of the differential dn in the obtained formula is on the order of $D_x/(n \ln n)$. When the number n is large, this ratio is small. Therefore we conclude that the differential $dF_{x*}(x*)$ is affected mostly by the variation in the variance D_x. The variation in the number n is of much lesser influence.

Substituting the probability density function (9.21) with its value calculated for $x* = \langle x* \rangle$, putting $\langle x* \rangle^2 /D_x \approx 2 \ln n$ is the above formula for the differential $dF_{x*}(x*)$ and substituting the differentials dD_x and dn with finite differences ΔD_x and Δn, we obtain the following formula which enables one to estimate the variation in the cumulative probability function for the given variations in the variance D_x and the number n of oscillations:

$$\Delta F_{x*} \simeq -\frac{1}{e}\left(\frac{\Delta D_x}{D_x} + \ln n \frac{\Delta n}{n}\right)$$

Example 9.6 Evaluate the convolution of the EVDv (9.21) with the normally distributed random variable X.

Solution Let X be a random variable whose probability density function is $f_x(x)$ and cumulative distribution function is $F_x(x)$. Then the random variables $V = X + Z*$ and $W = X - Z*$ have the following probability density and cumulative probability distribution functions:

$$f_v(v) = \int_{-\infty}^{\infty} f_x(x) f_{z*}(v-x)\, dx$$

$$F_v(v) = \int_{-\infty}^{v} f_x(x)\, dx \int_{0}^{v-x} f_{z*}(z*)\, dz* = \int_{-\infty}^{v} f_x(x) F_{z*}(v-x)\, dx$$

$$f_w(w) = \int_{-\infty}^{\infty} f_x(x) f_{z*}(x-w)\, dx$$

$$F_w(w) = \int_{-\infty}^{w} f_x(x)\, dx \int_{0}^{x-w} f_{z*}(z*)\, dz* = \int_{-\infty}^{w} f_x(x) F_{z*}(x-w)\, dx$$

In these formulas, the actual limits of integration for the variable X are defined by the range in which the function $f_x(x)$ is positive. If the variable $Z*$ is an extreme value of the variable $Z(t)$, the argument $z*$ enters the function $F_{z*}(z*)$ as $(z*)^2$ and therefore the distributions of the sum V of the variables X and $Z*$ and of the difference W of these variables coincide. With

$$f_x(x) = \frac{1}{\sqrt{2\pi D_x}} e^{-(x-\langle x \rangle)^2/(2D_x)}, \qquad -\infty < x < \infty$$

we have

$$F_v(v) = \frac{1}{\sqrt{2\pi D_x}} \int_{-\infty}^{v} e^{-(x-\langle x \rangle)^2/(2D_x)} [\exp(-ne^{-(v-x)^2/(2D_x)}) - e^{-n}]\, dx \quad (9.37)$$

Introducing a new variable

$$\xi = \frac{v-x}{\sqrt{2D_x}}$$

we obtain

$$F_v(v) = \frac{1}{\sqrt{\pi}} \int_0^\infty e^{-(\xi-\gamma)^2}[\exp(-ne^{-\delta\xi^2}) - e^{-n}]\,d\xi \qquad (9.38)$$

where the following notation is used:

$$\gamma = \eta - \gamma_x, \qquad \eta = \frac{v}{\sqrt{2D_x}}, \qquad \gamma_x = \frac{\langle x \rangle}{\sqrt{2D_x}}, \qquad \delta = \frac{D_x}{D_z}$$

Example 9.7 Evaluate the convolution of the EVD (9.21) with the Rayleigh distribution.

Solution In the case when the random variable X follows the Rayleigh distribution

$$f_x(x) = \frac{x}{x_0^2} e^{-x^2/(2x_0^2)}, \qquad 0 < x < \infty$$

where x_0 is the most likely x value, we obtain

$$F_v(v) = \frac{1}{x_0^2} \int_0^v x e^{-x^2/(2x_0^2)}[\exp(-ne^{-(v-x)^2/(2D_x)}) - e^{-n}]\,dx$$

$$= 2 \int_0^\eta (\eta - \xi)e^{-(\eta-\xi)^2}[\exp(-ne^{-\delta\xi^2}) - e^{-n}]\,d\xi \qquad (9.39)$$

Here the following notation is used:

$$\eta = \frac{v}{x_0\sqrt{2}}$$

Example 9.8 Find the convolution of a normal law (with zero mean) and the Rice law for extreme values. Use order statistics, based on the expression (9.2), and the extreme value distribution, expressed, for large n numbers, by the formula (9.23).

Solution The cumulative probability function for the Rice distribution, expressed by Eq. (3.31), is as follows:

$$P(z) = \frac{1}{\sigma} \int_z^\infty f(\eta)\,d\eta$$

$$= \frac{1}{2}\left\{1 + \Phi\left(\frac{z}{\varepsilon\sigma\sqrt{2}}\right) - \sqrt{1-\varepsilon^2}\left[1 + \Phi\left(\frac{z\sqrt{1-\varepsilon^2}}{\varepsilon\sigma\sqrt{2}}\right)\right]e^{-z^2/(2\sigma^2)}\right\}, \qquad 0 \le \varepsilon \le 1$$

where σ is the standard deviation of the process. The cumulative probability distribution function for the extreme response can be found, in accordance with the formula (9.2), as follows:

$$F_n(z) = P^n(z).$$

TABLE 9.1 Probability Distributions Obtained on the Basis of the EVD (Exact Values) and Extreme Value Statistics

$n = 50$

Dimensionless ordinate $z^*/\sqrt{D_z}$		2.00	2.50	2.75	3.00	3.50	4.00	5.00
Dimensionless probability density function $f_{z^*}(z^*)\sqrt{D_z}$	Exact value	0.0155826	0.6104653	1.0026645	0.9561751	0.3431495	0.0659642	0.9314889×10^{-3}
	Extreme value statistics	0.0096028	0.5924733	1.0093261	0.9722751	0.3499594	0.0672837	0.9501247×10^{-3}
Cumulative probability distribution function $F_{z^*}(z^*)$	Exact value	0.00115141	0.1111531	0.3199121	0.5738141	0.8963947	0.9833669	0.9998137
	Extreme value statistics	0.00069964	0.1057619	0.3157230	0.5720352	0.8962958	0.9833750	0.9998250

$n = 1000$

$z^*/\sqrt{D_z}$		3.00	3.50	3.75	4.00	4.25	4.50	5.00	6.00
$f_{z^*}(z^*)\sqrt{D_z}$	Exact value	0.499136×10^{-3}	0.8599112	1.3694844	0.9594312	0.4510415	0.1732131	0.0185639	0.9137801×10^{-4}
	Extreme value statistics	0.469551×10^{-3}	0.8579726	1.3704789	0.9604508	0.4515435	0.1734072	0.0185849	0.9147078×10^{-4}
$F_{z^*}(z^*)$	Exact value	0.149769×10^{-4}	0.1121979	0.4131988	0.7150074	0.8872640	0.9607269	0.9962805	0.9999848
	Extreme value statistics	0.140751×10^{-4}	0.1119503	0.4130858	0.7150522	0.8873642	0.9608857	0.9964066	~1

The probability density distribution function for a random variable Y, which obeys the normal distribution with zero mean, is

$$f_y(y) = \frac{1}{\sqrt{2\pi}\, s} e^{-y^2/(2s^2)}$$

Here s is the standard deviation of the process $Y(t)$. Then the cumulative probability distribution function of the random variable $X = Y + Z$ is as follows:

$$F_n(x) = \frac{1}{\sqrt{\pi}\, v} \int_0^\infty P_n(w\sigma\sqrt{2}) \exp\left[-\left(\frac{x}{s\sqrt{2}} - \frac{w}{v}\right)^2\right] dw$$

where $w = (x - y)/(\sigma\sqrt{2})$ is the variable of integration, and $v = s/\sigma$ is the ratio of the standard deviations of the normal and the Rice laws.

If the EVD (9.21) is used, then the cumulative distribution function for the extremes of the variable z is

$$F_n(z) = e^{-n(1-P_n)}$$

and the convolution $X = Y + Z$ can be obtained as follows:

$$F_n(x) = \frac{1}{\sqrt{\pi}\, v} \int_0^\infty \exp\left\{-n\left[1 - P_n(w\sigma\sqrt{2})\right] - \left(\frac{x}{s\sqrt{2}} - \frac{w}{v}\right)^2\right\} dw$$

Example 9.9 Compare the numerical results obtained on the basis of the exact solution for the probability density and the cumulative distribution functions

$$f_{x^*}(x^*) = \frac{x^*}{D_x} \exp\left[-\frac{(x^*)^2}{2D_x} - ne^{-x^{*2}/(2D_x)}\right]$$

$$F_{x^*}(x^*) = \exp\left[-ne^{-(x^*)^2/(2D_x)}\right] - e^{-n}$$

based on EVD with the results obtained on the basis of the extreme value statistics

$$f_{x^*}(x^*) = \frac{x^*}{D_x}(n+1)e^{-(x^*)^2/(2D_x)}\left[1 - e^{-(x^*)^2/(2D_x)}\right]^n$$

$$F_{x^*}(x^*) = \left[1 - e^{-(x^*)^2/(2D_x)}\right]^{n+1}$$

for $n = 50$ and $n = 1000$.

Solution The calculated extreme value distribution (EVD) probability density functions and the cumulative probability distribution functions obtained on the basis of the exact (double exponential) EVD and the extreme value statistics predicted are shown in Table 9.1 for $n = 50$ and $n = 1000$.

The calculated data show that the predictions based on the extreme value statistics underestimate the values of the functions $f_{x^*}(x^*)$ and $F_{x^*}(x^*)$ for small x^* values and overestimate them for large x^* values. The agreement between the predictions based on the extreme value statistics and the double-exponential EVD are the better, the larger is the number n of observations (oscillations).

9.4. Hermite Polynomials

The integrals representing cumulative probability distribution functions of the convolutions of the extreme value distributions cannot be found in a closed form. This statement can be illustrated by the integrals obtained in the solutions to the problems in Examples 9.5 and 9.6 of the previous section. Obtaining a closed-form analytical solution turns out to be impossible, even for the simplest case of a convolution of an extreme value distribution with a uniform distribution. At the same time, one often encounters considerable difficulties when trying to apply the conventional rules of numerical integration to obtain numerical results. These difficulties are due primarily to the fact that the integrands are high and narrow peak-like functions. On one hand, there is plenty of "unnecessary" additional computational work to be done to determine the actual limits of integration. On the other hand, it is hard to achieve satisfactory accuracy because of the necessity of choosing small increments for numerical integration for the established limits of integration. These difficulties can be successfully overcome by using Hermite polynomials.

Let us show, for instance, how Hermite polynomials can be applied to evaluate the integral in the solution to Example 9.5 of the previous section:

$$F_v(v) = \frac{1}{\sqrt{\pi}} \int_0^\infty e^{-(\xi-\gamma)^2}[\exp(-ne^{-\delta\xi^2}) - e^{-n}]\, d\xi$$

Substituting the variable ξ in this formula with the variable $-\xi$, we obtain

$$F_v(v) = \frac{1}{\sqrt{\pi}} \int_{-\infty}^0 e^{-(\xi+\gamma)^2}[\exp(-ne^{-\delta\xi^2}) - e^{-n}]\, d\xi$$

Summing up the two equations, we have:

$$F_v(v) = \int_{-\infty}^\infty e^{-\xi^2} f(\xi)\, d\xi \qquad (9.40)$$

where the function $f(\xi)$ is

$$f(\xi) = \frac{1}{2\sqrt{\pi}} e^{-\gamma^2}(e^{2\gamma\xi} + e^{-2\gamma\xi})[\exp(-ne^{-\delta\xi^2}) - e^{-n}] \qquad (9.41)$$

In accordance with the quadrature rule of numerical integration, using algebraic polynomials, the integral (9.40) can be computed by the formula

$$\int_{-\infty}^{\infty} e^{-\xi^2} f(\xi)\, d\xi = \sum_{k=1}^{m} A_k f(\xi_k) + R_m \qquad (9.42)$$

Choosing the Hermite polynomials

$$H_n(\xi) = (-1)^n e^{\xi^2} \frac{d^n}{d\xi^n}(e^{-\xi^2})$$

as suitable orthogonal polynomials, one can find that the knots ξ_k of the quadrature rule (9.42) can be determined as the roots of the equation

$$H_m(\xi_k) = 0 \qquad (9.43)$$

The first seven Hermite polynomials are

$$H_0(\xi) = 1$$
$$H_1(\xi) = \xi$$
$$H_2(\xi) = \xi^2 - 1$$
$$H_3(\xi) = \xi^3 - 3\xi$$
$$H_4(\xi) = \xi^4 - 6\xi^2 + 3$$
$$H_5(\xi) = \xi^5 - 10\xi^3 + 15\xi$$
$$H_6(\xi) = \xi^6 - 15\xi^4 - 45\xi^2 - 15$$

The tabulated values of Hermite polynomials are given in Appendix F.

The Hermite polynomials possess the following properties:

1. $$\frac{dH_n(\xi)}{d\xi} = nH_{n-1}(\xi)$$

2. $$\int_{-\infty}^{\infty} H_m(\xi) H_n(\xi) \alpha(\xi)\, d\xi = \begin{cases} n!, & \text{if } m = n \\ 0, & \text{if } m \neq n \end{cases}$$

    where

    $$\alpha(\xi) = \frac{1}{\sqrt{2\pi}} e^{-\xi^2/2}$$

is the standardized normal density function. Hence, Hermite polynomials are orthogonal with respect to the standardized normal probability density function. It is this property that make their use of convenience.

3. Using Hermite polynomials, the probability density function $f(\xi)$ of a standardized random variable (i.e. a variable with zero mean and unit variance) can be written as

$$f(\xi) = \alpha(\xi)\left[1 + \frac{m_3}{3!}H_3(\xi) + \frac{m_4 - 3}{4!}H_4(\xi)\right.$$
$$\left. + \frac{m_5 - 10m_3}{5!}H_5(\xi) + \frac{m_6 - 15m_4 + 30}{6!}H_6(\xi) + \cdots\right]$$

where m_i, $i = 0, 1, 2, \ldots$, are the moments of the standardized random variable in question. This expansion is known as the *Gram-Charlier series*. Using the Hermite polynomials, the coefficients A_k in the expansion (9.42) and the remainder R_m can be computed as

$$A_k = 2^{n+1}m!\sqrt{\pi}\,[H'_m(\xi_k)]^{-2}, \qquad k = 1, 2, \ldots, m \qquad (9.44)$$

$$R_n = \frac{m\sqrt{\pi}}{2^m(2m)!}\,f^{(2m)}(\xi) \qquad (9.45)$$

where m is the power of the polynomial. The values of ξ_k and A_k, calculated from Eqs (9.43) and (9.44), are given in Appendix G for different m values.

9.5. Application of Extreme Value Distributions for Reducing Casualties at Sea

The objective of the examples that follow is to illustrate how the application of the extreme value distributions can be used to establish guidelines for avoiding accidents and casualties at sea. Similar problems can be encountered in many other areas of engineering, such as fatigue strength, earthquakes, some problems in avionic engineering, various military situations, etc. Three practical problems are examined in this section:

1. The required underkeel clearance (UKC) for a ship passing a shallow waterway when entering a harbor
2. The required strength of a helicopter undercarriage while the helicopter lands on a ship's deck in rough sea conditions, and
3. The probability of occurrence of a ship slamming in heavy seas

In all of these cases the random process of the wave-induced ship motion, as well as the finite duration of the operations (a finite number of wave encounters or a finite number of oscillations), are considered.

9.5.1. Ship passing a shallow waterway

Choosing adequate underkeel clearance (UKC) is one of the most crucial and most difficult among the problems that occur in connection with the navigation of large ships, especially very large crude oil carriers. Ship operators, naval architects, hydraulic engineers, economists and managers encounter this problem when establishing whether or not the given vessel will safely pass a shallow waterway (such as, say, a harbor approach or a channel), when alloting the maximum loaded draft for a ship under design, when defining the proper depths of port approaches, etc. Since the majority of the comprehensive factors affecting UKC are of a probabilistic nature, the problem of choosing UKC guidelines should be treated from the probabilistic point of view.

Let the maximum ship draft be $T = T(t_1, t_2, \ldots, t_n)$, where t_i, $i = 1, 2, \ldots, n$, are random variables, affecting this draft (static draft error, water density uncertainty, wave-induced motion, maneuvering, ship hull flexibility, etc.), and the depth of water be $H = H(h_1, h_2, \ldots, h_k)$, where h_j, $j = 1, 2, \ldots, k$, are random variables due to uncertainties in predicting the water depth (uncertainties in the charted depth, relief of the sea bed, tide, siltation, etc.). Then the probability of safe passing a shallow waterway can be expressed as

$$P_s = P(H > T) = P(M = H - T > 0)$$

where M, representing the UKC, is, in effect, a margin of safety. The majority of the factors affecting the ship's draft and water depth do not usually change during the relatively short time of a single passing, and can therefore be considered "static" variables, while some of them (such as wave-induced ship motion and, quite often, also the relief of the sea bed) might vary substantially during this time and should be treated as "dynamic" variables. For manmade fairways having more or less uniform bottoms (i.e., port approach channels), the time-dependent uncertainties in sea bed relief may not be particularly distinguished, while for natural shallows they should apparently be accounted for.

In the following analysis, for the sake of simplicity, the random component due to the changes in the sea bed relief is not considered and only the wave-induced ship motion is taken into account. Thus, the margin M can be represented as a sum of the static and the dynamic components:

$$M = M_s + M_D(t)$$

where the time-independent random variable M_s is responsible for the ship safe passing a shallow waterway in the still water condition and the time-dependent random variable $M_D(t)$ reflects the additional safety margin due to the wave-induced ship motion. Both the static and the dynamic ship responses can be described by the normal law of the probability distributions:

$$f_s(m_s) = \frac{1}{\sqrt{2\pi D_s}} \, e^{-(m_s - \langle m_s \rangle)^2/(2D_s)}$$

$$f_D(m_D) = \frac{1}{\sqrt{2\pi D_D}} \, e^{-m_D^2/(2D_D)}$$

(9.46)

Here D_s and D_D are the variances of the random variables M_s and M_D and $\langle m_s \rangle$ is the mean value of the variable M_s. The mean of the dynamic variable M_D is assumed to be zero. The significant difference in the rate of changing of the random variables M_s and M_D in the time domain leads to the fact that the ordinates M_s of the "slow" random process $M_s(t)$ can be considered unchangeable, compared to the ordinates M_D of the "swift" process $M_D(t)$ which can make hundreds of oscillations, while the process $M_s(t)$ does not practically change its value. Therefore the problem of adding up the steady-state and the dynamic components of the safety margin reduces, in effect, to synthesizing the laws of the probability distributions for the random variable M_s and the maximum values of the random process $M_D(t)$.

The probabilistic warrant that no bottom touching occurs within the time the ship transients the given shallow waterway can be obtained by meeting at least one of the following requirements:

1. The probability of exceeding the still water UKC by the extreme vertical motion of the ship bottom needs to be small.
2. The probability of exceeding the interval Δt between two adjacent exceedings of the still water UKC by the time that the ship passes the waterway needs to be small.

For equal initial premises, both requirements should, generally speaking, lead to similar results.

The mathematical formulation of the first requirement can be based on the following extreme value distribution:

$$P_s = \exp\left[-n e^{-(m^*)^2/(2D_z)}\right] - e^{-n}$$

(9.47)

where m^* is the level of the steady-state (still water condition) margin of safety M_s, D_z is the variance of the process $Z(t)$ of the wave-induced vertical motions of the ship bottom, and n is the number of oscillations during the time Δt required to pass the waterway. For sufficiently large n values, the second term in the above formula is small and can be omitted. The formula (9.47) determines the probability of the event "the still water UKC will not be exceeded during the time Δt because of the ship's wave-induced motions." As for the variance D_z of these motions, it can be evaluated for the given sea condition either experimentally or theoretically. When the theoretical approach is applied, the variance D_z can be computed, using the Chapter 8 results, by the formula:

$$D_z = \int_0^\infty S_\zeta(\omega, h_{1/3}) a_z^2(\omega) \, d\omega \qquad (9.48)$$

Here $S_\zeta(\omega, h_{1/3})$ is the energy spectrum of the sea waves, ω is the wave frequency, $h_{1/3}$ is the significant wave height, and $a_z(\omega)$ is the response function for the ship's vertical motion. This function can be obtained by model tests in a towing tank or can be determined on the basis of the equations of the ship motion theory. The n value in the formula (9.47) can be computed as the ratio of the time Δt required for the ship to pass the dangerous area of the waterway to the effective period τ_e of the ship's motion. In this connection it should be pointed out that ocean-going ships sailing in irregular seas behave, like many other engineering systems, as narrow-band filters, selectively magnifying the "inputs" with frequencies close to those of the ship's natural motion (i.e., her oscillation frequencies in still water) and suppressing all other "inputs". For this reason one may assume that the effective period τ_e of oscillations is the natural period of ship motion in heave and pitch (these are typically quite close).

The second requirement (i.e., when the time, required to pass the dangerous area, is used as an appropriate safety criterion), can be written as follows:

$$P_x = P_0(\Delta t) = \exp\left[-n e^{-m_s^2/(2D_x)}\right] \qquad (9.49)$$

where m_s is the given "level" of the time duration.

The most probable extreme value of the ship's vertical motion can be calculated, for a sufficiently large n number, by the formula (9.34):

$$z_0 = \sqrt{2 D_z \ln n} \qquad (9.50)$$

By using this formula, the probability P_s of safe passing a shallow waterway can be written as

$$P_s = \exp\left[-e^{-(m_s^2 - z_0^2)/(2D_z)}\right] = \exp\left(-n^{1 - m_s^2/z_0^2}\right) \qquad (9.51)$$

When the mean value m_s of the still water UKC is equal to the most probable value z_0 of the ship's vertical displacement during her longitudinal motions, the probability P_s of safe passing is only $P_s = e^{-1} = 0.368$. This probability is definitely too low. It rapidly increases, however, with an increase in the m_s/z_0 ratio. When the m_s/z_0 ratio is equal to 1.5, then, in the case of $n = 1000$, the probability of safe passing becomes as high as $P_s = 0.9998$.

As has been shown in Sec. 9.2, the EVD is mostly affected by the $m_s/\sqrt{D_z}$ ratio. The variation in the number n of oscillations is of substantially lesser influence, especially when this number is large. This means that the length of the waterway and the time required to pass it (these, for the given ship speed, are proportional to the n value) have a lesser effect on the probability P_s of safe passing than the UKC and the wave conditions. For the given (accepted) probability P_s of safe passing, the required mean value m_s of the UKC can be calculated as

$$m_s = \sqrt{2D_z[\ln n - \ln(-\ln P_s)]} \qquad (9.52)$$

As evident from this formula, the required UKC in rough seas is affected primarily by the variance D_z of vertical ship motions. The plots in Fig. 9.4 illustrate the effect of the number of n of oscillations and the level of the assumed probability P_s on the $m_s/\sqrt{D_z}$ ratio. These plots were obtained assuming that m_s is a deterministic value.

In the conditions, when the steady-state margin of safety M_s is a normally distributed random variable and the wave-induced vertical

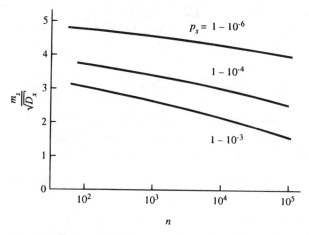

Figure 9.4 Factor of safety vs. number of oscillations for a ship passing a shallow waterway of deterministic depth in rough seas.

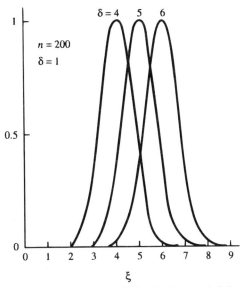

Figure 9.5 The integrands for the integral defining the probability of safe passing a shallow waterway.

ship motions, responsible for the dynamic factor of safety, follow the extreme value distribution, the probability P_s of safe passing a shallow waterway can be expressed by the formula (9.38)]:

$$P_s = \frac{1}{\sqrt{\pi}} \int_0^\infty e^{-(\xi-\gamma)^2}[\exp(-ne^{-\delta\xi^2}) - e^{-n}]\, d\xi \qquad (9.53)$$

where

$$\gamma = \frac{u - \langle m_s \rangle}{\sqrt{2D_s}} = \frac{u}{\sqrt{2D_s}} - \gamma_s$$

is the total factor of safety and

$$\delta = \frac{D_s}{D_z}$$

is the ratio of the variance D_s of the still water margin of safety to the variance D_z of the dynamic safety margin.

The formula (9.53) is based on the double-exponential extreme value distribution. If the extreme value statistics are used, then the probability P_s would be expressed as follows:

$$P_s = \frac{1}{\sqrt{2\pi D_s}} \int_{-\infty}^{u} e^{-(m_s - \langle m_s \rangle)^2/(2D_s)} [1 - e^{-(u-m_s)^2/(2D_z)}]^{n+1} \, dm_s$$

$$= \frac{1}{\sqrt{\pi}} \int_{0}^{\infty} e^{-(\xi - \gamma)^2} (1 - e^{-\delta \xi^2})^{n+1} \, d\xi \tag{9.54}$$

The integrals in the formulas (9.53) and (9.54) cannot be found in a closed form and should be evaluated numerically. The integrands of the integral (9.53) calculated for $\gamma = 4$, 5, and 6 are plotted in Fig. 9.5 for the case $n = 200$ and $\delta = 1$.

The integral (9.53) was solved many times, using Hermite polynomials, for different input data. The calculated probabilities of safe passing a shallow waterway are shown, as an illustration, for the case $n = 200$ and $\delta = 1$, in Fig. 9.6. The calculated data indicate that if the safety factor is increased above the value of, say, $\gamma_s = 2.630$ (which corresponds to the probability $P_s = 0.9999$), then the probability of safe passing of a shallow waterway will not increase appreciably. Therefore the level $\gamma_s = 2.630$ can be chosen as an allowable ("standardized") value for still water conditions. The ratios $v = \sqrt{\delta} = \sqrt{D_s/D_z}$ of the standard deviations of the still water margin of safety and the dynamic safety margin were calculated for the "standardized" probability $P_t = 1 - P_s = 1 - 0.9999 = 10^{-4}$ of touch-

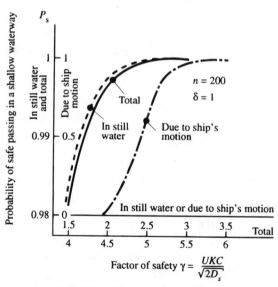

Figure 9.6 Probability of safe passing a shallow waterway vs. the factor of safety.

TABLE 9.2 The ratio of the standard deviation of the still water margin of safety to the dynamic safety margin

u	\multicolumn{5}{c}{n}				
	200	500	1000	10 000	20 000
3.00	6.589	7.070	7.420	8.470	8.750
3.25	3.962	4.245	4.450	5.070	5.270
3.50	2.847	3.045	3.188	3.625	3.750
4.00	1.840	1.960	2.050	2.316	2.390
5.00	1.104	1.166	1.212	1.358	1.400
6.00	0.805 3	0.844 5	0.874 2	0.970 2	0.998 3
8.00	0.538 1	0.559 8	0.576 2	0.629 7	0.645 6
12.00	0.334 05	0.345 65	0.354 28	0.382 16	0.390 35
20.00	0.193 87	0.200 14	0.204 78	0.219 57	0.223 87
30.00	0.127 97	0.132 02	0.135 00	0.144 51	0.147 26
50.00	0.076 395	0.078 785	0.080 548	0.086 15	0.087 765
100.00	0.038 118	0.039 305	0.040 179	0.042 959	0.043 762
200.00	0.019 049	0.019 642	0.020 078	0.021 465	0.021 865
500.00	0.007 647 4	0.007 855 5	0.008 029 9	0.008 584 2	0.008 744 4
1000.00	0.003 809 1	0.003 932 5	0.004 014 9		0.004 372 0

ing the bottom. The obtained data are shown in Table 9.2. The available results of full-scale observations indicate that this level is in satisfactory agreement with the existing maritime practice.

Since the consequences of touching the see bottom in waves are somewhat less dangerous than ship grounding in still water, the allowable cumulative probability of safe passing a shallow waterway in wave conditions can be chosen lower, than in still water conditions, say, $P_s = 0.999$. For the situation considered in Table 9.1, this value corresponds to the total factor of safety of $\gamma_t = 4.800$. Hence, the additional factor of safety, which should be applied due to the ship motion, is, in the case in question, $\gamma_D = \gamma_t - \gamma_s = 4.800 - 2.630 = 2.170$. In the hypothetical case of a ship passing, in the wave conditions, a shallow waterway with a deterministic depth, such a safety factor would result in the probability of safe passing as low as $P_s = 0.170$. This is due to the fact that, for a given value of the safety factor, wave conditions are characterized by substantially lower P_s values than still water conditions.

In the performed analyses we assumed that the wave spectrum is a narrow-band one, and therefore the "basic" distribution of the ship's motions is a Rayleigh distribution. If the actual wave conditions are such that this assumption cannot be applied (this might be the case when sailing in extremely shallow seas), then a more general Rice law should be used (see Sec. 3.3.6). In such a case the vertical displacements $X(t)$ of the ship hull should be evaluated as the convolution of the still water draft Y, which follows the normal distribution, and vertical wave-induced motions $Z(t)$, distributed in accordance with the Rice law. Using the results obtained in Example 9.7 of Sec. 9.3, we find that the probability of touching the bottom can be computed as

$$P_t = 1 - \frac{1}{\sqrt{\pi}v} \int_0^\infty P_n(w\sigma\sqrt{2})e^{-(u-w/v)^2}\,dw \qquad (9.55)$$

if the order statistics are used, and as

$$P_t = 1 - \frac{1}{\sqrt{\pi}v} \int_0^\infty \exp\left\{-n\left[1 - P_n(w\sigma\sqrt{2}) - \left(u - \frac{w}{v}\right)^2\right]\right\} dw \qquad (9.56)$$

if the double-exponential EVD is applied. In these formulas, $v = \sqrt{D_s/D_z} = s/\sigma$ is the ratio of the standard deviation of the UKC in the still water conditions to the standard deviation of vertical ship motions in waves, $u = \text{UKC}/(s\sqrt{2})$ is the dimensionless UKC, n is the number of ship oscillations (wave encounters) while passing the waterway, and w is the variable of integration.

9.5.2. Helicopter undercarriage strength when landing on a ship deck

The strength of the helicopter undercarriage when landing on the ground is determined by the landing velocity. This velocity may be considered as a random variable, following the normal law of probability distribution:

$$f_v(v) = \frac{1}{\sqrt{2\pi D_v}}\, e^{-(v-\langle v\rangle)^2/(2D_v)} \qquad (9.57)$$

where $\langle v \rangle$ is the mean and D_v is the variance of the random landing velocity V. The cumulative probability distribution function of this velocity is

$$F_v(v) = \frac{1}{2}\left[1 + \Phi\left(\frac{v - \langle v\rangle}{\sqrt{2D_v}}\right)\right] \qquad (9.58)$$

where $\Phi(\alpha)$ is the Laplace function. The allowable level v^* of the landing velocity can be obtained from Eq. (9.58) if the probability

$$P_s(v^*) = \frac{1}{2}\left[1 + \Phi\left(\frac{v^* - \langle v\rangle}{\sqrt{2D_v}}\right)\right] \qquad (9.59)$$

is considered as an acceptable (allowable) probability of safe landing. When a helicopter is landing on a ship deck, the effect of the ship's vertical motions in waves on the velocity of the encounter of the helicopter with the deck should be considered.

Although irregular waves seem at first sight to be absolutely chaotic, there is a certain "regularity" in the irregular wave trains. After some time of lull, say, a minute or two, the waves grow in size; then, after the highest wave has passed, the wave heights begin to reduce and a new, more or less brief, calm ensues (Fig. 9.7). The highest safety, in terms of the helicopter undercarriage strength, when landing on a ship deck, can be expected if such landing occurs within the calm period of the seas. Typically, ship operators, using the data of the on-board surveillance systems, signal to the helicopter pilot when the lull is expected to start, or when the "wave window" is ending. The main thing is, however, to be able to forsee the duration of the incoming lull. If the time Δt^* needed to land the helicopter on the ship deck is less than the time interval Δt between two adjacent exceedings of the certain ship vertical velocity level, viewed as a dangerous one, then safe landing becomes possible.

The cumulative probability distribution function for the extreme vertical ship velocity due to her wave-induced motion can be expressed by the formula

$$F_{\dot{z}*}(\dot{z}*) = \exp\left[-n^* e^{-(\dot{z}*)^2/(2D_{\dot{z}})}\right] - e^{-n^*} \qquad (9.60)$$

Here $\dot{z}^*$ is the extreme value of the vertical velocity $\dot{z}$, $D_{\dot{z}}$ is its variance within the time of the sea lull, $n^* = \Delta t^*/\tau_e$ is the number of ship oscillations during the time Δt^* of landing, and τ_e is the effective period of the ship's motion. The cumulative distribution function for the relative velocity of landing

$$V_r = V + \dot{Z}^* \qquad (9.61)$$

can be evaluated by the formula

$$P_v(v_r) = F_{v_r}(v_r)$$

$$= \int_{-\infty}^{v_r} f_v(v) F_{\dot{z}*}(v_r - v)\, dv$$

$$= \frac{1}{\sqrt{2\pi D_v}} \int_{-\infty}^{v_r} e^{-(v_r - \langle v \rangle)^2/(2D_v)} \left[\exp\left(-n^* e^{-(v_r - v)^2/(2D_{\dot{z}})}\right) - e^{-n^*}\right] dv$$

$$= \frac{1}{\sqrt{\pi}} \int_0^\infty e^{-(\xi - \gamma)^2} \left[\exp\left(-n^* e^{-\delta \xi^2}\right) - e^{-n^*}\right] d\xi \qquad (9.62)$$

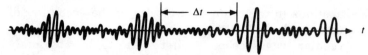

Figure 9.7 Typical sea elevation in time domain.

where

$$\gamma = \frac{v_r - \langle v \rangle}{\sqrt{2D_v}} = \gamma_t - \gamma_v \tag{9.63}$$

is the safety factor due to ship motion,

$$\gamma_v = \frac{\langle v \rangle}{\sqrt{2D_v}} \tag{9.64}$$

is the safety factor when landing on the ground,

$$\gamma_t = \frac{v_r}{\sqrt{2D_v}} \tag{9.65}$$

is the total safety factor, accounting for the relative helicopter-ship motion, and

$$\delta = \frac{D_r}{D_{\dot{z}}} \tag{9.66}$$

is the ratio of the variance of the relative velocity v_r of the encounter of the helicopter with the ship's deck to the variance of the ship's vertical velocity $\dot{z}$.

The following input data should be available when the calculations of the conditions of safety landing are carried out:

(a) mean value $\langle v \rangle$ of the ground landing velocity,
(b) variance D_r of this velocity,
(c) time Δt^* required for landing on the ship deck,
(d) effective period τ_e of ship motion (usually the ship's natural period in heave and pitch),
(e) variance $D_{\dot{z}}$ of the ship's vertical velocities, due to her motions during lull periods, and
(f) "standardized" (required) probabilities of safe landing on the ground (P_G) and on the ship deck (P_D),

The calculations can be carried out in the following sequence:

1. Define the required safety factor for landing on the ground

$$\bar{\gamma}_t = \frac{v^*}{\sqrt{2D_v}} \tag{9.67}$$

the safety factor $\gamma_v = \langle v \rangle / \sqrt{2D_v}$ due to ship motion and the probability $P_G = P_s(v^*)$ of safe landing on the ground. The letter can be computed by the formula (9.59).

2. Define the actual factor of safety

$$\gamma_t = \frac{v_r}{\sqrt{2D_v}} \tag{9.68}$$

for the given γ_v, $\delta = D_v/D_{\dot{z}}$, $n^* = \Delta t^*/\tau_e$ and $P_D = P_v(v_r)$ values, using the calculated values of the integral (9.62).

If the calculated total safety factor γ_t is not smaller than its required $\bar{\gamma}_t$ value, then helicopter landing may be permitted and supposedly will be safe. Clearly, it is assumed that reliable information about the beginning of a new lull period is available and can be continuously transmitted to the helicopter pilot.

If the allowable relative velocity value v^* of the helicopter landing and the actual ground landing velocity v_0, treated as a deterministic value, are given, then the calculation procedure becomes very simple. Indeed, in such a situation the equation (9.60) yields

$$\begin{aligned} \dot{z}^* &= \sqrt{2D_{\dot{z}}[\ln n^* - \ln(-\ln P_D + e^{-n^*})]} \\ &\cong \sqrt{2D_{\dot{z}}[\ln n^* - \ln(-\ln P_D)]} \end{aligned} \tag{9.69}$$

and the condition of safe landing becomes

$$\dot{z}^* \leq v^* - v_0 \tag{9.70}$$

If, for example, the number of ship oscillations during the time Δt^* of landing is $n^* = 30$, the required ("standardized") probability of safe landing of the helicopter on the ship's deck is $P_D = 0.9999$, the variance of the ship's vertical velocities due to her motions during lull periods is $D_{\dot{z}} = 0.030 \ m^2/s$, and the extreme value of the relative vertical velocity, calculated as the difference between the allowable vertical velocity v^* of the helicopter and the actual ground landing velocity v_0, is $v^* - v_0 = 1$ m/s, then the allowable level of the relative velocity at the moment of landing is

$$\dot{z}^* = \sqrt{2 \times 0.030[\ln 30 - \ln(-\ln 0.9999)]} = 0.870 \text{ m/s} < 1 \text{ m/s}$$

Hence, the helicopter landing can be permitted and should be expected to be safe.

9.5.3. Probability of occurrence of ship slamming

The occurrence of ship slamming in heavy irregular waves is characterized by the emergence of the ship's forebody from the sea during her resonance motion in heave and pitch, and a sufficiently high level of the relative velocity between the ship's bottom and the water surface at the time of impact. From a probabilistic point of view, ship slamming in rough seas is a random phenomenon, and the time intervals between the hydrodynamic impacts are random variables.

The relative vertical displacement $z(t)$ and the relative vertical velocity $\dot{z}(t)$ of the given ship's frame with an abscissa x, counted from the ship's mid-cross-section, is as follows:

$$z = -(\zeta - x\psi - \zeta_w), \qquad \dot{z} = -(\dot{\zeta} - x\dot{\psi} + n\psi - \dot{\zeta}_w)$$

where ζ and $\dot{\zeta}$ are the ship's absolute displacement and the absolute velocity in heave, ψ and $\dot{\psi}$ are the ship's absolute displacement and the velocity in pitch, v is the ship's forward speed, and ζ_w and $\dot{\zeta}_w$ are the vertical displacement and vertical velocity of the waves. If the parameters ζ and ψ of the ship's longitudinal motion for the given wave conditions are known (calculated or measured), then one can determine the variances of the random processes $Z(t)$ and $\dot{Z}(t)$:

$$D_z = \int_0^\infty S_w(\omega) a_z^2(\omega) \, d\omega, \qquad D_{\dot{z}} = \int_0^\infty S_w(\omega) a_{\dot{z}}^2(\omega) \, d\omega \qquad (9.71)$$

where $S_w(\omega)$ is the sea spectrum, and $a_z(\omega)$ and $a_{\dot{z}}(\omega)$ are the response functions for the relative displacement and the relative velocity, respectively.

A realization of the random process $Z(t)$ for a typical ship frame at her forebody can be written as

$$z(t) = z_a(t) \cos\left[\omega_e t + \varepsilon_z(t)\right] \qquad (9.72)$$

where $z_a(t)$ is the envelope of the process $Z(t)$, $\varepsilon_z(t)$ is the initial phase angle (which is also a function of time), and

$$\omega_e = \sqrt{\frac{D_{\dot{z}}}{D_z}} \qquad (9.73)$$

is the effective frequency of the random processes $Z(t)$ and $\dot{Z}(t)$. Then, by differentiation, we find the realization of the relative velocity as follows:

$$\dot{z}(t) \approx -\omega_e z_a(t) \sin[\omega_e t + \varepsilon_z(t)] \tag{9.74}$$

From Eqs (9.72) and (9.74) we have

$$z_a^2(t) = z^2(t) + \frac{\dot{z}^2(t)}{\omega_e^2} \tag{9.75}$$

It has been observed that slamming does not occur if the parameters of the ship motion are such that her relative velocity does not exceed a certain threshold value $\dot{z}^*$. It has also been found that in an approximate analysis the threshold level of the relative velocity can be determined as a function of the ship's length only (say, $\dot{z}^* \approx 0.1\sqrt{gL}$). If the ship's draft D_s at the stem in still water conditions is known, then the threshold value of the amplitude of the relative motion can be found from Eq. (9.75), by putting $z = D_s$ and $\dot{z} = \dot{z}^*$, as follows:

$$[(z_a)^*]^2 = D_s^2 + \frac{(\dot{z}^*)^2}{\omega_e^2} \tag{9.76}$$

Then, assuming that the spectrum of the random process $Z(t)$ is narrow and using the Rayleigh distribution for the amplitudes of this process, we find that the probability that the level $(z_a)^*$ is exceeded can be calculated as

$$P\{Z > (Z_a)^*\} = e^{-[(z_a)^*]^2/(2D_z)} \tag{9.77}$$

Introducing the relationship (9.76) into (9.77), we find that the probability of occurrence of slamming during the time equal to the effective period

$$\tau_e = \frac{2\pi}{\omega_e} = 2\pi\sqrt{\frac{D_z}{D_{\dot{z}}}}$$

of ship motions is

$$P = \exp\left[-\frac{1}{2D_z}\left(D_s^2 + \frac{(\dot{z}^*)^2}{\omega_e^2}\right)\right] = \exp\left[-\left(\frac{D_s^2}{2D_z} + \frac{(\dot{z}^*)^2}{2D_{\dot{z}}}\right)\right] \tag{9.78}$$

If the ship is operated in slamming conditions during the time T, one can assume that the number n of effective periods is $n = T/\tau_e$. Then, using the formula (9.2), we conclude that the probability that slamming occurs during the time T of sailing in heavy seas can be assessed as

$$P_{SL} = 1 - \exp\left[-n\left(\frac{D_s^2}{2D_z} + \frac{\dot{z}^{*2}}{2D_{\dot{z}}}\right)\right] \qquad (9.79)$$

If the formula (9.22) is used and the n value is large, then the probability of slamming can be found as

$$P_{SL} = 1 - \exp\left[-n\exp\left(\frac{D_s^2}{2D_z} + \frac{\dot{z}^{*2}}{2D_{\dot{z}}}\right)\right] \qquad (9.80)$$

For large enough n both formulas result in practically the same P_{SL} values. The formulas (9.79) and (9.80) indicate that large vessels characterized by low variances D_z and $D_{\dot{z}}$ and high $\dot{z}^*$ values, are less prone to slamming than small ships and that the probability P_{SL} of slamming can be reduced by keeping the draft D_s at the stem sufficiently high. Clearly, these formulas do not account for the effect of the voluntary speed reduction that ship masters usually take to avoid slamming. As a rule of thumb, one can assume that a typical ship master reduces the ship's speed or changes her course if slams occur more than three times out of 100 effective periods of wave encounter.

The level of the hydrodynamic load for flat bottom slamming can be assessed on the basis of the following elementary considerations. The application of the law of conservation of momentum results, for the case of a rapid immersion of a solid body into liquid, in the equation:

$$M\dot{z}_0 = (M + \mu)\dot{z}$$

Here M is the mass of the body, μ is the added mass of water, $\dot{z} = \dot{z}(t)$ is the speed of immersion, and $\dot{z}_0$ is the initial speed. From this equation we find

$$\dot{z} = \frac{M\dot{z}_0}{M + \mu}$$

and the hydrodynamic loading experienced by the body is

$$p = M(\dot{z}_0 - \dot{z}) = \frac{M}{M + \mu}\mu\dot{z}_0$$

Typically, during ship slamming, one can assume that $M \gg \mu$ (the ship's mass is significantly larger than the variable added mass of water), so that

$$p \simeq \mu\dot{z}_0$$

In an approximate analysis, one can assume also that the added mass μ can be calculated by the formula, obtained for the case of a direct impact of a plate on a water surface:

$$\mu = \frac{\pi}{2}\rho c^2$$

Here c is half the plate's width and ρ is the water density.

Let the velocity $\dot z_0$ of the impact be a random variable distributed in accordance with the trunketed Rayleigh law:

$$f_{\dot z_0}(\dot z_0) = C\frac{\dot z_0}{D}\exp\left[-\frac{(\dot z_0 - \dot z^*)^2}{2D}\right]$$

where the parameter D is related to the most probable value $\dot z_{max}$ of the velocity $\dot z_0$ as

$$D = \dot z_{max}(\dot z_{max} - \dot z^*)$$

The constant C can be determined from the condition of normalization

$$\int_0^\infty f_{\dot z_0}(\dot z_0)\,d\dot z_0 = 1$$

as follows:

$$C = \left\{e^{-\dot z^{*2}/(2D)} + \sqrt{\frac{\pi}{2D}}\,\dot z^*\left[1 + \Phi\left(\frac{\dot z^*}{\sqrt{2D}}\right)\right]\right\}^{-1} \tag{9.81}$$

The probability that the impact velocity $\dot z_0$ exceeds a level $\dot z^*$ is

$$\begin{aligned}P &= \int_v^\infty f_{\dot z_0}(\dot z_0)\,d\dot z_0 \\ &= \frac{C}{D}\int_{\dot z_0}^\infty \dot z_0 \exp\left[-\frac{(\dot z_0 - \dot z^*)^2}{2D}\right]d\dot z_0 \\ &= \sqrt{\frac{2}{D}}\,C\left(\sqrt{2D}\int_{(\dot z_0 - \dot z^*)/\sqrt{2D}}^\infty te^{-t^2}\,dt + \dot z^*\int_{(\dot z_0 - \dot z^*)/\sqrt{2D}}^\infty e^{-t^2}\,dt\right) \\ &= \sqrt{\frac{2}{D}}\,C\left\{\sqrt{\frac{D}{2}}\exp\left[-\frac{(\dot z_0 - \dot z^*)^2}{2D}\right] + \frac{\sqrt{\pi}}{2}\,\dot z^*\left[1 - \Phi\left(\frac{\dot z_0 - \dot z^*}{\sqrt{2D}}\right)\right]\right\} \\ &= C\left\{\exp\left(-\frac{\dot z_0 - \dot z^*}{2D}\right) + \sqrt{\frac{\pi}{2D}}\,\dot z^*\left[1 - \Phi\left(\frac{\dot z_0 - \dot z^*}{\sqrt{2D}}\right)\right]\right\} \tag{9.82}\end{aligned}$$

The obtained expression can be used to evaluate the probability that the impact pressure p during ship slamming in rough seas exceeds a certain level p^* if the velocity $\dot{z}_0$ in this expression is replaced by the value $\dot{z}_0 = p^*/\mu$, where μ is the added mass. This probability is determined assuming that ship slamming took place. The complete probability, that the impact load p exceeds a certain level p^*, should be calculated as a product of the probability of the occurrence of slamming [this can be determined by the formula (9.79) or (9.80)] and the probability computed by the formula (9.82).

Further Reading (see Bibliography): 6, 55, 56, 94.

Chapter 10

Reliability

> *"Theory discerns the underlying simplicity of phenomena. The nontheorist sees a crazy welter of phenomena; when he becomes a theorist, they fuse into a simple and dignified structure."*
>
> K. K. DARROW
> Member of Technical Staff
>   at Bell Laboratories in the mid-1950s

> *"All possible 'definitions' of probability fall far short of the actual practice."*
>
> WILLIAM FELLER
> "An Introduction to Probability Theory"

> *"One should always be able to say 'tables, chairs, glasses of beer', instead of 'points, straight lines, and planes'."*
>
> DAVID GILBERT

10.1. Major Definitions

Reliability is the ability of an item (system) to perform a required function under stated operation and maintenance conditions for a specified period of time. Reliability is a complex property, which includes, depending on the item's function and operation conditions, its dependability, durability, maintainability, repairability, availability, etc. Reliability engineering studies failure modes and mechanisms, the causes of occurrence of various failures, and ways to eliminate failures or, when a certain failure is acceptable, to bring its rate down to an allowable level.

The following major terms and definitions are used in reliability engineering.

Failure. The termination of the ability of an item to perform a required function.

Failure rate. For a stated period of the life of an item, the ratio of the total number of failures in a sample to the cumulative time of that sample.

Mean time-between-failures (MTBF). The mean of the duration of time between consecutive failures. For a stated period of the life of an item, the mean time between consecutive failures can be computed as the ratio of the cumulative observed time to the number of failures under stated conditions.

Mean time-to-failure (MTTF). The mean time of the item's operation until it fails, i.e., the mean time of failure-free operation. For non-repairable items, for a stated period in the life of an item, the mean time to failure can be computed as the ratio of the cumulative time for a sample to the total number of failures in the sample during the stated period and stated operation conditions.

Dependability. The ability of an item to perform a required function without failure under stated conditions for a stated period of time. This property is especially important for objects whose failures can lead to loss of human lives, large economic losses, etc.

Maintenance. The actions performed in an attempt to retain an item in a specified condition by providing systematic inspections, detection and prevention of an incipient failure (preventive maintenance), or the actions performed, as a result of failure, to restore an item to a specified condition (corrective maintenance).

Maintainability. The ability of an item, under stated conditions of use, to be retained in, or restored to, a state in which it can perform its required functions, provided that maintenance is conducted under specified conditions and using prescribed procedures and resources.

Availability. The ability of an item to perform its required function at a stated instant of time or over a stated period of time under combined aspects of its reliability, maintainability, and maintenance support.

Mean time-to-repair (MTTR). For repairable items, the total corrective maintenance time divided by the total number of corrective maintenance actions during a given period of time.

Various reliability-related problems have been and will be addressed as examples of the application of different probabilistic methods throughout the book. In this chapter we will address some major problems and approaches of reliability engineering in a more or less organized fashion.

10.2. Deterministic and Probabilistic Approaches in Reliability Engineering

Reliability depends on the concept that the item or the component has a certain resistance to loading. Failure occurs when the load induced by the operating conditions exceeds this resistance. If a deterministic approach is used, the reliability is ensured by introducing a sufficiently high "*safety factor.*" This is the ratio of the mean *capacity* ("*strength*") C to the mean *demand* ("*load*") D:

$$\eta = \frac{C}{D} \tag{10.1}$$

The magnitude of the safety factor is chosen depending on the consequences of failure, acceptable risks, the available information about the capacity and demand, the accuracy of predictions of these parameters, possible costs and social benefits, variability of materials parameters, construction (manufacturing) procedures, anticipated inspection schedules, etc. In a particular problem, the capacity and demand can be different from the strength and load, and their roles can be placed by, say, traffic capacity and anticipated traffic flow, culvert size and the quantity of water, critical (leading to buckling) and actual compressive stresses, acceptable and actual electrical resistance, water depth and ship draft, etc. Traditionally, the safety factors in engineering are being established from previous experiences for the considered system in its anticipated environment (operation conditions).

Another, also deterministic, approach attempts to consider the worst case by accounting for the variability in the capacity and demand. This is done by multiplying the safety factor computed by the formula (10.1) by the factor $(D + \Delta D)/(C - \Delta C)$ to consider that the demand can be somewhat higher and the capacity can be somewhat lower than their mean values. This leads to the following formula for the safety factor:

$$\eta = \frac{C}{D} \frac{D + \Delta D}{C - \Delta C} = \frac{1 + \Delta D/D}{1 - \Delta C/C} \tag{10.2}$$

It is clear, however, that, whatever the approach, as long as it is deterministic it is unable to assess the probability of failure.

The substance of the probabilistic approach to the assessment of the system's reliability or safety can be formulated as follows. Let $C = C(c_1, c_2, \ldots, c_k)$ be the system's (bearing) *capacity*, which is a random variable depending on various random parameters c_i, $i = 1, 2, \ldots, k$, and $D = D(d_1, d_2, \ldots, d_n)$ be the *demand* (loading), which is also a random variable depending on random parameters d_j, $j = 1, 2, \ldots, n$. The safety factor

$$\eta = \frac{C}{D} \tag{10.3}$$

is, in this case, also a random variable, and so is the *function of non-failure* (safety margin)

$$\psi = C - D \tag{10.4}$$

The *probability and nonfailure* ("reliability") is the probability that the system's capacity will be greater than the demand:

$$P = P\{C > D\} \tag{10.5}$$

If the laws of the distributions of the random variables c_i and d_j (which determine the capacity and the demand) are known, then, using methods of the theory of probability, one can find the probability densities $f_\psi(\psi)$ and $f_\eta(\eta)$ of the random variables ψ and η. Then the probability of nonfailure can be evaluated as

$$P = \int_0^\infty f_\psi(\psi)\, d\psi = \int_1^\infty f_\eta(\eta)\, d\eta \tag{10.6}$$

Although this probability is never zero, it can be made, if a probabilistic approach is used, as low as necessary. As one can see from Fig. 10.1,

$$P\{D > D_0\} = \omega_1$$

and

$$P\{C < C_0\} = P\{C < D_0\} = \omega_2$$

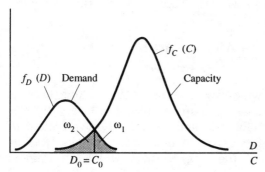

Figure 10.1 The relation location of the demand and capacity curves.

Then the probability of failure can be computed as

$$\omega_1\omega_2 = P\{C < D_0 < D\} = P\{C < D\} = P\{\psi < 0\}$$

The probability of nonfailure can be evaluated by the formula

$$P = P\{\psi > 0\} = 1 - \omega_1\omega_2 \tag{10.7}$$

Direct use of the probability of nonfailure is not always convenient, since, for highly reliable items, this probability is expressed by a fraction that is very close to unity. For this reason, even substantial changes in the system's design, having an appreciable impact on its reliability, may have a very minor effect on the probability of nonfailure. Therefore, another characteristic of the adequacy of a design, the "*safety index*" (*reliability index*) δ_R, is often used as a reliability criterion. This characteristic is substantially more sensitive to the changes in the design, materials, and loading conditions. If, for instance, the function of nonfailure (safety margin), ψ, is a random variable distributed in accordance with the normal law, so that

$$f_\psi(\psi) = \frac{1}{\sqrt{2\pi D_\psi}} \exp\left[-\frac{(\psi - \langle\psi\rangle)^2}{2D_\psi}\right] \tag{10.8}$$

then the probability P of nonfailure can be evaluated as follows:

$$\begin{aligned}P &= \int_0^\infty f_\psi(\psi)\,d\psi = \frac{1}{\sqrt{2\pi D_\psi}} \int_0^\infty \exp\left[-\frac{(\psi - \langle\psi\rangle)^2}{2D_\psi}\right] d\psi \\ &= \frac{1}{\sqrt{\pi}} \int_{-\delta_R}^\infty e^{-t^2}\,dt = \frac{1}{2}\left(\frac{2}{\sqrt{\pi}} \int_0^\infty e^{-t^2}\,dt + \frac{2}{\sqrt{\pi}} \int_0^{\delta_R} e^{-t^2}\,dt\right) \\ &= \frac{1}{2}[1 + \Phi(\delta_R)]\end{aligned} \tag{10.9}$$

Here

$$\delta_R = \frac{\langle\psi\rangle}{\sqrt{2D_\psi}} = \frac{\langle c \rangle - \langle d \rangle}{\sqrt{2(D_c + D_d)}} \tag{10.10}$$

is the safety index. As evident from this formula, the safety index is the reciprocal, up to a constant factor, of the coefficient of variation (COV)

$$v_\psi = \frac{\sqrt{D_\psi}}{\langle\psi\rangle}$$

and is related to this coefficient as follows:

$$\delta_R = \frac{1}{\sqrt{2}\,v_\psi} = \frac{\bar{\eta}-1}{\sqrt{2(\bar{\eta}^2 v_c^2 + v_d^2)}} \qquad (10.11)$$

Here $\bar{\eta} = \langle c \rangle / \langle d \rangle$ is the *nonrandom safety factor* and $v_c = \sqrt{D_c}/\langle c \rangle$ and $v_d = \sqrt{D_d}/\langle d \rangle$ are coefficients of variation of the capacity and demand, respectively. The formula (10.11) indicates how the safety index is related to the nonrandom safety factor and the coefficients of variation of the capacity and demand.

Another, more simple safety characteristic is the safety index δ_E, introduced as

$$P = \Phi(\delta_E) \qquad (10.12)$$

The indices δ_R and δ_E are related as follows:

$$\Phi(\delta_R) = 2\Phi(\delta_E) - 1 \qquad (10.13)$$

The calculated probabilities P of nonfailure and the indices δ_R are shown, for the given indices δ_E, in Table 10.1.

It is noteworthy that there are certain important factors that can have an appreciable effect on the system's reliability although they cannot be accounted for (quantified) on the basis of a probabilistic analysis. These are various human factors, possible violations of the maintenance schedule, unpredictable or inappropriate service conditions, etc. Such factors are usually addressed in engineering practice in a deterministic way, by introducing additional safety margins within the probabilistic scheme.

Example 10.1 Using the Chebyshev inequality from Sec. 5.3, assess the lower limit of the mean value $\langle \eta \rangle$ of the random safety margin η.

Solution The Chebyshev inequality

$$P(|X| \le \alpha) \ge 1 - \frac{D_x}{\alpha^2}$$

or

$$P(-\alpha \le X \le \alpha) \ge 1 - \frac{D_x}{\alpha^2}$$

TABLE 10.1

δ_E	P	δ_R
2.5	0.9995930	2.453
3.0	0.9999779	2.991
3.5	0.9999993	3.485
4.0	0.9999999	3.993
∞	1	∞

states that the probability that the random variable X can be found within an interval $(-\alpha, \alpha)$ is always higher than the number

$$P^* = 1 - \frac{D_x}{\alpha^2}$$

Hence, P^* is the minimum value of the probability P. Since the random variable X and the number α can be chosen in an arbitrary way, one can put

$$X = \eta - \xi\langle\eta\rangle, \qquad \alpha = \xi\langle\eta\rangle - 1$$

where ξ is a nonrandom variable. Then the Chebyshev inequality yields

$$P(1 \le \eta \le 2\xi\langle\eta\rangle - 1) \ge P^*$$

The left part of the obtained inequality is the probability that the random safety margin, η, is higher than one and represents, therefore, the probability of nonfailure. Hence, P^* is the lower limit of the probability of nonfailure.

The variance D_x can be expressed as

$$D_x = D_\eta + (\xi\langle\eta\rangle - \langle\eta\rangle)^2 = \langle\eta\rangle^2[v_\eta^2 + (\xi - 1)^2]$$

where $v_\eta = \sqrt{D_\eta}/\langle\eta\rangle$ is the coefficient of variation of the random variable η. Thus, the lower limit, P^* of the probability of nonfailure can be expressed as

$$P^* = 1 - \langle\eta\rangle^2 \frac{(\xi - 1)^2 + v_\eta^2}{(\xi\langle\eta\rangle - 1)^2}$$

The minimum value of the function $P^* = P^*(\xi)$ takes place for

$$\xi = 1 + \frac{\langle\eta\rangle}{\langle\eta\rangle - 1} v_\eta^2$$

and is equal to

$$P^* = \frac{1}{1 + [v_\eta\langle\eta\rangle/(\langle\eta\rangle - 1)]^2}$$

Solving this equation for the $\langle\eta\rangle$ value, we obtain the following formula for the lower limit of the safety margin:

$$\langle\eta\rangle = \frac{1}{1 - v_\eta[\sqrt{P^*/(1 - P^*)}]}$$

10.3. Dependability: Reliability Function

The simplest (and basic) objects in reliability engineering are elements (items) that do not lend themselves to any restoration (repair) and have to be replaced after the first failure. The reliability of such items is due completely to their *dependability*. This property can be measured by the probability that no failure occurs during the given period of time t. In reliability engineering this probability is called *reliability function*.

As with any other probability, the dependability of a sufficiently large population of unrepairable elements can be substituted by the frequency, and therefore the reliability function can be computed as

$$R(t) = \frac{N_s(t)}{N_0} \qquad (10.14)$$

where N_0 is the total number of items being tested and $N_s(t)$ is the number of items that are still sound by the time t. Differentiating this relationship with respect to the time t, we have

$$\frac{dR(t)}{dt} = \frac{1}{N_0}\frac{dN_s(t)}{dt} = -\frac{1}{N_0}\frac{dN_f(t)}{dt} \qquad (10.15)$$

where $N_f(t) = N_0 - N_s(t)$ is the number of items that failed by the time t.

Introduce a function

$$\lambda(t) = \frac{1}{N_s(t)}\frac{dN_f(t)}{dt} \qquad (10.16)$$

This function is known as the *failure rate*. As evident from this formula, the failure rate is the ratio of the number of failed items by the time t to the number of items that remained sound by this time. Introducing the function (10.16) into (10.15), we obtain

$$\frac{dR(t)}{dt} = -\lambda(t)\frac{N_s(t)}{N_0} = -\lambda(t)R(t) \qquad (10.17)$$

The function $R(t)$ must satisfy the following initial condition: $R(0) = 1$ (the probability of failure at the initial moment of time is zero). Then the equation (10.17) yields

$$R(t) = \exp\left[-\int_0^t \lambda(t)\,dt\right] \qquad (10.18)$$

This formula explains the physical meaning of the reliability function: this function expresses the probability of nonfailure during the time t or, more accurately, the probability that an item that has been operating successfully for the time t will fail within the time interval $(t, t + dt)$. The formula (10.18) expresses the probabilistic definition of the reliability function, while the formula (10.14) reflects its statistical definition.

When the failure rate λ is constant, the formula (10.18) yields

$$R(t) = e^{-\lambda t} \qquad (10.19)$$

This relationship is known as the *exponential formula of reliability*. The function

$$f(t) = -\frac{dP_D(t)}{dt} = \lambda(t) \exp\left[-\int_0^t \lambda(t)\, dt\right] \qquad (10.20)$$

is the probability density function for the flow of failures, or the *failure frequency*. The probability of failure during the time t can be evaluated as

$$Q(t) = 1 - R(t) = \int_0^t f(t)\, dt \qquad (10.21)$$

The failure rate $\lambda(t)$ characterizes the change in the dependability of an item in the course of its lifetime. While the function $R(t)$ never increases, the function $\lambda(t)$ can either decrease, or increase, or remain constant. At the initial moment of time $t = 0$, $R(0) = 1$ and $\lambda(t) = f(t)$. The approximate shape of the function $\lambda(t)$ is shown in Fig. 10.2. This graph is known as the "*bathtub curve*". A typical "bathtub curve" has three major portions, characterized by the behavior of the function $\lambda(t)$:

1. $d\lambda(t)/dt < 0$ ("infant mortality" portion, breaking-in period, debugging period)
2. $d\lambda(t)/dt \approx 0$ ("steady-state" period, the useful life of the system), and

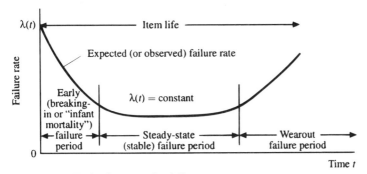

Figure 10.2 "Bathtub" curve for failure rates.

3. $d\lambda(t)/dt > 0$ (excessive "wear-out" portion).

It is worthy to point out that the type of curve in Fig. 10.2 is typical not only for engineering systems but also for many other systems (say, humans) as well. In practice, the function $\lambda(t)$ is determined experimentally.

The integral

$$\Lambda(t) = \int_0^t \lambda(t) \, dt$$

expresses the consumed lifetime of the item. The lifetime consumed within the time interval (t_1, t_2) can be calculated as

$$\Lambda(t_1, t_2) = \int_{t_1}^{t_2} \lambda(t) \, dt$$

Quite often, if the distribution functions are not available one can effectively employ numerical characteristics of the random variables of interest. Here are several mathematical (quantitative) formulations of some numerical characteristics used in the theory of dependability (see also Sec. 10.1):

1. *Mean time-to-failure* (*MTTF*) is the mean time of the item's operation until it fails (mean time of dependable, i.e., failure-free, operation):

$$\langle t \rangle = \int_0^\infty t f(t) \, dt$$

Since, as follows from Eq. (10.21), $f(t) = -dR(t)/dt$, then

$$\langle t \rangle = -\int_0^\infty t \frac{dR(t)}{dt} \, dt = -[tR(t)]_0^\infty + \int_0^\infty R(t) \, dt$$

where integration by parts is used. The first term in this formula is equal to zero, and therefore

$$\langle t \rangle = \int_0^\infty R(t) \, dt \qquad (10.22)$$

2. *Variance of the time-to-failure* is defined as

$$D_t = \int_0^\infty (t - \langle t \rangle)^2 f(t) \, dt = \int_0^\infty t^2 f(t) \, dt - \langle t \rangle^2$$

and is related to the dependability $R(t)$ as follows:

$$D_t = 2 \int_0^\infty tR(t) \, dt - \langle t \rangle^2 \qquad (10.23)$$

3. *Mean failure frequency* is defined as

$$\langle f(t) \rangle = \frac{1}{t} \int_0^t f(t) \, dt = \frac{1}{t}[1 - R(t)] \qquad (10.24)$$

4. *Time-to-failure* t_γ with the given probability $\gamma\%$ is defined by the relationship

$$R(t_\gamma) = \frac{\gamma\%}{100} \qquad (10.25)$$

Dependability of a system of n elements connected in series can be obtained as a product of the constituent dependabilities of the elements:

$$P_c(t) = \prod_{i=1}^n P_i(t)$$

The system's dependability reduces with an increase in the number of elements in it and is always lower than the dependability of the least reliable item. Therefore it is important to be able to identify this item and to increase its reliability. This is the most effective way to improve the reliability of a system. Even a minor improvement in the reliability of an unreliable item (component) can lead to an appreciable improvement in the system's reliability.

If the times till failure of different components in a complex system are statistically independent and follow the same law of the probability distribution, then the relationship between the *intensity* $\omega(t)$ *of the failure flow* and the probability density function $f(t)$ for the time between failures can be described by the following *equation of renewal*:

$$\omega(t) = f(t) + \int_0^t f(t - \xi)\omega(\xi) \, d\xi \qquad (10.26)$$

Example 10.2 The failure rate of an object is $\lambda(t) = at$, i.e., increases linearly in time. What is the probability that it will fail at the time t? What is the failure frequency at this time?

Solution Using the formula (10.18) we find

$$R(t) = \exp\left(-\int_0^t at \, dt\right) = \exp\left(-\frac{at^2}{2}\right)$$

The failure frequency can be determined by the formula (10.19):

$$f(t) = at \exp\left(-\frac{at^2}{2}\right)$$

Example 10.3 The failure of an object is a linear function of time: $\lambda(t) = at$. What is the dependability (the probability of failure-free operation) of this object during the mean time-to-failure?

Solution Using the formula for the reliability function $R(t)$ obtained in the previous example and applying the formula (10.22), we determine the mean time-to-failure as follows:

$$\langle t \rangle = \int_0^\infty \exp\left(-\frac{at^2}{2}\right) dt = \sqrt{\frac{\pi}{2a}}$$

The sought dependability is

$$R(\langle t \rangle) = \exp\left(-\frac{\pi}{4}\right) = 0.456$$

Example 10.4 An electronic device has an exponential distribution of the mean time-to-failure $\langle t \rangle$. What is the probability of the failure-free operation of the device during the time $t = \langle t \rangle$?

Solution The reliability function (the probability of failure-free operation) is

$$R(t) = e^{-\lambda t} = e^{-t/\langle t \rangle}$$

In the case $t = \langle t \rangle$, we obtain

$$R(\langle t \rangle) = e^{-1} = 0.634$$

Example 10.5 It has been established that the time-to-failure for a given object follows the Weibull distribution, so that the reliability function (the dependability of the object) is expressed as

$$R(t) = \exp\left[-\left(\frac{t}{b}\right)^a\right]$$

Determine the failure rate and the mean time-to-failure during the time $t = 100$ hours if the dependability of the object is $R = 0.95$ and the parameter a in the Weibull distribution is $a = 2$.

Solution From $R = 0.95$, we find

$$\frac{t}{b} = \sqrt{-\ln P_D} = 0.2265$$

and therefore the parameter b of the Weibull distribution is

$$b = \frac{100}{0.2265} = 441.5$$

The failure rate is

$$\lambda = \frac{a}{b}\left(\frac{t}{b}\right)^{a-1} = \frac{2}{441.5}(0.2265)^{2-1} = 1.026 \times 10^{-3}\ \text{h}^{-1}$$

The mean time-to-failure is as follows:

$$\langle t \rangle = b\Gamma\left(1+\frac{1}{a}\right) = 441.5\Gamma(1.5) = 441.5 \times 0.886\,23 = 391.3\ \text{h}$$

Example 10.6 An object has an exponential distribution of the time-to-failure t. What should the mean time-to-failure $\langle t \rangle$ be, so that the probability of failure-free operation is not less than $R = 0.99$ during the time of $t = 300$ hours?

Solution The reliability function is

$$R(t) = \exp\left(-\frac{t}{\langle t \rangle}\right) \approx 1 - \frac{t}{\langle t \rangle}$$

In order that the dependability is not less than 0.99 during the operation time of 300 hours, one should have

$$\langle t \rangle \geq \frac{t}{1 - R(t)} = \frac{300}{1 - 0.99} = 30\,000\ \text{h}$$

Example 10.7 The time-to-failure consists of two stages of the duration T_1 and T_2, respectively. The failure occurs at the moment $T = T_1 + T_2$. The times T_1 and T_2 are independent and are characterized by the exponential distributions $f_1(t) = \lambda_1 e^{-\lambda_1 t}$ and $f_2(t) = \lambda_2 e^{-\lambda_2 t}$. Find the intensity of the failure flow.

Solution The failure frequency (the probability density function for the time-to-failure) can be found as the convolution of the distributions $f_1(t)$ and $f_2(t)$:

$$f(t) = \int_0^t f_1(t-\xi) f_2(\xi)\, d\xi = \lambda_1 \lambda_2 e^{-\lambda_1 t} \int_0^t e^{(\lambda_1 - \lambda_2)\xi}\, d\xi = \frac{\lambda_1 \lambda_2}{\lambda_2 - \lambda_1}(e^{-\lambda_1 t} - e^{-\lambda_2 t})$$

Introducing this function into Eq. (10.26) of the renewal and seeking the solution to this equation in the form

$$\omega(t) = C(1 - e^{-(\lambda_1 - \lambda_2)t})$$

we obtain the following expression for the constant C:

$$C = \frac{\lambda_1 \lambda_2}{\lambda_1 + \lambda_2}$$

Thus, the intensity of the failure flow is expressed by the equation

$$\omega(t) = \frac{\lambda_1 \lambda_2}{\lambda_1 + \lambda_2}(1 - e^{-(\lambda_1 + \lambda_2)t})$$

Example 10.8 The distribution of the time-to-failure for the given item is given in the following table:

t, hours	0–1	1–2	2–3	3–4	4–5	5–6	6–7	7–8	8–9	9–10	>10
F(t)	0	0.03	0.08	0.20	0.45	0.65	0.80	0.90	0.95	0.98	1.00

Determine the main reliability indices of the item.

Solution The dependability (the probability of failure-free operation) for $t = 4$ hours is

$$R(4) = 1 - F(4) = 1 - 0.20 = 0.80$$

The probability of failure during the first 4 hours of operation is

$$Q(4) = F(4) = 0.10$$

The dependability for the time period from $t = 2$ hours to $t = 6$ hours is

$$P(2, 6) = \frac{1 - 0.65}{1 - 0.03} = 0.36$$

The probability of failure during this time is

$$Q(2, 6) = 1 - 0.36 = 0.64$$

The mean time-to-failure is

$$\langle t \rangle = \sum_{i=1}^{10} (1 - F(t_i)) = 1 + 0.97 + 0.92 + \cdots + 0.05 + 0.02 = 4.96$$

Example 10.9 An item has an exponential law of the probability distribution of the time-to-failure. The failure rate is $\lambda = 2.5 \times 10^{-5}$ h^{-1}. Determine the main probability indices of the time for the time $t_0 = 2000$ hours.

Solution The probability of failure-free operation during the time of 2000 hours is

$$R(2000) = e^{-\lambda t_0} = e^{-2.5 \times 10^{-5} \times 2000} = 0.9512$$

The approximate value of this probability can be found, for low λ values, as

$$R(2000) = 1 - \lambda t_0 = 1 - 2.5 \times 10^{-5} \times 2000 = 0.9500$$

The probability of failure is

$$Q(2000) = 1 - e^{-\lambda t_0} = 1 - 0.9512 = 0.0488$$

The probability of failure-free operation from 500 hours to 2500 hours can be found as

$$P(500, 2500) = e^{-2.5 \times 10^{-5}(2500 - 500)} = 0.9512$$

The mean time-to-failure is

$$\langle t \rangle = \frac{1}{\lambda} = \frac{1}{2.5 \times 10^{-5}} = 40\,000 \text{ h}$$

10.4. Experimental Evaluation of Dependability and the Failure Rate

The experimental evaluation of dependability and the failure rate involves testing of a large number of identical elements in identical (to the extent possible) test conditions. Let the number of tested elements be N and the times until failure for these elements be $t_1, t_2, \ldots, t_N$. The arithmetic mean $\langle t^* \rangle$ of the time-to-failure can be accepted as an estimate of the mean time-to-failure $\langle t \rangle$:

$$\langle t \rangle = \frac{t_1 + t_2 + \cdots + t_N}{N} \tag{10.27}$$

Often, however, because of the limited test time, one cannot obtain the times t_i, $i = 1, 2, \ldots, N$. In this case it is possible to obtain a low estimate for the mean time-to-failure by the formula

$$\langle t \rangle = \frac{t_1 + \cdots + t_n + (N - n)t}{N} \tag{10.28}$$

Here n is the number of failed elements. Similarly, the variance of the time-to-failure can be obtained by the formula

$$D_t = \frac{1}{N-1} \sum_{i=1}^{N} (t_i - \langle t \rangle)^2 \tag{10.29}$$

If the type of probabilistic law for the time-to-failure T is known or can be anticipated, then a knowledge of the parameters $\langle t \rangle$ and/or D_t is sufficient to establish this law. The exponential law is a single-parametric law and is completely determined by the mean value, the normal law is a two-parametric one and is determined by the mean value and the variance of the random variable, and so on. If, however, the probabilistic law of the distribution of the reliability function $R(t)$ is unknown, then it can be determined in the following way.

Let the time of testing be $t = t_0$. If the total number of elements that remained sound by the time t_0 is equal to N_s, an estimate of the dependability function can be computed as

$$P_N^*(t_0) = \frac{N_s}{N} \tag{10.30}$$

For sufficiently large N, one can assume $R(t_0) = P_N^*(t_0)$. For $t \leq t_0$, the reliability (dependability) function can be evaluated as

$$R(t) = \frac{N_s(t)}{N} \qquad (10.31)$$

The obtained experimental data can also be used to assess the function $f(t)$ of the failure frequency. Then the failure rate $\lambda(t)$ can be found as

$$\lambda(t) = \frac{f(t)}{R(t)} \qquad (10.32)$$

This formula can be obtained from (10.18) by solving this relationship for the function $\lambda(t)$ and considering the fact that the function $R(t)$ decreases with time; therefore the minus sign in the obtained formula can be omitted. The most typical reliability (dependability) laws are the exponential law, the Weibull law, the gamma distribution, and the normal law.

10.5. Goodness-of-Fit Criteria

10.5.1. Goodness-of-fit tests

In the previous section we discussed how the statistical distribution of a random variable, such as, for instance, time-to-failure, can be obtained from the experimental data. Quite often, one can intuitively guess, based on these data, how this distribution can be approximated by one of the known theoretical distributions, such as exponential, normal, Weibull, etc. There are, however, formalized statistical tests, known as *goodness-of-fit tests*, that enable one to decide whether the obtained sample data are consistent with a preconceived analytical distribution. This is done by assessing the probability that the test sample does not contradict (fits) the anticipated theoretical distribution.

The goodness-of-fit tests are carried out as follows. One chooses a certain random variable, κ (*goodness-of-fit criterion*), as a measure of the discrepancy between the observed ("statistical") and the theoretical ("probabilistic") laws of distribution. Then one evaluates such a magnitude (realization) κ_a of the value κ, for which the probability that the α value is larger than κ_a is equal to a sufficiently small number α: $P(\kappa > \kappa_a) = \alpha$. The desirable (acceptable) level of the α value should be established, based on the substance of the problem. If the experimentally obtained $\kappa = \kappa_q$ values exceed κ_α, then the deviation from the theoretical law of distribution cannot be explained by random causes only and is considered significant. In such a case the hypothesis about the law of distribution must be rejected. If the κ_q value does not exceed κ_α ($\kappa_q \leq \kappa_\alpha$), the deviation is considered acceptable, i.e., the experimental data are considered consistent with the hypothesis about the type of the law of the probability distribution. One can also check the hypothesis about the type of probability distribution in a different way: determine the probability $\alpha_q = P(\kappa > \kappa_q)$ from the κ_q value. If the obtained value α_q is smaller than α, then the derivation of the experimental data from the theoretical law is considered significant; if $\alpha_q \geq \alpha$, then the deviations are considered insignificant. The values α_q, which are very close to unity (very good agreement), correspond to an event of a very low probability. This could be an indication of the low quality of sampling (for instance, elements that would have resulted in substantial deviations from the mean were groundlessly eliminated).

Different values (criteria) can be chosen as a measure of the discrepancy between the statistical and the theoretical laws of distribution. Typically, these criteria are being chosen in such a way that the probabilistic law of distribution be simple enough and, for a sufficiently large amount of experimental data, be independent from the theoretical probability distribution function. In this section we describe the use (not necessarily with application to the reliability problems) of two of the most widely applied goodness-of-fit criteria: χ^2 (Pearson's) criterion and Kolmogorov's criterion.

10.5.2. Pearson's criterion

If *Pearson's criterion* (χ^2-criterion) is used, one determines the experimental value

$$\chi_q^2 = \sum_{i=1}^{k} \frac{n}{p_i} (p_i^* - p_i)^2 = \sum_{i=1}^{k} \frac{(m_i - np_i)^2}{np_i}$$

where k is the number of orders (segments) of the experimental values of the random variable X (the entire range of the observed data is broken into k segments), n is the sample size (clearly, k is always smaller than n), m_i is the size of the ith order (the number of observations for each ith order), $p_i^* = m_i/n$, $i = 1, 2, \ldots, k$, are the experimental frequencies, and p_i is the theoretical probability that the random variable X will be found within the ith order. It should be pointed out that although the theoretical law of the probability distribution is supposed to be known, the parameters of this law (mean, variance, etc.) are not. Therefore, when computing the theoretical probabilities, one has to use the statistically obtained characteristics for the theoretical laws. For $n \to \infty$, the law of the distribution of the χ_q^2, no matter what the law of the probability distribution for the random variable X is, tends to the χ^2-distribution with $r = k - s$ degrees of freedom. In this formula, s is the number of applied bonds, which can be calculated as the number of parameters in the theoretical law of distribution plus one. This additional "bond" is the condition of normalization

$$\sum_{i=k}^{k} p_i^* = 1$$

of the experimental frequencies p_i^*. The probabilities $P(\chi^2 \geq \chi_q^2)$ are tabulated in Appendix D as functions of χ_q^2 and r values. The Pierson criterion can be used when the sample size n and the order sizes m_i are sufficiently large ($n \geq 50 - 60$, $m_i \geq 5 - 8$). The calculations can be carried out using the following sequence of action:

1. Determine the quantity χ_q^2 for the given experimental sample.
2. Determine the number r of degrees of freedom as the number of orders (segments) minus the number of applied bonds: the number of applied bonds is the number of parameters in the theoretical law plus one.
3. Using the table for the χ^2 distribution (Appendix D) determine, from r and χ^2, the probability that a random variable, distributed in accordance with the χ^2 law with r degrees of freedom, exceeds the χ^2 value. If this probability is small (lower than α), the hypothesis should be rejected; otherwise it is considered consistent with the experimental data and should be accepted.

Example 10.10 A radioactive substance has been observed during 2608 equal (7.5 s each) time intervals. For each of these intervals the number of particles was recorded. The numbers m_i of the time intervals during which i particles were recorded by the counter are shown in Table 10.2. Using χ^2 criterion, check the goodness-of-fit of the observed data with the Poisson law:

$$P(i, a) = \frac{e^{-a} a^i}{i!}$$

Assume $\alpha = 5$ percent.

Solution The $\tilde{a}$ estimate for the Poisson law is

$$\tilde{a} = \frac{1}{n} \sum_{i=0}^{\infty} i m_i = 3.870$$

TABLE 10.2 Time intervals, m_i, during which i particles were recorded

i	m_i	i	m_i
0	57	6	273
1	203	7	139
2	383	8	45
3	525	9	27
4	532	More than 10	16
5	408		
		Total	$n = \sum m_i = 2608$

where

$$n = \sum_{i=0}^{\infty} m_i = 2608$$

We compute the theoretical probabilities p_i that i particles will hit the counter, using the tabulated values $P(i, \tilde{a}) = p_i$ for the Poisson law (Appendix A). As a result of integration between $a = 3$ and $a = 4$ we obtain the data shown in the Table 10.3. The number of the degrees of freedom is determined by the formula $r = k - p - 1$. The total number of time intervals is $k = 11$ and the number p of the parameters, determined on the basis of the observed data, is $p = 1$ (the parameter a). Then the number of degrees of freedom is $r = 11 - 1 - 1 = 9$.

Using the tabulated data for the χ^2 distribution with $r = 9$ and the calculated χ_q^2 value $\chi_q^2 = 13.05$, we find the probability $P(\chi^2 \geq \chi_q^2)$ that the number χ^2 exceeds the χ_q^2 value. Then we have

$$\alpha_q = P(\chi^2 \geq \chi_q^2) = 0.166$$

Since $\alpha_q > \alpha = 0.05$, the deviations from Poisson's law are insignificant.

TABLE 10.3 Evaluation of the experimental χ^2 criterion

i	p_i	np_i	$m_i - np_i$	$(m_i - np_i)^2$	$(m_i - np_i)^2/(np_i)$
0	0.021	54.8	2.2	4.84	0.088
1	0.081	211.2	−8.2	67.24	0.318
2	0.156	406.8	−23.8	566.44	1.392
3	0.201	524.2	0.8	0.64	0.001
4	0.195	508.6	23.4	547.56	1.007
5	0.151	393.8	14.2	201.64	0.512
6	0.097	253.0	20.0	400.00	1.581
7	0.054	140.8	−1.8	3.24	0.023
8	0.026	67.8	−22.8	519.84	7.667
9	0.011	28.7	−1.7	2.89	0.101
10	0.007	18.3	−2.3	5.29	0.289
$\sum$	1.000				$\chi_q^2 = 13.049$

TABLE 10.4 Experimental order statistics

Range of time-of-failure, h	Number of failures within this time	Frequency $\hat{p}_i$
0–50	5	0.12
50–100	6	0.14
100–150	11	0.26
150–200	8	0.19
200–250	7	0.17
250–300	5	0.12

Example 10.11 The order series shown in Table 10.4 was obtained as a result of the reliability testing of certain equipment. Find the mean time-to-failure. Using χ^2 criterion, determine whether the experimental data can be approximated by the uniform distribution.

Solution The mean time-to-failure can be found as

$$\langle t \rangle = \frac{50 + 0}{2} 0.12 + \frac{100 + 50}{2} 0.14 + \frac{100 + 150}{2} 0.26 + \cdots = 150.5 \text{ h}$$

The theoretical probabilities of the uniform distribution are

$$p_1 = p_2 = \cdots = p_6 = \frac{1}{6}$$

The χ^2 function is

$$\chi^2 \simeq \sum_{i=1}^{k} \frac{n}{p_i} (\hat{p}_i - p_i)^2 = \sum_{i=1}^{6} \frac{42}{1/6} \left(\hat{p}_i - \frac{1}{6} \right)^2$$
$$= 252[(0.12 - 0.17)^2 + (0.14 - 0.17)^2 + \cdots + (0.12 - 0.17)^2] = 3.63$$

The number of the degrees of freedom is

$$r = k - 1 = 6 - 1 = 5$$

For $r = 5$ and $\chi^2 = 3.63$, we find, using Appendix D data, that $p \simeq 0.6$. Hence, it is likely that the hypothesis that the experimental data can be approximated by a uniform distribution is consistent with these data.

10.5.3. Kolmogorov's criterion

The Russian mathematician A. N. Kolmogorov suggested that the maximum value D of the absolute value of the difference between the statistical and the theoretical distribution functions be used as a suitable measure of the discrepancy of these functions:

$$D_q = \max |F^*(x) - F(x)|$$

Here F^* is the statistical and F is the theoretical distribution function. Kolmogorov showed that when the number n of observations is

TABLE 10.5 Evaluation of the Kolmogorov's criterion

y	K(y)	y	K(y)	y	K(y)	y	K(y)
0.30	9×10^{-6}	0.70	0.289	1.10	0.822	1.50	0.978
0.35	3.03×10^{-4}	0.75	0.373	1.15	0.858	1.60	0.988
0.40	2.81×10^{-3}	0.80	0.456	1.20	0.888	1.70	0.994
0.45	0.013	0.85	0.535	1.25	0.912	1.80	0.997
0.50	0.016	0.90	0.607	1.30	0.932	1.90	0.99854
0.55	0.077	0.95	0.672	1.35	0.948	2.00	0.999329
0.60	0.136	1.00	0.730	1.40	0.960	2.10	0.999705
0.65	0.208	1.05	0.780	1.45	0.970	2.20	0.999874

large ($n \to \infty$), the law of the distribution of the value $\lambda = \sqrt{n}\,D$ tends to what is known as the Kolmogorov law of the probability distribution, regardless of the law of the distribution of the random variable X. In other words, if the condition $\lambda \geq y$ is fulfilled, the probability $P(\lambda \geq y) = P(y)$ tends to the limit $P(y) = 1 - K(y)$, for any probability distribution function of a continuous random variable, if the number n of observations tends to infinity. If Kolmogorov's criterion is used, the parameters of the theoretical law of the probability distribution do not have to be determined from the measured data. The probabilities

$$\alpha_q = P(D \geq D_q) = P(y) = 1 - K(y)$$

are tabulated in Table 10.5.

The calculations based on Kolmogorov's criterion can be carried out using the following steps:

1. Build the statistical distribution function $F^*(x)$ and choose the theoretical function $F(x)$.

2. Determine the maximum D value of the absolute value of the difference between the distributions $F^*(x)$ and $F(x)$.

3. Calculate the quantity $y = \sqrt{n}\,D$ and, using Table 10.5, find the probability $P(y)$ that, if the random variable X is indeed distributed in accordance with the law $F(x)$, the maximum discrepancy between the functions $F^*(x)$ and $F(x)$ due to random causes will not be smaller than the observed one. If the probability $P(y)$ is small, the hypothesis should be rejected. If this probability is large, the hypothesis that the given theoretical law is consistent with the experimental data should be accepted.

Example 10.12 The measured dimensions of 1000 parts, rounded to 0.5 mm, are shown in Table 10.6 (m_i is the number of measurements resulting in the dimensions x_i). Using Kolmogorov's criterion, check whether the observed data agree with a supposition that the number X obeys the law of normal

TABLE 10.6 Number of measurements, m_i, resulting in the dimensions x_i during i-th measurement

i	x_i	m_i	i	x_i	m_i
1	98.0	21	6	100.5	201
2	98.5	47	7	101.0	142
3	99.0	87	8	101.5	97
4	99.5	158	9	102.0	41
5	100.0	181	10	102.5	25

distribution with the mean $\langle x \rangle = 100.25$ mm and standard deviation $\sigma_x = 1$ mm.

Solution The theoretical probability distribution function is

$$F(x) = \frac{1}{2}[1 + \Phi(x - \langle x \rangle)]$$

The statistical distribution function can be calculated as

$$F^*(x_k) = \frac{1}{1000}\left(\sum_{i=1}^{k} m_i + 0.5 m_k\right)$$

The calculations are carried out in Table 10.7. The largest absolute value of the difference D is $D_q = 0.0089$. Then we obtain

$$\lambda = \sqrt{n} D_q = \sqrt{1000} \times 0.0089 = 0.281$$

Using the tabulated $P(\lambda)$ values we find: $\alpha_q = P(\lambda) = 1.000$. This value is large. Hence, the deviations are insignificant and one can conclude that the hypothesis on the goodness-of-fit of the observed data is in good agreement with the law of normal distribution with parameters $\langle x \rangle = 100.25$

TABLE 10.7 Calculated absolute differences between statistical and theoretical distributions

i	$x_i - \langle x \rangle$	$\frac{1}{2}\Phi(x_i - \langle x \rangle)$	$F(x_i)$	$F^*(x_i)$	$\|F^*(x_i) - F(x_i)\|$
1	-2.25	-0.4877	0.0123	0.0105	0.0018
2	-1.75	-0.4599	0.0401	0.0445	0.0044
3	-1.25	-0.3944	0.1056	0.1115	0.0059
4	-0.75	-0.2734	0.2266	0.2340	0.0074
5	-0.25	-0.0987	0.4013	0.4035	0.0022
6	0.25	0.0987	0.5987	0.5945	0.0042
7	0.75	0.2734	0.7734	0.7660	0.0074
8	1.25	0.3944	0.8944	0.8855	0.0089
9	1.75	0.4599	0.9599	0.9545	0.0054
10	2.25	0.4877	0.9877	0.9875	0.0002

mm, and $\sigma_x = 1.0$ mm. However, the high α value can cause doubts about the quality of sampling.

Example 10.13 It has been found that in Example 10.12 the maximum of the absolute value of the difference between the statistical distribution and the uniform distribution is $D = 0.07$. Using Kolmogorov's criterion, determine whether the experimental observations agree well with the hypothesis of the uniform distribution.

Solution For $n = 42$, we find

$$y = D\sqrt{n} = 0.07\sqrt{42} = 0.45$$

Using Table 10.5 for the Kolmogorov function we find

$$P(y) = P(0.45) = 1 - K(0.45) = 1 - 0.013 = 0.987.$$

Hence it is likely that the uniform distribution can be used as a suitable approximation of the experimental data. Note that the application of Kolmogorov's criterion leads to higher probabilities of likelihood than Pierson's criterion.

10.6. Reliability of Repairable Items

The following three major classes of engineering products can be distinguished, as far as the requirements to this reliability are concerned:

Class I. The product has to be made as reliable as possible. Such products are encountered in outer space engineering, in many military applications, and in other cases when cost effectiveness is not viewed as an important factor and the product has to be made reliable by all means. Such products are seldom manufactured in large quantities, and their failures are typically viewed as a catastrophe.

Class II. The product has to be made as reliable as possible, but only for a certain level of demand. If the actual demand happens to be larger than the design demand, the product might fail, although the probability of such failure should be made very small. Examples are: civil engineering structures, passenger elevators, ocean-going vessels, offshore structures, commercial aircrafts, railroad carriages, medical equipment, etc. These are typically highly expensive products, which, at the same time, are produced in relatively large quantities, and therefore applications of class I requirements to such products, responsible as these products might be, will lead to unjustifiable and unacceptable expenses and are therefore deemed to be economically unfeasible. The failure of such products is often associated with loss of human lives and is also viewed as a catastrophe.

Class III. The product is relatively inexpensive, produced in massive numbers, and its failure is not viewed as a catastrophe, i.e., a certain level of failures during normal operation is considered acceptable. Examples are: various household items, consumer products, agricultural equipment, etc.

The reliability of class II and, especially, class III products depends not only on their dependability but on their repairability as well. It is important that class II products are designed in such a way that their gradual and potential failures can be easily detected and eliminated in due time, and that the detected damages (say, fatigue cracks) can be effectively removed. The reliability of class III products is characterized, first of all, by their availability—the ability of an item to perform its required function at the given time or over a stated period of time under combined aspects of its reliability (dependability), maintainability, and maintenance support. A high level of the reliability (availability) of class III items can be achieved either by sufficiently high levels of dependability or by high levels of repairability, or by the most feasible combination of both.

The *mean time-between-failures* (MTBF) of a repairable item is defined for a stated period of its life as the mean value of the time between consecutive failures (see also Section 10.1). This time is evaluated as a ratio of the cumulative observed time to the number of failures under stated conditions. The stationary *failure flow* is characterized by the mean number of failures per unit time:

$$\lambda = \frac{1}{\langle t \rangle}$$

Here $\langle t \rangle$ is the mean time-to-failure.

The *mean time-to-repair* (MTTR) for a repairable item is defined as the total corrective maintenance time divided by the total number of corrective maintenance actions during a given period of time.

The non-steady-state *operational availability index* $K_a(t)$ is defined as the probability that the item will be available (workable) at the given moment t of time and will operate failure free during the given time T, beginning with the moment t. The steady-state operational availability index K_a is the probability that the item will operate failure free during the time T, beginning with an arbitrary and sufficiently "remote" time t. In other words, it is the probability of the event that the item is available at an arbitrary moment of time t and will then operate failure free during the required time T. The most often used availability characteristic of the class III items, whose normal operation includes regular repairs, is the *availability index*

K_a^*, defined as the steady-state probability that the item will be available (workable) at an arbitrary moment t of time taken between the preplanned preventive maintenance activities. The availability index can be calculated by the formula:

$$K_a^* = \frac{1}{1 + \sum_{i=1}^{n} \langle t_r \rangle_i / \langle t_f \rangle_i} \qquad (10.33)$$

Here $\langle t_f \rangle_i$ is the mean time between successive failures for the ith element of the product and $\langle t_r \rangle_i$ is the mean time-to-repair for this element. The availability *index* K_a^* enables one to make assessments of the unforeseen idle times and to consider these times at early stages of the design of the product.

The operational availability index $K_a(t)$ can be calculated in a situation, where the probability of the failure-free operation during the time interval t is independent of the beginning of this interval, by the formula

$$K_a(t) = K_a^* R(t)$$

where $R(t)$ is the dependability of the item. This formula determines the probability that two events occur: (1) the item is available at an arbitrary moment of time with the probability K_a^* and (2) will operate failure free during the time period of the duration t. The problem of the availability of an item will be revisited in Sec. 11.3.7 on the basis of the theory of Markovian processes.

10.7. Choosing the Appropriate Reliability Indices

In previous sections, various reliability characteristics (indices) were introduced and discussed:

1. The probability, $R(t)$, of nonfailure during the time t (reliability function or the dependability).
2. Mean time-to-failure, $\langle t \rangle$ (MTTF), which is the mean time of dependable (failure-free) operation of nonrepairable items.
3. Mean time-to-repair, $\langle t_r \rangle$ (MTTR), which is the mean time required for a corrective maintenance (repair) of a repairable item.
4. Mean time-between-failures, $\langle t_f \rangle$ (MTBF), which is the mean time between consecutive failures of a repairable item.
5. Failure rate, $\lambda(t)$, which is the ratio of the failed items by the time t to the number of items that remained sound by this time.

6. Failure frequency, $f(t)$, which is the probability density function of the flow of failures.
7. Time-of-failure, t_γ, with the given probability $\gamma\%$. This time is defined by the relationship

$$R(t_\gamma) = \frac{\gamma\%}{100}$$

i.e., determines the time after which the failure will occur with the given probability R.
8. Availability index, K_a^*, which is the probability that the item will be available at an arbitrary moment of time taken between the preplanned maintenance activities.
9. Operational availability index, K_a, which is the probability that the item is not only available at an arbitrary moment of time sufficiently remote from the beginning of the operations but will also operate failure free during the given time t.

The following additional indices can be introduced, when necessary, by analogy with the above indices for items that can fail during storage:

10. The probability, $R_s(t)$, of nonfailure during storage for the time t.
11. Mean time of nonfailure $\langle t_s \rangle$ during storage.
12. Mean time $\langle t_m \rangle$, of maintenance, and others.

All these indices belong, as a rule, to one of the following two types: "time-dependent" indices (MTTF, MTBF, MTTR, etc.) and "probability-dependent" indices (dependability, repairability, etc.). In the "time-dependent" indices the observed random variables are various random times: time-to-failure, time-between-failures, time-to-repair, etc. In the "probability-dependent" indices, the observed random variables are the numbers of various events: the number of failures, repairs, etc. The selection of the appropriate reliability indices should be done depending on the type of item, its function, general operation requirements, consequence of failure, etc. Reliability indices must have a clear physical meaning, must lend themselves to simple experimental evaluation and verification, as well as to standardization, when necessary. The total number of reliability indices in a given problem should be minimal.

The following general guidelines can be helpful in choosing the appropriate reliability index or indices. This should be preceded by selection of the failure criteria and the "*ultimate condition*" of the product. The ultimate condition can be defined as the condition (state) that makes further use (operation) of the product (item) impossible or unfeasible. For nonrepairable items, typical "ultimate conditions" are: failure of the item; a situation where the failure rate of the items increases dramatically and can lead to a rapid failure or to violation of the safety requirements; a situation where the item becomes obsolete. For repairable items, the ultimate conditions might be somewhat different: dramatic loss of the effectiveness of the item; a drastic increase in the failure rate, so that another repair is considered economically unfeasible; the obsolescence of the item, etc.

The choice of failure criteria can be carried out using the following guidelines:

1. Based on customer requirements and the item operation conditions, establish the nomenclature of the parameters and the allowable tolerance limits, with consideration of the state-of-the-art in the given area of engineering.
2. Develop a list of technical characteristics that determine the performance of the item, considering the customer requirements and the possibility of meeting these requirements.
3. Establish the allowable limits for the chosen technical characteristics, so that exceeding these limits would be considered a failure. These limits should be included in the specifications as failure criteria.
4. Establish the "code number" of the item (Such an approach that enables one to formalize the selection of an appropriate reliability criteria, has been recommended by the Russian State Committee of Standardization in the early 70's). This number consists of four figures. The first figure is "1" if the item is non-repairable and is "0" for repairable items. The second figure characterizes the conditions making the further operation of the item impossible or unfeasible. For nonrepairable items (i.e., if the first figure is "1"), the second figure is also "1" if the item is operated until it fails; the second figure is "2" if the item is operated either until it fails or until it reaches the "ultimate condition"; the second figure is "3" if the item is operated until it fails or until it completes its mission. For repairable items (i.e., if the first figure is "2"), the second figure is "1" if the item is operated until the first failure; the second figure is "2" if the item is operated either until the first failure or until the "ultimate condition" is reached; the second figure is "3" if the item is operated either until its first failure or until its function is fulfilled; the second figure is "4" if the "ultimate condition" is reached; this figure is "5" if the "ultimate

condition" is reached in the idle state of the item, or if the item fails, or when its function is completed, although it might still be in the operating state.

The third figure characterizes the operation conditions of the item, depending on whether it is operated continuously ("1"), with regular interruptions ("2"), or with irregular interruptions ("3"). These factors are concerned with the loading conditions, thereby affecting the reliability criteria.

The fourth figure in the code number characterizes the consequences of failure. For nonrepairable items, this figure is "1" if the prevailing factor on which the consequences of failure are judged is simply the very fact of failure. The fourth figure for nonrepairable items is "2" if the item is judged depending on whether it does or does not fulfill the expected functions in the given capacity. For repairable items, the fourth figure is "1" if the prevailing factor on which the consequences of failure are judged is the items failure, regardless of the duration of idling; the fourth figure is "2" if the consequences of failure are judged depending on whether the item does or does not fulfill the expected functions in the given capacity; the fourth figure is "3" if the prevailing factor on which the consequences of failure are judged is the fact that the item is found to be inactive (idle); the fourth figure is equal to "4" if the prevailing factor on which the consequences of failure are judged is the items failure and its idling; the fourth figure is equal to "5" if the prevailing factor on which the consequences of failure are judged is the fact that the item does or does not filfill its functions in the given capacity, for an arbitrary moment at the beginning of the "working conditions."

5. After the "code number" of the item is established, the appropriate reliability index can be designated based on Table 10.8 recommendations. Note that the steady-state failure rate λ (or the effective time-to-failure $t_c = 1/\lambda$) is designed for items whose further use is deemed unfeasible after the specified (preplanned) service period. Such a situation occurs when the actual mean time-to-failure exceeds significantly the specified lifetime of the item.

Example 10.14 What does the "code number" 1312 mean?

Solution This is a nonrepairable item (1), which is expected to be operated either until failure or until its mission is fulfilled (3). The item operates in a continuous fashion (1), and the prevailing factor to be used to judge the consequences of its failure is whether the item does or does not perform satisfactorily in the given capacity at the given time (2).

Example 10.15 What are the prevailing factors that should be used to judge the consequences of failure for the following items (products): a household

TABLE 10.8 Selection of the appropriate reliability criterion (index) based on a code number reflecting the reliability requirements and conditions

Code number		Reliability index (indices)
1111	2111	Mean time-to-failure $\langle t \rangle$
1121	2121	
1131	2131	
1211	2211	Steady-state failure rate λ
1221	2221	and time-to-failure t_γ
1231	2231	(or time-to-failure and service time)
1222	2232	Reliability function $R(t)$
1232	2422	and time-to-failure t_γ
2222	2432	(or time-to-failure and service time)
1312		Reliability function $R(t)$
2312		
2411		Mean time-to-failure $\langle t \rangle$
2421		and time-to-failure t_γ
2431		(or time-to-failure and service time)
2413		Time-to-failure t_γ (or time-to-failure and service time)
2423		Availability index K_a^*
2433		and time-to-failure t_γ (or time-to-failure and service time)
2414		Mean time-to-failure $\langle t \rangle$ and time-to-failure t_γ (or time-to-failure and service time)
2424		Availability index K_a^*
2434		mean time-to-failure $\langle t \rangle$, and time-to-failure t_γ (or time-to-failure and time)
2415	2515	Operational availability
2425	2525	index K_a and time-to-
2435	2535	failure t_γ (or time-to-failure and service time)

TV set, an electronic system of a strategic ballistic missile, an automatic line for manufacturing commercial products, technological equipment for a good production factory, a fire engine?

Solution For a household TV set, the prevailing factor that should be used to judge the consequences of failure is simply the very fact of failure. The downtime of a household TV is typically not expected to cause serious consequences. As far as an electronic system of a ballistic missile is concerned, the prevailing factor upon which its reliability should be judged is the fact of the fulfillment or nonfulfillment of the missile's task in the given situation at the given time. For an automatic line designed for manufacturing commercial products, it is the undesirable idling time that is the prevailing factor which should be used for assessing the consequences of failure, since it is this downtime that will be responsible for the shortcut in the production volume. In the case of a technological equipment for a food production factory both the fact of failure and the forced downtime of the equipment should be taken into account when evaluating the consequences of failure. This case is typical for items whose functioning is equally affected by the repair losses and the losses due to the forced idling. As to the fire engine, the major factor that should be considered, when assessing the consequences of failure, is whether this engine will or will not be able to perform adequately in the "working condition," i.e., when it is needed to extinguish fire. This case is typical for items that operate during their service time in two major conditions: the "waiting" condition and the "working" one.

Example 10.16 Determine the reliability index that should be used to characterize the reliability of a household electric bulb.

Solution The bulb is a nonrepairable item (1), it is operated until failure (1), it can be operated, depending on the service conditions, in continuous (1), cyclicly regular (2), or cyclicly irregular (3) conditions. As far as the consequences of failure are concerned, the prevailing factor for the assessment of these consequences is the very fact of failure. These considerations lead to the following "code numbers:" 1111, 1121, 1131. Whatever the code number, the appropriate reliability index is the mean time-to-failure.

Example 10.17 Determine the reliability index that should be used to characterize the reliability of a rope intended for lifting and descending a passenger elevator in a tower.

Solution The rope is a nonrepairable item (1), which is operated until failure or until reaching an "ultimate state" (2). The load acting on the rope changes in a cyclicly regular fashion (2). The consequences of failure are catastrophic; therefore the prevailing factor when assessing such consequences is the fact that the rope does or does not fulfill the designed function (2). Hence, the code number for this rope is 1222. In accordance with Table 10.8, the appropriate reliability indices are: (1) the dependability (reliability) function, i.e., the probability of nonfailure during the time t of the elevator lifting or descending and (2) the time-to-failure t_y (or the time-to-failure and service time).

Example 10.18 Determine the appropriate reliability indices for the items listed in Example 10.15.

Solution The code numbers for these items are: 2411, 2421, or 2431 for a household TV set; 2312 for an electronic system of a ballistic missile; 2424 or 2434 for an automatic line producing commercial products; 2414 for technological equipment for a food line; and 2415, 2425, or 2435 for a fire engine. The corresponding reliability indices of these items are: failure rate and time-to-failure (or time-to-failure and service life); the reliability function (the probability of nonfailure during the given time); the availability index, failure rate, and time-to-failure (or time-to-failure and service time); failure rate, time-to-failure (or time-to-failure and service time); and the operational availability index and time-to-failure and service time).

10.8. Updating Reliability: Conjugate Distributions

The estimated probability P of a system's dependability can be treated itself as a random variable. Its probabilistic characteristics can be determined from an appropriate probability distribution. Let us assume that one knows the mean $\langle p \rangle$ and the variance D_p of the probability P. With four parameters $P_{min} = 0$, $P_{max} = 1$, $\langle p \rangle$, and D_p, the principle of maximum entropy suggests that the beta distribution should be taken as an appropriate probability distribution, since it is the one with the maximum entropy (see Sec. 3.3.11 and Chapter 6). Using the formulas (3.47) and (3.48), we obtain the following expression for the probability density function of the random variable P:

$$f(p) = \frac{(\alpha + \beta + 1)!}{\alpha! \beta!} p^\alpha (1-p)^\beta \qquad (10.34)$$

where α and β are integers. Comparing this formula with the formula (2.4) for the binomial distribution

$$f(x) = C(n, x) p^x (1-p)^{n-x} = \frac{n!}{(n-x)! x!} p^x (1-p)^{n-x} \qquad (10.35)$$

we conclude that the value $x = \alpha$ can be viewed as the number of successes in n trials in a binomial process, while the value $n - x = \beta$ is the number of failures. In terms of the parameters of the beta distribution, the binomial distribution can be written as

$$f(x) = \frac{(\alpha + \beta)!}{\alpha! \beta!} p^\alpha (1-p)^\beta \qquad (10.36)$$

Some probability distributions possess the following important property. The formula

$$P(H_i | A) = P(H_i) P(A | H_i) \qquad (10.37)$$

for the conditional probability (see Sec. 1.3) states that the posterior (post-experimental) probability $P(H_i | A)$ can be calculated as a product of the prior (pre-experimental) probability $P(H_i)$ and the new information provided by the distribution $P(A|H_i)$. Let a certain distribution be selected as the prior distribution and another one be chosen as the distribution providing new information ("additional knowledge"). If the calculated posterior distribution is of the same type as the prior distribution, then the distribution's characterizing prior information and the new information are said to be *conjugate distributions*. In other words, if a pair of distributions is such that the operation, defined by the formula (10.37), does not change the type of that one which is chosen as the prior distribution, then this pair of distributions are conjugate distributions. The beta and the binomial distributions are examples of conjugate distributions: if the prior distribution is a beta one and the new information is provided by a binomial distribution, then the posterior distribution is also a beta distribution.

The mean and the variance of the beta distribution (with $a = P_{min} = 0$ and $b = P_{max} = 1$) are

$$\langle p \rangle = \frac{\alpha + 1}{\alpha + \beta + 2}, \quad D_p = \frac{(\alpha + 1)(\beta + 1)}{(\alpha + \beta + 2)^2(\alpha + \beta + 3)} \quad (10.38)$$

Example 10.19 Determine the beta distribution for four successes and one failure. How will this distribution change with no prior knowledge of the item's performance?

Solution For four successes and one failure ($\alpha = 4$, $\beta = 1$) the formulas (10.38) yield $\langle p \rangle = 0.71$ and $D_p = 0.0256$. With $\alpha > \beta$, the distribution skews to the lower D_p values. This is due to the fact that if there are more successes than failures, the probability of nonfailure should increase. The distribution of the probability of nonfailure of $\alpha = 4$ and $\beta = 1$ is shown in Fig. 10.3. With no prior knowledge of the item's performance ($\alpha = \beta = 0$), the resulting distribution would be $f(p) = 1$. This is a uniform distribution, as it is supposed to be, in accordance with the principle of maximum entropy. For one success and no failures ($\alpha = 1$, $\beta = 0$) one obtains a triangular distribution: $f(p) = 2p$.

Example 10.20 An experiment is repeated a number of times and no failures were observed. Determine the expected number of successes as a function of the probability of non-failure.

Solution Solving the first equation in (10.38) for the number α of successes and putting, for the case of no failures, $\beta = 0$, we obtain:

$$\alpha = \frac{2\langle p \rangle - 1}{1 - \langle p \rangle} \quad (10.39)$$

If, for instance, $\langle p \rangle = 0.71$, then $\alpha = 1.448)$. Since the α value should be an integer, we have $\alpha = 1$. This corresponds to a triangular distribution $f(p) = 2p$.

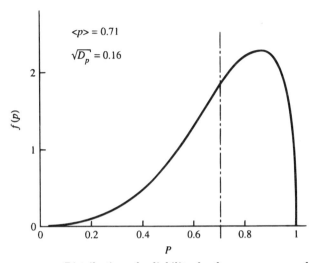

Figure 10.3 Distribution of reliability for four successes and one failure. (From M. E. Harr, "*Reliability-Based Design in Civil Engineering,*" McGraw-Hill, New York, 1987).

Example 10.21 Suppose that an item failed, despite the relatively high level $\langle p \rangle = 0.95$ of the anticipated probability of non-failure. What is the revised estimate of this probability?

Solution Let the prior measure β_0 of β in Eq. (10.34) be very small in comparison with the prior estimate α_0 of α. Assuming that β_0 was zero in the initial design and using Eq. (10.39) with $\langle p \rangle = 0.95$, we obtain $\alpha_0 = 18$. Hence, for the new posterior failure, $\alpha = 18$ and $\beta = 1$ produces, from the first equation in (10.38), the revised mean probability $\langle p \rangle = 0.905$ of non-failure. An additional failure ($\alpha = 18$, $\beta = 2$) reduces this probability to $\langle p \rangle = 0.864$. Note that, after the first failure, as many as 19 additional continuous successes (37 successes and 1 failure) would have to be registered in order to return the reliability to its original estimate of 95 percent.

10.9. Confidence Intervals

Suppose that one deals with a normal random variable X and wishes to obtain a general expression for the upper, b, and lower, a, limits that will satisfy the condition (Fig. 10.4)

$$P(a \leq x \leq b) = \frac{\alpha}{2}$$

or $\qquad P(\langle x \rangle - k\sigma_x \leq x \leq \langle x \rangle + k\sigma_x) = 1 - \alpha \qquad (10.40)$

The range (a, b) is called the *confidence interval* of the variable X and $1 - \alpha$ is the *confidence level*. The probability $1 - \alpha$ provides a measure

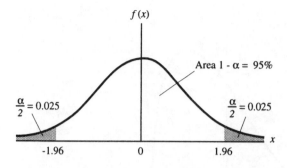

Confidence level (1 - α), %	$k_{\alpha/2}$
90.00	1.64
95.00	1.96
95.45	2.00
98.00	2.33
99.00	2.58
99.50	2.81
99.73	3.00
99.90	3.29
99.99	3.89
99.994	4.00

Figure 10.4 Confidence interval for normal distribution.

of belief that the confidence limits (a, b) will contain the realization x. The choice of the confidence level $1 - \alpha$ is a matter of the risk that one is willing to take in accepting an erroneous value of the random variable under consideration. Values of the factor k can be obtained from the tables for the normal distribution (Appendix B). We will revisit the problem of the evaluation confidence intervals twice: in Chapter 14, in connection with the random geometry of machine parts, and in Chapter 16, in connection with processing of random data.

Example 10.22 The flow rate in a channel is a normally distributed random variable with the mean of $\langle q \rangle = 14.5$ ft^3/s and a coefficient of variation of 40 percent. Obtain the 98 percent confidence limits for the flow rate.

Solution With $\langle q \rangle = 14.5$ ft^3/s and $v_q = 0.4$ we have $\sigma_q = 0.40 \times 14.5 = 5.8$ ft^3/s. For $1 - \alpha = 98$ percent, from the table in Fig. 10.4 we obtain $k = 2.33$. Hence, the lower confidence limit is $a = 14.5 - 2.33 \times 5.80 = 0.99$ and the upper one is $b = 14.5 + 2.33 \times 5.80 = 28.01$.

Example 10.23 The thickness of the adhesive layer in an adhesively bonded joint is a normally distributed random variable with the mean $\langle t \rangle = 1$ mil

(25.4 μm) and the variance $D_t = 0.0625$ mil^2. Determine the 99.9 percent confidence limits for the thickness of the adhesive.

Solution For $1 - \alpha = 99.9$ percent the table in Fig. 10.4 yields $k = 3.29$. The confidence limits are $a = 1 - 3.29\sqrt{0.0625} = 0.1775$ mils and $b = 1 + 3.29\sqrt{0.0625} = 1.8225$ mils. Thus, the probability that the actual thickness of the adhesive layer will be found within the interval 0.1775 mils $< h <$ 1.8225 mils is 0.999. If the confidence level is reduced to 90 percent, the confidence limits are $a = 1 - 1.64\sqrt{0.0625} = 0.5900$ mils and $b = 1 + 0.64\sqrt{0.0625} = 1.1600$ mils. The confidence interval is substantially narrower in this case.

10.10. Accelerated Testing

With today's highly reliable products, reliability tests at normal use conditions would take too long to obtain meaningful failure characteristics. Therefore one has to force products to fail by testing them at much more severe conditions than the anticipated actual conditions, i.e., to *accelerate* the product failures by enhancing one or several failure mechanisms. This enables one to obtain reliability characteristics at practical test times and with relatively small numbers of test specimens. Clearly, since one has to "distort" at least one of the actual operating parameters, theoretical modeling is needed to predict the performance of the product at these conditions from the observed performance during accelerated tests. These are known as accelerated test models.

The basic requirement (hypothesis) underlying accelerated test modeling and experimental design is that failure mechanisms and modes should be the same at accelerated and real-life conditions. The only difference is that in accelerated tests "things happen faster" because of higher temperatures, larger stresses, elevated humidities, greater accelerations, higher voltages, etc.

Example 10.24 Determine the acceleration factor, assuming that the probability distribution function is the same at accelerated and actual environmental conditions. Consider exponential, Weibull, and log-normal distributions.

Solution When exponential law is applied the probability of failure is

$$F(t) = 1 - e^{-\lambda t}$$

where λ is the steady-state failure rate and t is time. If this probability is required to be the same in actual and accelerated conditions, then the following relationship between the actual, λ_r, and accelerated, λ_a, failure rates must take place:

$$\lambda_a = (k_A)\lambda_r$$

where k_A is the acceleration factor. In other words, if the accelerated tests indicate that the failure rate is λ_a and the mean time-to-failure is t_a, then, at use conditions, the failure rate can be evaluated as

$$\lambda_r = \lambda_a \frac{t_a}{t_r}$$

where t_r is the mean time-to-failure for use conditions.

In the case of the Weibull distribution,

$$F(t) = \exp\left[-\left(\frac{r}{b}\right)^a\right]$$

and the acceleration factor is

$$k_A = \frac{t_r}{t_a} = \left(\frac{a}{b}\right)_a \div \left(\frac{a}{b}\right)_r$$

If one intends to keep the shape parameter a the same, then the scale parameter b_r at use conditions should be related to the scale parameter b_a at accelerated tests as

$$b_r = (k_A) b_a$$

In a log-normal distribution, the probability of failure is expressed by the formula

$$F(t) = \frac{1}{2}\left[1 + \Phi\left(\frac{\log t/y_0}{\sqrt{2D}}\right)\right]$$

Making the acceleration transformation in the time scale and assuming that the variance D is the same at both actual and accelerated test conditions, we obtain

$$k_A = \frac{(y_0)_r}{(y_0)_a}$$

where $(y_0)_r$ and $(y_0)_a$ values at actual and accelerated conditions, respectively, are related to the mean time-to-failure t for these conditions as

$$y_0 = t e^{-D/(2M^2)}$$

where $M = 0.43043$ is the coefficient of transfer of the natural logarithms to the decimal ones. Clearly,

$$k_A = \frac{t_r}{t_a}$$

Thus, the linear acceleration of the environmental conditions does not change the type of distribution and its shape parameter. Only the scale parameter is changed and should be multiplied by the acceleration factor. Then one can easily convert the results of the accelerated test data analysis to the failure rate projections in the use condition.

Example 10.25 The thermal expansion (contraction) mismatch strain ε during accelerated testing of a solder joint interconnection in an electronic package is a random variable, distributed in accordance with the normal law

$$f_\varepsilon(\varepsilon) = \frac{1}{\sqrt{2\pi D_\varepsilon}} \exp\left[-\frac{(\varepsilon - \langle \varepsilon \rangle)^2}{2D_\varepsilon}\right]$$

The number of cycles to failure N_f is related to the strain ε by the formula

$$N_f = \left(\frac{\varepsilon^*}{\varepsilon}\right)^m$$

where ε^* is a factor of the angular strain in the joint and the exponent m characterizes the particular solder material. Find the probability density distribution function for the random variable N_f and the probability that this variable exceeds a certain value N.

Solution Using the formula

$$f_N(N_f) = f_\varepsilon(\varepsilon) \frac{d\varepsilon}{dN_f}$$

("elementary probabilities" of the number-of-cycles-to-failure, N_f, and the induced strain, ε, should be the same) we find

$$f_N(N_f) = \frac{N_f^{-(m+1)/m}}{\sqrt{2\pi D_N}} \exp\left[-\frac{m^2(N_f^{-1/m} - \langle N_f \rangle)^2}{2D_N}\right]$$

where the following notation is used:

$$\langle N_f \rangle = \frac{\varepsilon^*}{\langle \varepsilon \rangle}, \quad D_N = \frac{m^2}{(\varepsilon^*)^2} D_\varepsilon$$

The probability that the strain ε will exceed the level $\bar{\varepsilon}$ can be found as

$$P(\varepsilon > \bar{\varepsilon}) = \int_{\bar{\varepsilon}}^{\infty} f_\varepsilon(\varepsilon)\, d\varepsilon$$

$$= \frac{1}{\sqrt{2\pi D_\varepsilon}} \int_{\bar{\varepsilon}}^{\infty} \exp\left[-\frac{(\varepsilon - \langle \varepsilon \rangle)^2}{2D_\varepsilon}\right] d\varepsilon$$

$$= \frac{1}{\sqrt{\pi}} \int_{(\bar{\varepsilon} - \langle \varepsilon \rangle)/\sqrt{2D_\varepsilon}}^{\infty} e^{-t^2}\, dt$$

$$= \frac{1}{\sqrt{\pi}} \left(\int_0^{\infty} e^{-t^2}\, dt - \int_0^{(\bar{\varepsilon} - \langle \varepsilon \rangle)/\sqrt{2D_\varepsilon}} e^{-t^2}\, dt\right)$$

$$= \frac{1}{2} - \frac{1}{2}\Phi\left(\frac{\bar{\varepsilon} - \langle \varepsilon \rangle}{\sqrt{2D_\varepsilon}}\right)$$

where $\Phi(\alpha)$ is the Laplace function. Thus, the probability that the number of cycles to failure will exceed a certain level N^* can be calculated as

$$P(N_f > N^*) = \frac{1}{2}\left[1 - \Phi\left(\frac{\varepsilon^* N^* - \langle \varepsilon \rangle}{\sqrt{2D_\varepsilon}}\right)\right]$$

10.11. Technical Diagnostics

10.11.1. Objective of technical diagnostics

Technical diagnostics emerged in connection with the increased requirements for safety, dependability, and durability of complex and costly engineering systems as one of the important directions of reliability engineering. The main objective of technical diagnostics is to determine the state (*diagnosis*) of the system from the standpoint of its propensity to failure.

Technical diagnostics deals with a wide variety of problems, associated with obtaining, processing, and interpretation of the diagnostic information, including the development and analysis of diagnostic models and decision algorithms. Using technical diagnostics methods, the reliability engineer can detect early incipient flaws and damages, and predict and prevent potential failures. Quite often this can be done during the regular (preventive) maintenance of the system, even without interrupting its normal operation. The amount of information necessary for carrying out diagnostic analyses is never complete, and therefore a probabilistic evaluation should be conducted to draw a sufficiently reliable conclusion.

Technical diagnostics is sometimes viewed as a part of a more general theory of image recognition. This theory is aimed at the recognition of images of any nature, such as, for instance, geometric figures, sounds, or speech. Technical diagnostics studies recognition algorithms in application to diagnostic problems. These can be viewed, from the standpoint of the image recognition theory, as classification problems.

The main objective of the diagnostic models and algorithms is to establish the relationships between the state (diagnosis) of the technical system and its manifestation (*signature*) in the *space of diagnostic signals*. Based on the solution to a diagnostic problem, the reliability engineer can judge whether the given item should be considered faulty or sound. Clearly, such a conclusion is always associated with the risk of a "false alarm" ("seller's risk") or with the risk of "missing a target" ("buyer's risk"). Technical diagnostics uses the theory of statistical decisions to minimize these risks.

The general probabilistic diagnostic problem can be formulated as follows. It is known that the system can possibly be in one of n random states (diagnoses) D_i, $i = 1, 2, \ldots, n$. One knows also the signatures (parameters) of these states, as well as their probabilities. The problem is to develop a decision rule (algorithm) that would enable an engineer to attribute these signatures to one of the possible states of the system. In addition, it is desirable to assess the confidence of the taken decision, as well as the risk of taking an erroneous decision.

10.11.2. Statistical methods of recognition

10.11.2.1. Application of the Bayes formula.
The Bayes method (see Sec. 1.6), because of its simplicity and effectiveness, is widely used for the purpose of technical diagnostics. This method is advisable in all cases where the amount of statistical information is large.

The Bayes formula can be written, for diagnostics applications, as follows:

$$P(D_i/k_j) = P(D_i)\frac{P(k_j/D_i)}{P(k_j)}, \quad i = 1, 2, \ldots, m, j = 1, 2, \ldots, n \quad (10.41)$$

In this formula, $P(D_i)$ is the probability of the state (diagnosis) D_i that is evaluated on the basis of the statistical data (prior probability of the diagnosis), $P(k_j/D_i)$ is the probability of occurrence of the signature k_j in the system found in the state D_i, $P(D_i/k_j)$ is the probability of the state (diagnosis) D_i after the signature k_j has been observed (posterior probability of the diagnosis), and $P(k_j)$ is the probability of the occurrence of the signature k_j in any component of the system, regardless of the state (diagnosis) of the particular component. The relationship (10.41) follows from the fact that the probability that a certain state (diagnosis) D_i takes place along with its signature k_j can be obtained either as $P(D_i)P(k_j/D_i)$ or as $P(k_j)P(D_i/k_j)$. If N components are evaluated and in N_i of them the state D_i takes place, then the (prior) probability of the diagnosis D_i is

$$P(D_i) = \frac{N_i}{N}$$

If N_i components are found in the state (diagnosis) D_i and in N_{ij} of them the signature k_j is observed, then the probability of the occurrence of the signature k_j in the system found in the state D_i is

$$P(k_j/D_i) = \frac{N_{ij}}{N_i}$$

Finally, if the signature k_j is observed in N_j components out of the total number of N components, then the probability of occurrence of the signature k_j in the system is

$$P(k_j) = \frac{N_j}{N}$$

Typically, the observations are carried out for a *complex signature* K. This includes "elementary" signatures $k_1, k_2, \ldots, k_n$, each of which may consist of a certain number of "subsignatures." As a result of these observations (inspections) one determines the realization K^* of the complex signature K. Then the Bayes formula can be written as

$$P(D_i/K^*) = P(D_i)\frac{P(K^*/D_i)}{P(K^*)}, \quad i = 1, 2, \ldots, m \quad (10.42)$$

Here $P(D_i/K^*)$ is the corrected probability of the diagnosis D_i after the results of the inspection of the complex signature K become available and $P(D_i)$ is the prior probability of the diagnosis D_i (based on the preceding statistical analyses). The formula (10.42) can be applied to each of the m possible states (diagnoses) of the system. If the system can be found only in one of such states, then the following normalization conditions must be fulfilled:

$$\sum_{S=1}^{m} P(D_S) = 1$$

In practice, there is often a possibility that several states exist simultaneously, and some of these states can be in combination with the others. Then one has to view separate states and separate combinations of states as particular diagnoses D_i. If the complex signature K consists of n signatures, then

$$P(K^*/D_i) = P(k_1/D_i)P(k_2/k_1 D_i) \cdots P(k_n/k_1 \cdots k_{n-1} D_i) \quad (10.43)$$

or, for diagnostically independent signatures,

$$P(K^*/D_i) = P(k_1/D_i)P(k_2/D_i) \cdots P(k_n/D_i) \quad (10.44)$$

In many practical problems, especially when the number of signatures is large, the signatures can be considered independent, even if there are essential correlations between them. The probability $P(K^*)$ of occurrence of a complex signature K^* is

$$P(K^*) = \sum_{S=1}^{m} P(D_S)P(K^*/D_S)$$

The generalized Bayes formula can be written in this case as

$$P(D_i/K^*) = \frac{P(D_i)P(K^*/D_i)}{\sum_{s=1}^{n} P(D_s)P(K^*/D_s)} \tag{10.45}$$

where the probability $P(K^*/D_i)$ can be calculated by the formula (10.44). As evident from (10.45),

$$\sum_{i=1}^{m} P(D_i/K^*) = 1$$

It has to be this way, since one of the signatures has to be realized in any event, while simultaneous realization of two independent diagnoses is impossible. It should be pointed out that the denominator in the formula (10.45) is the same for all the diagnoses. This enables one to determine first the probabilities of the simultaneous occurrence of the ith diagnosis and the given realization of the complex signature

$$P(D_i K^*) = P(D_i)P(K^*/D_i)$$

and then evaluate the posterior probability of the diagnosis

$$P(D_i/K^*) = \frac{P(D_i K^*)}{\sum_{s=1}^{m} P(D_s K^*)}$$

When the Bayes method is used, the *decision rule* can be formulated as follows: the object (system) with the complex signature K^* has a state (diagnosis) D_i if the posterior probability $P(D_i/K^*)$ is the largest in comparison with the other states (diagnoses). This rule is usually associated with the consideration of a *threshold probability* of a diagnosis:

$$P(D_i/K^*) \geq P_{S,i}$$

Here $P_{S,i}$ is the chosen level of recognition for the diagnosis D_i (say, $P_{S,i} = 0.9$ or higher). If this condition is not fulfilled, additional investigations are necessary to establish the diagnosis. The selection of the threshold probability $P_{S,i}$ should be carried out considering the consequences of failure, because of an erroneous diagnosis, the effort required to conduct additional investigations, etc.

The two major shortcomings of the Bayes method are: the necessity to obtain a large amount of the preliminary statistical information and the "suppression" of rarely encountered states (diagnoses). In order to overcome the second shortcoming it is advisable to carry out additional diagnostics analysis, for the case of equally possible diagnoses, by putting $P(D_i) = 1/m$, i.e., by assuming that all the prior probabilities of the different states are the same. Then the diagnosis D_i, for which the probability $P(K^*/D_i)$ is the largest, will have the largest posterior probability:

$$K^* \in D_i \quad \text{if} \quad P(K^*/D_i) > P(K^*/D_j), \qquad j = 1, 2, \ldots, m; \; i \neq j$$

In other words, one establishes the diagnosis D_i if the given set of signatures is encountered more often in the diagnosis D_i than in other diagnoses (the *method of maximum likelihood*). In the method of maximum likelihood, "typical" and "atypical" diagnoses (states) are equally represented in the diagnostic analysis. The appropriate threshold in such an approach can be selected based on the condition

$$P(K^*/D_i) \geq P_{S,i}$$

Example 10.26 Two signatures, k_1 and k_2, have been recorded during the operation of a gas turbine: the increase in the gas temperature behind the turbine by more than 50 °C (k_1) and the increase in the acceleration time (i.e., the time required to achieved the maximum r/min) by more than 5 s (k_2). It is known that, for the given turbine type, the occurrence of these signatures has to do either with a damage in the fuel supply system (D_1) or with an increase in the radial gap in the turbine (D_2). It is also known that, if the turbine operates normally (state D_0), the signature k_1 does not occur and the signature k_2 is observed in 5 percent of all cases; that in the state D_1 the probabilities of occurrence of the signatures k_1 and k_2 are 20 and 30 percent, respectively; and that in the state D_2 these probabilities are 40 and 50 percent. Finally, it is also known, based on statistical data, that by the end of the operation time (lifetime) 80 percent of the turbines of the type in question are in the condition D_0, 5 percent are in the condition D_1, and 15 percent are in the condition D_2. Determine the probabilities of the states in which both signatures k_1 and k_2 occur, assuming that they are statistically (diagnostically) independent. Determine the probabilities of the states in which the signature k_1 will not occur, and the signature k_2 will, and the probability that none of the signatures k_1 or k_2 occur.

Solution For the sake of convenience, let us present the input information in the form of a *diagnostic matrix* (Table 10.9). In accordance with the formula (10.43), the probability of the same D_0 (the turbine operates

TABLE 10.9 Diagnostic matrix

i	D_i	$P(k_1/D_i)$	$P(k_2/D_i)$	$P(D_i)$
0	D_0	0	0.05	0.80
1	D_1	0.20	0.30	0.05
2	D_2	0.40	0.50	0.15

normally), when the signatures k_1 and k_2 occur, can be calculated as

$$P(D_0/k_1k_2) = \frac{0 \times 0.05 \times 0.80}{0 \times 0.05 \times 0.80 + 0.20 \times 0.30 \times 0.05 + 0.4 \times 0.5 \times 0.15}$$

$$= \frac{0}{0.033} = 0$$

Similarly, the probabilities that the state D_1 or the state D_2 take place are

$$P(D_1/k_1k_2) = \frac{0.20 \times 0.30 \times 0.05}{0.033} = 0.909$$

$$P(D_2/k_1k_2) = \frac{0.40 \times 0.50 \times 0.15}{0.033} = 0.091$$

Since the absence of signature k_1 is the signature $\bar{k}_1$, the probability $P(k_1/D_i)$ can be calculated as $P(\bar{k}_1/D_i) = 1 - P(k_1/D_i)$. Having this in mind and using the formula (10.43), we obtain

$$P(D_0/\bar{k}_1k_2) = \frac{1 \times 0.05 \times 0.8}{1 \times 0.05 \times 0.8 + 0.8 \times 0.3 \times 0.05 + 0.6 \times 0.5 \times 0.15}$$

$$= \frac{0.04}{0.097} = 0.412$$

$$P(D_1/\bar{k}_1k_2) = \frac{0.8 \times 0.3 \times 0.05}{0.097} = 0.123$$

$$P(D_2/\bar{k}_1k_2) = \frac{0.6 \times 0.5 \times 0.15}{0.097} = 0.464$$

$$P(D_0/\bar{k}_1\bar{k}_2) = \frac{1 \times 0.95 \times 0.8}{1 \times 0.95 \times 0.8 + 0.8 \times 0.7 \times 0.05 + 0.6 \times 0.5 \times 0.15}$$

$$= \frac{0.760}{0.833} = 0.912$$

$$P(D_1/\bar{k}_1\bar{k}_2) = \frac{0.8 \times 0.7 \times 0.05}{0.833} = 0.034$$

$$P(D_2/\bar{k}_1\bar{k}_2) = \frac{0.6 \times 0.5 \times 0.15}{0.833} = 0.054$$

As evident from the calculated data, if neither of the signatures k_1 and k_2 is detected, the normal state, D_0, of the turbine is the most likely one (91.2 percent). The probability of the state D_2, characterized by an elevated radial gap, is the second most likely one. Its likelihood is, however, only 5.4 percent. The probability of the state D_1 (malfunction of the fuel supply system) is only 3.4 percent. On the other hand, if both the signatures k_1

and k_2 are observed, it is the state (diagnosis) D_1 that is the most likely one (90.9 percent). Although the state D_2 is possible as well, its probability is only 9.1 percent. In a situation where the signature k_1 (increase in the gas temperature) is not observed while the signature k_2 (elevated acceleration time) is, the difference in the probabilities of the states D_2 (the radial gap is too large) and D_0 (normal operation conditions) is insignificant. This means that additional investigations are necessary to establish the actual cause of the signature k_2.

10.11.2.2. Wald's method. In the Bayes method the total number of observations is decided upon beforehand. In Wald's method, the observations are performed in a sequential fashion and the number of these observations is as large as it is necessary to make a decision with the given risk level. Wald's method is concerned with the *optimal stopping rule*. The "choosy bride" problem (Example 1.13, Sec. 1.1) is an example of a situation when a decision on optimal stopping should be made.

Let the diagnoses D_1 and D_2 be judged based on a signature k_1. It has been found, for instance, using the Bayes method, that the probability $P(k_1/D_2)$ is larger than the probability $P(k_1/D_1)$, so that

$$\frac{P(k_1/D_2)}{P(k_1/D_1)} > A, \qquad K^* \in D_2 \qquad (10.46)$$

where A is the upper boundary of the decision to be made. Clearly, the decision should be made in this case in favor of the diagnosis D_2. In the opposite situation, where the signature k_1 is encountered substantially more often for the diagnosis D_1 than for the diagnosis D_2, the decision should be made in favor of the diagnosis D_2:

$$\frac{P(k_1/D_2)}{P(k_1/D_1)} < B, \qquad K^* \in D_1 \qquad (10.47)$$

where B is the lower boundary of the decision to be made. In the case where

$$B < \frac{P(k_1/D_2)}{P(k_1/D_1)} < A \qquad (10.48)$$

(the "*likelihood relationship*"), additional information is needed for making a decision. Such additional information, in accordance with Wald's method, can be obtained by further inspection of the signature k_2. If, for instance, this signature is not found for the object under observation, then the following product of the two likelihood relationships is formed:

$$\frac{P(k_1/D_2)}{P(k_1/D_1)} \frac{P(\bar{k}_2/D_2)}{P(\bar{k}_2/D_1)} \qquad (10.49)$$

If this product exceeds the upper boundary A, then a decision $K^* \in D_2$ should be made. In a similar fashion, the lower boundary B of the decision making can be considered. Additional inspections should be carried out until, for the chosen boundaries A and B, a highly likely (well-substantiated) decision can be made.

Quite often it is more convenient to consider the natural logarithm of the likelihood ratio rather than the likelihood ratio itself:

$$\ln\left[\frac{P(k_1/D_2)}{P(k_1/D_1)} \frac{P(\bar{k}_2/D_2)}{P(\bar{k}_2/D_1)}\right] > \ln A \qquad (10.50)$$

Such a representation is usually used if the signatures are normally distributed.

The procedure described above can be formally generalized for independent signatures as follows. Let $m - 1$ observations be conducted; it turns out that they were insufficient for making a decision:

$$B < \frac{P(k_1/D_2)}{P(k_1/D_1)} \cdots \frac{P(k_r/D_2)}{P(k_r/D_1)} < A, \qquad r = 1, 2, \ldots, m - 1 \quad (10.51)$$

After the mth observation, however, the condition

$$\frac{P(k_1/D_2)}{P(k_1/D_1)} \cdots \frac{P(k_m/D_2)}{P(k_m/D_1)} > A \qquad (10.52)$$

is fulfilled. Then one can conclude that the diagnosis D_2 is more likely than the diagnosis D_1. If, after the mth observation, the condition

$$\frac{P(k_1/D_2)}{P(k_1/D_1)} \cdots \frac{P(k_m/D_2)}{P(k_m/D_1)} < B \qquad (10.53)$$

is satisfied, then the diagnosis D_1 should be accepted. In practical diagnostic procedures the most informative signatures must be analyzed first to reduce the total amount of inspections.

Two types of error are likely during the recognition process. The decision can be made in favor of the diagnosis D_2, while, in effect, it is the state D_1 that takes place, and vice versa. If, for instance, the state D_1 of a system is a sound one and the state D_2 is a faulty one, then the error of the first type is a "false alarm" and the error of the second type is associated with "missing the fault" (the "target"). In market analyses, these errors are known as "seller's risk" and "buyer's risk" (or "customer's risk"), respectively.

Denoting the errors associated with a "false alarm" and "missing the target" as α and β, respectively, one can write the relationships (10.51) and (10.52) as follows:

$$\frac{1-\beta}{\alpha} \geq A, \qquad \frac{\beta}{1-\alpha} \leq B \qquad (10.54)$$

In practical applications, the levels α and β are established at a sufficiently low level (say, $\alpha = \beta = 0.05$) based on the possible consequences of making an erroneous decision. Wald's method is also known as *chain sampling* (see Sec. 16.5.3) and is used in statistical quality control and in the general theory of choice and decision making.

Example 10.27 The mean value of the variable stress in a sound gas turbine is $\langle x_1 \rangle$, while in a faulty one it is substantially higher than $\langle x_2 \rangle$. The variances of the stress in either state of the turbine can be assumed to be the same: $\sigma^2 = \sigma_2^2 = \sigma^2$. The diagnostics is carried out by measuring the variable stress in the turbine blades. The random process of the induced stress can be considered to be normal. Using Wald's method, assess the state of the turbine based on the tensometric data.

Solution In accordance with the procedure of Wald's method, one measures first the stress in the first blade and forms the ratio of the probability density functions, calculated for $x = x_1$:

$$\frac{f(x_1/D_2)}{f(x_1/D_1)} = \exp\left\{-\frac{1}{2\sigma^2}[(x_1 - \langle x_1 \rangle)^2 - (x_1 - \langle x_2 \rangle)^2]\right\}$$

After the stress in the nth blade has been measured, the following quantity can be computed:

$$\eta_n = \ln \frac{f(x_1/D_2) \cdots f(x_n/D_2)}{f(x_1/D_1) \cdots f(x_n/D_1)} = \frac{1}{2\sigma^2} \sum_{i=1}^{n}[(x_i - \langle x_1 \rangle)^2 - (x_i - \langle x_2 \rangle)^2]$$

$$= \frac{\langle x_2 \rangle - \langle x_1 \rangle}{\sigma^2}\left[\sum_{i=1}^{n} x_i - \frac{\langle x_2 \rangle + \langle x_1 \rangle}{2} n\right]$$

If this quantity falls within the limits $\ln A$ and $\ln B$, so that

$$\ln B < \eta_n < \ln A$$

then there are no sufficient grounds to decide whether the turbine should be considered sound or faulty. This condition can also be written as

$$b_1 + an < \sum_{i=1}^{n} x_i < b_2 + an \qquad (10.55)$$

where

$$b_1 = \frac{\sigma^2}{\langle x_2 \rangle - \langle x_1 \rangle} \ln \frac{\beta}{1-\alpha}$$

$$b_2 = \frac{\sigma^2}{\langle x_2 \rangle - \langle x_1 \rangle} \ln \frac{1-\beta}{\alpha}$$

and

$$a = \frac{\langle x_2 \rangle + \langle x_1 \rangle}{2}$$

The condition (10.55) for different numbers n of observations corresponds (Fig. 10.5) to the area between the two parallel lines. If the sum $\sum_{i=1}^{n} x_i$ is found within the area between these lines, the state of the turbine is unclear and the tests should be continued. However, if this sum is found outside the "corridor" between the parallel lines, then it is possible to make a proper decision about the state of the turbine. The width of this "corridor" increases with the decrease in the parameters α and β, the decrease in the difference $\langle x_2 \rangle - \langle x_1 \rangle$, and the increase in the variance σ^2 of the measured stress in the turbine blades.

10.11.2.3. Methods of statistical decisions. Unlike the Bayes and Wald methods, the methods of statistical decisions are based on conditions of optimization, such as, for instance, the condition of minimum risk. Methods of statistical decisions have found extensive application in radars, communication systems, and in other areas of radio and telecommunications engineering as well as in numerous economics and management problems. We will examine, however, a mechanical problem.

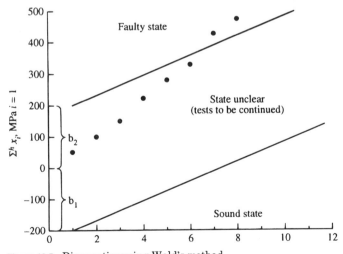

Figure 10.5 Diagnostics using Wald's method.

Let, for instance, the diagnostics of an engine be based on the random amount X of the iron detected in the lubrication oil. The problem is in selecting the critical value x_0 of the parameter X. If the actual amount X of the detected iron is larger than its critical value x_0, further operation of the engine cannot be permitted. In the case $X < x_0$, the engine is considered sound. In mathematical terms, this requirement can be written as follows:

$$X \in D_1 \quad \text{for } X < x_0 \quad \text{and} \quad X \in D_2 \quad \text{for } X > x_0 \quad (10.56)$$

where D_1 is the sound and D_2 is the faulty state of the engine.

Let H_{ij} ($i, j = 1, 2$) be possible statistical decisions. The first subscript in the notation H_{ij} identifies the accepted diagnosis and the second subscript identifies the actual state. Hence, H_{11} and H_{22} are the correct decisions and H_{12} ("missing the defect") and H_{21} ("false alarm") are erroneous decisions. Let the probability density functions for the variable X be $f_1(x/D_1)$ in the sound condition and $f_2(x/D_2)$ in the faulty condition of the engine.

The probability of a "false alarm" can be calculated as the probability of two events: (1) the engine is sound and (2) the critical value x_0 is chosen in such a way that $X > x_0$. Then we have

$$P(H_{21}) = P(D_1)P(X > x_0/D_1) = P_1 \int_{x_0}^{\infty} f_1(x/D_1)\, dx$$

where $P_1 = P(D_1)$ is the prior probability of the diagnosis D_1. This probability is supposed to be available from the preliminary statistical data. The probability of "missing the defect" can be determined in a similar way:

$$P(H_{12}) = P(D_2)P(X < x_0/D_2) = P_2 \int_{-\infty}^{x_0} f_2(x/D_2)\, dx$$

where $P_2 = P(D_2)$ is the prior probability of the diagnosis D_2.

The probability of making an erroneous decision can be obtained as the sum of the "weighed" probabilities of the "false alarm" and "missing the defect:"

$$R = C_{21}P_1 \int_{x_0}^{\infty} f_1(x/D_1)\, dx + C_{12}P_2 \int_{-\infty}^{x_0} f_2(x, D_2)\, dx$$

Here C_{21} and C_{12} are the "costs" of making erroneous decisions. The level of such "costs" should be estabished beforehand with consideration of the consequences of making an erroneous decision. Typically, the cost of "missing a defect" is higher than the cost of a "false alarm" ($C_{12} \gg C_{21}$). Sometimes the costs C_{11} and C_{22} of making correct decisions H_{11} and H_{22} are introduced, in addition to the costs C_{21} and C_{12} of the erroneous decisions. The costs C_{11} and C_{22} are negative.

In the general case, the mean risk can be expressed as follows:

$$R = C_{11}P_1 \int_{-\infty}^{x_0} f_1(x/D_1)\,dx + C_{21}P_1 \int_{x_0}^{\infty} f_1(x/D_1)\,dx$$
$$+ C_{12}P_2 \int_{-\infty}^{x_0} f_2(x/D_2)\,dx + C_{22}P_2 \int_{x_0}^{\infty} f_2(x/D_2)\,dx \quad (10.57)$$

In order to determine the *minimum risk*, one should differentiate this equation by x_0 and equate the obtained derivative to zero. This results in the relationship

$$\frac{f_1(x_0/D_1)}{f_2(x_0/D_2)} = \frac{C_{12} - C_{22}}{C_{21} - C_{11}} \frac{P_2}{P_1} \quad (10.58)$$

To make sure that the x_0 value determined from this equation corresponds to a minimum (not to a maximum) risk, one should select, from two roots given by the condition (10.56), the root x_0, which satisfies the condition

$$\frac{f'_1(x_0/D_1)}{f'_2(x_0/D_2)} < \frac{C_{12} - C_{22}}{C_{21} - C_{11}} \frac{P_2}{P_1} \quad (10.59)$$

If the distributions $f_1(x/D_1)$ and $f_2(x/D_2)$ are unimodal ones, i.e., if they have only one maximum, then, for $\langle x_1 \rangle \langle x_0 \rangle < \langle x_2 \rangle$, the condition (10.57) is always fulfilled.

From the conditions (10.56) and (10.58), one can conclude that the engine is sound ($X \in D_1$) if the condition

$$\frac{f_1(x/D_1)}{f_2(x/D_2)} > \frac{C_{12} - C_{22}}{C_{21} - C_{11}} \frac{P_2}{P_1} = \lambda \quad (10.60)$$

is fulfilled; otherwise the engine should be considered to be faulty ($X \in D_2$). The quantity

$$\lambda = \frac{C_{12} - C_{22}}{C_{21} - C_{11}} \frac{P_2}{P_1} \quad (10.61)$$

is the threshold value (the likelihood ratio). If "bonuses" for making correct decisions are not considered ($C_{11} = C_{22} = 0$), the condition

(10.59) can be simplified:

$$\lambda = \frac{C_{12}}{C_{21}} \frac{P_2}{P_1} \tag{10.62}$$

The actual cost of the consequences of making an erroneous decision are often unknown and cannot be easily determined. At the same time, it is always desirable to minimize the probability of a "false alarm" for the given (allowable) level of "missing the defect," and vice versa. If the allowable (desirable) level of the probability of a "false alarm" is known and one wants to minimize the probability of "missing a target," then the following *Neumann–Pierson condition* can be used:

$$P_1 \int_{x_0}^{\infty} f_1(x/D_1)\, dx \leq A \tag{10.63}$$

In this condition, A is the allowable level of the probability of a "false alarm" and P_1 is the probability that the object is in the sound state. The minimum value of the probability of "missing a target" corresponds to the x_0 value. This value can be found from the relationship (10.63) with the equality sign used.

In practical applications, one can assume $A = kP_2$, where k is the factor of safety with respect to the probability P_2 of "missing a target." This factor can be taken as, say, $k = 1$ to 3 when the anticipated consequences of failure are not very dangerous and as $k = 3$ to 10 in more responsible cases.

When solving technical diagnostics problems, one can also choose an approach based on the evaluation of the limiting x_0 values. In this case the governing equation, similar to the condition (10.63), is

$$P_2 \int_{-\infty}^{x_0} f_2(x/D_2)\, dx = B$$

and the B value can be evaluated as

$$B = \frac{1}{kN}$$

Here N is the total number of products and k is the factor of safety ($k = 1$ to 10).

Example 10.28 The diagnostics of a mechanical transmission line is carried out based on the detected content of iron in the lubrication oil. If the transmission is sound, the mean value $\langle x_1 \rangle$ and the standard deviation σ_1 of the amount of iron are $\langle x_1 \rangle = 5$ (5 g of iron per 1 ton of oil) and $\sigma_1 = 2$, respectively. Because of the defects in the bearings, the actual values of these characteristics are $\langle x_2 \rangle = 12$ and $\sigma_2 = 3$. According to the available statistical data, the faulty condition in the transmissions of the type in question were observed in 10 percent of the inspected transmissions. Assuming that the amount of lubrication oil is a normally distributed random variable, determine the maximum allowable amount of iron in the oil. Use the method of minimum risk. The correct decisions are not encouraged ($C_{11} = C_{22} = 0$) and the ratio of the cost of missing the defect to the cost of a false alarm is $C_{12}/C_{21} = 20$.

Solution Using the condition (10.58), we obtain the following value for the probability density ratio:

$$\frac{f_1(x_0/D_1)}{f_2(x_0/D_2)} = 20 \frac{0.1}{0.9} = 2.22$$

The probability density functions for the amount of iron in the lubrication oil for the sound and the faulty conditions are

$$f_1(x/D_1) = \frac{1}{2\sqrt{\pi}} \exp\left[-\frac{(x-5)^2}{8}\right]$$

and

$$f_2(x/D_2) = \frac{1}{3\sqrt{2\pi}} \exp\left[-\frac{(x-12)^2}{18}\right]$$

respectively. Introducing these distributions into the probability density ratio, we obtain the following equation for the critical x_0 value:

$$-\frac{(x_0-5)^2}{8} + \frac{(x_0-12)^2}{18} = \ln \frac{2 \times 2.22}{3} = 0.3920$$

The root of this equation is $x_0 = 7.456$.

The probability of a "false alarm" can be found as

$$P(H_{21}) = P_1 \int_{x_0}^{\infty} f_1(x/D_1) \, dx$$

$$= 0.90 \int_{7.456}^{\infty} \frac{1}{2\sqrt{\pi}} \exp\left[-\frac{(x-5)^2}{8}\right] dx = 0.0984$$

The probability of "missing the target" is as follows:

$$P(H_{12}) = P_2 \int_{-\infty}^{x_0} f_2(x/D_2) \, dx$$

$$= 0.10 \int_{-\infty}^{7.456} \frac{1}{3\sqrt{2\pi}} \exp\left[-\frac{(x-12)^2}{18}\right] dx = 0.0065$$

The mean risk can be computed as

$$R = C_{21}P_1 \int_{x_0}^{\infty} f_1(x/D_1)\, dx + C_{12}P_2 \int_{-\infty}^{x_0} f_2(x/D_2)\, dx$$

$$= C_{21}(0.0984 + 20 \times 0.0065) = 0.2284 C_{21}$$

If, for instance, the cost of a "false alarm" is $C_{21} = 1$, then $R = 0.2284$.

Example 10.29 Using the input data from the previous example and the Neumann–Pierson condition, determine the probability of a "false alarm," the probability of "missing a target," and the mean risk. The factor of safety can be assumed to be $k = 1$.

Solution With $k = 1$ and $P_2 = 0.10$, we have $A = kP_2 = 0.10$. Then the condition (10.63) yields

$$0.90 \int_{x_0}^{\infty} \frac{1}{2\sqrt{\pi}} \exp\left[-\frac{(x-5)^2}{8}\right] dx = 0.10$$

This equation has the following solution: $x_0 = 7.44$. The sought probabilities of a "false alarm" and "missing a target" are $P(H_{21}) = 0.100$ and $P(H_{12}) = 0.0064$, respectively. The calculated mean risk is $R = 0.230$.

10.12. Structural Reliability

10.12.1. Ultimate and fatigue strength

A structure or a structural element subjected to static loading must possess three major qualities: strength, rigidity, and stability. From the reliability point of view, a structure is a complex system with redundancy. Such a redundancy is due to the employment of duplicating elements and is characterized by how these elements are integrated into the system, as well as by the statical indeterminacy of the structure. In a statical indeterminate structure, failure of one or even several elements does not necessarily change the configuration of the structure and may not lead to its failure.

Structural elements are usually considered as items that operate until the first failure, i.e., as items that are not intended to be restored after they fail. If it happens, these elements are simply replaced by new ones. Such an approach is typically taken when strength calculations are conducted or when structural design is being carried out, although it might not be necessarily the case in actual practice. As to the structure as a whole, its strength, especially when the structure belongs to class II or III (see Sec. 10.6), is evaluated assuming that the structure lends itself to repair. Repairability is therefore an important property of complex engineering structures. This property, required in civil, aerospace, ocean, and other structures, might not be needed in microelectronic structures at low levels of packaging (chip-substrate assemblies, electronic components, printed circuit boards with or without surface-mounted components, etc.), but might be required at higher levels (TV set, computer, various electronic equipments, etc.).

In any structure one can usually identify the most crucial elements, the failure of which will lead to the failure of the entire system (e.g., the main girder in a ship's hull, the chip in an electronic package, the lightwave guide in a photonic system, etc.), while the damage in, or even the failure of, an element of lesser importance (e.g., a crack in a ship's bulkhead or a fracture in a solder joint or in the molding compound of a plastic electronic package) does not usually result in the system's failure. Therefore an assessment of the structural reliability is based on the reliability of the most crucial elements of the structure.

Structural reliability should be evaluated and ensured on both short- and long-term scales, i.e., a structure or a structural element should possess both high dependability and high durability. High dependability means that the structure should be able to withstand high-level loading for a short period of time. High durability means that the structure should be able to withstand relatively low level loading, but for a long time. Dependability is concerned with the ultimate strength of a structural element (structure) when the material is overstressed and failure occurs at a high stress level. Durability is concerned with fatigue and/or brittle strength of the material and is associated with the accumulation of damages and the initiation and propagation of fatigue and/or brittle cracks. In the case of an ultimate failure, the material utilizes completely its qualities ("bearing capacity") established by the stress–strain curve. Fatigue and brittle failures occur at relatively low stress levels.

All of the three major structural qualities—strenth, rigidity, and stability—depend on the structural design, materials properties, and the magnitude and distribution of the external or internal (say, thermal) load. These factors can be either constant or vary in time, and can be treated, depending on the approach taken, as deterministic (nonrandom) or random parameters. If a deterministic approach is considered, the material characteristics are treated as deterministic values, while the random nature of loading is accounted for by multiplying the "actual" (nominal) load by a certain safety factor.

The fatigue or brittle cracks, as well as extensive tear, creep, etc., are most likely to occur in areas of high stress concentration. Therefore it is the analysis of the state of stress in the areas of stress concentration that is the basis of the assessement of the durability of the entire structure. Methods of prediction and measurement of stress concentration in different structural joints are addressed in stress–

strength-related disciplines: theories of elasticity, plasticity, and creep; strength of materials; structural analysis; theory of plates and shells; etc.

Although both ultimate strength (dependability) and long-term strength (durability) are important for any structure, the role of these qualities may be different for different structures and different loading conditions. Let us assume, for the sake of simplicity, that structural reliability can be assessed by the time-to-failure T_s in the case of the evaluation of short-term strength and the time-to-failure T_l in the case of long-term strength. Then, considering both short- and long-term reliabilities, the reliability function can be written as

$$R(t) = R_1(t) R_2(t) \qquad (10.64)$$

where $R_1(t) = P(T_s > t)$ is the reliability function with respect to failures due to the system's dependability and $R_2(t) = P(T_l > t)$ is the reliability function with respect to failures due to the system's durability. The formula (10.64) reflects an assumption that the qualities of dependability and durability may be considered to be statistically independent. Typically, the accumulation of structural damages begins to have an effect on the long-term reliability function $R_2(t)$ only after a certain time T_f of the operation of the system. At the initial time interval, $0 < t < T_f$, only the dependability function $R_1(t)$ is important from the standpoint of the total reliability of the structure.

Quite often, however, the functions $R_1(t)$ and $R_2(t)$ cannot be considered independently. This is due to the fact that both dependability and durability of the structure are affected by the same qualities—strength, rigidity, and structural stability—and the measures, aimed at the improvement of these qualities, will automatically influence both short- and long-term reliability of the structure. If, for instance, the function $R_2(t)$ can be expressed through the function $R_1(t)$, then the structural reliability function $R(t)$ becomes a function of the dependability $R_1(t)$ only:

$$R(t) = f(R_1(t)) \qquad (10.65)$$

Depending on the actual measures taken, this function can be either an increasing or a decreasing one, i.e., by improving the ultimate strength of the structure, one may or may not improve its durability and, hence, its general reliability. For example, an increase in the yield stress of steel will improve the dependability of the structure, although it may not result in an improvement of its fatigue strength. However, if the designer intends to utilize the increase in the allowable ultimate stress by building a lighter structure, he should do this with great caution: the increase in the stress level can worsen the fatigue strength of the structure unless the fatigue limit is increased as well. Thus, if the dependability and the durability are of equal importance, the structural and materials improvements must be aimed at the improvement of both functions $R_1(t)$ and $R_2(t)$. At the same time, it should be pointed out that no structure can have a satisfactory durability without having a high enough dependability, i.e., only a structure with a sufficiently high ultimate strength can have a satisfactory long-term reliability. Therefore the primary task of a structural engineer is to ensure high dependability, i.e., high ultimate strength.

10.12.2. Dependability of a structural element

In structural reliability, the dependability of a structural element cannot be expressed directly through the probability density function of the "flow" of failures, since this function cannot be determined experimentally and is usually unknown. Therefore time-to-failure, which is a suitable dependability criterion in situations where failure rates can be easily determined (say, in radio and electronic systems), is not used as an adequate criterion of structural dependability. Instead, one typically uses an approach described in Sec. 10.2. In accordance with this approach, one determines first the probability density function $f_x(x)$ of the random variable X of interest (stress, deflection, compressive force, etc.). This can be done by treating this variable as a nonrandom (deterministic) function of the random loading whose probability distribution is known or can be easily determined. Then one calculates the probability of nonfailure:

$$\Gamma(x_1, x_2) = P(x_2 < X < x_1) \tag{10.66}$$

This is the probability that the random variable X of interest will not fall beyond the safe interval (x_1, x_2). Then the material of a structural element and its geometry are established in such a way that the condition (10.66) is fulfilled with sufficiently high probability Γ. If the probability density function $f_t(t)$ for the structural failure rate is known, then, requiring that

$$P(x_2 < X < x_1) = P(T > t)$$

one would have

$$\int_{x_2}^{x_1} f_x(x)\, dx = \int_{t}^{\infty} f_t(t)\, dt \tag{10.67}$$

so that

$$\Gamma_x(x_1, x) = R(t) \tag{10.68}$$

The relationship establishes the link between the approach taken in structural reliability and the approach based on the concept of a failure rate. It enables one to find the time-to-failure t for the given probability of nonfailure Γ. Thus, the dependability level established on the basis of the formula (10.66) is not in any contradiction to the dependability level that can be found on the basis of the probability density function for the failure rate flow. These two concepts will be in complete agreement if Eq. (10.67) is fulfilled.

Example 10.30 A single-span simply supported beam of length l is subjected to a uniform, time-dependent, normally distributed load $q(t)$. Find the required section modulus S. The probability that the maximum bending stress exceeds the allowable stress level $[\sigma]$ and should not be higher than the given probability P.

Solution The maximum bending moment in the mid-cross-section of the beam is

$$M_{\max} = \frac{l^2}{8} q(t)$$

Then the maximum stress can be computed as

$$\sigma_{\max} = \frac{M_{\max}}{S}$$

where S is the section modulus. Hence, the section modulus is related to the applied load by the equation

$$S = \frac{M_{\max}}{\sigma_{\max}} = \frac{l^2}{8[\sigma]} q(t)$$

Since the load $q(t)$ is distributed in accordance with the normal law

$$f_q(q) = \frac{1}{\sqrt{2\pi D_q}} \exp\left[-\frac{(q - \langle q \rangle)^2}{2D_q}\right]$$

the probability that the maximum bending stress $\sigma_{\max}$ exceeds the $[\sigma]$ value can be found as

$$P = P(\sigma_{\max} > [\sigma]) = P\left(q > \frac{8[\sigma]}{l^2} S\right)$$

$$= \int_{(8[\sigma]/l^2)S}^{\infty} f_q(q) \, dq = \frac{1}{\sqrt{\pi}} \int_a^{\infty} e^{-t^2} \, dt,$$

where

$$a = \frac{(8[\sigma]/l^2)S - \langle q \rangle}{\sqrt{2D_q}}$$

The obtained relationship can be rewritten as

$$P = \frac{1}{\sqrt{\pi}} \left(\int_0^\infty e^{-t^2} dt - \int_0^a e^{-t^2} dt \right) = \frac{1}{2}[1 - \Phi(a)]$$

so that

$$\Phi(a) = 1 - 2P$$

In order to determine the required section modulus S for the given probability P and the given allowable stress $[\sigma]$, one should compute the Laplace function $\Phi(a)$ and then, using the tabulated values of this function (see Appendix B), determine the parameter a and compute the required section modulus

$$S = \frac{l^2}{8[\sigma]} (\langle q \rangle + a\sqrt{2D_q})$$

Example 10.31 In the previous example, it is the length of the beam l that is not known beforehand and should be treated as a normally distributed random variable, while the load q is a given nonrandom time-dependent variable. Find the section modulus that would ensure that the probability P of the event "the maximum bending stress σ_{max} exceeds the allowable $[\sigma]$ level" is very small.

Solution The maximum stress is expressed as

$$\sigma_{max} = \frac{l^2}{8S} q(t)$$

The length l of the beam is distributed in accordance with the normal law

$$f_l(l) = \frac{1}{\sqrt{2\pi D_l}} \exp\left[-\frac{(l - \langle l \rangle)^2}{2D_l}\right]$$

Then the probability that this length exceeds a certain value

$$l^* = \sqrt{\frac{8S\sigma_{max}}{q}}$$

is

$$P(l > l^*) = \int_{l^*}^\infty f_l(l)\, dl = \frac{1}{\sqrt{2\pi D_l}} \int_{l^*}^\infty \exp\left[-\frac{(l - \langle l \rangle)^2}{2D_l}\right] dl$$

$$= \frac{1}{\sqrt{\pi}} \int_{(l^* - \langle l \rangle)/\sqrt{2D_l}}^\infty e^{-t^2} dt = \frac{1}{\sqrt{\pi}} \left(\int_0^\infty e^{-t^2} dt - \int_0^{(l^* - \langle l \rangle)/\sqrt{2D_l}} e^{-t^2} dt \right)$$

$$= \frac{1}{2}[1 - \Phi(a)], \quad \text{for } a = \frac{l^* - \langle l \rangle}{\sqrt{2D_l}}$$

Thus, in order to find the appropriate section modulus S, one should calculate first the Laplace function

$$\Phi(a) = 1 - 2P,$$

then, using the tabulated values of this function (Appendix B), determine the a value, calculate the corresponding length

$$l^* = \langle l \rangle + a\sqrt{2D_l}$$

and, finally, compute the section modulus

$$S = \frac{q^{*2}}{8[\sigma]}$$

Example 10.32 In Example 10.30, both the external load q and the allowable stress (say, the yield stress of the material) σ_Y are normally distributed random variables. Determine the section modulus S of the beam's cross section which would ensure a low probability P that the maximum stress in the beam will exceed the yield stress.

Solution The load q results in the stress

$$\sigma_{max} = \frac{l^2}{8S} q$$

Hence, the yield stress σ_Y corresponds to the load

$$q_Y = \frac{8S\sigma_Y}{l^2}$$

The probability of nonfailure can be defined, in accordance with the formula (10.9), as follows:

$$P_n = \frac{1}{2}[1 + \Phi(\delta_R)]$$

Here $\Phi(\delta_R)$ is the Laplace function and

$$\delta_R = \frac{\langle \psi \rangle}{\sqrt{2D_\psi}}$$

is the safety index. The function ψ of nonfailure can be found as

$$\psi = q_Y - q = \frac{8S}{l^2}(\sigma_Y - \sigma_{max})$$

Then we have

$$\langle \psi \rangle = \frac{8S}{l^2}\left(\langle \sigma_Y \rangle - \frac{l^2}{8S}\langle q \rangle\right) = \frac{8S}{l^2}\langle \sigma_Y \rangle - \langle q \rangle$$

and

$$D_\psi = \left(\frac{8S}{l^2}\right)^2\left[D_Y + \left(\frac{l^2}{8S}\right)^2 D_q\right] = \left(\frac{8S}{l^2}\right)^2 D_Y - D_q$$

The safety index is expressed by the equation

$$\delta_R = \frac{(8S/l^2)\langle\sigma_Y\rangle - \langle q\rangle}{\sqrt{2[(8S/l^2)^2 D_y - D_q]}}$$

Solving this equation for the section modulus S, we find

$$S = \frac{l^2}{8} \frac{\langle q\rangle\langle\sigma_Y\rangle}{2\delta_R^2 D_Y - \langle\sigma_Y\rangle^2} \left[\sqrt{1 + \frac{(2\delta_R^2 D_Y - \langle\sigma_Y\rangle^2)(2\delta_R^2 D_q + \langle q\rangle^2)}{\langle q\rangle^2 \langle\sigma_Y\rangle^2}} - 1\right]$$

Thus, in order to determine the required section modulus S, one should first evaluate the function

$$\Phi(\delta_R) = 2P_n - 1 = 1 - 2P$$

compute the δ_R value for the given probability P, and then calculate the modulus S.

10.12.3. Solder glass attachment in a ceramic electronic package

As an example of a probabilistic design of a structural element in electronics, examine a solder ("seal") glass attachment in a ceramic ("Cerdip/Cerquad") package of an integrated circuit (IC) device.

During the last decade, the glass and microelectronics industries have actively pursued the development and use of glass solders in various "hermetic" (i.e., helium leak-proof) semiconductor packages. The mechanical performance of solder glass in a package subjected to the temperature change during package fabrication, testing, or storage is due to the stresses occurring primarily because of the thermal expansion (contraction) mismatch of the seal glass and the body of the package. Solder glass is a brittle material and breaks easily when stretched or bent. The thermally induced stresses in the solder glass can be minimized if all of the ceramic components of the package (the window frame, the base, and the lid) have the same coefficients of thermal expansion. In such a case the shearing stresses concentrating at the end portions of the glass seal are expected to be small, and the normal stresses in the glass layer are due exclusively to its thermal mismatch with the ceramics. If, for some reason, ceramics with different coefficients of expansion are used, it is imperative that these coefficients, being temperature dependent, match well at least at low-temperature conditions, when the thermally induced stresses are the highest. Since solder glasses, being brittle materials, are able to withstand much greater stresses in compression than in tension, it is desirable for compressive stresses in the glass attachment that the glass has a smaller coefficient of expansion than the ceramics.

The maximum interfacial shearing stress in a thin solder glass layer can be assessed by the formula

$$\tau_{max} = kh_g \sigma_{max}$$

where h_g is the thickness of the glass layer,

$$k = \sqrt{\frac{\lambda}{\kappa}}$$

is the parameter of the assembly compliance,

$$\lambda = \frac{1-v_c}{E_c h_c} + \frac{1-v_g}{E_g h_g}$$

is the in-plane compliance of the assembly, h_c is the total thickness of the ceramics parts, E_c and E_g are Young's moduli of the ceramics and the glass, respectively, v_c and v_g are Poisson's ratios of these materials,

$$\kappa = \frac{2(1+v_c)}{3E_c} h_c + \frac{4(1+v_g)}{3E_g} h_g$$

is the interfacial compliance of the assembly,

$$\sigma_{max} = \frac{\Delta\alpha \, \Delta t}{\lambda h_g}$$

is the maximum normal stress in the midportion of the glass layer, Δt is the change in temperature, $\Delta\alpha = \bar{\alpha}_c - \bar{\alpha}_g$ is the difference in the "effective" coefficients of thermal expansion for the ceramics and the glass,

$$\bar{\alpha}_c = \frac{1}{\Delta t} \int_t^{t_0} \alpha_c(t) \, dt, \quad \bar{\alpha}_g = \frac{1}{\Delta t} \int_t^{t_0} \alpha_g(t) \, dt$$

are these coefficients for the given temperature t, t_0 is the annealing ("zero stress," "setting up") temperature, and $\alpha_c(t)$ and $\alpha_g(t)$ are the "instantaneous" values of the coefficients of expansion. In an approximate analysis, one can assume that the axial compliance λ is due to the seal glass only [that is, $\lambda \simeq (1-v_g)/(E_g h_g)$], and therefore the maximum normal stress σ_{max} in the glass can be computed by a simplified formula:

$$\sigma_{max} = \frac{E_g}{1-v_g} \Delta\alpha \, \Delta t$$

While the geometrical characteristics of the assembly, the temperature change, and the elastic constants of the ceramics and the glass can be determined with high accuracy, the reliability in the prediction of the difference in the coefficients of thermal expansion of the glass and the ceramics is not as good. This is due, first of all, to the fact that these coefficients exhibit strong temperature dependence and are very sensitive to variations in the glass composition, thermal treatment, etc. However, what is more important is that, because of the obvious incentive to minimize the thermal contraction mismatch of the glass and the ceramics, such a mismatch is characterized by a small difference in close numbers. This leads to an appreciable uncertainty in the evaluation of this difference, and justifies the application of a probabilistic approach.

Treating the coefficients of thermal expansion as random varibles, we evaluate the probability P that the stress in the glass is compressive and does not exceed an allowable value σ^* as a probability that the difference

$$\psi = \alpha_c - \alpha_g$$

falls within the interval

$$0 \leq \psi \leq \psi^* = \frac{\sigma^*}{E_g} \frac{1-v_g}{\Delta t}$$

Assuming that the random variables α_c and α_g are normally distributed, i.e., that their probability density functions are expressed as

$$f_c(\alpha_c) = \frac{1}{\sqrt{2\pi D_c}} \exp\left[-\frac{(\alpha_c - \langle\alpha_c\rangle)^2}{2D_c}\right]$$

$$f_g(\alpha_g) = \frac{1}{\sqrt{2\pi D_g}} \exp\left[-\frac{(\alpha_g - \langle\alpha_g\rangle)^2}{2D_g}\right]$$

we conclude that the function ψ of nonfailure (safety margin) is distributed in accordance with Eq. (10.8), in which the mean $\langle\psi\rangle$ and the variance D_ψ of the random variable ψ are $\langle\psi\rangle = \langle\alpha_c\rangle + \langle\alpha_g\rangle$ and $D_\psi = D_c + D_g$, respectively. Then the probability $P = P\{0 \leq \psi \leq \psi^*\}$

can be computed as

$$P = \int_0^\psi f_\psi(\psi)\, d\psi$$

$$= \frac{1}{\sqrt{2\pi D_\psi}} \int_0^\psi \exp\left[-\frac{(\psi - \langle\psi\rangle)^2}{2D_\psi}\right] d\psi$$

$$= \frac{1}{\sqrt{2\pi}} \int_{-\gamma}^{\gamma^* - \gamma} e^{-t^2/2}\, dt$$

$$= \frac{1}{\sqrt{2\pi}} \int_{-\infty}^{\gamma^* - \gamma} e^{-t^2/2}\, dt - \frac{1}{\sqrt{2\pi}} \int_{-\infty}^{-\gamma} e^{-t^2/2}\, dt$$

$$= \Phi_1(\gamma^* - \gamma) - [1 - \Phi_1(\gamma)] = \Phi_1(\gamma^* - \gamma) + \Phi_1(\gamma) - 1$$

where

$$\Phi_1(t) = \operatorname{erf} t = \frac{1}{\sqrt{2\pi}} \int_{-\infty}^{t} e^{-t^2/2}\, dt = \frac{1}{2}[1 + \Phi(t\sqrt{2})]$$

and the reliability (safety) indices γ and γ^* are expressed by the formulas

$$\gamma = \frac{\langle\psi\rangle}{\sqrt{D_\psi}}, \qquad \gamma^* = \frac{\psi^*}{\sqrt{D\psi}}$$

If the P value is close to unity, then one can be confident that the normal stresses in the glass layer are compressive and do not exceed the allowable stress level σ^*, while the shearing stresses, which concentrate at the ends of the glass layer, are small and do not exceed the $\tau^* = kh_g \sigma^*$ value.

Let, for instance, the elastic constants of the solder glass be $E_g = 0.66 \times 10^6$ kg/cm^2 and $v_g = 0.27$, the seal temperature be 485°C, the lowest (testing) temperature be -65°C (so that the change in temperature is $\Delta t = 550$°C), the calculated "effective" coefficients of expansion at this temperature be $\bar{\alpha}_g = 6.75 \times 10^{-6}$ °C^{-1} and $\bar{\alpha}_c = 7.20 \times 10^{-6}$ °C^{-1}, the standard deviations of these coefficients be $\sqrt{D_c} = \sqrt{D_g} = 0.25 \times 10^{-6}$ °C^{-1}, and the experimentally obtained compressive strength of the glass be $\sigma_u = 5500$ kg/cm^2. With the safety factor of, say, $\eta = 4$, we have $\sigma^* = \sigma_u/4 = 1375$ kg/cm^2. Therefore the required level of the function of nonfailure is

$$\psi^* = \frac{(1 - 0.27)1375}{0.66 \times 10^{-6} \times 550} = 2.765 \times 10^{-6}$$

its mean value is

$$\langle \psi \rangle = \bar{\alpha}_c - \bar{\alpha}_g = 0.450 \times 10^{-6} \, °C^{-1}$$

and the variance is

$$D_\psi = D_c + D_g = 0.125 \times 10^{-12} \, (°C)^{-2}$$

Then the calculated reliability indices are $\gamma = 1.2726$ and $\gamma^* = 7.8201$, and the probability of nonfailure is

$$P = \Phi_1(6.5475) + \Phi_1(1.2726) - 1 = 0.898$$

If the standard deviations $\sqrt{D_c}$ and $\sqrt{D_g}$ of the coefficients of thermal expansion are only $0.1 \times 10^{-6} \, °C^{-1}$, then the reliability indices would become $\gamma = 3.1825$ and $\gamma^* = 19.5559$, respectively, and the probability P of nonfailure would increase to $P = 0.999$. Hence, the standard deviations, reflecting the degree of uncertainty in the prediction of the coefficients of expansion, have a strong impact on the probability criterion and must be kept sufficiently small to ensure high reliability of the package. In this connection, it should be pointed out that it is advisable that vendors of packaging materials provide information not only about the mean values of the coefficients of expansion but about their variances as well. This would enable a designer to establish the adequate stress level, so as not to compromise the long-term reliability of the package.

10.12.3. Integrated Circuit in a "Smart Card" Design
Basic Assumptions

As another illustration of the application of a probabilistic approach in electronic packaging, we examine the use of such an approach to evaluate improvements in the reliability of an integrated circuit (IC), i.e., a silicon chip, in a "Smart Card" design. Chips mounted on flexible substrates, such as "Smart Cards," often break because of the excessive bending of the cards. In the analysis which follows we examine a simple probabilistic model for the assessment of the probability of failure of the IC in a chip-card assembly due to the card flexure.

Based on the developed model, we intend to establish the variance of the in-service curvature of the card from the reported failure statistics in the field and the additional experimental data obtained by bending the cards on circular mandrels. The mean value of the in-service curvature is considered zero. The developed model will be then applied for the assessment of the effect of the structural improvements (thinning and polishing the chip, improvements in the chip's location and orientation, application of compliant adhesives, etc) on the mean value of the assembly curvature at the point of the IC fracture and on the probability of fracture. This will enable the designer to judge whether the introduced improvements are sufficient to achieve the desired (required) low failure rate level of the IC's.

The following major assumptions are used in the analysis:

1. The level of the external loading on the chip can be characterized by the curvature κ of the mandrel on which the chip-card assembly is bent. The bearing capacity (ultimate strength) of the chip can be characterized by the curvature K of the mandrel at the point of the IC fracture.
2. Both the above curvatures are random variables which follow the normal law of the probability distribution, and the mean value of the curvature κ of the substrate in service conditions is zero:

$$f_\kappa(\kappa) = \frac{1}{\sqrt{2\pi D_\kappa}} \exp\left(-\frac{\kappa^2}{2D_\kappa}\right) \qquad (10.69)$$

$$f_K(K) = \frac{1}{\sqrt{2\pi D_K}} \exp\left[-\frac{(K - \langle K \rangle)^2}{2D_K}\right] \qquad (10.70)$$

In these formulas, D_κ and D_K are the variances of the curvatures κ and K, and $\langle K \rangle$ is the mean value of the curvature K.

3. The variance of the card curvature in the field conditions is independent of the structural improvements in the design at the IC location, and the value of this variance, determined for the existing design, can be used for the prediction of the probability of failure (nonfailure) of the IC in the improved design as well.
4. The statistical data for the curvature K can be obtained on the basis of a multi-step loading of the assembly on circular mandrels of the decreasing radii.

With the expressions (10.69) and (10.70), we find, that the function of nonfailure (safety margin) $\psi = K - \kappa$, expressing the difference between the curvature at failure and the actual curvature, has the following probability density:

$$f_\psi(\psi) = \frac{1}{\sqrt{2\pi D_\psi}} \exp\left[-\frac{(\psi - \langle \psi \rangle)^2}{2D_\psi}\right]$$

where $\langle \psi \rangle = \langle K \rangle$ is the mean value of the function ψ and

$$D_\psi = D_\kappa + D_K \qquad (10.71)$$

is its variance. The probability that the function ψ is positive, i.e., the probability of nonfailure of the IC, is

$$P_R = P\{\psi > 0\} = P\{K > \kappa\}$$
$$= \int_0^\infty f_\psi(\psi)\, d\psi = \frac{1}{2}[1 + \Phi(\delta_R)] \tag{10.72}$$

where

$$\Phi(\delta_R) = \frac{2}{\sqrt{\pi}} \int_0^{\delta_R} e^{-t^2}\, dt$$

is the probability integral, and

$$\delta_R = \frac{\langle \psi \rangle}{\sqrt{2D_\psi}} \tag{10.73}$$

is the safety index.

It should be pointed out that, if the failure rate λ is low, this rate can be easily determined from the probability of failure

$$P_f = 1 - e^{-\lambda t} \simeq 1 - (1 - \lambda t) = \lambda t$$

as

$$\lambda = \frac{P_f}{t}$$

where t is the given period of time.

Expected in-Service Curvatures

It is very difficult to determine the in-service curvatures of chip-card assemblies by direct measurements. At the same time, it is important to know the variance of these curvatures, otherwise no quantitative probabilistic assessment would be possible at all. The assessment of the variance in the in-service substrate curvature can be carried out on the basis of an indirect method. This method uses the reported failure statistics and the additional information obtained from the measured curvatures at IC fracture during bending of the chip-substrate assembly on a circular mandrel. The suggested approach is based on the following simple considerations.

If the reported percentage of the cracked chips in the field is P_f, then the corresponding safety index δ_R can be found from the calculated probability of nonfailure $P_R = 1 - P_f$, using the relationship (10.72). Then the anticipated variance D_κ of the substrate (assembly) curvature in the field can be computed on the basis of the formulas (10.71) and (10.73) as follows:

$$D_\kappa = \frac{1}{2}\left(\frac{\langle K \rangle}{\delta_R}\right)^2 - D_K \qquad (10.74)$$

The Required IC Strength

The required (desired) mean and variance of the substrate (assembly) curvature at the IC failure can be established based on the following reasoning.

Let the acceptable percentage of the failed assemblies per month, because of the chip cracking, be P_f. Then, using (19.73) we conclude that the mean $\langle K \rangle$ and the variance D_K of the assembly curvature at the IC fracture should satisfy the inequality:

$$\frac{\langle K \rangle}{\sqrt{2(D_K + D_\kappa)}} > \delta_R \qquad (10.75)$$

If the variance D_κ of the curvature K is assumed the same in the existing and the improved designs, then the required increase in the IC strength should result in the increase in the mean value $\langle K \rangle$ of the assembly (mandrel) curvature at the point of the IC fracture. The new $\langle K \rangle$ value should not be smaller than

$$\langle K \rangle = \delta_R \sqrt{2 D_\psi}. \qquad (10.76)$$

Experimental Procedure and Numerical Example

In accordance with the considered experimental approach, we fabricated seven mandrels with a rather wide range of radii shown in the Table 10.10.

We started to test the given card at the largest radii by bending it at the given radius twice, each time in the opposite direction, and measuring the functioning of the card electrically before proceeding to the next (smaller) radius. Testing was stopped after electrical failure was observed. As a rule, this corresponded to the chip fracture. Some experimental data, obtained for the radii $r = 1/K$ of the mandrels at the point of the chip fracture, are shown in Table 10.11.

TABLE 10.10 Mandrel Radii

Mandrel	Radii, inches						
1	1.0	1.1	1.2	1.3	1.4	1.5	5.0
2	1.5	1.6	1.7	1.8	1.9	2.0	5.0
3	2.0	2.1	2.2	2.3	2.4	2.5	5.0
4	2.5	2.6	2.7	2.8	2.9	3.0	5.0
5	3.0	3.1	3.2	3.3	3.4	3.5	5.0
6	3.5	3.6	3.7	3.8	3.9	4.0	5.0
7	1.0	2.0	3.0	4.0	5.0	6.0	7.0

The calculated mean $\langle K \rangle$ and variance D_K of the mandrel curvature at the point of the IC failure, obtained from these data, are $\langle K \rangle = 0.956$ 1/in, and $D_K = 0.0208$ 1/in^2. Note that, because of the limited amount of experimental data, we calculated the variance D_K by the formula

$$D_K = \frac{1}{n-1} \sum_{i=1}^{n} (K_i - \langle K \rangle)^2,$$

where n is the number of the performed tests.

Let us estimate the confidence intervals (see Sections 14.2 and 16.2) for the obtained $\langle K \rangle$ value for different probabilities P of error. From the tables for Student's distribution (Appendix E), and for the number of the degrees of freedom $n - 1 = 11 - 1 = 10$ ($n = 11$ is the number of measurements in Table 10.11) we calculate, for different probabilities P of error the t value and the deviation $\Delta = t\sqrt{D_K}/\sqrt{n}$ from the mean. The lower and the upper confidence limits are $\langle K \rangle - \Delta$ and $\langle K \rangle + \Delta$, respectively. The computations are carried out in a tabular form (Table 10.12). Hence, if the desired (required) probability of error is, say, $P = 5\%$, the $\langle K \rangle$ value can be found within the interval between 0.859 1/in and 1.053 1/in.

Let the reported percentage of failure for the existing chip-card assemblies, because of the IC fracture, be as high as $P_f = 1.4\%$ per month. Then the corresponding probability of nonfailure is: $P_R = 1 - P_f = 0.986$, the probability integral is $\phi(\delta_R) = 0.972$, and the safety index is $\delta_R = 1.096$. From the relationship (10.74) we find that the variance of the in-service curvature is $D_\kappa = 0.360$ 1/in^2. If one requires that the percentage of the failed assemblies due to the IC fracture be only, say, 0.5%, per month, then the probability of nonfailure should be increased to $P_R = 0.995$. In this case the formula (10.72) yields: $\Phi(\delta_R) = 0.990$, and the safety index becomes substantially higher: $\delta_R = 1.390$.

If the variance D_K remains unchanged, the formula (10.75) results in the following mean value of the required substrate (mandrel) cur-

TABLE 10.11 Radii of the Mandrels at which Chip Fracture Occurred

r, in	1.6	1.2	1.1	1.1	1.1	1.0	1.0	1.0	0.9	0.9	0.9
$K = 1/r$, 1/in	0.625	0.833	0.909	0.909	0.909	1.0	1.0	1.0	1.111	1.111	1.111
$(K - \langle K \rangle)^2$	0.1096	0.0151	0.0022	0.0022	0.0022	0.0019	0.0019	0.0019	0.0240	0.0240	0.0240

TABLE 10.12 Calculated Confidence Limits

P	0.01	0.02	0.05	0.10
t	3.169	2.764	2.228	1.812
Δ	0.1378	0.1202	0.0969	0.0788
$\langle K \rangle - \Delta$	0.818	0.836	0.859	0.877
$\langle K \rangle + \Delta$	1.094	1.076	1.053	1.035

vature at the IC fracture $\langle K \rangle = 1.213$ 1/in. This corresponds to the curvature radius of $\langle r \rangle = 1/\langle K \rangle = 0.825$ in. The mean radius $\langle r \rangle$ of the mandrel causing IC fracture in the existing design is, as follows from the Table 10.11, somewhat higher: $\langle r \rangle = 1/\langle K \rangle = 1.046$ in. Hence, the reduction in the mean value of the mandrel radius at the point of the IC fracture by a factor of only 1.27 is expected to result in the decrease in the probability of IC fracture by a factor of $1 - 0.986/1 - 0.995 = 2.80$. Thus, even a relatively minor improvement in the fabrication and encapsulation of the IC is expected to lead to an appreciable increase in its reliability. Such a favorable situation is typical for systems of low reliability.

10.12.5. Kinetic Approach to Material Failure

According to the kinetic (thermofluctuation) approach to the strength of materials, in any solid body there are always some particles whose energies are large enough to break the chemical bonds with the adjacent particles. The minimum value U_0 of the energy required to break the bond is called the *activation energy*. The concept of the activation energy enables one to assess the durability (long-term strength) of a solid body (say, a polymeric material) subjected to an elevated temperature. This can be done based on the following probability density function (*Boltzmann's statistics*) for the kinetic energy E of a particle:

$$f(E) = \lambda e^{-\lambda E}, \quad \lambda = \frac{1}{kT}$$

where T is the absolute temperature, k is Boltzmann's constant, and the product kT is the mean value of the energy E. This distribution holds for both solids and gases.

Ludwig Boltzmann (1844–1906) was an Austrian physicist who concerned himself throughout his career with the physical and mathematical implications of an atomic view of matter. Boltzmann's statistics (see Example 1.36), a fundamental law of classical statistical mechanics, simply states that the kinetic energy E of a particle is a random variable that follows the exponential distribution.

The probability that the energy E exceeds the level U_0 is

$$P = P(E > U_0) = \int_{U_0}^{\infty} f(E)\,dE = e^{-\lambda U_0}$$

The mean number n_a of activated particles (i.e., the particles whose energy exceeds the U_0 level) can be evaluated as

$$n_A = Ne^{-\lambda U_0}$$

where N is the total number of particles in the system.

The *mean time-to-fluctuation*, τ, i.e., the mean value of the time required for the energy E of an activated particle to deviate substantially from its mean value $\langle E \rangle = kT$, can be assessed by assuming that the ratio of the time τ_0 of a single "attempt" of this particle to overcome the energy barrier U_0 to the mean time-to-fluctuation τ is equal to the ratio n_A/N of the number n_A of activated particles to the total number N of particles:

$$\frac{\tau_0}{\tau} = \frac{n_A}{N} = e^{-\lambda U_0}$$

Hence,

$$\tau = \tau_0 e^{-\lambda U_0} = \tau_0 \exp\left(\frac{U_0}{kT}\right) \tag{10.77}$$

This relationship is known as the *Boltzmann–Arrhenius equation* and has numerous applications in physics and reliability of materials. Svante Arrhenius (1859–1927), a Swedish chemist, is the founder of the "ionist" school of physical chemistry.

If the material is also subjected, in addition to an elevated temperature T, to a tensile stress σ, then the effect of this stress can be accounted for by modifying the formula (10.77) as follows:

$$\tau = \tau_0 \exp\left(\frac{U_0 - \gamma\sigma}{kT}\right) \tag{10.78}$$

where γ is the stress sensitivity factor, which depends on the structure of the material and the degree of the accumulated damage. The *Bueche–Zhurkov formula* (10.78) underlies the kinetic approach to the evaluation of the strength of materials. This formula was suggested independently by F. Bueche in the United States and S. N. Zhurkov in Russia in the early 1950s. In accordance with the kinetic approach, it is the random thermal fluctuations of particles (atoms) that are primarily responsible for the material's strength (failure), while the role of the external stress is simply in reducing the activation energy. This stress leads to a lower durability of the material and a higher probability of its failure:

$$P = \exp\left(-\frac{U_0 - \gamma\sigma}{kT}\right) \quad (10.79)$$

The decrease in the strength of solid bodies with an increase in thermal fluctuations was first explained by A. Smekal back in 1936. As evident from the formulas (10.78) and (10.79), the stress-at-failure can be computed as

$$\sigma = \frac{U_0}{\gamma} \quad (10.80)$$

The kinetic approach to material failure is currently widely applied, in addition to, or as a modification of, the fracture mechanics approach, to the evaluation of fracture of polymeric materials. According to the kinetic approach, the fracture of polymeric materials in the nonbrittle state (taking place at relatively high temperatures and/or during low-level static loading) is due to the gradual development and accumulation of microdamages caused by thermal fluctuations. Fracture in the brittle state (taking place at low-temperature conditions and/or during high-level dynamic loading) is due to a rapid growth of microcracks, leading to bond rupture. Such a growth can also be caused by thermal fluctuations of the material's particles. The time τ_0 of a single "attempt" of a particle to overcome an energy barrier (activation energy) U_0 can be assumed to be equal to the period of the vibrations of atoms and, for C—C bonds in a polymeric material, is about $\tau_0 = 3 \times 10^{-14}$ s if all the bonds in the given cross section are being broken simultaneously. It can be found to be within the range between 8×10^{-7} and 3×10^{-14} s if the bonds are being broken one at a time, depending on the conditions of the experiment. The activation energy U_0 for various polymeric materials lies within the range 100 to 300 kJ/mol. The stress sensitivity factor for a nonoriented condition of a polyimide is about $\gamma = 1.3 \times 10^{-27}$ m^3. For an oriented condition it can be significantly lower (by a factor of 0.15 to 0.20). In many applications, it is only the governing relationship of the type (10.78) or (10.79) that is considered, while the numerical values of the parameters are evaluated experimentally for each particular application.

Further Reading (*see Bibliography*): 2, 3, 8, 10, 14, 16, 18, 25, 36, 37, 41, 50, 54, 63, 71, 76, 82, 83, 87, 100, 106, 107, 113, 118, 127, 128, 133, 135, 144.

Chapter

11

Markovian Processes

"Mathematics for an engineer is a means, a tool, which is similar to a caliper, chisel, hammer, file for a craftsman, or a foot rule, bench axe and a saw for a carpenter."
ALEXEY KRYLOV,
Russian Mechanician
and Naval Architect

"Life is the art of drawing sufficient conclusions from insufficient premises."
SAMUEL BUTLER

"Any equation longer than three inches is most likely wrong."
UNKNOWN ENGINEER

11.1. Flows of Events

A random sequence of homogeneous events is known in the applied probability theory as a *flow of events*. Examples are: requests at a telephone exchange; random shock loadings on a structure; the occurrence of breakdowns during the operation of a machine or a computer; the arrival of cars at toll booths on highways, etc. A flow of events is *stationary* if its probability characteristics are time independent. In such a case the probability that a particular number of events will be found within the given time interval depends only on the length of this interval, and is independent of the location of this interval on the time axis. A flow of events is *ordinary* if the probability that an elementary time interval Δt contains more than one event is negligibly small compared to the probability that only one event is found within this interval. Events in an ordinary flow occur one at a time. A flow of events is referred to as a *flow without an aftereffect* if the number of events within the given time interval is independent of the number of events on any other (nonoverlapping) time interval. In a flow without an aftereffect, the events occur independently of each other. A flow of events is *elementary* if it possesses all three of the above properties, i.e., if it is stationary, ordinary, and has no aftereffects. A flow of events is a *Poisson flow* if it is ordinary and has no aftereffects. A stationary Poisson flow in an elementary one.

The time interval T between two successive events in an elementary flow has an exponential distribution

$$f(t) = \lambda e^{-\lambda t} \qquad (11.1)$$

where $\lambda = 1/\langle t \rangle$ is the inverse of the mean value $\langle t \rangle$ of the time T. The value λ is known as the *intensity* of the flow of events. It is constant for a stationary flow and is a function of time for a nonstationary flow.

An ordinary flow of events is a *Palm flow* (or recurrent flow, or a flow with a limited aftereffect) if the time intervals between successive events are independent random variables having the same probability distributions. A Palm flow is always stationary. If the probability distributions of the time intervals between the events in a Palm flow are exponential, this flow becomes an elementary one.

An elementary flow in which every kth event is retained and the intermediate events are removed is an *Erlang flow of the order k* (see Sec. 3.3.8). In the Erlang flow, the time interval between the successive events has the following probability distribution:

$$f_k(t) = \frac{\lambda(\lambda t)^{k-1}}{(k-1)!} e^{-\lambda t} \qquad (11.2)$$

The mean, variance, and the coefficient of variation of this distribution are

$$\langle t \rangle = \frac{k}{\lambda}, \qquad D_t = \frac{k}{\lambda^2}, \qquad v_t = \frac{1}{\sqrt{k}} \qquad (11.3)$$

Let us show how the formula (11.2) can be obtained for a series of exponential distributions with parameter λ. Let us form first a convolution of two exponential distributions, i.e., find the distribution of the sum of two independent random variables X_1 and X_2, whose probability densities are

$$f_1(x_1) = \lambda e^{-\lambda x_1}, \qquad f_2(x_2) = \lambda e^{-\lambda x_2}$$

Using the general formula for the convolution of distributions (Section 5.2) we have

$$g(z) = \int_{-\infty}^{\infty} f_1(x_1) f_2(z - x_1)\, dx_1 = \int_0^z f_1(x_1) f_2(z - x_1)\, dx_1$$

$$= \lambda^2 \int_0^z e^{-\lambda x_1} e^{-\lambda(z - x_1)}\, dx_1 = \lambda^2 z e^{-\lambda z}$$

This is Erlang's distribution of order 2.

The convolution of n exponential distributions with parameter λ can be determined by consecutively finding the convolutions of two, three, etc., distributions. This leads to the formula

$$g_n(x) = \lambda e^{-\lambda x} \sum_{i=1}^{n} (-1)^{i-1} C(n, i)$$

which, for $n \to \infty$, results in the formula (11.2). It is simpler, however, to obtain the formula (11.2) by proceeding from the elementary flow, retaining every nth point in it, and deleting the intermediate points. Let the random variable X be a sum of random variables X_i:

$$X = \sum_{i=1}^{n} X_i$$

where random variables X_i have exponential distributions. We evaluate the probability density $f_n(x)$ of the random variable X, first finding the element of probability $f_n(x)\, dx$. This is the probability that the random variable X is found within the elementary interval $(x, x + dx)$. For the value X to be found within this interval, it is necessary that exactly $n - 1$ events fall within the interval x and one event within the interval dx. The probability of such an occurrence is $f_n(x)\, dx = P_{n-1} \lambda\, dx$, where P_{n-1} is the probability that $n - 1$ events fall within the interval x. Since, in the elementary flow, the number of events falling on the interval x has Poisson's distribution with the parameter $a = \lambda x$, we have

$$f_n(x)\, dx = \frac{(\lambda x)^{n-1}}{(n - 1)!} \lambda e^{-\lambda x}\, dx$$

This leads to the formula (11.2). Sometimes a "shift" in the flow of events is applied, and the $n - 1$ number is substituted by a n number (see Sec. 3.3.8).

11.2. Markovian Processes and Chains

A random (stochastic) process is a *Markovian process*, or a *process without an aftereffect*, if the probability of any state of the system characterized by this process in the "future" ($t > t_0$) depends only on its state in the "present" ($t = 0$) and is independent of how the system reached the "present" state, i.e., independent of the "prehistory", the "past" of the process.

Markovian processes occupy a special and a very important place in the theory of random processes. This is due to the fact that the behavior of many engineering and physical systems can be adequately modeled using Markovian processes. At the same time, the mathematical "equipment" required to solve many practically important engineering and physical problems, which can be approximated as Markovian processes, is well developed. Here are examples of some applied problems of a general nature that can be addressed on the basis of the theory of such processes:

1. There are no exact methods for the evaluation of the complete probabilistic characteristics (distribution functions) of the responses of dynamic systems (linear or nonlinear, discrete or continuous) subjected to random excitations. The only exception is a linear dynamic system subjected to normal excitation. As long as the complete probabilistic characteristics of the dynamic responses cannot be found directly, one has to use various labor and time-consuming approaches for computing the moments of the system's response and, using these moments, to "guess" the probability density function of the process. However, if the input process is a Markovian one, there exist general and powerful methods for obtaining probability density functions for the response ("output") process, even for nonlinear systems.

2. In many applied problems there is a need to evaluate the probability that the parameters of a system will remain within the given (desirable) boundaries, whether fixed or moving. Quite often, the normal operation of a system is violated if such boundaries are exceeded. For instance, stresses in a mechanical system should not exceed the allowable level, control systems have a finite zone of stable operation, a sufficient underkeel clearance for a ship approaching a port is necessary for the ship's safety, etc. Although some elementary problems of this type can be solved using the theory of random variables (the random variable of interest representing the "output" process is treated, in this case, as a nonrandom function of the random variable representing the "input" process), the complete information about the behavior of dynamic systems in regions with boundaries can be obtained only for Markovian systems.

3. Optimization of dynamic systems (mechanical, electrical, telemetric, communications, and others), when subjected to random

excitations, can be carried out only for those systems that can be modeled as Markovian ones.

Broad application of the theory of Markovian processes in engineering practice is due to the fact that many actual random processes are or can be "made" Markovian ones if one increases, in an appropriate fashion, the number of variables describing the state of the system. For instance, the random displacement $Z(t)$ of a one-degree-of-freedom dynamic system is governed by a differential equation of the second order, and therefore the displacement z_n at the moment t_n depends not only on the displacement z_{n-1} at the previous moment of time t_{n-1} but on the system's velocity $\dot{z}_{n-1}$ at the moment t_{n-1} as well. Hence, one has to know, in addition to z_{n-1}, also the displacement z_{n-2} at the moment t_{n-2} preceding the moment t_{n-1}, in order to be able to evaluate the velocity $\dot{z}_{n-1}$. Thus, the function $Z(t)$ is determined by the values of this function at two, not one, preceding moments of time, and therefore the process $Z(t)$ is not a Markovian one. However, if one increases the number of random variables under consideration and examines the coupled processes (responses) $Z(t)$ and $\dot{Z}(t)$, then such a "two-dimensional" random process, as is obvious from the previous discussion, becomes a Markovian process.

Many real dynamic systems transforming broad-band excitations are characterized by low dissipations. Therefore the response functions of these systems have a strongly pronounced peak in the area of their fundamental frequency. In such a situation, the spectrum of the excitation can be assumed constant, with sufficient accuracy, within the portion of the response function of interest. Therefore such an excitation process can be approximated by a "white noise," thereby making the problem suitable for the application of the theory of Markovian processes. It is noteworthy also that normal random processes with a power spectrum expressed by an algebraic fraction can also be treated as Markovian processes. This can be done by expressing this spectrum through partially constant spectra, i.e., through a series of excitations of the "white noise" type (see Sec. 8.3).

Application of the methods of the theory of Markovian processes enabled one to solve numerous practically important problems in many areas of engineering (electrical, mechanical, civil, aerospace, radio- and telecommunications, control, reliability and quality, queueing, etc.), applied science (physics, chemistry, biology, medicine, etc.), and even in such areas as business, economics, and finance.

A Markovian process with discrete states and discrete time is a *Markov chain*. For such a process it is convenient to consider the moments of time $t_1, t_2, \ldots$, where the system S can change its state, as successive steps in the process, from the initial state $S(0)$ to the states $S(1)$, $S(2)$, ..., $S(k)$, ... The event $\{S(k) = s_i\}$ that the system is found in the state s_i immediately after the kth step ($i = 1, 2, ...$) is a random event. Therefore, the sequence of states $S(0)$, $S(1)$, $S(2)$, ..., $S(k)$, ... can be viewed as a sequence of random events. The initial state $S(0)$ can be either deterministic or accidental (random). A Markov chain formed by the sequence of events $S(0)$, $S(1)$, $S(2)$, ..., $S(k)$, ... can be interpreted as an integer-valued random process $X(t)$ whose possible values are $1, 2, ..., n$ and determine n different states (realizations) $s_1, s_2, ..., s_n$. This process "jumps" from one integer value to another at the moments of time $t_1, t_2, ...$.

An illustration of a process described by a Markov chain is the motion of a man who tosses a coin at certain time intervals and, depending on the outcome, makes a step to the right or to the left. Another example is a random motion of a particle that can move one unit to the right or to the left.

Let $p_i(k)$ be the probability that the system S is in the state s_i, $i = 1, 2, ..., n$, after taking the kth step and prior to the $(k+1)$th step. The set of probabilities $p_i(k)$ is the *probabilities of the states* of a Markov chain. For any k, the condition of normalization

$$\sum_{i=1}^{n} p_i(k) = 1, \qquad k = 0, 1, 2, ... \tag{11.4}$$

must be fulfilled.

A Markov chain is characterized by the *initial probability distribution* $p_i(0)$, $i = 1, 2, ..., n$, at the beginning of the process and the *probability of transition* P_{ij}. The probability of transition is the conditional probability that the system S will be found in the state s_j after taking the kth step, provided that it was in the state s_i after the previous $(k-1)$th step was taken.

A Markov chain is *homogeneous* if the transition probabilities are independent from the ordinal number of the step and depend only on the number of steps from which and to which the system passes:

$$P\{S(k) = s_j \,|\, S(k-1) = s_i\} = P_{ij} \tag{11.5}$$

In other words, the transition probability P_{ij} is a conditional probability that the system S will be found in the state s_j after the kth step, provided that the system was in the state s_i after the previous $(k-1)$th step. The transition probabilities P_{ij} of a homogeneous Markov chain form an $n \times n$ square matrix (*transition matrix*, or *probability matrix*):

$$\|P_{ij}\| = \begin{Vmatrix} P_{11} & P_{12} & \cdots & P_{1j} & \cdots & P_{1n} \\ P_{21} & P_{22} & \cdots & P_{2j} & \cdots & P_{2n} \\ \cdots & \cdots & \cdots & \cdots & \cdots & \cdots \\ P_{i1} & P_{i2} & \cdots & P_{ij} & \cdots & P_{in} \\ P_{n1} & P_{n2} & \cdots & P_{nj} & \cdots & P_{nn} \end{Vmatrix} \quad (11.6)$$

The first index in the notation P_{ij} indicates the state of the system at the earlier moment of time and the second one determines the state of the system at the later moment of time. In accordance with the condition of normalization, the sum of the transition probabilities in any row of the matrix (11.6) must be equal to unity:

$$\sum_{j=1}^{n} P_{ij} = 1, \quad i = 1, 2, \ldots, n \quad (11.7)$$

A matrix that possesses this property is called a *stochastic matrix*. The knowledge of the initial probability distribution $p_i(0)$ and the transition matrix $\|P_{ij}\|$ enables one to evaluate the probability $p_i(k)$ at any step k. This can be done on the basis of the recursion formula

$$p_i(k) = \sum_{j=1}^{n} p_j(k-1) P_{ji}, \quad i, j = 1, 2, \ldots, n \quad (11.8)$$

In an homogeneous Markov chain, the transition probabilities in the matrix $\|P_{ij}\|$ and in the recursion formula (11.8) are independent from the ordinal number k of the step.

Example 11.1 Let a system X be in one of the three states, x_1, x_2, x_3, and the process in this system is developed in accordance with a homogeneous Markov chain. The possible transitions of the system from one state to another are given by the transition matrix

$$P = \begin{vmatrix} P_{11} & P_{12} & P_{13} \\ P_{21} & P_{22} & P_{23} \\ P_{31} & P_{32} & P_{33} \end{vmatrix} = \begin{vmatrix} \frac{1}{3} & \frac{1}{3} & \frac{1}{3} \\ \frac{1}{2} & \frac{1}{6} & \frac{1}{3} \\ \frac{1}{4} & 0 & \frac{3}{4} \end{vmatrix}$$

Is the transition from the state x_3 to the state x_2 possible?

Solution As evident from the information contained in the transition matrix, the system can move from the state x_1 to the state x_2, to the state x_3, or remain in the state x_1 with the same probability 1/3. The second line of the matrix indicates that the system will remain in the state x_2 with the probability 1/6, can move to the state x_1 with the probability 1/2, or move to the state x_3 with the probability 1/3. Finally, if the system is in the state x_3, it is most likely that it will remain in this state (the probability of this event is 3/4); in one of the four cases it will move to the state x_1. The transition from the state x_3 to the state x_2 is zero, i.e., this transition is impossible.

Example 11.2 A particle is moving in a random fashion within two absorbing barriers. What is the transition matrix for this motion if the total number of the possible positions of the particle is $N = 5$?

Solution Let the extreme positions of the particle be x_1 and x_5. For absorbing barriers, $P_{11} = P_{55} = 1$. As far as the internal states $i = 2, 3, 4$ are concerned, the system can move in the direction of the state x_1 with the probability q and in the direction of the state x_5 with the probability $p = 1 - q$. Since $p + q = 1$, the probabilities P_{ii} are zero for $i = 2, 3, 4$. Hence, the transition matrix is

$$P = \begin{vmatrix} 1 & 0 & 0 & 0 & 0 \\ q & 0 & p & 0 & 0 \\ 0 & q & 0 & p & 0 \\ 0 & 0 & q & 0 & p \\ 0 & 0 & 0 & 0 & 1 \end{vmatrix}$$

Example 11.3 A particle is moving in a random fashion within two reflecting barriers. Write the transition matrix for this motion if the total number of the possible positions is $N = 5$.

Solution If the particle, after having reached the barrier, is reflected to its previous position, only the states x_2 and x_4 can be such positions. Then we have

$$P = \begin{vmatrix} 0 & 1 & 0 & 0 \\ q & 0 & p & 0 \\ 0 & q & 0 & 0 \\ 0 & 0 & q & p \\ 0 & 0 & 0 & 0 \end{vmatrix}$$

If the conditions of reflection are such that the particle can occupy, after being reflected, any position, including x_3, then

$$P = \begin{vmatrix} 0 & 0 & 1 & 0 & 0 \\ q & 0 & p & 0 & 0 \\ 0 & q & 0 & p & 0 \\ 0 & 0 & q & 0 & p \\ 0 & 0 & 1 & 0 & 0 \end{vmatrix}$$

Example 11.4 A factory produces TV sets. Depending on the demand by the end of each year, the factory may be in one of two states: state 1, if its order book is full, and state 2, if the order book is empty. The probability that the factory is kept in state 1 throughout the next year is 4/5. The probability of leaving state 2 by improving the product is 3/5. Write the transition matrix for the production process.

Solution The transition matrix for the production process, treated as a Markov chain, is

$$P = \begin{vmatrix} \frac{4}{5} & \frac{1}{5} \\ \frac{3}{5} & \frac{2}{5} \end{vmatrix}$$

A markovian process with discrete states but continuous time is a *continuous Markov chain*. For such a process, the probability of transition from the state s_i to the state s_j is zero for any moment of time. For a continuous Markov chain, the *transition probability density* λ_{ij} is used instead of the transition probability P_{ij} to characterize the process. This density is defined as the limit

$$\lambda_{ij} = \lim_{\Delta t \to 0} P_{ij}(t + \Delta t, t)$$

of the ratio of the transition probability P_{ij} from the state s_i to the state s_j during a small time interval Δt after a moment t to the duration Δt of this interval as it tends to zero. The transition probability density can be either constant (in this case the continuous Markovian process is a homogeneous one) or time dependent (in this case the process is nonhomogeneous).

When considering random processes with discrete states and continuous time, it is assumed that the transitions of a system S from one state to another is governed by some flow of events. In such a case the densities of transition probabilities assume the values of intensities λ_{ij} of the governing flow of events: the system "jumps" from the state s_i to the state s_j, as soon as the first event occurs in the flow with an intensity λ_{ij}. If the governing flow is a Poisson one, the random process in the system is a markovian process. The *probabilities of state* $p_i(t) = P\{S(t) = s_i\}$, $i = 1, 2, \ldots, n$, of such a process, i.e., the probability that the system S is in the state s_i at the moment of time t, can be found from the following *Kolmogorov equation*:

$$\frac{dp_i(t)}{dt} = \sum_{j=1}^{n} \lambda_{ji} p_j(t) - p_i(t) \sum_{j=1}^{n} \lambda_{ij}, \quad i = 1, 2, \ldots, n \quad (11.9)$$

For any t, the condition of normalization

$$\sum_{i=1}^{n} p_i(t) = 1 \quad (11.10)$$

must be fulfilled.

If all the flows of events, which bring the system from one state to another, are elementary, i.e., are stationary Poisson processes with constant intensities λ_{ij}, then the *limiting probabilities of states*

$$p_i = \lim_{t \to \infty} p_i(t), \quad i = 1, 2, \ldots, n$$

are independent from the initial state of the system. In such a *limiting stationary condition* the probabilities of the system, passing from state to state, no longer change, the system becomes an *ergodic* one, and the corresponding random process is an *ergodic process* (see Sec. 7.3).

Example 11.5 The road traffic in the given direction is an elementary flow with intensity λ. A hitch-hiker tries to stop a car. Find the distribution of the time T during which he has to wait for a car, and the mean value and the variance of this time.

Solution Since an elementary flow does not have aftereffects, its "future" is "past" independent, and the distribution of the waiting time T is the same as the distribution of the time intervals between successive passing cars, i.e., exponential with parameter λ:

$$f(t) = \lambda e^{-\lambda t}$$

The mean and the variance of the waiting time T are

$$\langle t \rangle = \frac{1}{\lambda}, \quad D_t = \frac{1}{\lambda^2}$$

Example 11.6 A passenger arrives at a bus stop at an arbitrary moment of time. The buses are run regularly with a time interval τ. What is the distribution of the time T for which the passenger will have to wait for a bus? What is the distribution of the time T for which the passenger will have to wait for a bus? Determine also the mean and the variance of this time.

Solution The moment of time at which the passenger arrives at the bus stop is a random variable which is distributed uniformly within the interval τ between two buses. The distribution density of the waiting time T is also constant, and therefore

$$f(t) = \frac{1}{\tau} = \lambda$$

Then the mean value of the waiting time is $\langle t \rangle = 1/(2\lambda)$ and its variance is $D_t = 1/(12\lambda^2)$.

11.3. Some Often-Encountered Markovian Processes

11.3.1. Random telegraph signal

Let a process $X(t)$ start at the value $x(0) = 1$ at the initial point of time $t = 0$ and change sign at random moments of time t_k, $k = 1, 2, \ldots$, constituting a Poisson process with the rate λ. The number n of changes of sign between any two moments of time has a Poisson distribution with the rate λ. Such a process is known as the *random telegraph signal*.

In the case of an even number of changes of sign at the interval $(0, t)$, $X(t) = +1$. The probability of this event is

$$P(X(t) = +1) = \left(1 + \frac{(\lambda t)^2}{2!} + \frac{(\lambda t)^4}{4!} + \cdots\right) e^{-\lambda t} = e^{-\lambda t} \cosh \lambda t$$

In the case of an odd number of changes of sign, $X(t) = -1$. The probability of this event can be calculated as

$$P(X(t) = -1) = e^{-\lambda t} \sinh \lambda t$$

The mean (expected) value of the process is therefore

$$\langle x \rangle = 1 \times P(X(t) = 1) + (-1)P(x(t) = -1)$$
$$= e^{-\lambda t}(\cosh \lambda t - \sinh \lambda t) = e^{-2\lambda t}$$

The correlation function of the random telegraph signal can be obtained as

$$K_x(t_1, t_2) = e^{-2\lambda(t_2 - t_1)}$$

11.3.2. Shot noise

If a light falls on a heated cathode in a vacuum-tube diode, the cathode ejects photoelectrons. Every time a photoelectron is ejected, a current pulse is generated. If photoelectrons are emitted at random moments of time t_k, $-\infty < k < \infty$, then the total current in the electric circuit can be represented as

$$X(t) = \sum_{k=-\infty}^{\infty} h(t - t_k)$$

where $h(t)$ is the function describing the shape of the current pulse. If the moments of time t_k of the emission of the photoelectrons form a Poisson flow of events, the process $X(t)$ is stationary and is known as *shot noise*.

In general, a random process is said to be a shot noise process if it is induced by a train of impulses acting on a system at random moments of time t_k. The mean value of the shot noise process $X(t)$ is

$$\langle x \rangle = \lambda \int_{-\infty}^{\infty} h(t - \xi)\, d\xi = \lambda \int_{-\infty}^{\infty} h(u)\, du$$

The correlation function of this process can be found as (*Campbell's theorem*)

$$K_x(\tau) = \lambda \int_{-\infty}^{\infty} h(u)h(\tau + u)\, du$$

where λ is the intensity of the Poisson process. If, for instance, the shot pulses are rectangles of the height a and the duration T, then

$$K_x(\tau) = \lambda a^2 (T - |\tau|), \quad \text{for } |\tau| < T$$

The correlation function is zero for $|\tau| > T$.

11.3.3. Random pulses

If an aircraft flying over the ocean sends out a radar pulse, it receives a multitude of echoes. These are Doppler-shifted copies of the transmitted pulse which is reflected from numerous points on the surface of the water. Those points reflect the pulse with random amplitudes and random delays which, for a turbulent wind-blown surface of the ocean, can be considered statistically independent. The echoes constitute a random signal known as *clutter*:

$$X(t) = \sum_{k=-\infty}^{\infty} A_k h(t - t_k)$$

Here $h(t)$ represents the shape of the reflected radar pulse, t_k is the time when the kth pulse reaches the receiver, and A_k is the amplitude of the pulse. If the process $X(t)$ lasts from $-\infty$ to ∞, the times t_k constitute a Poisson process, and the times t_k and the amplitudes A_k are statistically independent, then the process $X(t)$ is called a *random pulse process*.

The mean (expected) value of the clutter process is

$$\langle x \rangle = \lambda \langle A \rangle \int_{-\infty}^{\infty} h(t - \tau) \, d\tau = \lambda < \langle A \rangle \int_{-\infty}^{\infty} h(u) \, du$$

where $\langle A \rangle$ is the mean value of the amplitudes A_k. The correlation function of this process is

$$K_x(\tau) = \lambda \langle A^2 \rangle \int_{-\infty}^{\infty} h(u) h(\tau + u) \, du$$

where $\langle A^2 \rangle$ is the mean value of the pulse amplitude squared.

11.3.4. Poisson impulse process

The *Poisson impulse process* is defined as

$$X(t) = \sum_{k=-\infty}^{\infty} A_k \delta(t - t_k)$$

where A_k are random "amplitudes" of instantaneous impulses occurring at random moments of time t_k and $\delta(t)$ is Dirac's delta function. The correlation function of this process is

$$K_x(\tau) = \lambda \langle A^2 \rangle \delta(\tau)$$

and its spectral density is

$$S_x(\omega) = \lambda \langle A^2 \rangle = \text{constant}$$

Hence, the Poisson impulse process is a "white noise."

Let us revisit the process consisting of random pulses with random amplitudes examined in Sec. 11.3.3. Denoting this process as $Y(t)$, we obtain the following expression for its spectrum:

$$\begin{aligned} S_y(\omega) &= \lambda \langle A^2 \rangle \int_{-\infty}^{\infty} \int_{-\infty}^{\infty} h(u) h(\tau + u) e^{-i\omega\tau} \, du \, d\tau \\ &= \lambda \langle A^2 \rangle \int_{-\infty}^{\infty} \int_{-\infty}^{\infty} h(u) e^{i\omega u} h(\tau + u) e^{-i\omega(\tau + u)} \, du \, d\tau \\ &= \lambda \langle A^2 \rangle \int_{-\infty}^{\infty} h(u) e^{i\omega u} \, du \int_{-\infty}^{\infty} h(v) e^{-i\omega} \, dv \\ &= \lambda \langle A^2 \rangle a^2(\omega) = S_x(\omega) a^2(\omega) \end{aligned}$$

where

$$a(\omega) = |\Phi(i\omega)| = \int_{-\infty}^{\infty} h(u) e^{-i\omega u} \, du$$

is the spectrum of the individual pulse $h(t)$. Thus, the spectral density of the random pulse process $Y(t)$ is what it would be if it were the output of a linear dynamic system ("filter") with a transfer function $a(\omega)$ if the input were a random process $X(t)$ with a uniform spectral density $S_x(\omega) = \lambda \langle A^2 \rangle$.

11.3.5. Gaussian process

Let us put the mean value of the Poisson impulse process, examined in the previous section, equal to zero ($\langle A \rangle = 0$) and assume that the intensity of the process tends to infinity ($\lambda \to \infty$) and the mean square amplitude tends to zero ($\langle A^2 \rangle \to 0$) in such a way that the spectral density $S_x(\omega) = \lambda \langle A^2 \rangle$ of the process remains finite and unchanged. The delta functions $A_k \delta(t - t_k)$, whether positive or negative, are now coming thick and fast, but their absolute values $|A_k|$ are vanishingly small. A great many of these impulses enter the system during any time interval whose duration is of the order of magnitude of the response time of the system. Out of the system emerges a dense train of small, overlapping series $A_k h(t - t_k)$ of its impulse response. The output

$$Y(t) = \int_{-\infty}^{\infty} h(u)X(t - u)\, du = \sum_{k=-\infty}^{\infty} A_k h(t - t_k)$$

of the system, sampled at the time t, is a Gaussian (normal) random variable, consisting of the sum of a large number of small, random constituents.

The joint probability density function

$$f_y(y_1, t_1; y_2, t_2; \ldots; y_n, t_n)$$

of n samples $y_1 = y(t_1)$, $y_2 = y(t_2)$, $\ldots$, $y_n = y(t_n)$ of the output of the system will also be normal, i.e., will be a *gaussian random (stochastic) process*. In the examined example it arose as a result of filtering a dense Poisson impulse process. A gaussian random process $X(t)$ with zero mean is called *gaussian noise*.

11.3.6. Brownian process

The shot noise process examined in Sec. 11.3.2 can be viewed as an output of a linear system ("filter") whose impulse response is $h(t)$ and whose input is a Poisson impulse process of unit amplitude: $\langle A \rangle = \langle A^2 \rangle = 1$. When this special Poisson impulse process is integrated, the result is also a Poisson process. The integral

$$X(t) = \int_0^t Y(u)\, du$$

of a Poisson impulse process ("white noise") $Y(t)$ of unit amplitude is called the *brownian process*, because it represents in one dimension the position of a particle undergoing brownian motion. So many impulses contribute, during any perceptible time interval t, to the above integral that $X(t)$ is a gaussian random process. A typical realization (sample function) consists of numerous vanishingly small, random, positive or negative jumps. The brownian process is a "white noise." Its mean value is zero and the variance is

$$D_x = \langle [x(t)]^2 \rangle = \int_0^t \int_0^t Y(t_1) Y(t_2)\, dt_1\, dt_2$$
$$= \lambda \langle A^2 \rangle \int_0^t \int_0^t \delta(t_1 - t_2)\, dt_1\, dt_2 = \lambda \langle A^2 \rangle \int_0^t dt_1 = N_0 t$$

where $N_0 = \lambda \langle A^2 \rangle$ is the spectral density of the "white noise" $Y(t)$. The probability density function of the process $X(t)$ is

$$f_x(x, t) = \frac{1}{\sqrt{2\pi Dt}} e^{-x^2/(2Dt)}$$

where $D = N_0 = \lambda \langle A^2 \rangle$ is the *diffusion constant*. The width $\sqrt{Dt}$ of this function is proportional to the square root of the time. The correlation function of the brownian process is

$$K_x(t_1, t_2) = Dt_1, \qquad t_2 > t_1$$

The brownian motion process, when it is generalized and used in other areas of engineering and applied science (quantum mechanics, thermal noise in electric circuits, etc.), is known also as *Wiener–Lévy process*. Being a "white noise," the brownian process is a markovian process. Its behavior for times $t > t_2$ depends only on the $X(t_2)$ value and not on how the process $X(t)$ reached that value.

The brownian process can be treated as a limiting case of the following mechanism. Every t seconds a coin is tossed, and successive tosses are statistically independent. A particle is moved s units to the right if the coin falls "heads" and s units to the left if the coin falls "tails." At the time $n\tau$, the probability that the particle has progressed a total of ks units to the right and $(n - k)s$ units to the left [i.e., the probability that it is found at the position $X = (2k - n)s$] can be calculated on the basis of the binomial distribution

$$P\{x(n\tau) = (2k - n)s\} = C(k, n)p^k q^{n-k}$$

where p and $q = 1 - p$ are the probabilities of the occurrences of "heads" and "tails," respectively. The mean value of the position of the particle at the time $n\tau$ is

$$\langle x \rangle = x \langle 2k - n \rangle = ns(2p - 1) = ns(p - q)$$

where k is the binomial random variable. The variance of its position is

$$\text{var } [x(n\tau)] = 4s^2 \text{ var } k = 4ns^2 pq$$

If $p > q$, the particle "drifts" to the right; if $p > q$, it drifts to the left. The position of the particle is randomly dispersed about the $\langle x \rangle$ value and the variance of the particle's position is proportional to the number n of steps.

If the intervals τ between tosses are very brief and the length s of the steps is very small, then the probabilities q and p differ only slightly from 1/2:

$$p = \frac{1}{2}(1 + \varepsilon), \qquad q = \frac{1}{2}(1 - \varepsilon), \qquad \varepsilon \ll 1$$

and
$$t = n\tau, \qquad D = \frac{s^2}{\tau}, \qquad v = \frac{\varepsilon s}{\tau}$$

At any finite time t, the number n will be very large, and the random variable $X(t)$ will have a gaussian distribution in the limit $\tau \to 0$, with s and ε decreasing in proportion to $\sqrt{\tau}$. Then

$$\langle x \rangle = ns\varepsilon \to vt, \qquad D_x = \text{var } x(t) = Dt$$

and the probability density function of $x(t)$ will be

$$f_x(x, t) = \frac{1}{\sqrt{2\pi Dt}} \exp\left[-\frac{(x - vt)^2}{2Dt}\right]$$

The resulting random process is called *brownian motion with drift*. The v value is the drift velocity and D is the diffusion constant.

11.3.7. Flows of failures and restorations: availability of an item

In the theory of reliability of repairable items (see Sec. 10.6) one can consider failures and restorations as flows of events. These events occur or start at random moments of time, and the duration of the repair process is also a random variable. In many applications, the flows of failures and restorations can often be treated as stationary, ordinary flows without an aftereffect. This enables one to use the methods of the theory of markovian processes for their description.

The number of failures and restorations (repairs) of one item during a certain period of operation has to be the same; however, the functions of the flow of failures and the flow of restorations can be different, since the operation times and restoration times are different. The probability of occurrence of n failures during the time t, assuming that these are rare events, follows the Poisson law:

$$P_n(t) = \frac{(\lambda t)^n}{n!} e^{-\lambda t}$$

where λ is the failure rate. The restoration time t is a random variable that typically can be assumed to obey the exponential law:

$$f_r(t) = \mu e^{-\mu t}$$

Here μ is the intensity of the restoration of the workability of the item. Such a distribution is especially suitable in the cases where the restorations are carried out swiftly and the number of restorations reduces with an increase in their duration.

The flows of failures and restorations are considered when evaluating the availability of a repairable item. Availability is a complex reliability index which characterizes both the dependability and repairability of a repairable item (see also Sec. 10.6). Let $K(t)$ be the probability that the item is in the working condition and $k(t)$ is the probability that the item is in the idle condition. Using the Kolmogorov equation (11.9) for a Markov chain, we obtain the following equations for the probabilities $K(t)$ and $k(t)$:

$$\frac{dK(t)}{dt} = \mu k(t) - \lambda K(t)$$

$$\frac{dk(t)}{dt} = \lambda K(t) - \mu k(t)$$

Clearly, for any moment of time t, the condition of normalization of probabilities requires that $K(t) + k(t) = 1$. With this relationship, the above equations can be written in the form

$$\frac{1}{\lambda + \mu} \frac{dK(t)}{dt} + K(t) = \frac{\mu}{\lambda + \mu}$$

$$\frac{1}{\lambda + \mu} \frac{dk(t)}{dt} + k(t) = \frac{\lambda}{\lambda + \mu}$$

in which the probabilities $K(t)$ and $k(t)$ are separated. These equations have the following solutions:

$$K(t) = Ce^{-(\lambda + \mu)t} + \frac{\mu}{\lambda + \mu}$$

$$k(t) = Ce^{-(\lambda + \mu)t} + \frac{\lambda}{\lambda + \mu}$$

where the constant of integration C is determined from the initial conditions, depending on whether the item is in the workable condition or not at the initial moment $t = 0$. If it is, then $K(0) = 1$ and $k(0) = 0$. If it is not, then $K(0) = 0$ and $k(0) = 1$. Using these conditions, we find that the constant C is equal to $C = \lambda(\lambda + \mu)$ if the item is in a workable condition at the initial moment of time, and is equal to $C = \mu/(\lambda + \mu)$ if it is not.

The *availability function* $K(t)$ can therefore be expressed as

$$K(t) = \frac{\mu}{\lambda+\mu} + \frac{\lambda}{\lambda+\mu} e^{-(\lambda+\mu)t}$$

if the item is in the workable condition at the initial moment of time, and as

$$K(t) = \frac{\mu}{\lambda+\mu} - \frac{\mu}{\lambda+\mu} e^{-(\lambda+\mu)t}$$

in a situation where the item is not in the workable condition at the initial moment of time. As one can see, the availability function consists of the constant (steady state) and the transitional parts. The constant part is called the *availability index* (see Sec. 10.6):

$$K_a = \frac{\mu}{\lambda+\mu} = \frac{1}{1+\lambda/\mu} = \frac{1}{1+\langle t_r\rangle/\langle t_f\rangle}$$

where $\langle t_r \rangle = 1/\mu$ is the mean time-to-repair and $\langle t_f \rangle = 1/\lambda$ is the mean time-between-successive-failures. The availability index indicates the percentage of time in which the item is in a workable ("available") condition. If the system consists of many items, then the obtained formula can be generalized in accordance with the formula (10.33).

11.3.8. Linear filtering: Wiener–Hopf equation

Let a receiving unit in a telecommunication channel be subjected to a random input process $X(t)$ which is a known (in the general case, nonlinear) nonrandom function of a random *signal* $s[t, \lambda(t)]$ and a random corrupting error (*noise*) $\xi(t)$:

$$X(t) = F[s(t, \lambda(t)), \xi(t)]$$

The signal s depends on the time t, both explicitly and implicitly, through one or more parameters $\lambda(t)$. The statistical characteristics of the random processes s and ξ, including their interaction, are supposed to be known as well. The objective is to find out, using the available information, which realization of the information (signal) s is contained in the received random process (oscillation) x, and to develop, on this basis, recommendations for the design of an optimal receiver. This should be able to minimize, using certain optimization criteria, the adverse influence of the noise. The body of knowledge that deals with the optimal *detection and estimation of signals* is known as *filtering of random signals*. In many practically important cases, the random processes s and ξ can be considered as stationary ones, and the process $X(t)$ can be treated as a sum of these processes:

$$X(t) = s[t, \lambda(t)] + \xi(t), \qquad 0 \le t \le T$$

Here T is the time of observation. Additional simplification can be achieved if the process $\xi(t)$ is a Markovian one ("white noise"):

$$X(t) = s[t, \lambda(t)] + n(t), \qquad 0 \le t < T$$

Because of the noise, the estimated signal $\hat{s}(t)$ will not coincide with the transmitted signal $s(t)$. This leads to filtration errors. The minimization of these errors can be achieved using various optimization criteria: the least mean square error, the maximum signal-to-noise ratio, or the maximum *a posteriori* probability. If the criterion of the least mean square error is used, an optimal receiver is designed by minimizing the quantity

$$\langle \varepsilon^2 \rangle = \langle [\hat{s}(t) - s(t + \Delta)]^2 \rangle$$

In this equation we introduced a time lag Δ. If $\Delta > 0$, the estimate $\hat{s}(t)$ of the obtained information will *predict* the value of the signal $s(t)$ in advance. If $\Delta = 0$, the problem is reduced to the *extraction* ("*smoothening*") of the signal $s(t)$ from the process $X(t)$. If $\Delta < 0$, the estimate $\hat{s}(t)$ provides information of the input signal that took place at the moment $t - \Delta$, i.e., it *restores* the signal.

In the simplest case, the process $X(t)$ can be treated as a linear function of stationary and normal processes $s(t)$ and $n(t)$. The rigorous mathematical solution to the problem of filtering of random signals, known as *linear filtering*, was obtained by a Russian mathematician A. N. Kolmogorov, in 1941, and an American mathematician N. Wiener, in 1949. The main results of their theory can be summarized as follows.

Let the processes $s(t)$ and $\xi(t)$ be normal random processes with known correlation functions

$$K_s(\tau) = \langle s(t)s(t+\tau) \rangle = \sigma_s^2 R_s(\tau)$$

$$K_\xi(\tau) = \langle \xi(t)\xi(t+\tau) \rangle = \sigma_\xi^2 R_\xi(\tau)$$

$$K_{s\xi}(\tau) = \langle s(t)\xi(t+\tau) \rangle$$

If the dynamic system (filter), characterized by the impulse characteristic $h(t)$, $t \ge 0$, is subjected to a stationary random process $X(t)$, then the output process can be evaluated as

$$Y(t) = \hat{s}(t) = \int_0^\infty h(\tau)X(t-\tau)\,d\tau$$

The mean square error is

$$\langle \varepsilon^2 \rangle = \left\langle \left[\int_0^\infty h(\tau)X(t-\tau)\,d\tau - s(t+\Delta) \right] \right\rangle$$

$$= K_s(0) - 2\int_0^\infty h(\tau)K_{sx}(\tau+\Delta)\,d\tau$$

$$+ \int_0^\infty \int_0^\infty h(\tau_1)h(\tau_2)K_x(\tau_2 - \tau_1)\,d\tau_1\,d\tau_2$$

where $K_{sx}(\tau)$ is the bivariate correlation function of the processes $s(t)$ and $X(t)$:

$$K_{sx}(\tau) = \langle s(t)x(t+\tau) \rangle$$

and $K_x(\tau)$ is the correlation function of the random process $X(t)$:

$$K_x(\tau) = \langle x(t)x(t+\tau) \rangle$$

The impulse response $h_0(t)$ of the optimal filter can be determined, based on the following approach from the theory of variations. The function $h(t)$ is sought in the form

$$h(t) = h_0(t) + \mu g(t)$$

where μ is a time-independent parameter and $g(t)$ is an arbitrary function of time t. The condition of minimization

$$\left. \frac{d\langle \varepsilon^2 \rangle}{d\mu} \right|_{\mu=0} = 0$$

yields

$$\int_0^\infty g(\tau) \left[\int_0^\infty h_0(u)K_x(\tau-u)\,du - K_{sx}(\tau+\Delta) \right] d\tau = 0$$

Since $g(\tau)$ is an arbitrary function, the following equation must be fulfilled:

$$\int_0^\infty h_0(u)K_x(\tau-u)\,du = K_{sx}(\tau+\Delta), \qquad \tau \geq 0$$

This Fredholm integral equation of the first kind is the basic equation of the theory of linear filtering. It is known as the *Wiener–Hopf equation*.

If the solution to the Wiener–Hopf equation is found, then one can try to build a physically realizable linear filter (this is known as *Wiener filter*) which performs optimal smoothing ($\Delta = 0$) or forecasting ($\Delta > 0$). The mean square error of the optimal linear filtering is expressed as

$$\langle \varepsilon^2 \rangle = \sigma_x^2 + \int_0^\infty h_0(\tau) \, d\tau \left[\int_0^\infty K_x(\tau - u) h_0(u) \, du - 2K_{sx}(\tau + \Delta) \right]$$

The described linear theory of filtering has, however, several major shortcomings. First of all, in the majority of cases the Wiener–Hopf equation does not lend itself to a direct solution. In addition, the linear theory of filtering encompasses a relatively small number of cases of practical interest. This is due primarily to the fact that linear filters are able to extract only that information that is linearly related to the received signal and, in addition, obeys the normal law of the probability distribution. However, in most typical problems encountered in telecommunications engineering, the received information $X(t)$ is related nonlinearly to the signal, and, although the process $X(t)$ can be indeed in many cases approximated by a normal distribution, the most typical signals are not normal ones. Finally, the formal solutions to the Wiener–Hopf equation often result in responses (optimal filters) that physically cannot be realized. In the example that follows, the optimal filter should have an impulse characteristic that contains the exponents $e^{\gamma t}$ and $e^{-\gamma t}$, and therefore cannot be built easily.

Example 11.7 Design, using the Wiener–Hopf equation, a linear filter which would minimize the mean square error in the received normal stationary process $X(t)$ with the correlation function

$$K_x(\tau) = \sigma_x^2 e^{-\alpha|\tau|}$$

Solution For the case in question, the Wiener–Hopf equation is

$$\int_0^t [\sigma_x^2 e^{-\alpha|\tau - u|} + S_0 \delta(\tau - y)] h_0(u) \, du = \sigma_x^2 e^{-\alpha|\tau|}$$

or

$$e^{-\alpha\tau} \int_0^t e^{\alpha u} h_0(u) \, du + e^{\alpha\tau} \int_0^t e^{-\alpha u} h_0(u) \, du = e^{-\alpha\tau} - \frac{S_0}{\sigma_x^2} h_0(\tau)$$

By multiplying this integral equation by $e^{\alpha\tau}$ and then differentiating it twice by the variable τ, we obtain the following differential equation for the Green function $h_0(u)$:

$$\frac{d^2 h_0(\tau)}{d\tau^2} - \gamma^2 h_0(\tau) = 0$$

where the eigenvalue γ is expressed as

$$\gamma^2 = \alpha^2 \left(1 + \frac{2\sigma_x^2}{\alpha S_0}\right)$$

The solution to the obtained equation is

$$h_0(\tau) = C_1 e^{-\gamma \tau} + C_2 e^{\gamma \tau}$$

The constants

$$C_1 = \frac{(\alpha + \gamma)^2 (\alpha - \gamma)}{(\alpha + \gamma)^2 - (\alpha - \gamma)^2 e^{-2\gamma t}}$$

$$C_2 = \frac{(\alpha + \gamma)(\alpha - \gamma)^2}{(\alpha + \gamma)^2 e^{2\gamma t} - (\alpha - \gamma)^2}$$

can be found by introducing the solution for the function $h_0(\tau)$ into the integral equation for this function and by equating the coefficients in front of the functions $e^{-\alpha \tau}$ and $e^{-\alpha(t-\tau)}$.

The least mean square error due to the application of the filter in question is as follows:

$$\langle \varepsilon^2 \rangle = S_0 (\gamma - \alpha)$$

11.4. Queues

The way this section is set forth is different of the other sections or chapters in this book. We simply put together some formulas and ready-to-use results from the Queueing Theory.

11.4.1. Queueing systems

A *queueing system* is a system designed to serve customers (demands) arriving at random moments of time. Examples are: telephone exchange, aircraft waiting to land, a garage, cars waiting at a traffic light, a storage facility, a booking office, a hairdresser's establishment, industrial machines awaiting repair, an interactive computer system, etc. Originally, the specific probabilistic problems that lead to the development of the theory of queues emerged in connection with the operation of telephone systems. Later it was found that similar problems arise in trade, in product equipment management, in scheduling of events and processes, in calculating the traffic capacity of highways, bridges, tolls, intersections, airfields, canal locks, berths at seaports, etc. The main definitions used in the queueing theory are set forth below.

A facility that serves customers (demands) is called a *server* (*channel*). There are *single-server* (one-channel) and *multiserver* (multichannel) queueing systems. A hairdressing establishment with one hairdresser is an example of a single-server system, while an establishment with several people is an example of a multiserver system. The theory of queues establishes relationships between the random flow of demands, the capacity of the given channel, the number of channels, and the efficiency of the service system.

One differentiates between *congestion systems* (*systems with refusals*) and *delay queueing systems*. A customer arriving at a congestion system when all the servers are busy is refused service and departs without taking part in any further proceedings. In a delay queueing system, a customer arriving when all the servers are busy does not leave the system but drives up and waits for the server to become free. The number of places in the queue may be limited or unlimited. If the number of places is zero, a delay system turns into a congestion system. A queue can be *bounded* both in terms of the number of customers in it (the size or length of the queue) and in terms of the queueing time ("systems with impatient customers"). Delay systems are also subdivided in terms of the queue discipline, i.e., customers may be served either in the order of arrival in a random fashion, or some customers may be able to obtain service before others (a *priority service*). A priority serivce may have several gradations, or ranks, of prontos (service without delay).

The analyses of queueing systems can be carried out in a simple way when all the flows of events transferring it from the given state are elementary and can be treated as stationary Poisson's flows. This means that both the arrival and the service processes are stationary Poisson's processes, and therefore the service time is exponential. Such queueing systems are called *elementary* queueing systems. If all the flows of events are stationary Poisson flows, then the process in a queueing system is a markovian random process with discrete states and continuous time.

The problems the queueing theory deals with include finding the probabilities of various states of a queueing system and establishing relationships between the parameters (the number of servers, the intensity of the arrival process, the distribution of the service time, etc.) and the characteristics of the efficiency of the service system. For instance, one can consider: the average number of customers served by a queueing system per unit time, or the *absolute capacity* of the system for service; the probability of serving an arriving customer, or the *relative capacity* of the system for service; the probability of a refusal, i.e., the probability that an arriving customer will not be served; the average number of customers present in a system (both being served and in the queue); the average number of customers in the queue; the average waiting time of a customer (the average time the customer is present in the system); the average queueing time of a customer (the average time the customer spends in the queue); the average number of busy servers. In general, all these characteristics are time dependent. However, in many practically important cases it can be assumed that the actual service system operates under the same conditions for a long time and, therefore, a simpler situation, which can be assumed to be stationary, is considered.

A queueing system is said to be *open* if the intensity of the arrival process is independent from the state of the system itself. For an open system, which operates under limiting stationary conditions, the average waiting time $\langle t_w \rangle$ of a customer is related to the average number $\langle z \rangle$ of customers in the system by *Little's formula*:

$$\langle t_w \rangle = \frac{\langle z \rangle}{\lambda} \qquad (11.11)$$

where λ is the intensity of the arrival process. A similar formula relates the average queueing time $\langle t_q \rangle$ of a customer and the average number $\langle r \rangle$ of customers in the queue:

$$\langle t_q \rangle = \frac{\langle r \rangle}{\lambda} \qquad (11.12)$$

It should be pointed out that Little's formulas (11.11) and (11.12) hold for any open queueing system, as long as the arrival and the service processes are stationary. The average number $\langle k \rangle$ of busy servers in an open queueing system is related to the absolute capacity A of the system as

$$\langle k \rangle = \frac{A}{\mu} \qquad (11.13)$$

where $\mu = 1/\langle t_{ser} \rangle$ is the intensity of the service process and $\langle t_{ser} \rangle$ is the average service time.

Below we will give, without derivations, some practical formulas for the limiting probabilities of states and the characteristics of the efficiency for several often-encountered queueing systems.

11.4.2. An elementary congestion system (Erlang's problem)

Let customers arrive at an n-server system with refusals, in accordance with a stationary Poisson process with an intensity λ. The service time is exponential with a parameter $\mu = 1/\langle t_{ser} \rangle$. The states of the system are numbered in accordance with the number of customers in the system (since there is no queue, the number of customers coincides with the number of busy servers). Thus, we have the following states of the system:

s_0 = {the system is idle}

s_1 = {one server is busy and the other servers are idle}

. . . .

s_k = {k servers are busy and the other $n - k$ servers are idle}

. . . .

s_n = {all n servers are busy}

The limiting probabilities of the above states are expressed by Erlang's formula:

$$p = \left(1 + \sum_{k=1}^{n} \frac{\rho^k}{k!}\right)^{-1}, \quad \rho = \frac{\lambda}{\mu} \qquad (11.14)$$

$$p_k = \frac{\rho^k}{k!} p_0, \quad k = 1, 2, \ldots, n \qquad (11.15)$$

The characteristics of the system's efficiency are as follows: the average number A of customers served per unit time (absolute capacity) is

$$A = \lambda(1 - p_n)$$

the probability Q of serving an arriving customer (relative capacity) is

$$Q = 1 - p_n$$

the probability of a refusal P_{ref} is

$$P_{\text{ref}} = p_n$$

and the average number of busy servers is

$$\langle k \rangle = \rho(1 - p_n)$$

For large values of the number n of servers, the probabilities p_k of states can be evaluated by the formula

$$p_k = \frac{P(k, \rho)}{R(n, \rho)}, \quad k = 0, 1, 2, \ldots, n \tag{11.16}$$

where, for a Poisson distribution,

$$P(m, a) = \frac{a^m}{m!} e^{-a} = R(m, a) - R(m-1, a) \tag{11.17}$$

and

$$R(m, a) = \sum_{k=0}^{m} \frac{a^k}{k!} e^{-a} \tag{11.18}$$

are tabulated functions.

11.4.3. An elementary single-server system with an unbounded queue

Consider a single-server system at which customers arrive in a stationary Poisson process with intensity λ. The service time is exponential with the parameter $\mu = 1/\langle t_{\text{ser}} \rangle$. The queue has an unlimited length. The limiting probabilities exist only for $\rho = \lambda/\mu < 1$. For $\rho \geq 1$ the queue increases indefinitely. The states of the system are numbered in accordance with the number of customers queueing or being served. These are the states:

$s_0 = \{\text{the system is idle}\}$

$s_1 = \{\text{the server is busy and there is no queue}\}$

$s_2 = \{\text{the server is busy and one customer is waiting to be served}\}$

. . . .

$s_k = \{\text{the server is busy and } k-1 \text{ customers are in the queue}\}$

The limiting probabilities of the states are expressed as

$$p = 1 - \rho, \quad p_k = \rho^k (1-\rho), \quad k = 1, 2, \ldots \tag{11.19}$$

where

$$\rho = \frac{\lambda}{\mu} < 1 \tag{11.20}$$

The characteristics of the system's efficiency are

$$A = \lambda, \quad Q = 1, \quad P_{\text{ref}} = 0$$

$$\langle z \rangle = \frac{\rho}{1-\rho}, \quad \langle r \rangle = \frac{\rho^2}{1-\rho}$$

$$\langle t_w \rangle = \frac{\rho}{\lambda(1-\rho)}, \quad \langle t_q \rangle = \frac{\rho_2}{\lambda(1-\rho)}$$

$$\langle k \rangle = \frac{\lambda}{mu} = 0 \qquad (11.21)$$

11.4.4 An elementary single-server system with a bounded queue

Consider a single-server queueing system at which customers arrive at a stationary Poisson process with intensity λ. The service time is exponential with the parameter $\mu = 1/\langle t_{\text{ser}} \rangle$. There are n places in the queue. If a customer arrives when all the places are occupied, he or she is refused service and departs. The states of the system are:

$s_0 = \{$the system is idle$\}$

$s_1 = \{$the server is busy and there is no queue$\}$

$s_2 = \{$the server is busy and one customer is in the queue$\}$

....

$s_k = \{$the server is busy and $k - 1$ customers are in the queue$\}$

....

$s_{m+1} = \{$the server is busy and m customers are in the queue$\}$

The limiting probabilities of states exist for any $\rho = \lambda/\mu$ and are as follows:

$$p_0 = \frac{1-\rho}{1-\rho^{m+2}}, \quad p_k = \rho^k p_0, \quad k = 1, 2, \ldots, m+1 \qquad (11.22)$$

The characteristics of the system's efficiency are

$$A = \lambda(1 - p_{m+1}), \quad Q = 1 - p_{m+1}, \quad P_{\text{ref}} = p_{m+1} \qquad (11.23)$$

The average number of busy servers is

$$\langle k \rangle = 1 - p_0 \qquad (11.24)$$

The average number of customers in the queue is

$$\langle r \rangle = \frac{\rho^2[1 - \rho^m(m + 1 - m\rho)]}{(1 - \rho^{m+2})(1 - \rho)} \qquad (11.25)$$

The average number of customers in the system is $\langle z \rangle = \langle r \rangle + \langle k \rangle$. Then Little's formula (11.11) yields

$$\langle t_w \rangle = \frac{\langle z \rangle}{\lambda}, \qquad \langle t_q \rangle = \frac{\langle r \rangle}{\lambda} \qquad (11.26)$$

11.4.5. An elementary multiserver system with an unbounded queue

Let customers arrive at an n-server queueing system in a stationary Poisson process with intensity λ. The service time of a customer is exponential with the parameter $\mu = 1/\langle t_{\text{ser}} \rangle$. Limiting probabilities exist only for $\rho_n = \kappa < 1$, where $\rho = \lambda/\mu$. The states of the system are numbered in accordance with the number of customers in the system:

$s_0 = \{\text{the system is idle}\}$

$s_1 = \{\text{one server is busy}\}$

....

$s_k = \{k \text{ servers are busy and there is no queue}\}$

....

$s_n = \{\text{all } n \text{ servers are busy}\}$

$s_{n+1} = \{\text{all } n \text{ servers are busy and there is one customer in the queue}\}$

$s_{n+r} = \{\text{all } n \text{ servers are busy and there are } r \text{ customers in the queue}\}$

The limiting probabilities of states are given by the formulas:

$$\left.\begin{aligned} p_0 &= \left(1 + \sum_{k=1}^{n} \frac{\rho^k}{k!} + \frac{\rho^{n+1}}{n \times n!} \frac{1}{1 - \kappa}\right)^{-1} \\ p_k &= \frac{\rho^k}{k!} p_0, \qquad 1 \leq k \leq n \\ p_{n+r} &= \frac{\rho^{n+r}}{n^r \times n!} p_0, \qquad r \geq 1 \end{aligned}\right\} \qquad (11.27)$$

Using the functions $P(m, a)$ and $R(m, a)$ one can reduce these equations to the form

$$p_k = \frac{P(k, \rho)}{R(n, \rho) + P(n, \rho)[\kappa/(1-\kappa)]}, \quad k = 0, 1, \ldots, n$$

$$P_{n+r} = \kappa^r p_n, \quad t = 1, 2, \ldots \quad (11.28)$$

The characteristics of the efficiency of the system are

$$\left.\begin{aligned}
\langle r \rangle &= \frac{\rho^{n+1} p_0}{n \times n!(1-\kappa)^2} = \kappa p_n (1-\kappa)^2 \\
\langle z \rangle &= \langle r \rangle + \langle k \rangle + \rho \\
\langle t_w \rangle &= \frac{\langle z \rangle}{\lambda} \\
\langle t_q \rangle &= \frac{\langle r \rangle}{\lambda}
\end{aligned}\right\} \quad (11.29)$$

11.4.6. An elementary multiserver system with a bounded queue

The conditions and the numbering of the states are the same as in the previous item, except that the number m of places in the queue is limited. Limiting probabilities of states exist for any λ and μ, and are expressed by the formulas:

$$\left.\begin{aligned}
p_0 &= \left(1 + \sum_{k=1}^{n} \frac{\rho^k}{k!} + \frac{\rho^{n+1}}{n \times n!} \frac{1-\kappa^m}{1-\kappa}\right)^{-1} \\
p_k &= \frac{\rho^k}{k!} p_0, \quad 1 \le k \le n \\
p_{n+r} &= \frac{\rho^{n+r}}{n^r \times n!} p_0, \quad 1 \le r \le m
\end{aligned}\right\} \quad (11.30)$$

where

$$\kappa = \frac{\rho}{n} = \frac{\lambda}{n\mu} \quad (11.31)$$

The characteristics of the efficiency of the system are

$$\left.\begin{array}{l} A = \lambda(1 - p_{n+m}), \quad Q = 1 - p_{n+m}, \quad P_{\text{ref}} = p_{n+m} \\ \langle k \rangle = \rho(1 - p_{n+m}) \\ \langle r \rangle = \dfrac{\rho^{n+1} p_0}{n \times n!} \dfrac{1 - (n+1)\kappa^m + m\kappa^{m+1}}{(1 - \kappa)^2} \\ \langle z \rangle = \langle r \rangle + \langle k \rangle \\ \langle t_q \rangle = \dfrac{\langle r \rangle}{\lambda}, \quad \langle t_w \rangle = \dfrac{\langle z \rangle}{\lambda} \end{array}\right\} \quad (11.32)$$

11.4.7. A multiserver congestion system for a stationary Poisson arrival and an arbitrary service time

The solution obtained in Erlang's problem remains valid when the arrival process is a stationary Poisson one and the service time T_{ser} has an arbitrary distribution with expectation $\langle t_{\text{ser}} \rangle = 1/\mu$.

11.4.8. A single-server system with an unbounded queue for a stationary Poisson arrival and an arbitrary service time

If the arrivals at a single-server queueing system come in a stationary Poisson flow with intensity λ and the service time T_{ser} has an exponential distribution with expectation $1/\mu$ and the coefficient of variation v_μ, then the average number of customers in the queue is expressed by the *Pollaczek–Khinchin formula*:

$$\langle r \rangle = \frac{\rho^2 (1 + v_\mu)^2}{2(1 - \rho)} \quad (11.33)$$

where

$$\rho = \frac{\lambda}{\mu}$$

and the average number of customers in the system is

$$\langle z \rangle = \frac{\rho^2 (1 + v_\mu)^2}{2(1 - \rho)} + \rho \quad (11.34)$$

Using Little's formula we obtain

$$\langle t_q \rangle = \frac{\rho^2 (1 + v_\mu)^2}{2\lambda(1 - \rho)}, \quad \langle t_w \rangle = \frac{\rho^2 (1 + v_\mu^2)}{2\lambda(1 - \rho)} + \frac{1}{\mu} \quad (11.35)$$

11.4.9. A single-server queueing system for an arbitrary (Palm) arrival process and an arbitrary service time

There are no exact formulas for this case, and an approximation for the length of the queue is given by the formula:

$$\langle r \rangle \approx \frac{\rho^2(v_\lambda^2 + v_\mu^2)}{2(1-\rho)} \tag{11.36}$$

where v_λ is the coefficient of variation of the interarrival time, $\rho = \lambda/\mu$, λ is the inverse of the expectation of that time, $\mu = 1/\langle t_{\text{ser}} \rangle$ is the inverse of the average service time, and v_μ is the coefficient of variation of the service time. The average number of customers in the queueing system

$$\langle z \rangle = \frac{\rho^2(v_\lambda^2 + v_\mu^2)}{2(1-\rho)} + \rho \tag{11.37}$$

and the average queueing and waiting times of a customer are, respectively,

$$\langle t_q \rangle \approx \frac{\rho^2(v_\lambda^2 + v_\mu^2)}{2\lambda(1-\rho)} \tag{11.38}$$

and

$$\langle t_w \rangle \simeq \frac{\rho^2(v_\lambda^2 + v_\mu^2)}{2\lambda(1-\rho)} + \frac{1}{\mu} \tag{11.39}$$

Example 11.8 A Poisson flow of items arrives at a repair shop with an intensity of four items per hour. Determine the probability that $k = 0, 1, 2, 3, 4$ items will arive within one hour.

Solution The probability that k items will arrive at the shop within the time τ is

$$P(k, \tau) = \frac{(\gamma\tau)^k}{k!} e^{-\gamma\tau}$$

In this case, $\gamma = 4$, $\tau = 1$, $k = 0, 1, 2, 3, 4$. Thus,

$$P(0, 1) = \frac{1}{1} e^{-4} = 0.0183, \qquad P(1, 1) = \frac{4}{1!} e^{-4} = 0.0733$$

$$P(2, 1) = 0.147, \qquad P(3, 1) = 0.195$$

$$P(4, 1) = 0.195$$

Example 11.9 Using the condition of the previous example, what is the probability that at least three items will arrive for the time $\tau = 30$ min that are required to have one item repaired?

Solution The sought probability is

$$P(k \geq 3, 1/2) = \sum_{k=3}^{\infty} P(k, 1/2) = 1 - \sum_{k=0}^{2} P(k, 1/2) = 0.323$$

Example 11.10 A gas station has two pumps ($n = 2$), but its lot can accommodate not more than four cars ($m = 4$). The arrival of cars at the station is a stationary Poisson's process with intensity $\lambda = 1$ car/min. The service time for a car follows an experimental distribution with the mean value $\langle t \rangle = 2$ min. Find the limiting probabilities of the states of the gas station and its characteristics.

Solution We have $\lambda = 1$, $\mu = 1/2 = 0.5$, $\rho = 2$, $\kappa = \rho/n = 1$. Using the general formulas for the birth and death scheme and designating $\rho = \lambda/\mu$, we obtain the following formulas for the probabilities of the states:

$$p_0 = \left(1 + \frac{\rho}{1!} + \frac{\rho^2}{2!} + \cdots + \frac{\rho^{n+m}}{n^m + n!}\right)^{-1}$$

$$= \left\{1 + \frac{\rho}{1!} + \frac{\rho^2}{2!} + \frac{\rho^n}{n!}\left[\frac{\rho}{n} + \left(\frac{\rho}{n}\right)^2 + \cdots + \left(\frac{\rho}{n}\right)^m\right]\right\}^{-1}$$

For $\kappa = \rho/n = 1$ we have

$$p_0 = \left(1 + \frac{\rho}{1!} + \frac{\rho^2}{2!} + \cdots + \frac{\rho^n}{n!} + \frac{m\rho^n}{n!}\right)^{-1}$$

$$p_k = \frac{\rho^k}{k!} p_0 \quad (1 \leq k \leq n)$$

$$p_{n+r} = \frac{\rho^n}{n!} p_0 \quad (1 \leq r \leq m)$$

$$P = p_{n+m}$$

$$Q = 1 - p_{n+m} = 1 - \frac{\rho^n}{n!} p_0$$

$$A = \lambda Q = \lambda\left(1 - \frac{\rho^n}{n!} p_0\right)$$

$$\langle k \rangle = \frac{A}{\mu} = \rho\left(1 - \frac{\rho^n}{n!} p_0\right)$$

$$\langle r \rangle = \sum_{r=1}^{m} r p_{n+r} = \sum_{r=1}^{m} r \frac{\rho^n}{n!} p_0 \sum_{r=1}^{m} r = \frac{\rho^n}{n!} \frac{m(m+1)}{2} p_0$$

$$\langle z \rangle = \langle r \rangle + \langle k \rangle$$

$$\langle t_w \rangle = \frac{\langle z \rangle}{\lambda}, \quad \langle t_q \rangle = \frac{\langle r \rangle}{\lambda}$$

With our input data, we find

$$p_0 = \left(1 + 2 + \frac{2^2}{2!} + \frac{2^2}{2!} 4\right)^{-1} = \frac{1}{13}$$

$$p_1 = p_2 = \cdots = p_6 = \frac{2}{13}$$

$$P = \frac{2}{13}$$

$$Q = 1 - P_{\text{ref}} = \frac{11}{13}$$

$$A = \lambda Q = \frac{11}{13} \approx 0.85 \text{ car/min}$$

$$\langle k \rangle = \frac{A}{\mu} = \frac{22}{13} = 1.69 \text{ pumps}$$

$$\langle r \rangle = \frac{2^2}{2!} \frac{4(4+1)}{2} \frac{1}{13} \approx 1.54 \text{ cars}$$

$$\langle z \rangle = \langle r \rangle + \langle k \rangle \approx 3.23 \text{ cars}$$

Example 11.11 A team of r mechanics services n machines of the same type ($r \leq n$). Each of the machines may require a mechanic's attention at a random moment of time. The machines fail independently of one another. The probability of failure of a machine during the time $(t, t + \tau)$ is $\lambda \tau$, where λ is the failure rate. The probability that a failed machine will be put back to operation during the time $(t, t + \tau)$ is $\mu \tau$, where μ is the restoration rate. Each mechanic can handle only one machine at a time and each machine is handled by one mechanic only. The rates λ and μ are time independent. They are also independent from the number n of machines in operation and from the number of machines under repair. Determine the probability that in such a steady-state situation a certain number of machines will be idle at a given moment of time.

Solution Let E_k be the event that k machines are out of operation at a given moment of time. The system in question can be in one of the following states: $E_0, E_1, \ldots, E_n$. This is a "birth-and-death" process for which $\lambda_k = (n-k)\lambda$ for $0 \leq k \leq n$, $\lambda_k = 0$ for $k \geq n$, $\mu_k = k\mu$ for $1 \leq k \leq r$, $\mu_k = r\mu$ for $k \geq r$. It could be found that

$$p_k = \begin{cases} \dfrac{n!}{k!(n-k)!} \rho^k p_0, & 1 \leq k \leq r \\ \dfrac{n!}{r^{n-k} r!(n-k)!} \rho^k p_0, & r \leq k \leq n \end{cases}$$

where $\rho = \lambda/\mu$ and

$$p_0 = \left[\sum_{k=0}^{r} \frac{n!}{k!(n-k)!} \rho^k + \sum_{k=r+1}^{n} \frac{n!}{r! r^{n-k}(n-k)!} \rho^k \right]^{-1}$$

In particular, for one server ($r = 1$) we obtain

$$p_k = \frac{n!}{(n-k)!} \rho^k p_0$$

and

$$p_0 = \left[\sum_{k=0}^{n} \frac{n!}{(n-k)!} \rho^k \right]^{-1}$$

TABLE 11.1 The results for $n = 8, r = 2$

Number of idle machines	Number of machines awaiting servicing	Number of idle mechanics	p_k
0	0	2	0.2048
1	0	1	0.3277
2	0	0	0.2294
3	1	0	0.1417
4	2	0	0.0687
5	3	0	0.0275
6	4	0	0.0083
7	5	0	0.0017
8	6	0	0.0002

Let, for instance, eight machines ($n = 8$) be serviced by two mechanics ($r = 2$). What is the best way to organize their work: (a) to put them both in charge of all the machines, so that the one who is free at the moment takes care of the machine that has just stopped, or (b) to assign four specific machines to each mechanic? The results are tabulated in Table 11.1, assuming $\rho = 0.2$.

The number of machines that are idle because the mechanics are busy with other machines is

$$\sum_{k=2}^{\infty} (k-2)p_k = 0.3045$$

The total idling time of the machines (under repair of waiting for repair) is

$$\sum_{k=2}^{\infty} kp_k = 1.6875 \text{ days}$$

The mean duration of free time for one mechanic is

$$2 \times 0.2048 + 1 \times 0.3277 = 0.7373 \text{ days}$$

Hence, each mechanic is idle for 0.3686 working days. If four specific machines are assigned to each mechanic then the number of idle machines, the number of machines waiting to be serviced, the number of idle mechanics, and the probabilities p_k that the system is found in the state E_k at the time t can be evaluated in the form of Table 11.2.

TABLE 11.2 The results for $n = 4, r = 1$

Number of idle machines	Number of machines awaiting servicing	Number of idle mechanics	p_k
0	0	1	0.3984
1	0	0	0.3189
2	1	0	0.1914
3	2	0	0.0760
4	3	0	0.0153

The mean time of unproductive idling of the machines (waiting for repair) is

$$1 \times 0.1914 + 2 \times 0.0760 + 3 \times 0.153 = 0.3893 \text{ days}$$

Hence, the whole group of eight machines will be idle for 0.7886 of the working day. Thus, the loss of working time of the machines because of waiting for repair is twice as large as in the case when both mechanics are in charge of all the machines. The total loss of time for the four machines because of waiting for repair and the repair time itself is

$$1 \times 0.3189 + 2 \times 0.1914 + 3 \times 0.0760 + 4 \times 0.0153 - 0.9909 \text{ days}$$

Thus, all eight machines will be out of service for 1.9818 working days. A mechanic will be idle on average for 0.3984 days, while the machines are expected to be idle for a longer time.

11.5. Continuous Markovian Processes: Smoluchowski's Equation

A continuous markovian process is characterized by a *transition function* $P(x, t | x_0, t_0)$. This function expresses the conditional probability that at the moment of time t the value of the random process $X(t)$ will be below level x, provided that at the moment t_0 preceding the moment t the value of the function $X(t)$ was x_0:

$$P(x, t | x_0, t_0) = P\{X(t) < x | X(t_0) = x_0\}$$

The function

$$p(x, t | x_0, t_0) = \frac{\partial P(x, t | x_0, t_0)}{\partial x}$$

is the *transition probability density*.

Let a transition $x_0, t_0 \to x, t$, characterized by the probability $p(x, t | x_0, t_0) \, dx$, be realized through some intermediate state ξ, τ. The relationship

$$p(x, t | x_0, t_0) = \int_{-\infty}^{\infty} p(x, t | \xi, \tau) p(\xi, \tau | x_0, t_0) \, d\xi \qquad (11.40)$$

which follows from the rule of the addition of probabilities, is known as *Smoluchowski's equation*. In mathematical literature this equation is usually referred to as the *Chapman–Kolmogorov equation*. This integral equation provides the fundamental property of the markovian process and expresses an obvious fact, that the transition probabilities for any three sequential moments of time must be in a certain way related to each other. Unfortunately, the Smoluchowski equation is an integral equation and does not lend itself to readily obtainable solutions. However, there exists a differential equation whose solution is equivalent to that of the Smoluchowski equation. This is the Fokker–Planck equation.

11.6. Fokker–Planck Equation

The intermediate moment of time τ in Smoluchowski's equation can be chosen in an arbitrary fashion. Let us choose this moment in such a way that it precedes the moment t by a short time $\Delta t = t - \tau$. Multiply this equation by an arbitrary function $R(x)$ and then integrate over the x axis:

$$\int_{-\infty}^{\infty} R(x) p(x, t | x_0, t_0)\, dx$$
$$= \int_{-\infty}^{\infty} p(\xi, t - \Delta t | x_0, t_0)\, d\xi \int_{-\infty}^{\infty} p(x, t | \xi, t - \Delta t) R(x)\, dx$$

The function $R(x)$ has the following expansion in Taylor series around the point $x = \xi$:

$$R(x) = R(\xi) + R'(\xi)(x - \xi) + \tfrac{1}{2} R''(\xi)(x - \xi)^2 + \cdots$$

Retaining the first three terms in this expansion, introducing it into the right-hand part of the above equation, and using the condition of normalization

$$\int_{-\infty}^{\infty} p(x, t | \xi, t - \Delta t)\, dx = 1$$

we obtain

$$\int_{-\infty}^{\infty} R(x) p(x, t | x_0, t_0)\, dx = R(\xi) \int_{-\infty}^{\infty} p(\xi, t - \Delta t | x_0, t_0)\, d\xi$$
$$+ \left[R'(\xi) \int_{-\infty}^{\infty} (x - \xi) p(x, t/\xi, t - \Delta t)\, dx \right.$$
$$\left. + \frac{1}{2} R''(\xi) \int_{-\infty}^{\infty} (x - \xi)^2 p(x, t | \xi, t - \Delta t)\, dx \right]$$
$$\times \int_{-\infty}^{\infty} p(\xi, t - \Delta t | x_0, t_0)\, d\xi$$

Taking the term containing the function $R(\xi)$ to the left-hand part and substituting in it the variable ξ for the variable x, we have

$$\int_{-\infty}^{\infty} R(x) p(x, t | x_0, t_0)\, dx - R(x) \int_{-\infty}^{\infty} p(x, t - \Delta t | x_0, t_0)\, dx$$
$$= \left[R'(x) \int_{-\infty}^{\infty} (x - \xi) p(x, t | \xi, t - \Delta t)\, dx \right.$$
$$\left. + \frac{1}{2} R''(\xi) \int_{-\infty}^{\infty} (x - \xi)^2 p(x, t | \xi, t - \Delta t)\, dx \right]$$
$$\times \int_{-\infty}^{\infty} p(\xi, t - \Delta t | x_0, t_0)\, d\xi$$

Dividing this equation by Δt and calculating the limit for $\Delta t \to 0$, we obtain the following equation:

$$\int_{-\infty}^{\infty} R(x) \frac{\partial}{\partial t} p(x, t | x_0, t_0) \, dx$$

$$= \int_{-\infty}^{\infty} p(\xi, t | x_0, t_0) \left[R'(\xi) A(\xi, t) + \frac{1}{2} R''(\xi) B(\xi, t) \right] d\xi \quad (11.41)$$

where the following notation is introduced:

$$A(\xi, t) = \lim_{\Delta t \to 0} \frac{\langle x - \xi \rangle}{\Delta t} = \lim_{\Delta t \to 0} \left[\frac{1}{\Delta t} \int_{-\infty}^{\infty} (x - \xi) p(x, t | \xi, t - \Delta t) \, dx \right]$$

$$B(\xi, t) = \lim_{\Delta t \to 0} \frac{\langle (x - \xi)^2 \rangle}{\Delta t} = \lim_{\Delta t \to 0} \left[\frac{1}{\Delta t} \int_{-\infty}^{\infty} (x - \xi)^2 p(x, t | \xi, t - \Delta t) \, dx \right]$$

As one can see from these equations, $A(\xi, t)$ is the mean value of the velocity of the change in the state of the process in the point ξ at the moment of time t, and $B(\xi, t)$ is the measure of scattering of the points x with respect to the point ξ. This scattering increases in accordance with the diffusion law: when the time t changes by the amount Δt (from $t - \Delta t$ to t) the scattering increases proportionally to $\tau = t - \Delta t$.

Let us substitute x for ξ in the right-hand part of Eq. (11.41) and carry out the integration by parts, assuming that

$$\lim_{x = \pm \infty} R = 0, \qquad \lim_{x = \pm \infty} R' = 0$$

Then we obtain

$$\int_{-\infty}^{\infty} R(x) \frac{\partial}{\partial t} p(x, t | x_0, t_0) \, dx = - \int_{-\infty}^{\infty} R(x) \frac{\partial}{\partial x} [p(x, t | x_0, t_0) A(x, t)] \, dx$$

$$+ \frac{1}{2} \int_{-\infty}^{\infty} R(x) \frac{\partial^2}{\partial x^2} [p(x, t | x_0, t_0) B(x, t)] \, dx$$

Since the function $R(x)$ is chosen in an arbitrary way, the obtained equation will be fulfilled if the following relationship takes place:

$$\frac{\partial p(x)}{\partial t} = - \frac{\partial}{\partial x} [A(x) p(x)] + \frac{1}{2} \frac{\partial^2}{\partial x^2} [B(x) p(x)] \quad (11.42)$$

Here $p(x) \equiv p(x, t | x_0, t_0)$, $A(x) \equiv A(x, t)$, and $B(x) \equiv B(x, t)$. The obtained differential equation in partial derivatives is known as the *Fokker–Planck equation* or the *Kolmogorov forward equation*. It describes the evolution of the probability density function of a one-dimensional continuous markovian process $X(t)$ and is equivalent to the integral Smoluchowski equation (11.40). Equation (11.42) is sometimes referred to as the *diffusion equation*.

The Fokker–Planck equation (11.42) can be generalized for the case of a multidimensional process as follows:

$$\frac{\partial}{\partial t} p(x_1, x_2, \ldots, x_n) = -\sum_{k=1}^{n} \frac{\partial}{\partial x_k} [A_k(x_1, x_2, \ldots, x_n) p(x_1, x_2, \ldots, x_n)]$$
$$+ \frac{1}{2} \sum_{k=1}^{n} \sum_{i=1}^{n} \frac{\partial^2}{\partial x_k \partial x_l}$$
$$\times [B_{kl}(x_1, x_2, \ldots, x_n) p(x_1, x_2, \ldots, x_n)] \quad (11.43)$$

where

$$A_k = \lim_{\Delta t \to 0} \frac{\langle x_{k\tau} - x_k \rangle}{\Delta t}$$

$$B_{kl} = \lim_{\Delta t \to 0} \frac{\langle (x_{k\tau} - x_k)(x_{l\tau} - x_l) \rangle}{\Delta t} \quad (11.44)$$

The solution to the Fokker–Planck equation must satisfy the condition of normalization

$$\int_{-\infty}^{\infty} p(x, t | x_0, t_0) \, dx = 1$$

and, in the case of nonrandom initial conditions, the conditions

$$p(x, t_0 | x_0, t_0) = \prod_{k=1}^{n} \delta(x_k - x_{0k})$$

and the appropriate boundary conditions. These are determined by the substance of the physical problem. In the case of random initial conditions, it is not the initial conditions themselves but their probability density functions that are given. In such a case the sought solution to the Fokker–Planck equation must satisfy the condition of normalization:

$$\int_{-\infty}^{\infty} p(x, t | t_0) \, dx = 1$$

and the initial condition:

$$p(x, t_0 | t_0) = p_0(x_0)$$

The Fokker–Planck equation, when applied, for instance, to the random process of the Brownian movement, can be interpreted as follows. Let there be a large number of particles, generated as the initial moment of time t_0 from the position x_0. This situation occurs in the case of delta-like (i.e., deterministic) initial conditions. The concentration (density) of these particles at the moment t in the point x is proportional to $p(x, t)$. The total flow $G(x, t)$ of the Brownian particles is due to the *systematic flow*

$$A(x)p(x, t)$$

and the "*random*" (*diffusion*) *flow*

$$-\frac{1}{2}\frac{\partial}{\partial x}[B(x)p(x, t)]$$

and can be calculated as

$$G(x, t) = A(x)p(x, t) - \frac{1}{2}\frac{\partial}{\partial x}[B(x)p(x, t)] \qquad (11.45)$$

Therefore the coefficients A and B are called the *drift* and the *diffusion* coefficients, respectively. The diffusion flow for $B(x, t) = B = $ constant is proportional to the gradient of the concentration of the gas and is directed from the higher concentration to the lower one. Therefore the diffusion flow is negative. The Fokker–Planck equation is, in effect, the equation of continuity of the flow $G(x, t)$:

$$\frac{\partial}{\partial t}p(x, t) + \frac{\partial}{\partial x}G(x, t) = 0 \qquad (11.46)$$

If the random process $X(t)$ can assume any value on the t axis, then Eq. (11.46) is fulfilled anywhere from $-\infty$ to ∞. Quite often the following boundary conditions can be applied:

$$G(-\infty, t) = G(\infty, t) = 0$$

and

$$p(-\infty, t) = p(\infty, t) = 0$$

If the drift and the diffusion coefficients $A(x)$ and $B(x)$ are time independent, the probability density function tends to a stationary distribution $p(x)$, independent from the initial conditions. Therefore, for a stationary state,

$$\frac{\partial}{\partial t}p(x) = 0$$

and, hence, $G(x) = G$ = constant. Then the Fokker–Planck equation (11.45) becomes

$$\frac{d}{dx}[B(x)p(x)] - 2A(x)p(x) = -2G \tag{11.47}$$

In a situation when the constants A and B are x independent, this liner equation is, in effect, the generalization of the empiric *Feek law* for diffusion processes (in this case G is the density of the flow of the material and B is the diffusion coefficient) and the *Fourier law* for heat conduction (in this case G is the density of the heat flow and B is thermal conductivity):

$$G(x) = Ap(x) - \frac{1}{2}B\frac{dp(x)}{dx}$$

For $G = 0$, Eq. (11.47) has the following solution:

$$p(x) = \frac{C}{B(x)} \exp\left[2\int_0^x \frac{A(x)}{B(x)}dx\right] \tag{11.48}$$

where the constant of integration C can be obtained from the condition of normalization. The physical interpretation of Eq. (11.47) can be as follows. As is known from field theory, the flow is a vector that is perpendicular to a certain plane through which the liquid or gas flows. The total flow through the cross section will be zero only in the case where the convection and the diffusion flows are equal in magnitude and opposite in signs. Thus, having evaluated the coefficients $A(x)$ and $B(x)$, one can often write down the expression for a one-dimensional stationary probability density without even solving the Fokker–Planck equation.

Example 11.12 The process $X(t)$ is described by a stochastic differential equation

$$\dot{x} = kx(a^2 - x^2) + n(t), \qquad k > 0$$

where a = constant and $n(t)$ is a "white noise." Determine the probability density function for this process.

Solution From Eqs (11.44) we find

$$A(x) = kx(a^2 - x^2), \qquad B(x) = \pi S_0 = \text{constant}$$

where S_0 is the spectrum of the "white noise" $n(t)$. Then the formula (11.48) yields

$$p(x) = C \exp\left[\frac{k}{2\pi S_0}(2a^2x^2 - x^4)\right]$$

where the constant of integration can be found from the condition of normalization and is as follows:

$$C = \left\{ \int_{-\infty}^{\infty} \exp\left[\frac{k}{2\pi S_0}(2a^2x^2 - x^4)\right] dx \right\}^{-1}$$

Example 11.13 Determine the coefficients of drift and diffusion and write and solve the Fokker–Planck equation for th markovian process $x(t)$, described by the following *Langevin equation* for the velocity $x(t)$ of a particle:

$$\dot{x} + \alpha x = n(t)$$

This equation was first obtained by the French physicist P. Langevin in connection with the study of brownian motion.

Solution The solution of the Langevin equation results in the following formula for the velocity x of the particle:

$$x(t) = xe^{-\alpha t} + \alpha e^{-\alpha t} \int_0^t e^{\alpha \xi} n(\xi) \, d\xi$$

When there is a need, the position of the particle can be evaluated by integrating this equation. The first moment of the increment

$$x(\tau) - x = x(e^{-\alpha \tau} - 1) + \alpha e^{-\alpha \tau} \int_0^\tau e^{\alpha \xi} n(\xi) \, d\xi$$

with $\langle n(t) \rangle = 0$, is

$$\langle x(\tau) - x \rangle = x(e^{-\alpha \tau} - 1)$$

or, for small τ, putting $e^{-\alpha \tau} \approx 1 - \alpha \tau$,

$$\langle x(\tau) - x \rangle = -\alpha x \tau$$

The second moment of the increment $x(\tau) - x$ is

$$\langle [x(\tau) - x]^2 \rangle = \left\langle x^2(e^{-\alpha \tau} - 1)^2 + 2x(e^{-\alpha \tau} - 1)e^{-\alpha \tau} \int_0^\tau e^{\alpha \xi} n(\xi) \, d\xi \right.$$

$$\left. + e^{-2\alpha \tau} \int_0^\tau \int_0^\tau e^{\alpha(t_1 + t_2)} n(t_1) n(t_2) \, dt_1 \, dt_2 \right\rangle$$

$$= x^2 \tau^2 + e^{-2\alpha \tau} \int_0^\tau \int_0^\tau e^{\alpha(t_1 - t_2)} \langle n(t_1) n(t_2) \rangle \, dt_1 \, dt_2$$

$$= x^2 \tau^2 + e^{-2\alpha \tau} \frac{\pi}{2\alpha} S_0 (e^{2\alpha \tau} - 1)$$

$$= x^2 \tau^2 + \pi S_0 \tau$$

The coefficients of drift and diffusion are

$$A(x, t) = \lim_{\tau \to 0} \frac{1}{\tau} \langle x(\tau) - x \rangle = -\alpha x$$

$$B(x, t) = \lim_{\tau \to 0} \frac{1}{\tau} \langle [x(\tau) - x]^2 \rangle = \pi S_0$$

and the Fokker–Planck equation can be written as

$$\frac{\partial p(x,\,t)}{\partial t} = \alpha\frac{\partial}{\partial x}[xp(x,\,t)] + \frac{\pi}{2}S_0\frac{\partial^2 p(x,\,t)}{\partial x^2}$$

This equation has the following solution:

$$p(x,\,t) = \frac{1}{\sqrt{2\pi D_x(t)}}\exp\left[-\frac{[x - \langle x(t)\rangle]^2}{2D_x(t)}\right]$$

where the mean velocity is

$$\langle x(t)\rangle = x_0 e^{-\alpha t} + \alpha e^{-\alpha t}\int_0^t e^{\alpha\xi}\langle n(\xi)\rangle\,d\xi = x_0 e^{-\alpha t}$$

and its variance is

$$D_x(t) = \frac{\pi}{2}\frac{S_0}{\alpha}(1 - e^{-2\alpha t})$$

The markovian process $x(t)$ with the coefficients of drift and diffusion of the type $A(x) = -\alpha x$ and $B(x) = \pi S_0 = $ constant is known as the *Ornstein–Uhlenbeck process* and the steady-state ($t \to \infty$) distribution

$$p(x) = \frac{1}{\pi}\sqrt{\frac{\alpha}{S_0}}\exp\left[-\frac{\alpha(x - x_0)^2}{\pi S_0}\right]$$

is known as the *Maxwell distribution*.

Example 11.14 A thermoelectric device is used to measure the variance of a stationary fluctuating electrical current $\xi(t)$. The thermoelectric device consists of a galvanometer and a "quadratic" thermocouple. If the time constant α of the galvanometer is significantly smaller than the time constant β of the thermocouple, the deviation $\phi(t)$ of the galvanometer hand can be described by the equation

$$\dot\phi(t) + \alpha\phi(t) = a\xi^2(t)$$

where the factor a characterizes the sensitivity of the device and the factor α characterizes its inertia. The current $\xi(t)$ is a normal random process with zero mean and the correlation function

$$K_\xi(\tau) = De^{-\beta|\tau|}$$

Determine the probability density function of the random $\phi(t)$ of the deviation of the hand of the galvanometer.

Solution (Tyhonov) Since $\beta \gg \alpha$, the Fokker–Planck equation can be used to solve the problem. Introduce a new variable

$$\zeta(t) = \xi^2(t) - D$$

If a normal stationary process $\xi(t)$ has a zero mean and a correlation function

$$K_\xi(\tau) = DR(\tau)$$

then the correlation function of the process $\xi^2(t)$ can be found as

$$K_{\xi^2}(\tau) = \langle \xi^2(t)\xi^2(t+\tau)\rangle = D^2[1 + 2R^2(\tau)]$$

In the case in question, $R(\tau) = e^{-\beta|\tau|}$, so that

$$K_\zeta(\tau) = 2D^2 e^{-2\beta|\tau|}$$

With the new variable ζ the equation for the deviation $\phi(t)$ can be written as

$$\dot\phi(t) = -\alpha\phi + aD + a\zeta(t)$$

It can be shown that if an "output" process $\eta(t)$ is related to the "input" process $\xi(t)$ by the nonlinear stochastic differential equation

$$\dot\eta(t) = f[\eta(t)] + g[\eta(t)]\xi(t)$$

where f and g are given nonrandom functions that characterize the system, then the coefficients of drift and diffusion in the Fokker–Planck equation can be found as

$$A(\eta) = f[\eta], \qquad B = N_0 g^2[\eta]$$

where

$$N_0 = \int_{-\infty}^{\infty} \langle \xi(t)\xi(t+\tau)\rangle \, d\tau$$

In the problem in question the coefficients A and B can be found as

$$A(\phi) = -\alpha\phi + aD, \qquad B(\phi) = \frac{2a^2}{\beta} D^2$$

Introducing these formulas into the solution

$$f(\phi) = \frac{C}{B(\phi)} \exp\left[2 \int_0^\phi \frac{A(\xi)}{B(\xi)} \, d\xi\right]$$

of the stationary Fokker–Planck equation, we obtain

$$f(\phi) = \frac{C\beta}{2a^2 D^2} \exp\left[2 \int_0^\phi \frac{\beta}{2a^2 D^2} (-\alpha\xi + aD) \, d\xi\right]$$

$$= \frac{C\beta}{2a^2 D^2} \exp\left(-\frac{1}{2} \frac{\alpha\beta}{a^2 D^2} \phi^2 + \frac{\beta}{aD} \phi\right)$$

Evaluating the constant C from the condition of normalization

$$\int f(\phi) \, d\phi = 1$$

we finally obtain

$$f(\phi) = \frac{1}{\sqrt{2\pi D_\phi}} \exp\left[-\frac{(\phi - \langle\phi\rangle)^2}{2D_\phi}\right]$$

where

$$\langle\phi\rangle = \frac{aD}{\alpha}, \quad D_\phi = \frac{a^2 D^2}{\alpha\beta}$$

The relative error of the measurements can be found as

$$\delta = \frac{\sqrt{D_\phi}}{\langle\phi\rangle} = \sqrt{\frac{\alpha}{\beta}}$$

Thus, the accuracy of the measurements increases with an increase in the ratio of the time constant of the device to the correlation time of the current $\xi(t)$.

11.7. Kolmogorov Backward Equation

In the previous section the intermediate moment of time τ was chosen as $\tau = t - \Delta t$. This means that the differentiation in Eq. (11.42) was carried out with respect to the "future" time $t = \tau + \Delta t$. Let us choose now the intermediate moment of time τ as $\tau = t + \Delta t$. Hence, in this case we address the "past" time $t = \tau - \Delta t$, and Smoluchowski's equation (11.40) can be written as

$$p(x, t | x_0, t_0) = \int_{-\infty}^{\infty} p(x, t | \xi, t + \Delta t) p(\xi, t + \Delta t | x_0, t) \, d\xi \quad (11.49)$$

The expansion of the function $p(x, t | \xi, t + \Delta t)$ in a Taylor series around the point $\xi = x_0$ is

$$p(x, t | \xi, t + \Delta t) = p(x, t | x_0, t + \Delta t)$$
$$= \frac{\partial p(x, t | x_0, t + \Delta t)}{\partial x_0} (\xi - x_0)$$
$$+ \frac{1}{2} \frac{\partial^2 p(x, t | x_0, t + \Delta t)}{\partial x_0^2} (\xi - x_0)^2 + \cdots$$

Introducing this expansion into Eq. (11.49), considering the condition of normalization

$$\int_{-\infty}^{\infty} p(x, t | x_0, t + \Delta t) \, dx = 1$$

and dividing both parts of the obtained equation by Δt, we obtain

$$\frac{p(x, t|x_0, t_0) - p(x, t|x_0, t + \Delta t)}{\Delta t}$$

$$= \frac{\partial p(x, t|x_0, t + \Delta t)}{\partial x_0} \left[\frac{1}{\Delta t} \int_{-\infty}^{\infty} (\xi - x_0) p(\xi, t + \Delta t|x_0, t_0) \, d\xi \right]$$

$$+ \frac{1}{2} \frac{\partial^2 p(x, t|x_0, t + \Delta t)}{\partial x_0^2} \left[\frac{1}{\Delta t} \int_{-\infty}^{\infty} (\xi - x_0)^2 p(\xi, t + \Delta t|x_0, t_0) \, d\xi \right]$$

By putting $\Delta t \to 0$ and assuming that the limits

$$A(x_0, t_0) = \lim_{\Delta t \to 0} \frac{1}{\Delta t} \int_{-\infty}^{\infty} (\xi - x_0) p(\xi, t + \Delta t|x_0, t_0) \, d\xi$$

and

$$B(x_0, t_0) = \lim_{\Delta t \to 0} \frac{1}{\Delta t} \int_{-\infty}^{\infty} (\xi - x_0)^2 p(\xi, t + \Delta t|x_0, t_0) \, d\xi$$

take finite values, we obtain the following equation for the transitional probability density function $p(x, t|x_0, t_0)$:

$$-\frac{\partial p(x)}{\partial t} = A(x) \frac{\partial p(x)}{\partial x} + \frac{1}{2} B(x) \frac{\partial^2 p(x)}{\partial x^2} \qquad (11.50)$$

where $p(x) \equiv p(x, t|x_0, t_0)$, $A(x) = A(x, t)$, and $B(x) = B(x, t)$. This equation is known as the *Kolmogorov backward equation*. The name is due to the fact that it is an equation for the probability function for the pair of initial variables x_0 and t_0 moving backward from t.

11.8. Barriers

As has been mentioned earlier, one of the main merits of the theory of markovian processes is the possibility of using this theory to obtain the stationary distribution of the output processes for dynamic systems, both linear and nonlinear, if the input process can be approximated by a "white noise." Another class of practically important problems that can be solved on the basis of the theory of markovian processes are various problems associated with reaching barriers (boundaries). The simplest types of barriers are so-called absorbing and reflecting barriers, although there are many other more complex types of barriers. In the case of an *absorbing barrier*, the process abrupts at the moment of time when the barrier is reached. In the case of a *reflecting barrier*, the particle, after having hit the barrier, instantaneously returns into an internal point of the interval. The process then resumes from this point and, being a markovian one, is independent from the "prehistory" of the motion. In the analysis that follows, we examine the case of absorbing barriers.

Let a random markovian process $X(t)$ describing the location of a particle be such that all its realizations take place within the absorbing barriers a and b. Let $q(x_0, x, t)$ be the probability density function for the location of particles that have never reached the barriers. This density function should be dependent on the initial position x_0 of the particle, its current position x, and time t, and should obey the boundary conditions:

$$q(x_0, a, t) = q(x_0, b, t) = 0 \qquad (11.51)$$

The probability $Q(x_0, t)$ that the particle, initially located at the point x_0, will always stay within the interval (a, b), i.e., will never reach the barriers, is

$$Q(x_0, t) = \int_0^b q(x_0, x, t)\, dx \qquad (11.52)$$

At the initial moment of time, when none of the realizations could reach a barrier, this function is $Q(x_0, t_0) = 1$. On the other hand, for $t \to \infty$, all the realizations will sooner or later reach the boundaries, and therefore $Q(x_0, \infty) = 0$. It should be pointed out that the function $q(x_0, a, t)$ does not satisfy the condition of normalization, since we deliberately excluded the probabilities $P(a, t)$ and $P(b, t)$ of the absorption at the barriers a and b. The "complete" probability density function can be written as

$$w(x_0, x, t) = q(x_0, x, t) + P(a, t)\delta(x - a) + P(b, t)\delta(x - b) \qquad (11.53)$$

where $\delta(x)$ is the Dirac delta function.

Although the function $q(x_0, a, t)$ is not normalized, the physical substance of the problem is such that one can nonetheless apply the Fokker–Planck equation for the evaluation of the evolution of this function. Since for delta-like (deterministic) initial conditions the equation for the one-dimensional distribution coincides with the equation for the probability transfer density, one can use, instead of the function $q(x_0, x, t)$, the nonnormalized probability transfer density function $p(x, t | x_0, t_0)$. This function can be determined from the Fokker–Planck equation and should satisfy the following boundary conditions:

$$p(b, t | x_0, t_0) = p(a, t | x_0, t_0) = 0$$

$$p(x, t | a, t_0) = p(x, t | b, t_0) = 0$$

These indicate that the probability density function $p(x, t|x_0, t_0)$ is referred to realizations that never reach the barriers. The function $p(x, t|x_0, t_0)$ obeys not only the Fokker–Planck equation but the Kolmogorov backward equation as well. The function $Q(x_0, t)$ can be expressed through the probability density function $p(x, t|x_0, t_0)$ as follows:

$$Q(x_0, t) = Q(x_0, t_0, t) = \int_a^b p(x, t|x_0, t_0)\, dx \tag{11.54}$$

Integrating both parts of Eq. (11.50) with respect to x within the limits from a to b, we obtain

$$-\frac{\partial Q}{\partial t} = A(x)\frac{\partial Q}{\partial x} + \frac{1}{2}B(x)\frac{\partial^2 Q}{\partial x^2} \tag{11.55}$$

Since for a homogeneous process,

$$p(x, t|x_0, t_0) = p(x, t - t_0|x_0)$$

and

$$\frac{\partial p(x, t - t_0|x_0)}{\partial t_0} = \frac{\partial p(x, t - t_0|x_0)}{\partial t}$$

Eq. (11.55) results in the following equation for the probability Q that the barriers will not be reached:

$$\frac{\partial Q}{\partial t} = A(x)\frac{\partial Q}{\partial x} + \frac{1}{2}B(x)\frac{\partial^2 Q}{\partial x^2} \tag{11.56}$$

The probability

$$P(x, t) = 1 - Q(x, t)$$

that the process $X(t)$ will reach, during the time t, either of the barriers a or b can be found from the equation

$$\frac{\partial P(x)}{\partial t} = A(x)\frac{\partial P}{\partial x} + \frac{1}{2}B(x)\frac{\partial^2 P}{\partial x^2} \tag{11.57}$$

The solution to this equation must satisfy the initial condition $P(x_0, 0) = 0$ and the boundary conditions $P(a, t) = P(b, t) = 1$. In addition,

$$\lim_{t \to \infty} P(x_0, t) = 1$$

The probability $P(x_0, t)$ is the cumulative probability distribution function for the time t, when the process $X(t)$ reaches for the first time the barrier a or b from the initial position x_0.

Although the equation (11.57) cannot be solved, it can be used for the evaluation of the mean time $\langle t \rangle$ of reaching the barrier:

$$\langle t \rangle = \int_0^\infty t w(t)\, dt = \int_0^\infty t \frac{\partial P(x_0, t)}{\partial t}\, dt$$

$$= -\int_0^\infty t \frac{\partial Q(x_0, t)}{\partial t}\, dt = \int_0^\infty Q(x_0, t)\, dt \qquad (11.58)$$

Here the integration by parts is used and the condition $Q(x_0, \infty) = 0$ is considered. Integrating Eq. (11.56) by t from zero to infinity, and considering the conditions $Q(x_0, t_0) = 1$ and $Q(x_0, \infty) = 0$ and the relationship (11.58), one can obtain the following equation for the time $\langle t \rangle$:

$$-1 = A(x) \frac{d\langle t \rangle}{dx} + \frac{1}{2} B(x) \frac{d^2\langle t \rangle}{dx^2} \qquad (11.59)$$

Introducing a new variable

$$u = \frac{d\langle t \rangle}{dx}$$

we obtain the following linear equation of the first order:

$$\frac{1}{2} B(x) \frac{du}{dx} + A(x) u + 1 = 0$$

This equation has the following solution:

$$u = \left[-\int_a^{x_0} \frac{2}{B(x)} e^{\phi(x)}\, dx + C_1 \right] e^{-\phi(x)}$$

where

$$\phi(x) = 2 \int \frac{A(x)}{B(x)}\, dx$$

Then, returning to the variable $\langle t \rangle$, we have

$$\langle t \rangle = \int_a^{x_0} \left[-\int_a^{\xi_0} \frac{2}{B(\xi)} e^{\phi(\xi)}\, d\xi \right] e^{-\phi(\xi)}\, d\xi + C_1 \int_a^{x_0} e^{-\phi(\xi)}\, d\xi + C_2 \qquad (11.60)$$

The constants of integration C_1 and C_2 can be found from the boundary conditions $T(a) = T(b) = 0$. Putting $x_0 = a$, we obtain $C_2 = 0$.

Then, substituting $x_0 = b$ into the obtained solution, we have

$$C_1 = \frac{\int_a^b \left[\int_a^\xi \frac{2}{B(\xi)} e^{\phi(\xi)} d\xi\right] e^{-\phi(\xi)} d\xi}{\int_a^b e^{-\phi(\xi)} d\xi}$$

and the solution (11.60) yields

$$\langle t \rangle = \int_{x_0}^b \left[\int_a^\xi \frac{2}{B(\xi)} e^{\phi(\xi)} d\xi\right] e^{-\phi(\xi)} d\xi \int_a^{x_0} e^{-\phi(\xi)} d\xi$$
$$+ \int_{x_0}^a \left[\int_a^\xi \frac{2}{B(\xi)} e^{\phi(\xi)} d\xi\right] e^{-\phi(\xi)} d\xi \int_{x_0}^b e^{-\phi(\xi)} d\xi \left(\int_a^b e^{-\phi(\xi)} d\xi\right)^{-1}$$
(11.61)

This formula was obtained by V. J. Tyhonov.

Example 11.15 The stochastic equation of the process $X(t)$ is

$$\dot{x} = A + \alpha n(t)$$

where A = constant and $n(t)$ is a "white noise." The process is characterized by the diffusion coefficient

$$B = \frac{1}{2} \alpha^2 N_0$$

Determine the mean time $\langle t \rangle$ when the process $X(t)$ will reach a barrier a or b.

Solution Using the formula (11.61), we obtain

$$\langle t \rangle = \frac{(b - x_0)(e^{-\lambda a} - e^{-\lambda x_0}) - (x_0 - a)(e^{-\lambda x_0} - e^{-\lambda b})}{A(e^{-\lambda a} - e^{-\lambda b})}$$
(11.62)

where

$$\lambda = \frac{2A}{B}$$

If $x_0 = 0$ and $a = b$, the obtained solution can be simplified:

$$\langle t \rangle = \frac{b}{A}\left(\frac{\cosh \lambda b - 1}{\sinh \lambda}\right) = \frac{b}{A} \tanh\left(\frac{A}{B} b\right)$$

If $A = 0$, then the solution (11.62) yields

$$\langle t \rangle = \frac{(b - x_0)(x_0 - a)}{B}$$

Example 11.16 The stochastic equation of the process $X(t)$ is

$$\dot{x} + \alpha x = n(t)$$

where $n(t)$ is a "white noise." Determine, for symmetric barriers $a = -b$, the mean time $\langle t \rangle$ for reaching the barriers from the initial position $x_0 = 0$.

Solution The coefficients of drift and diffusion are

$$A = -\alpha x_0, \qquad B = \tfrac{1}{2} N_0$$

and the function $\phi(x)$ is

$$\phi(x) = 2 \int \frac{A}{B} \, dx = -\frac{2\alpha}{N_0} x^2 = -\frac{x^2}{2 D_x}$$

where

$$D_x = \frac{N_0}{4\alpha}$$

is the variance of the process in a stationary condition. Then, using the formula (11.61), we obtain

$$\langle t \rangle = \frac{1}{\alpha} \sqrt{\pi} \int_0^{b/\sqrt{(2 D_x)}} \Phi(\xi) e^{\xi^2} \, d\xi$$

where

$$\Phi(\xi) = \frac{2}{\sqrt{\pi}} \int_0^{\xi} e^{-t^2} \, dt$$

is the probability integral.

11.9. Method of Stochastic Differential Equations

Some of the results obtained previously on the basis of the Fokker–Planck equation could also be obtained directly from the differential equations of the random process. Such an approach was introduced first by P. Langevin in connection with the theory of brownian motion. Examine this approach using, as an example, the Langevin equation for the velocity x of the particle (see Example 11.13, Sec. 11.6):

$$\dot{x} + \alpha x = n(t).$$

This equation has the following solution:

$$x = x_0 e^{-\alpha t} + e^{-\alpha t} \int_0^t e^{\alpha \xi} n(\xi) \, d\xi$$

where x_0 is the initial velocity. The mean of this velocity can be easily found as

$$\langle x \rangle = x_0 e^{-\alpha t}$$

If the process $n(t)$ is a "white noise," its values for any two moments of time t and t' are uncorrelated. This property can be written as follows:

$$\langle n(t)n(t') \rangle = B\, \delta(t - t')$$

where $B = \pi S_0 =$ constant is a measure of the intensity of the random impulses acting on the particle. With this relationship, the expression for the variance of the random velocity $X(t)$ can be found as

$$D_x = \langle (x - \langle x \rangle)^2 \rangle = e^{-2\alpha t} \int_0^t \int_0^t e^{\alpha(t+t')} \langle n(t)n(t') \rangle\, dt\, dt'$$

$$= Be^{-2\alpha t} \int_0^t \int_0^t e^{\alpha(t+t')} \delta(t - t')\, dt\, dt'$$

$$= Be^{-2\alpha t} \int_0^t e^{2\alpha t}\, dt = \frac{B}{2\alpha}(1 - e^{-2\alpha t})$$

The obtained values of the mean $\langle x \rangle$ and variance D_x coincide with those obtained in the Example 11.13, Sec. 11.6, on the basis of the Fokker–Planck equation. The probability $p(x, t)$ can be found by assuming that the random variable $X(t)$ has a normal distribution with the mean $\langle x \rangle$ and variance D_x.

The obtained solution is applicable to electric circuits as well. In this case all the results concerning the velocity $x(t)$ should be referred to the current $I(t)$, and the results concerning the displacement $s(t) = \int_0^t x(t)\, dt$ of the particle are true for the charge transmitted through the cross section of the circuit:

$$q(t) = \int_0^t I(t)\, dt$$

It should be pointed out that, although in the above example the method of stochastic differential equations was applied to the case where the input process is a "white noise," i.e., a delta-correlated random process, this method, unlike the Fokker–Planck equation, can also be used in problems for which the random excitation is not a "white noise," and, hence, the response $X(t)$ is not a markovian process.

Further Reading (see Bibliography): 5, 12, 14, 15, 27, 40, 42, 74, 109, 110, 143.

Chapter 12

Random Fatigue

> "A good theory of a complex system is merely a good 'caricature' of this system, which exaggerates the most typical properties of the system and deliberately ignores all the other, insignificant, properties."
> YAKOV FRENKEL,
> Russian Physicist

> "Experiment without a theory is blind. A theory without an experiment is dead."
> UNKNOWN ENGINEER

> "It is always better to be approximately right, than precisely wrong."
> UNKNOWN ENGINEER

12.1. Characteristics of Fatigue Failure

Failure of structural elements subjected to variable loading is usually due to the accumulation of microdamages which are transmitted into the developing fatigue cracks. The process, leading to failure, consists of two major stages: (1) accumulation of fatigue damages and (2) the growth and propagation of fatigue cracks.

The long-term reliability ("workability") of engineering materials is usually characterized by the *Wöhler curve*. This was introduced by a German railroad engineer, A. Wöhler, in 1860 and reflects the relationship between the level (amplitude) of the variable load and the number of cycles to failure. For a typical structural metal, the fatigue curve is as shown in Fig. 12.1. The kink on this curve is due to different prevailing mechanisms of the development of fatigue damages in the regions of *limited fatigue* and *steady-state fatigue*. Fatigue failure in the region of limited fatigue is due primarily to the accumulation of shearing plastic deformations, while the failure in

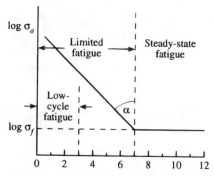

Figure 12.1 Stress/number of cycles to failure curve in logarithmic coordinates

the steady-state region is due, first of all, to the diffusion processes of the motion of dislocations. For typical structural metals, the $m = \tan \alpha$ value is $m = 6$ to 14. Many materials, especially nonmetals, do not have a steady-state region at all, although the slope of the fatigue curve in the region of high number of cycles is substantially smaller than in the region of limited fatigue. If the maximum stress at which the material does not fail, no matter how large the number of cycles, exists, it is known as the *fatigue limit*.

Fatigue tests are usually conducted for symmetric stress cycles. The fatigue limit is the largest in this case. The fatigue limit for an asymmetric cycle, with the coefficient of asymmetry $r = \sigma_{\min}/\sigma_{\max}$, can be estimated by using the following *Goodman law*:

$$\sigma_r = \sigma_{-1}\left(1 - \frac{\langle \sigma \rangle}{\sigma_u}\right) \tag{12.1}$$

where

$$\langle \sigma \rangle = \frac{1}{2}(\sigma_{\max} + \sigma_{\min}) \tag{12.2}$$

is the mean value of the stress cycle and σ_u is the strength limit.

The initiation of fatigue cracks is accelerated by stress concentration. This increases both the mean value of the stress cycle and the stress amplitude by a factor of k, which is equal to the stress concentration factor. This, as evident from the Goodman formula, can be accounted for by reducing the fatigue limit of the asymmetric cycle by a factor

$$a = \frac{1}{1 - k(\langle \sigma \rangle/\sigma_u)} \tag{12.3}$$

12.2. Linear Accumulation of Damages

Fatigue cracks propagate slowly. If such a crack could be easily detected and removed in due time, it does not provide a serious threat to the reliability of the structure. The real danger occurs when there is a high possibility that a fatigue crack can turn into a brittle crack, leading to a rapid and catastrophic failure.

The change in the structure of the material or structural performance under the action of variable loading is known as *fatigue damage*. For the assessment of fatigue life, one usually applies the linear theory of the accumulation of damages—the *theory of summation of damages*. In accordance with this theory, the fatigue damages are independent of the degree of the consumption of the fatigue lifetime at a given moment of time. The damages are independent also from the "prehistory" of loading. Therefore the damages caused by the current loading cycle can simply be added to the previous damages. Such an assumption seems to be particularly justified in the case of random loading, when, due to the sequential action of cycles with high and low stresses, material weakenings caused by high stress cycles interchange with its strengthenings caused by low stress cycles, so that the effect of the "prehistory" of loading is continuously smoothed down.

Thus, in accordance with the linear theory of the accumulation of damages, the total fatigue damage can be evaluated as

$$\xi = \sum_{i=1}^{k} \xi_i = \sum_{i=1}^{k} \frac{n_i}{N_i}$$

Here $\xi_i = n_i/N_i$ is the damage from the ith level of loading, k is the number of loading levels, n_i is the number of cycles of loading of the ith level $(\sigma_i = \sigma_a)$, and N_i is the number of cycles of loading of the ith level leading to a fatigue failure. Such failure can either be caused by a single loading of the ith level or be due to the entire spectrum of loading. Assuming that the work W of the external loading leading to a fatigue failure is the same in both cases, we have

$$W = \frac{N_i}{n_i} W_i = \sum_{i=1}^{k} W_i$$

where W_i is the work of n_i cycles of a variable loading of the ith level. From this equation we obtain

$$W_i = \frac{n_i}{N_i} \sum_{i=1}^{k} W_i$$

Summing up the elementary works W_i in both parts of this equation with respect to i, we obtain

$$\zeta = \sum_{i=1}^{k} \frac{n_i}{N_i} = 1 \tag{12.4}$$

This is the basic formula of the linear theory of summation of fatigue damages.

12.3. Probabilistic Assessment of Fatigue Life

The probabilistic approach to the assessment of the fatigue life of structures and structural elements enables one to account for the actual loading conditions in the most complete way. The simplest approach to the assessment of the fatigue life of a structural element subjected to a continuous random loading can be based on the condition:

$$\int_{N_s}^{N_f} \frac{dN}{N} = 1 \tag{12.5}$$

which is equivalent to Eq. (11.4) in a discrete case. In this condition, N_s is the threshold of cycle sensitivity (the maximum number of cycles for which the probability of fatigue failure can still be considered zero), N_f is the number of cycles corresponding to achievement of the fatigue curve, and N is the number of cycles till fatigue failure.

The number N of cycles to failure can be determined from the logarithmic fatigue curve:

$$N = N_f \left(\frac{\sigma_f}{\sigma_a}\right)^m, \qquad m = \tan \alpha \tag{12.6}$$

The total number of cycles accumulated for the time T can be computed as $N_T = T/\tau_e$, where τ_e is the effective period of random loading. Assuming that the loading cycles are distributed uniformly within the interval $0 < N < N_T$, we conclude that the probability density function for the number of cycles N is

$$f_N(N) = \frac{1}{N_T} = \frac{\tau_e}{T} \tag{12.7}$$

The condition of the equality of elementary probabilities for the amplitudes σ_a of loading and the number N of cycles is

$$f_\sigma(\sigma_a)\, d\sigma_a = f_N(N)\, dN$$

where $f_\sigma(\sigma_a)$ is the probability density distribution function for the amplitudes of loading. Then we have

$$dN = f_\sigma(\sigma_a) \frac{d\sigma_a}{f_N(N)} = \frac{T}{\tau_e} f_\sigma(\sigma) \, d\sigma_a \qquad (12.8)$$

Introducing Eqs. (12.6) and (12.8) into the condition (12.5) we obtain the following formula for the time-to-failure:

$$T = \beta \tau_e N_f \qquad (12.9)$$

where the factor

$$\beta = \frac{\sigma_f^m}{\displaystyle\int_{\sigma_f}^{\sigma_u} \sigma_a^m f(\sigma_a) \, d\sigma_a} \qquad (12.10)$$

considers the effect of material properties and the loading level.

If, for instance, the stress amplitudes σ_a are distributed in accordance with the Rayleigh law, that is,

$$f(\sigma_a) = \frac{\sigma_a}{D_\sigma} \exp\left[-\left(\frac{\sigma_a^2}{2D_\sigma}\right)\right] \qquad (12.11)$$

then the formula (12.10) yields

$$\beta = \frac{D_\sigma \sigma_f^m}{\displaystyle\int_{\sigma_f}^{\sigma_u} \sigma_a^{m+1} e^{-\sigma_a^2/(2D_\sigma)} \, d\sigma_a} \qquad (12.12)$$

In a situation when the loading cycle is asymmetric and, in addition, there is a stress concentration, the following substitutions should be made in the formula (12.12): $\sigma_a \to k\sigma_a$, $\sigma_f \to \sigma_f/a$, $\sigma_u \to \sigma_u/a$. Then this formula yields

$$\beta = \frac{x_0^m}{k^{m+2} \displaystyle\int_{x_0}^{x_1} x^{m+1} e^{-x^2/2} \, dx} \qquad (12.13)$$

where

$$x = \frac{\sigma_a}{\sqrt{D_\sigma}}, \qquad x_0 = \frac{\sigma_f}{\sqrt{D_\sigma}}, \qquad x_1 = \frac{\sigma_u}{\sqrt{D_\sigma}}$$

The integral in the formula (12.13) can be taken numerically or expressed through a tabulated function. Since the integrand in the denominator of this formula does not make physical sense for $\sigma_a > \sigma_u$ anyway (the integrand is zero for $\sigma_a > \sigma_u$), one can put the upper limit in the integral equal to ∞. Then we have

$$\int_{x_0}^{\infty} x^{m+1} e^{-x^2/2}\, dx = 2^{m/2} \Gamma\left(\frac{m+2}{2}\right) P(x_0^2, m+2)$$

where $\Gamma(\alpha)$ is the gamma function (3.38) and $P(\chi^2, m)$ is the Pierson function (3.41). The coefficient in front of the Pierson function can be calculated as

$$2^{m/2} \Gamma\left(\frac{m+2}{2}\right) = \begin{cases} \sqrt{\dfrac{\pi}{2}}\, m!! & \text{for odd } m \\ 2^{m/2}\left(\dfrac{m}{2}\right)! & \text{for even } m \end{cases}$$

where

$$m!! = 1 \times 3 \times 5 \times \cdots \times (m-2) m$$

The long-term distribution of the stress amplitudes can also be exponential or distributed in accordance with the log-normal law or with the Weibull law, depending on the particular application.

It should be mentioned that the fatigue limit σ_f may not be constant. The experimental data indicate that for the majority of materials this limit gradually decreases with an increase in the number of cycles. In addition, the fatigue limit depends on the loading spectrum: the larger is the portion of this spectrum above the fatigue limit, the larger is the rate of its decrease. In an approximate analysis, one can assume that the rate of decrease of the fatigue limit depends on its initial value, the rate of the accumulation of the fatigue damages, and the total level of the accumulated damages at a given moment of time:

$$\frac{d\sigma_f}{dn} = -\alpha \frac{\sigma_f^0}{N}\left(1 - \frac{n}{N}\right)^{\alpha-1}$$

Here σ_f is the fatigue limit after n loadings, σ_f^0 is the initial value of the fatigue limit, N is the number of cycles to fatigue failure, and α is a parameter depending on the material properties. This equation has the following solution:

$$\sigma_f = \sigma_f^0 \left(1 - \frac{n}{N}\right)^{\alpha}$$

As follows from this formula, the decrease in the fatigue limit depends, for the given material, on the accumulated fatigue damage only, and becomes zero at the moment of failure.

More complicated models consider also the fact that the rate of decrease in the fatigue limit is larger for higher loading levels. This results in the relationship

$$\sigma_f = \sigma_f^0 (1 - v)^{\alpha(1-k)}$$

where $k = \sigma_f^0/\sigma_a$ is the ratio of the initial fatigue limit to the level σ_a of loading.

Example 12.1 A structural element experiences linear deformations, so that the induced stress can be evaluated by the formula

$$\sigma(t) = Kq(t)$$

where the factor K is determined by the element's geometry and $q(t)$ is a static random loading with zero mean value. The fatigue limit σ_f of the material is a known random variable. Determine the geometrical factor K, so that the durability T of the element would not be lower than the allowable (required) value T^*.

Solution Using a conservative approach, we assume that the fatigue limit σ_f is low enough, so that the parameter x_0 is such that the Pierson function $P(x_0^2, m+2)$ can be put equal to unity. This assumption will result in an overestimation of the stress level and is acceptable in an approximate and conservative analysis. Then the formulas (12.9) and (12.13) result in the following expression for the required fatigue limit:

$$\sigma_f = \left[\frac{T}{\tau_e N_f} 2^{m/2} k^{m+2} \Gamma\left(\frac{m+2}{2}\right) \right]^{1/m} \sqrt{D_\sigma}$$

If the fatigue limit is distributed in accordance with the Weibull law, then the probability that this limit is higher than a certain value σ_f is

$$P = \exp\left[-\left(\frac{(\sigma_f - \sigma_0)}{b}\right)^a \right]$$

where σ_0, a, and b are the parameters of the Weibull distribution (see Sec. 3.3.7). Introducing the formula for the required σ_f value into this formula we obtain

$$P = \exp\left(-\left\{ \frac{\left[\frac{T}{\tau_e N_f} 2^{m/2} k^{m+2} \Gamma\left(\frac{m+2}{2}\right) \right]^{1/m} \sqrt{D_\sigma} - \sigma_0}{b} \right\}^a \right)$$

Since $D_\sigma = K^2 D_q$, we have

$$K = \frac{\sigma_0 + b(-\ln P)^{1/a}}{\sqrt{D_q}\left[\dfrac{T}{\tau_e N_f} 2^{m/2} k^{m+2} \Gamma\left(\dfrac{m+2}{2}\right)\right]^{1/m}}$$

Example 12.2 An elongated 1 m wide rectangular steel plate clamped on the support contour experiences vibrations under the action of a time-dependent load $q(t)$, uniformly distributed over the plate's surface. The load is random, with zero mean and standard deviation $\sqrt{D_q} = 0.25$ MPa. The fatigue limit of the plate's material follows a Weibull distribution with parameters $\sigma_0 = 220$ MPa, $a = 6$, and $b = 50$ MPa. The fatigue curve of the material is characterized by the parameters $m = 10$ and $N_f = 10^7$. The coefficient of stress concentration can be assumed to be $k = 1$. The required service time of the plate structure is $T = 2$ years $= 63 \times 10^6$ s. Find the required thickness, h, of the plate for which the probability P that the fatigue limit σ_f will not be exceeded during the service time is 0.99. The effective period τ_e of loading can be assumed to be equal to the fundamental period of free vibrations of the plate.

Solution For an elongated steel clamped plate, the maximum bending stress σ_{max} and the frequency ω of the fundamental mode of vibrations can be calculated as

$$\sigma_{max} \simeq 0.497 \frac{a^2}{h^2} q, \qquad \omega \simeq 6.782 \frac{h}{a^2} \sqrt{\frac{E}{\rho}}$$

where a is the plate's width, h is its thickness, E is Young's modulus of the material, and ρ is the material's density. Then we have

$$K = \frac{\sigma_{max}}{q} = 0.497 \frac{a^2}{h^2}, \qquad \tau_e = \frac{2\omega}{\omega} = 0.926 \frac{a^2}{h} \sqrt{\frac{\rho}{E}}$$

The formula for the factor K obtained in the previous example yields

$$K\tau_e^{1/m} = \frac{\sigma_0 + b(-\ln P)^{1/a}}{\sqrt{D_q}\left[\dfrac{T}{N_f} 2^{m/2} k^{m+2} \Gamma\left(\dfrac{m+2}{2}\right)\right]^{1/m}}$$

$$= \frac{220 + 50(-\ln 0.99)^{1/6}}{0.25\left[\dfrac{63 \times 10^6}{10^7} 2^5 \Gamma(6)\right]^{1/10}} = 354.6$$

With $\rho = 795.6$ kgf s^2/m^4 and $E = 2.2 \times 10^6$ kgf/cm$^2 = 2.2 \times 10^{10}$ kgf/m^2, we obtain

$$K\tau_e^{1/m} = 0.497 \frac{a^2}{h^2} \left(0.926 \frac{a^2}{h} \sqrt{\frac{\rho}{E}}\right)^{1/m} = \frac{0.2094}{h^{2.1}}$$

Thus,

$$h^{2.1} = \frac{0.2094}{354.6} = 0.0005905$$

so that

$$h = 0.0290 \text{ m} = 29.0 \text{ mm}$$

Further Reading (*see Bibliography*): 6, 10, 14, 15, 36, 37, 41, 46, 87, 103, 128, 136, 147.

Chapter

13

Random Vibrations

"Everyone knows that we live in the era of engineering, however he rarely realizes that literally all our engineering is based on mathematics and physics."
B. L. VAN DER WAERDEN

"I never said it was possible. I only said it was true." CHARLES RICHET,
Nobel Laureate in Physiology

"Truth is rarely pure, and never simple."
OSCAR WILDE
"The Importance of Being Earnest"

13.1. Vibrations of Elastic Systems

Vibrations (oscillations) due to random excitations, random initial conditions, or random behavior of the system's parameters are also random. Examples are: electrical current fluctuations, noises in telecommunications systems, vibrations of structures caused by earthquakes, ocean wave atmospheric turbulence, etc.

Many examples of random vibrations of one-degree-of-freedom systems have been examined in previous chapters as illustrations (examples) of the application of various probabilistic methods (Chapters 7, 8, 9, and 11). In this chapter we address several general methods and approaches used for analyzing dynamic systems that experience random vibrations. Let us start with a typical example.

13.1.1. Equations of nonlinear vibrations of rectangular plates

Consider an elongated plate whose ends cannot get closer during vibrations. These vibrations can be described by the equation:

$$D\left(1 + 2\varepsilon \frac{\partial}{\partial t}\right) \frac{\partial^4 w}{\partial x^4} + m \frac{\partial^2 w}{\partial t^2} - \frac{Eh}{2(1-v^2)a} \frac{\partial^2 w}{\partial x^2} \int_{-a/2}^{-a/2} \left(\frac{\partial w}{\partial x}\right)^2 dx = q(x, t)$$

(13.1)

where $w(x, t)$ is the deflection function, $q(x, t)$ is the external loading, $D = Eh^3/[12(1-v^2)]$ is the flexural rigidity of the plate, E and v are the elastic constants of the material, h is the thickness of the plate, a is the plate's width, m is its mass per unit length, and ε is the coefficient of viscous damping.

Equation (13.1) can be obtained from the following von Kármán equations of large deflections of a rectangular plate:

$$\left. \begin{aligned} D\nabla^4 w + m \frac{\partial^2 w}{\partial t^2} &= hL(\phi, w) + q(x, y, t) \\ \frac{1}{E} \nabla^4 \phi &= -\frac{1}{2} L(w, w) \end{aligned} \right\}$$

Here $\phi = \phi(x, y, t)$ is the stress (Airy) function,

$$\nabla^4 = \frac{\partial^4}{\partial x^4} + 2 \frac{\partial^4}{\partial x^2 \partial y^2} + \frac{\partial^4}{\partial y^4}$$

is the biharmonic operator, and the operator $L(\phi, w)$ is expressed as

$$L(w, \phi) = \frac{\partial^2 w}{\partial x^2} \frac{\partial^2 \phi}{\partial y^2} - 2 \frac{\partial^2 w}{\partial x \partial y} \frac{\partial^2 \phi}{\partial x \partial y} + \frac{\partial^2 w}{\partial y^2} \frac{\partial^2 \phi}{\partial x^2}$$

The in-plane ("membrane") stresses in the plate can be expressed through the stress function ϕ as follows:

$$\sigma_x^2 = \frac{\partial^0 \phi}{\partial y^2}, \quad \sigma_y^0 = \frac{\partial^2 \phi}{\partial x^2}, \quad \tau_{xy}^0 = -\frac{\partial^2 \phi}{\partial x \partial y}$$

The displacements of the points, located in the midplane of the plate, in the directions x and y can be evaluated by the formulas:

$$\left. \begin{aligned} u(x, y) &= \int_0^x \left[\frac{1}{E}(\sigma_x^0 - v\sigma_y^0) - \frac{1}{2}\left(\frac{\partial w}{\partial x}\right)^2\right] dx \\ v(x, y) &= \int_0^y \left[\frac{1}{E}(\sigma_y^0 - v\sigma_x^0) - \frac{1}{2}\left(\frac{\partial w}{\partial y}\right)^2\right] dx \end{aligned} \right\}$$

where the terms containing the angles of rotation $\partial w/\partial x$ consider additional deflections due to these angles. In a situation where the edges of the plate's support contour cannot get closer during the vibrations of the plate, the displacements u and v must be zero at these edges:

$$u\left(\pm\frac{a}{2}, y\right) = v\left(x, \pm\frac{b}{2}\right) = 0$$

Then the above formulas for the displacements result in the following additional relationships between the deflection function $w(x, y)$ and the stress function $\phi(x, y)$:

$$\left.\begin{array}{l}\displaystyle\int_{\pm a/2}^{0}\left[\frac{1}{E}\left(\frac{\partial^2\phi}{\partial y^2} - v\frac{\partial^2\phi}{\partial x^2}\right) - \frac{1}{2}\left(\frac{\partial w}{\partial x}\right)^2\right]dx = 0 \\ \displaystyle\int_{\pm b/2}^{0}\left[\frac{1}{E}\left(\frac{\partial^2\phi}{\partial x^2} - v\frac{\partial^2\phi}{\partial y^2}\right) - \frac{1}{2}\left(\frac{\partial w}{\partial y}\right)^2\right]dy = 0\end{array}\right\}$$

In the above equations, the origin of the coordinates x, y is in the middle of the plate on its midplane, and a and b are the plate's dimensions in the directions x and y, respectively. In the case of an elongated plate ($b \to \infty$), we have

$$\nabla^4 w = \frac{\partial^4 w}{\partial x^4}, \quad L(\phi, w) = \frac{\partial^2 w}{\partial x^2}\frac{\partial^2 \phi}{\partial y^2} = \sigma_x^0 \frac{\partial^2 w}{\partial x^2}, \quad L(w, w) = 0$$

and the non-Kármán equations reduce to a single equation:

$$D\frac{\partial^4 w}{\partial x^4} + m\frac{\partial^2 w}{\partial t^2} = h\sigma_x^0\frac{\partial^2 w}{\partial x^2} + q(x, t)$$

The conditions of the nondeformability of the plate's contour yield

$$\sigma_x^0 = \frac{E}{(1-v^2)a}\int_0^{a/2}\left(\frac{\partial w}{\partial x}\right)^2 dx = \frac{E}{2(1-v^2)a}\int_{a/2}^{-a/2}\left(\frac{\partial w}{\partial x}\right)^2 dx$$

Excluding the stress σ_x^0 from the obtained relationships and supplementing the elastic and inertia forces with the forces of viscous damping, one obtains Eq. (13.1).

In accordance with the method of principal coordinates, the deflection function $w(x, t)$ can be sought in the form of the expansion

$$w(x, t) = \sum_{k=1}^{\infty} T_k(t)X_k(x) \tag{13.2}$$

where $X_k(x)$ is the (known) kth mode of vibrations and $T_k(t)$ is the (unknown) principal coordinate corresponding to this mode.

Equation (13.1) can be solved using the Galerkin method. In accordance with the procedure of this method, we introduce the expansion

(13.2) into Eq. (13.1), multiply the obtained expression by $X_j(x)$ and integrate this expression over the plate's width a. Considering the mutual orthogonality of the functions $X_k(x)$ and $X_j(x)$, we obtain the following equations for the unknown principal coordinates $T_k(t)$:

$$\ddot{T}_k(t) + 2r_k \dot{T}_k(t) + \omega_k^2 T_k(t) + T_k(t) \sum_{i=1}^{\infty} \alpha_{ik} T_i^2 = \frac{Q_k}{M_k}, \quad k = 1, 2, \ldots \quad (13.3)$$

Here the following notation is used:

$$\left.\begin{aligned}
M_k &= m \int_{-a/2}^{a/2} X_k^2(x)\,dx, \quad Q_k(t) = \int_{-a/2}^{a/2} q(x,t)X_k(x)\,dx \\
\omega_k^2 &= \frac{D}{M_k} \int_{-a/2}^{a/2} X_k(x) X_k^{iv}(x)\,dx = \frac{D}{M_k} \int_{-a/2}^{a/2} [X_k''(x)]^2\,dx \\
r_k &= \varepsilon \frac{D}{M_k} \int_{-a/2}^{a/2} X_k(x) X_k^{iv}(x)\,dx = \varepsilon \frac{D}{M_k} \int_{-a/2}^{a/2} [X_k''(x)]^2\,dx \\
\alpha_{ik} &= -\frac{Eh}{2(1-\nu^2)aM_k} \int_{-a/2}^{a/2} [X_k'(x)]^2\,dx \int_{-a/2}^{a/2} X_i''(x)X_i(x)\,dx \\
&= \frac{Eh\kappa_i\kappa_k}{2aM_k} \quad \kappa_k = \int_{-a/2}^{a/2} [X_k'(x)]^2\,dx
\end{aligned}\right\} \quad (13.4)$$

In these formulas, M_k is the generalized mass of the kth mode of vibrations, $Q_k(t)$ is the generalized force corresponding to the kth mode, ω_k is the linear frequency of undamped linear vibrations of the kth mode, r_k is the damping coefficient, and α_{ik} are the factors of nonlinearity. Equations (13.3) describe the vibrations of a multidegree-of-freedom nonlinear system.

13.1.2. Steady-state vibrations of linear elastic systems

In the linear approximation, the in-plane ("membrane") stresses are not considered, and the parameters α_{ik} of nonlinearity can be put equal to zero ($\alpha_{ik} \equiv 0$). This leads to the following uncoupled linear differential equations for the principal coordinates $T_k(t)$:

$$\ddot{T}_k(t) + 2r_k \dot{T}_k(t) + \omega_k^2 T_k(t) = \frac{Q_k(t)}{M_k}, \quad k = 1, 2, \ldots \quad (13.5)$$

The steady-state solution to this equation can be sought in the form of the Duhamel integral (see Sec. 7.5):

$$T_k(t) = \frac{1}{M_k} \int_0^{\infty} Q_k(t-\tau) h_k(\tau)\,d\tau \quad (13.6)$$

This solution is applicable starting at the moment of time that is sufficiently remote from the initial moment, i.e., at the moment when free vibrations fade away because of damping and need not be accounted for. Therefore the initial conditions do not enter the expression (13.6).

The bivariate correlation function for the random processes $T_k(t)$ and $T_l(t)$ is as follows:

$$K_{T_k T_l}(t_1, t_2) = \frac{1}{T} \int_0^T T_k(t_1) T_l(t_2) \, dt$$

$$= \frac{1}{M_k M_l} \int_0^\infty \int_0^\infty h_k(\tau_1) h_l(\tau_2)$$

$$\times \left[\frac{1}{T} \int_0^T Q_k(t_1 - \tau_1) Q_l(t_2 - \tau_2) \, dt \right] d\tau_1 \, d\tau_2$$

$$= \frac{1}{M_k M_l} \int_0^\infty \int_0^\infty h_k(\tau_1) h_l(\tau_2) K_{Q_k Q_l}(\tau + \tau_1 - \tau_2) \, d\tau_1 \, d\tau_2$$

where

$$05 K_{Q_k Q_l}(\tau + \tau_1 - \tau_2) = \frac{1}{T} \int_0^T Q_k(t_1 - \tau_1) Q_l(t_2 - \tau_2) \, dt$$

$$= K_{Q_k Q_l}[(t_2 - \tau_2) - (t_1 - \tau_1)]$$

is the bivariate correlation function for the generalized forces $Q_k(t)$ and $Q_l(t)$. Then the bivariate spectral density for the generalized principal coordinates T_k and T_l is

$$S_{T_k T_l}(\omega) = \frac{1}{\pi} \int_{-\infty}^\infty K_{T_k T_l}(\tau) e^{-i\omega\tau} \, d\tau$$

$$= \frac{2}{\pi M_k M_l} \int_0^\infty \int_0^\infty \int_0^\infty e^{-i\omega\tau} h_k(\tau_1) h_l(\tau_2) K_{Q_k Q_l}(\tau + \tau_1 - \tau_2) \, d\tau_1 \, d\tau_2 \, d\tau$$

$$= \frac{2}{\pi M_k M_l} \int_0^\infty \int_0^\infty \int_0^\infty e^{-\omega(\tau + \tau_1 - \tau_2)} e^{i\omega\tau_1} e^{-i\omega\tau_2}$$

$$\times h_k(\tau_1) h_l(\tau_2) K_{Q_k Q_l}(\tau + \tau_1 - \tau_2) \, d\tau_1 \, d\tau_2 \, d\tau$$

$$= \frac{2}{\pi M_k M_l} \int_0^\infty h_k(\tau_1) e^{i\omega\tau_1} \, d\tau_1 \int_0^\infty h_l(\tau_2) e^{-i\omega\tau_2} \, d\tau_2$$

$$\times \int_0^\infty K_{Q_k Q_l}(\tau + \tau_1 - \tau_2) e^{-i\omega(\tau + \tau_1 - \tau_2)} \, d\tau$$

$$= \frac{S_{Q_k Q_l}(\omega)}{M_k M_l} \int_0^\infty h_k(t) e^{i\omega t} \, dt \int_0^\infty h_l(t) e^{-i\omega t} \, dt$$

where

$$S_{Q_k Q_l}(\omega) = \frac{2}{\pi} \int_0^\infty K_{Q_k Q_l}(\tau + \tau_1 - \tau_2) e^{-i\omega(\tau + \tau_1 - \tau_2)} d\tau$$

$$= \frac{1}{\pi} \int_{-\infty}^\infty K_{Q_k Q_l}(\tau) e^{-i\omega\tau} d\tau$$

is the bivariate spectral density of the generalized forces. Since the complex frequency characteristic of the response function $h_k(t)$ is

$$\Phi_k(i\omega) = \int_{-\infty}^\infty h_k(t) e^{-i\omega t} dt = \int_0^\infty h_k(t) e^{-i\omega t} dt$$

the bivariate spectral density of the coordinates T_k and T_l can be found as

$$S_{T_k T_l}(\omega) = S_{Q_k Q_l}(\omega) \frac{\Phi_k(i\omega) \Phi_l(-i\omega)}{M_k M_l}$$

Introducing the formula (8.37) into this expression, we obtain

$$S_{T_k T_l}(\omega) = \frac{S_{Q_k Q_l}(\omega)}{M_k M_l (\omega_k^2 - \omega^2 + 2ir_k\omega)(\omega_l^2 - \omega^2 - 2ir_k\omega)} \quad (13.7)$$

Then the bivariate variances of the generalized principal coordinates T_k and T_l are

$$D_{T_k T_l} = \int_0^\infty S_{T_k T_l}(\omega) \, d\omega$$

$$= \frac{1}{M_k M_l} \int_0^\infty \frac{S_{Q_k Q_l}(\omega) \, d\omega}{(\omega_k^2 - \omega^2 + 2ir_k\omega)(\omega_l^2 - \omega^2 - 2ir_l\omega)}$$

Using the theorem of the spectrum of a derivative, expressed by the formula (8.26), we have

$$S_{\dot{T}_k T_l} = S_{T_k \dot{T}_l} = i\omega S_{T_k T_l}, \qquad S_{\dot{T}_k \dot{T}_l} = -\omega^2 S_{T_k T_l}$$

and therefore

$$D_{\dot{T}_k T_l} = D_{T_k \dot{T}_l} = \frac{i}{M_k M_l} \int_0^k \frac{\omega S_{Q_k Q_l}(\omega) \, d\omega}{(\omega_k^2 - \omega^2 + 2ir_k\omega)(\omega_l^2 - \omega^2 - 2ir_k\omega)} \quad (13.8)$$

$$D_{\dot{T}_k \dot{T}_l} = -\frac{1}{M_k M_l} \int_0^\infty \frac{\omega S_{Q_k Q_l}(\omega) \, d\omega}{(\omega_k^2 - \omega^2 + 2ir_k\omega)(\omega_l^2 - \omega^2 - 2ir_l\omega)}$$

If the generalized forces $Q_k(t)$ can be approximated by "white noises," so that

$$K_{Q_kQ_l}(\tau) = \pi S_{Q_kQ_l} \delta(\tau), \quad S_{Q_kQ_l} = \text{constant} \tag{13.9}$$

then the bivariate variances of the generalized principal coordinates and the principal velocities are expressed by the formulas:

$$\left.\begin{aligned}
D_{T_kT_l} &= \frac{4\pi S_{Q_kQ_l}}{M_k M_l} \frac{r_k + r_l}{(\omega_k^2 - \omega_l^2)^2 + 4(r_k + r_l)(r_k\omega_l^2 + r_l\omega_k^2)} \\
D_{\dot{T}_kT_l} &= D_{T_k\dot{T}_l} = \frac{2\pi S_{Q_kQ_l}}{M_k M_l} \frac{(r_k + r_l)(\omega_l^2 - \omega_k^2)}{(\omega_k^2 - \omega_l^2)^2(r_k + r_l) + r_k\omega_l^2 + r_l\omega_k^2} \\
D_{\dot{T}_k\dot{T}_l} &= \frac{4\pi S_{Q_kQ_l}}{M_k M_l} \frac{r_k\omega_l^2 + r_l\omega_k^2}{(\omega_k^2 - \omega_l^2)^2 + 4(r_k + r_l)(r_k\omega_l^2 + r_l\omega_k^2)}
\end{aligned}\right\} \tag{13.10}$$

As evident from these formulas, when the coefficients of energy dissipation r_k are small (which is typically the case for lower modes of vibrations of elastic systems), one can neglect the bivariate correlations of the generalized coordinates and generalized velocities. Such a correlation becomes appreciable and should be considered only for very close values of the frequencies ω_k and ω_l. This circumstance enables one to examine the random displacements of the particular generalized coordinate independently of other coordinates. For $k = l$ we have

$$D_{T_kT_k} = \frac{\pi S_{Q_kQ_l}}{2M_k^2 r_k \omega_k^2}, \quad D_{\dot{T}_kT_k} = D_{T_k\dot{T}_k} = 0, \quad D_{\dot{T}_k\dot{T}_k} = \frac{\pi S_{Q_kQ_l}}{2M_k^2 r_k} \tag{13.11}$$

Let us determine the probability that the process $T_k(t)$ exceeds a certain level T^*, assuming that this process is a narrow-banded, normal stationary process with a statistically independent derivative. The mean number of crossings the level T^* is

$$N_{T^*} = \frac{1}{\tau_e} \exp\left(-\frac{T^{*2}}{2D_{T_k}}\right) \tag{13.12}$$

For $T^* = 0$, we have

$$N_0 = \frac{1}{\tau_e} \tag{13.13}$$

This formula determines the mean "number of zeros," i.e., the number of crossings at the zero level $T = 0$ per unit time. Since the process

$T_k(t)$ is narrow-banded, the number of zeros is equal to the number of maxima. Hence, only N_{T*} "attempts" out of the total number of maxima N_0 are successful, i.e., exceed the level T^*. The probability of such exceedings can be determined as the ratio of the "successful attempts" N_{T*} to the total number N_0 of maxima:

$$P\{T_k > T^*\} = \frac{N_{T*}}{N_0} = \exp\left(-\frac{T^{*2}}{2D_{T_k}}\right) = \exp\left(-\frac{M_k^2 r_k \omega_k T^{*2}}{\pi S_{Q_k Q_l}}\right) \quad (13.14)$$

The same result could be obtained based on an assumption that the amplitudes of the random process $T_k(t)$ follow the Rayleigh law.

Example 13.1 The spectrum $S_x(\omega)$ of the excitation acting on a chassis of an electronic component is localized within the frequency range from ω_1 to ω_2. Determine, for the cases shown in Fig. 13.1 (cantilever fastening, clamped fastening, simply supported fastening), the spectral density $S_y(\omega)$ of the vibration amplitudes for a point that is the most remote from the support contour of the chassis. The chassis can be treated as an elongated rectangular plate. The response functions for these cases can be obtained

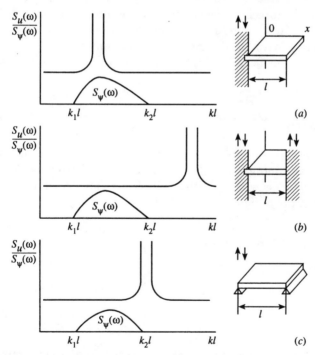

Figure 13.1 Relative locations of the response functions and excitation spectra for different support conditions

as

$$a(\omega, x) = \frac{(\cosh u + \cos u)(\cosh kx + \cos kx) - (\sinh u - \sin u)(\sinh kx + \sin kx)}{2(1 + \cosh u \cos u)}$$

$$a(\omega, x) = \frac{\sinh u/2 \cos [k(x - l/2)] + \sin u/2 \cosh [k(x - l/2)]}{\cosh u/2 \sin u/2 + \sinh u/2 \cos u/2}$$

and

$$a(\omega, x) = \frac{\cosh u/2 \cos [k(x - l/2)] + \cos u/2 \cosh [k(x - l/2)]}{2 \cosh u/2 \cos u/2}$$

respectively. In these equations, the u value is $u = kl$ and k is the frequency parameter expressed as

$$k = \sqrt[4]{\frac{\rho h \omega^2}{D}}$$

Here h is the chassis thickness, ρ is the density of its material, $D = Eh^3/[12(1 - v^2)]$ is the flexural rigidity of the chassis, l is the chassis width, E and v are the elastic constants of the material, and ω is the frequency of vibrations.

Solution The response functions for the locations, at which the amplitudes of vibrations are the largest, should be calculated for the point $x = 0$, in the case of a cantilever chassis, and for the point $x = l/2$, in the case of a chassis clamped or supported at the ends. The spectral densities of the vibration amplitudes at these points can be evaluated as

$$S_y(\omega) = a^2(\omega, 0)S_x(\omega) = \left(\frac{\cosh u + \cos u}{1 + \cosh u \cos u}\right)^2 S_x(\omega)$$

$$S_y(\omega) = a^2(\omega, l/2)S_x(\omega) = \left(\frac{\sinh u/2 + \sin u/2}{\cosh u/2 \sin u/2 + \sinh u/2 \cos u/2}\right)^2 S_x(\omega)$$

and

$$S_y(\omega) = a^2(\omega, l/2)S_x(\omega) = \left(\frac{\cosh u/2 + \cos u/2}{2 \cosh u/2 \cos u/2}\right)^2 S_x(\omega)$$

respectively.

The squares of the response functions $a^2(\omega) = S_y(\omega)/S_x(\omega)$, and the input spectra, are plotted in Fig. 13.1 as functions of the parameter $u = kl$, in the range from

$$u_1 = k_1 l = \sqrt[4]{\frac{\rho h \omega_1^2}{D}}$$

to

$$u_2 = k_2 l = \sqrt[4]{\frac{\rho h \omega_2^2}{D}}$$

As evident from the obtained plots, the second type of fastening ("clamped-clamped" chassis) is the best, since the entire spectrum of the input process is shifted from the region of resonance vibrations.

Example 13.2 An object of mass M, fixed to a base of a vibration measuring device by means of a spring with the spring constant K, is connected to a viscous damper and is subjected to random loading. The loading can be approximated by a "white noise," so that the equation of motion of the mass can be written as

$$M\ddot{x}(t) + R\dot{x}(t) + Kx(t) = n(t)$$

The frequencies of the "white noise" $n(t)$ are within the finite range $0 \leq \omega \leq \omega_1$. Determine the spectral functions and the variances of the induced displacement $x(t)$ and the induced velocity $\dot{x}(t)$ of the object, and the variance of the random dynamic excitation applied to the base in the case $\omega_1 \to \infty$.

Solution In order to determine the complex frequency characteristic of the system, we assume $P = n(t) = P_0 e^{i\omega t}$ and $x = x_0 e^{i\omega t}$, where P_0 and x_0 are the complex amplitudes of harmonic vibrations. Then we obtain:

$$(-M\omega^2 + iR\omega + K)x_0 e^{i\omega t} = P_0 e^{i\omega t}$$

The complex frequency characteristic of the system is

$$\Phi_x(i\omega) = \frac{X}{P} = \frac{1}{K - M\omega^2 - iR\omega} = \frac{1}{K} \frac{1}{1 - \omega^2/\omega_0^2 + i(2r/\omega_0)(\omega/\omega_0)}$$

where $\omega_0 = \sqrt{M/K}$ is the natural frequency of free undamped vibrations and $r = R/(2M)$ is the coefficient of damping. The spectral density of the random displacement $X(t)$ is

$$S_x(\omega) = |\Phi_x(i\omega)|^2 S_0 = \frac{S_0}{K^2} \frac{1}{(1 - \omega^2/\omega_0^2)^2 + (2r/\omega_0)^2(\omega/\omega_0)^2}$$

for $0 \leq \omega \leq \omega_1$. Outside this region, the spectrum $S_x(\omega)$ is equal to zero.
The complex frequency characteristic of the velocity $\dot{X}(t)$ is

$$\Phi_{\dot{x}}(i\omega) = i\omega \Phi_x(i\omega)$$

and therefore the spectral density of the velocity can be evaluated as

$$S_{\dot{x}}(\omega) = |i\omega \Phi_x(i\omega)|^2 S_0(\omega) = \omega^2 S_x(\omega)$$

Similarly, the spectral density of the acceleration is

$$S_{\ddot{x}}(\omega) = \omega^4 S_x(\omega)$$

The total force, transmitted to the base of the device, is due to the restoring and dissipation forces and can be calculated as

$$R_b = Kx + R\dot{x} = (K + iR\omega)x$$

The complex frequency characteristic of this force is

$$\Phi_R(i\omega) = (K + iR\omega)\Phi_x(i\omega)$$

Hence, the spectrum $S_R(\omega)$ of the force R_b transmitted to the base is

$$S_R(\omega) = |\Phi_R(i\omega)|^2 S_0 = K^2 S_x(\omega) + R^2 S_{\dot{x}}(\omega)$$

The variances D_x and $D_{\dot{x}}$ of the response process $X(t)$ and its time derivative $\dot{X}(t)$ can be found as follows:

$$D_x = \int_0^\infty S_x(\omega)\, d\omega = \int_0^\infty |\Phi_x(i\omega)|^2 S_0\, d\omega$$

$$= \frac{S_0}{K^2} \int_0^{\omega_1} \frac{d\omega}{(1 - \omega^2/\omega_0^2)^2 + (2r/\omega_0)^2(\omega/\omega_0)^2}$$

$$= \frac{S_0 \omega_0}{K^2} \left[\frac{1}{8\sqrt{1-\mu^2}} \ln \frac{\eta^2 + 2\eta\sqrt{1-\mu^2} + 1}{\eta^2 - 2\eta\sqrt{1-\mu^2} + 1} \right.$$

$$\left. + \frac{1}{4\mu}\left(\arctan \frac{\eta + \sqrt{1-\mu^2}}{\mu} + \arctan \frac{\eta - \sqrt{1-\mu^2}}{\mu} \right) \right]$$

and

$$D_{\dot{x}} = \int_0^\infty S_{\dot{x}}(\omega)\, d\omega$$

$$= \frac{S_0 \omega_0^3}{K^2} \left[-\frac{1}{8\sqrt{1-\mu^2}} \ln \frac{\eta^2 + 2\eta\sqrt{1-\mu^2} + 1}{\eta^2 - 2\eta\sqrt{1-\mu^2} + 1} \right.$$

$$\left. + \frac{1}{4\mu}\left(\arctan \frac{\eta + \sqrt{1-\mu^2}}{\mu} + \arctan \frac{\eta - \sqrt{1-\mu^2}}{\mu} \right) \right]$$

where the following notation is used:

$$\mu = \frac{r}{\omega_0}, \qquad \eta = \frac{\omega_1}{\omega_0}$$

For $\omega_1 \to \infty$ we find

$$D_x = \frac{\pi}{4} \frac{S_0 \omega_0}{K^2 \mu} = \frac{\pi}{4} \frac{S_0 \omega_0^2}{K^2 r}, \qquad D_{\dot{x}} = \frac{\pi}{4} \frac{S_0}{K^2} \frac{\omega_0^4}{r}$$

Thus, although the spectrum of the "white noise" loading $n(t)$ is unlimited, the variances D_x and $D_{\dot{x}}$ of the random processes of the displacement and the velocity remain finite.

The variance of the force R_b, transmitted to the device base, is

$$D_{R_b} = K^2 D_x + R^2 D_{\dot{x}} = \frac{\pi}{4} S_0 \omega_0 \left(\frac{\omega_0}{r} + \frac{4r}{\omega_0} \right)$$

Example 13.3 For the system considered in the previous example, determine the most advantageous value of damping that would minimize the standard deviation of the force applied to the system's base.

Solution As evident from the obtained formula for the variance D_{R_b} of the force R_b acting on the system's base, this variance is infinitely large for both $r \to 0$ and $r \to \infty$. Therefore, there should exist a value of the parameter r that would result in the minimum D_{R_b} value. Indeed, differentiating the equation

$$D_{R_b} = \frac{\pi}{4} S_0 \omega_0 \left(\frac{\omega_0}{r} + \frac{4r}{\omega_0} \right)$$

with respect to r and equating the obtained expression to zero, we find $r = \omega_0/2$. Hence, the minimum value of the force acting on the system's base is

$$D_{R_b} = \pi S_0 \omega_0$$

and takes place for the following value of the viscous damping force:

$$R = 2M \frac{\omega_0}{2} = M \omega_0$$

As evident from this formula, the optimal viscous damping force is quite large. The obtained formula indicates also that low-frequency systems (characterized by low ω_0 values) lend themselves more easily to vibration damping.

13.1.3. Free vibrations with random initial conditions

In the previous section we examined steady-state (stationary) vibrations caused by random excitations, characterized by their power spectra. Such vibrations can be analyzed without taking into the account the initial conditions. Indeed, the general solution to the linear nonhomogenous ordinary differential equation of the type (13.5) can be written as the sum of the general solution to the homogeneous equation:

$$\ddot{T}_k(t) + 2r_k \dot{T}_k(t) + \omega_k^2 T_k(t) = 0 \qquad (13.15)$$

and the particular solution (13.6). The solution to Eq. (13.15) when the ratio r_k/ω_k does not exceed unity (which is the case in the majority of practical situations) can be written as

$$T_k(t) = e^{-r_k t} \left[T_k(0) \left(\cos \lambda_k t + \frac{r_k}{\lambda_k} \sin \lambda_k t \right) + \dot{T}_k(0) \frac{\sin \lambda_k t}{\lambda_k} \right] \qquad (13.16)$$

Here

$$\lambda_k = \sqrt{\omega_k^2 - r_k^2} \qquad (13.17)$$

is the frequency of the kth mode of free vibrations with consideration of the effect of viscous damping, and $T_k(0)$ and $\dot{T}_k(0)$ are the initial conditions for the principal coordinate and its velocity. As obvious from the formula (13.16), the contribution of the transient vibrations, expressed by this formula, to the general solution of the nonhomogeneous equation (13.15) is small if the coefficient of damping r_k is appreciable and/or the time t is large. In the analysis that follows we consider the opposite extreme case, namely, the case where there are no continuous external forces acting on the system $[Q_k(t) \equiv 0]$ and the system's vibrations are due to the initial disturbance of its displacement and/or velocity.

In actual situations it is impossible to obtain or to observe motions with absolutely certain initial conditions. Nonetheless, in the "classical" theory of oscillations, transient vibrations in dynamic systems are treated as deterministic ones. "Give me the initial conditions, and I will predict the future of the universe," Laplace wrote. Consideration of the inevitable scattering in the initial conditions enables one to assess to what extent the actual transient vibrations might differ from those predicted by the deterministic solution.

From the formula (13.16) we have

$$\dot{T}_k(t) = T_k(0)\dot{f}_k(t) + \dot{T}_k(0)\dot{g}_k(t)$$
$$\ddot{T}_k(t) = T_k(0)\ddot{f}_k(t) + \dot{T}_k(0)\ddot{g}_k(t) \qquad (13.18)$$

where the functions $f_k(t)$ and $g_k(t)$ are expressed as follows:

$$f_k(t) = e^{-r_k t}\left(\cos \lambda_k t + \frac{r_k}{\lambda_k} \sin \lambda_k t\right), \qquad g_k(t) = e^{-r_k t} \frac{\sin \lambda_k t}{\lambda_k} \qquad (13.19)$$

Let the initial conditions $T_k(0)$ and $\dot{T}_k(0)$ be random variables with the means $\langle T_k(0) \rangle$ and $\langle \dot{T}_k(0) \rangle$, variances $D_{T_k(0)}$ and $D_{\dot{T}_k(0)}$, and the bivariate correlation moment $K_{T_k \dot{T}_k}$. Since, as evident from the solution (13.16), the principal coordinate $R_k(t)$ depends linearly on the initial conditions $T_k(0)$ and $\dot{T}_k(0)$, then the mean $\langle T_k(t) \rangle$ and the correlation function $K_{T_k}(t, t')$ of the principal coordinate $T_k(t)$ can be found as

$$\langle T_k(t) \rangle = \langle {}^*T_k(0) \rangle e^{-r_k t}\left(\cos \lambda_k t + \frac{r_k}{\lambda_k} \sin \lambda_k t\right) + \frac{1}{\lambda_k} \langle \dot{T}_k(0) \rangle e^{-r_k t} \sin \lambda_k t$$

and

$$K_{T_k}(t, t') = \langle T_k(t) T_k(t') \rangle$$
$$= D_{T_k(0)} e^{-r_k(t+t')} \left(\cos \lambda_k t + \frac{r_k}{\lambda_k} \sin \lambda_k t \right) \left(\cos \lambda_k t' + \frac{r_k}{\lambda_k} \sin \lambda_k t' \right)$$
$$+ \frac{1}{\lambda_k} K_{T_k(0)\dot{T}_k(0)} \left[e^{-r_k(t+t')} \left(\cos \lambda_k t + \frac{r_k}{\lambda_k} \sin \lambda_k t \right) \sin \lambda_k t' \right.$$
$$\left. + e^{-r_k(t+t')} \left(\cos \lambda_k t' + \frac{r_k}{\lambda_k} \sin \lambda_k t' \right) \sin \lambda_k t \right]$$
$$+ \frac{1}{\lambda_k^2} D_{\dot{T}_k(0)} e^{-r_k(t+t')} \sin \lambda_k t \sin \lambda_k t'$$

The variance of the principal coordinate $T_k(t)$ can be found from the obtained expression by putting $t' = t$:

$$D_{T_k}(t) = D_{T_k(0)} f_k^2(t) + 2 K_{T_k(0)\dot{T}_k(0)} f_k(t) g_k(t) + D_{\dot{T}_k(0)} g_k^2(t)$$

If the initial conditions are uncorrelated, i.e., if $K_{T_k(0)\dot{T}_k(0)} = 0$, then this variance can be evaluated as

$$D_{T_k}(t) = D_{T_k(0)} f_k^2(t) + D_{\dot{T}_k(0)} g_k^2(t)$$

Similarly, one can find the expressions for the means and the correlation functions of the velocities and accelerations:

$$\langle \dot{T}_k(t) \rangle = \langle T_k(0) \rangle \dot{f}_k(t) + \langle \dot{T}_k(0) \rangle \dot{g}_k(t)$$
$$\langle \ddot{T}_k(t) \rangle = \langle T_k(0) \rangle \ddot{f}_k(t) + \langle \dot{T}_k(0) \rangle \ddot{g}_k(t)$$
$$K_{\dot{T}_k}(t, t') = D_{T_k(0)} \dot{f}_k(t) \dot{f}_k(t') + K_{T_k(0)\dot{T}_k(0)} [\dot{f}_k(t) \dot{g}_k(t') + \dot{f}_k(t') \dot{g}_k(t)]$$
$$+ D_{\dot{T}_k(0)} \dot{g}_k(t) \dot{g}_k(t')$$
$$K_{\ddot{T}_k}(t, t') = D_{T_k(0)} \ddot{f}_k(t) \ddot{f}_k(t') + K_{T_k(0)\dot{T}_k(0)} [\ddot{f}_k(t) \ddot{g}_k(t') + \ddot{f}_k(t') \ddot{g}_k(t)]$$
$$+ D_{\dot{T}_k(0)} \ddot{g}_k(t) \ddot{g}_k(t')$$

The deterministic solution (13.16) corresponds to the case when $\langle T_k(0) \rangle = T_k(0)$ and $K_{T_k}(t, t') \equiv 0$, i.e., when $D_{T_k(0)} = 0$, $D_{\dot{T}_k(0)} = 0$, and $K_{T_k(0)\dot{T}_k(0)} = 0$.

Example 13.4 A simply supported beam of a circular cross section experiences an instantaneous uniform loading of the magnitude q. The loading q

is random and follows the Rayleigh law of the probability distribution:

$$f_q(q) = \frac{q}{q_0^2} e^{-q^2/(2q_0^2)}$$

where q_0 is the most likely value of the load q. Determine the probability that the induced dynamic stress in the beam will exceed the allowable level σ^*.

Solution The differential equation of the beam's vibrations can be written as

$$EI \frac{\partial^4 w}{\partial x^4} + m \frac{\partial^2 w}{\partial t^2} = q\, \delta(t)$$

where $\delta(t)$ is Dirac's delta function, EI is the flexural rigidity of the beam's cross section, and m is the beam's mass per unit length. The solution of this equation can be sought in accordance with the method of principle coordinates, in the form of an expansion

$$w(x, t) = \sum_{k=1}^{\infty} T_k(t) \sin \frac{k\pi x}{l}$$

where $T_k(t)$ is the principal coordinate of the kth mode and l is the beam's length. The induced velocities of the beam's cross sections are

$$\dot{w}(x, t) = \sum_{k=1}^{\infty} \dot{T}_k(t) \sin \frac{k\pi x}{l}$$

At the initial moment of time, $\dot{w}(x, 0) = q/m$ and therefore

$$\sum_{k=1}^{\infty} \dot{T}_k(0) \sin \frac{k\pi x}{l} = \frac{q}{m}$$

Multiplying both parts of this equation by $\sin(j\pi x/l)$, integrating the obtained expression over the beam's length l, and considering the fact that the functions $\sin(k\pi x/l)$ and $\sin(j\pi x/l)$ are mutually orthogonal ones, and therefore

$$\int_0^l \sin \frac{k\pi x}{l} \sin \frac{j\pi x}{l}\, dx = \begin{cases} l/2, & \text{for } j = k \\ 0, & \text{for } j \neq k \end{cases}$$

we have

$$\dot{T}_k(0) = \frac{2ql}{\pi m k}, \quad k = 1, 3, 5, \ldots$$

Clearly, for a uniformly distributed load, only the odd vibration modes occur.

In the presence of dissipating (damping) forces, the principal coordinates are expressed by the formula (13.16). This formula, considering the fact that the initial displacements are zero [$w(x, 0) \equiv 0$] and, hence,

$T_k(0) \equiv 0$, yields

$$T_k(t) = e^{-r_k t}\dot{T}_k(0)\frac{\sin \lambda_k t}{\lambda_k} = \frac{2ql}{\pi mk}e^{-r_k t}\frac{\sin \lambda_k t}{\lambda_k}, \qquad k = 1, 3, 5, \ldots$$

In order to evaluate the frequencies ω_k of undamped vibrations, one can seek the solution to the differential equation of the beam's response in the form:

$$w(x, t) = \sum_{k=1}^{\infty} \sin\frac{k\pi x}{l} \sin \omega_k t$$

Introducing the expansion into the equation of motion, we obtain

$$\omega_k = \left(\frac{k\pi}{l}\right)^2 \sqrt{\frac{EI}{m}}$$

For sufficiently low damping, one can put $\lambda_k \approx \omega_k$, so that the kth principal coordinate is

$$T_k(t) = \frac{2q}{\sqrt{mEI}}\left(\frac{l}{k\pi}\right)^3 e^{-r_k t}\sin \lambda_k t$$

The maximum bending stress occurs in the mid-cross-section ($x = l/2$) of the beam and, for a circular beam, can be calculated as follows:

$$\sigma_{max} = -E\frac{d}{2}\frac{\partial^2 w}{\partial x^2}\bigg|_{x=l/2} = E\frac{d}{2}\sum_{k=1,3,5,\ldots}^{\infty}\left(\frac{k\pi}{l}\right)^2 T_k(t)\sin\frac{k\pi x}{l} = C(t)q$$

where d is the beam's diameter and the factor $C(t)$ is expressed as

$$C(t) = \left(\frac{4}{\pi}\right)^2 \sqrt{\frac{E}{\rho}}\frac{l}{d^2}\sum_{k=1,3,5,\ldots}^{\infty}\frac{1}{k}e^{-r_k t}\sin \lambda_k t$$

where ρ is the material's density. Since the factor $C(t)$ is time-dependent, one has to determine first the moment t^* of time for which this factor is the largest. Then the probability that the induced stress σ_{max} will exceed the level σ^* can be evaluated as

$$P(\sigma_{max} > \sigma^*) = P(q > q^*) = \exp\left(\frac{q^{*2}}{2q_0^2}\right) = \exp\left[-\frac{\sigma^{*2}}{2C^2(t^*)q_0^2}\right]$$

where the factor $C(t^*)$ should be computed by the formula

$$C(t^*) = \left(\frac{4}{\pi}\right)^2 \sqrt{\frac{E}{\rho}}\frac{l}{d^2}\sum_{k=1,2,5,\ldots}^{\infty}\frac{1}{k}e^{-r_k t^*}\sin \lambda_k t^*$$

The required $C(t^*)$ value can be established, for the assumed (allowable) probability P, as

$$C(t^*) = \frac{\sigma^*}{q_0\sqrt{-\ln P}}$$

Here σ^* is the allowable stress level and P is the allowable (desired) probability that this level is exceeded. The required diameter of the beam can be found from the transcendental equation

$$\sum_{k=1,3,5,\ldots}^{\infty} \frac{1}{k} \exp\left[-\frac{r_k}{\lambda_k}(\lambda_k t^*)\right] \sin \lambda_k t^* = \left(\frac{\pi}{4}\right)^2 \frac{\sigma^*}{q_0} \sqrt{\frac{\rho}{-2E \ln P}} \frac{d^2}{l}$$

in which

$$\lambda_k t^* \simeq \omega_k t^* = \left(\frac{k\pi}{2}\right)^2 \sqrt{\frac{E}{\rho} \frac{d}{l^2}} t^*$$

In an approximate analysis of the beam's response, the fundamental mode only can be considered. In this case the moment of time t^* can be evaluated from the equation

$$\sin \lambda_1 t^* = 1$$

which yields

$$t^* = \frac{\pi}{2\lambda_1}$$

Then we have

$$d^2 = \left(\frac{4}{\pi}\right)^2 l \frac{q_0}{\sigma^*} \sqrt{\frac{-E \ln P}{\rho}} \exp\left(-\frac{\pi}{2}\frac{r_1}{\lambda_1}\right)$$

If, for instance, the length of the beam is $l = 5$ cm $= 0.05$ m, the intensity of the applied load is $q_0 = 0.1$ kgf s/m^2, the allowable level of the dynamic stress is $\sigma^* = 25 \times 10^6$ kgf/m^2, Young's modulus of the material is $E = 2 \times 10^{10}$ kgf/m^2, its density is $\rho = 780$ kgf s^2/m^4, the allowable probability that the stress σ^* is exceeded is $P = 10^{-4}$, and the ratio of the damping coefficient to the fundamental frequency of lateral vibrations is $r_1/\lambda_1 = 0.05$. With this input data, we obtain $d = 0.00255$ m $= 0.255$ cm.

13.2. Nonlinear Vibrations

13.2.1. Methods for analyzing nonlinear random vibrations

The choice of a method for obtaining a solution to an equation of motion of the type of Eq. (13.1), in which the excitation is a random variable or a random function, depends on the intensity of the excitation and the relationship between its correlation time and the system's time constant(s). It should be emphasized that the intensity of the excitation is determined not so much by the absolute magnitudes of the ordinates of the input process itself (characterized, say, by its variance) but, first of all, by the magnitudes of the ordinates of the output process (response function).

For analyzing random excitations of low intensity, one can use the method of statistical linearization, regardless of the relationship between the correlation time and the time constant (say, period) of the dynamic system. This method enables one to evaluate the mean and the correlation function of the output process in a relatively simple and straightforward way. In the case of a highly intensive excitation, the choice of the method for obtaining the solution to the problem of random vibrations depends on the relationship between the correlation time and the system's constant. If the correlation time is substantially greater than the system's constant, then the external excitation changes slowly in comparison with the time characteristic of the system and a "quasistatic" approximation can be applied. In such a case, the inertia and dissipation terms in the basic differential equation can be neglected and a relatively simple problem for a non-inertia excitation can be considered. In another extreme case, when the correlation time exceeds significantly the time constant of the system, one can use a "white noise" approximation and apply methods of the theory of Markovian processes.

13.2.2. Statistical linearization

The basic idea of this approach in application to elastic vibrations is as follows. The equations of motion are linearized in one way or another and then the theory of linear random vibrations is applied. Let us show the substance of one of the modifications of this method, using an example of a prismatic simply supported single-span beam subjected to bending moments and reactive tensile forces. These forces are due to the fact that the beam's supports cannot get closer during its vibrations.

The differential equation of vibrations of a single-span beam subjected, in addition to lateral inertia forces, to the reactive forces from the supports can be written as follows:

$$EI \frac{\partial^4 w}{\partial x^4} + m \frac{\partial^2 w}{\partial t^2} - S \frac{\partial^2 w}{\partial x^2} = q(x, t) \qquad (13.20)$$

Here $w = w(x, t)$ is the deflection function, EI is the flexural rigidity of the beam, m is its mass per unit length, and $q(x, t)$ is the external load per unit beam length. If the beam's supports cannot move together during the vibration process, the magnitude of the axial reactive forces S can be calculated as

$$S = \frac{EA}{2l} \int_0^l \left(\frac{\partial w}{\partial x} \right)^2 dx \qquad (13.21)$$

where l is the beam's length, E is Young's modulus of the material, and A is the cross-sectional area of the beam.

The solution to Eq. (13.20) can be sought in the form:

$$w(x, t) = T(t) \sin \frac{\pi x}{l} \qquad (13.22)$$

where $T(t)$ is the principal coordinate of vibrations. Using Galerkin's method, we substitute (13.22) into Eq. (13.20), multiply the obtained expression by $\sin \pi x/l$ and integrate over the beam's length l. Complementing the external loading with the dissipation (damping) forces, we obtain the following equation for the principal coordinate $T(t)$:

$$\ddot{T}(t) + 2r\dot{T}(t) + \omega_0^2 T(t) + \alpha T^3(t) = \frac{Q(t)}{M} \qquad (13.23)$$

Here the generalized mass M, the linear frequency ω_0 of the undamped vibrations, the parameter α of nonlinearity, and the generalized external force $Q(t)$ are expressed as follows:

$$\left. \begin{array}{l} M = m \int_0^l \sin^2 \frac{\pi x}{l} \, dx = \frac{ml}{2} \\[6pt] \omega_0^2 = \frac{EI}{M} \int_0^l \left[\left(\sin \frac{\pi x}{l} \right)'' \right]^2 dx = \frac{\pi^4 EI}{l^4 m} \\[6pt] \alpha = \frac{EA}{2Ml} \left\{ \int_0^l \left[\left(\sin \frac{\pi x}{l} \right)' \right]^2 dx \right\}^2 = \frac{\pi^4 EA}{4ml^4} \\[6pt] Q(t) = \int_0^l q(x, t) \sin \frac{\pi x}{l} \, dx \end{array} \right\}$$

The nonlinear equation (13.20) can be substituted by an equivalent, in terms of the vibration frequency, linear equation

$$\ddot{T}(t) + 2r\dot{T}(t) + \sigma^2 T(t) = \frac{Q(t)}{M} \qquad (13.24)$$

where the equivalent frequency σ can be found on the basis of one of the methods of nonlinear mechanics. Assuming that damping is small and does not affect the frequency of vibrations, one can determine this frequency from the equation:

$$\ddot{T}(t) + \omega_0^2 T(t) + \alpha T^3(t) = 0 \tag{13.25}$$

Seeking the solution to this equation in the form

$$T(t) = A \cos \sigma t \tag{13.26}$$

and introducing Eq. (13.26) into Eq. (13.25), we find

$$(-\sigma^2 + \omega_0^2 + \tfrac{3}{4}\alpha A^2) \cos \sigma t + \tfrac{1}{4}\alpha A^2 \cos 3\sigma t = 0$$

where the formula

$$4 \cos^3 \phi = 3 \cos \phi + \cos 3\phi$$

was used. Neglecting the third harmonic in the obtained solution, we have

$$\sigma^2 \simeq \omega_0^2 + \tfrac{3}{4}\alpha A^2 \tag{13.27}$$

Since the "variance" of harmonic vibrations with an amplitude A is $D_T = A^2/2$, this equation can be written as follows:

$$\sigma^2 = \omega_0^2 + \tfrac{3}{2}\alpha D_T \tag{13.28}$$

where D_T is the variance of the random process $T(t)$. If this process were deterministic, the formulas (13.27) and (13.28) would be equivalent. In the case of a random process, the frequency σ is, strictly speaking, a random variable. Since, however, this frequency is determined by the envelope of the process $T(t)$ and, as a consequence of that, is changing substantially slower than the process $T(t)$ itself, one can substitute the random variable σ by its mean value and use the formula (13.28).

Let us assume, for instance, that the process $Q(t)$ is a "white noise." Then, applying the first formula in (13.11) to the linearized equation (13.24), we have

$$D_T = \frac{\pi S_Q}{2M^2 r \sigma^2} = \frac{\pi S_Q}{2M^2 r(\omega^2 + \tfrac{3}{2}\alpha D_T)}$$

Solving this equation for the variance D_T we obtain

$$D_T = \frac{\omega^2}{3\alpha}\left(\sqrt{1 + \frac{3\pi\alpha S_Q}{M^2 r \omega^4}} - 1\right) \tag{13.29}$$

The method of statistical linearization does not provide any means of evaluating the probability density function of the output process, and therefore one has to assume, if necessary, a certain law of the probability distribution for this process. Let us assume that the response amplitudes follow the Rayleigh law. Then we obtain the following formula for the probability that the process $T(t)$ exceeds a certain level T^*:

$$P\{T > T^*\} = \exp\left[-\frac{(T^*)^2}{2D_T}\right]$$

$$= \exp\left[-\frac{3\alpha(T^*)^2}{2\omega^2}\left(\sqrt{1 + \frac{3\pi\alpha S_Q}{M^2 r \omega^4}} - 1\right)^{-1}\right] \quad (13.30)$$

When $\alpha = 0$, this formula leads to the formula (13.14). As one can see from (13.30), consideration of the nonlinearity reduces the probability of exceeding the level T^* and, therefore, as far as the vibration amplitudes are concerned, a linear approach is conservative.

Examine now a more general approach of statistical linearization. Consider a nonlinear relationship

$$Y(t) = f[X(t)]$$

where $X(t)$ and $Y(t)$ are random processes. One wishes to substitute the relationship with the following linear one:

$$Z(t) = f_0 + K\mathring{X}(t)$$

Here f_0 and K are nonrandom parameters and

$$\mathring{X}(t) = X(t) - \langle x(t) \rangle$$

is a centered random process. The characteristic f_0 of nonlinearity and the coefficient K have to be chosen in such a way that the processes $Y(t)$ and $Z(t)$ are statistically equivalent. One can require, for instance, that these two processes have equal means and variances:

$$\langle y(t) \rangle = \langle z(t) \rangle, \qquad D_y(t) = D_z(t)$$

Since $\langle z(t) \rangle = f_0$, then

$$f_0 = \langle y \rangle = \langle z \rangle = \int_{-\infty}^{\infty} f(x) p(x)\, dx$$

where $p(x)$ is the probability density function of the input process $X(t)$ for the given time t. Since

$$D_y = D_z = K^2 D_x$$

and, on the other hand,

$$D_y = \int_{-\infty}^{\infty} f^2(x) p(x)\, dx - \langle y \rangle^2$$

we obtain

$$K = \left\{ \frac{1}{D_x} \left[\int_{-\infty}^{\infty} f^2(x) p(x)\, dx - \langle y \rangle^2 \right] \right\}^{1/2}$$

Another approach for evaluating the K value can be based on the requirement that the mean square of the error due to the statistical approximation be minimum. Since

$$y - z = [\langle y \rangle + \mathring{Y}(t)] - [f_0 + K\mathring{X}(t)]$$

the mean square of the approximation error is

$$\langle \varepsilon^2 \rangle = \langle y \rangle^2 + D_y - 2\langle y \rangle f_0 - 2KK_{xy} + f_0^2 + K^2 D_x$$

where K_{xy} is the correlation moment of the variables X and Y. The necessary conditions of the extremum of the $\langle \varepsilon^2 \rangle$ value are expressed by the relationship:

$$\left. \begin{array}{l} \dfrac{\partial \langle \varepsilon^2 \rangle}{\partial f_0} = -2\langle y \rangle + 2f_0 = 0 \\[1em] \dfrac{\partial \langle \varepsilon^2 \rangle}{\partial K} = -2K_{xy} + 2KD_x = 0 \end{array} \right\}$$

so that

$$f_0 = \langle y \rangle = \int_{-\infty}^{\infty} f(x) p(x)\, dx, \qquad K = \frac{K_{xy}}{D_x}$$

Since

$$K_{xy} = \langle \mathring{X}[f(x) - \langle y \rangle] \rangle = \int_{-\infty}^{\infty} f(x)(x - \langle x \rangle) p(x)\, dx$$

we obtain

$$K = \frac{1}{D_x} \int_{-\infty}^{\infty} f(x)(x - \langle x \rangle)p(x)\, dx$$

It could be shown that the obtained extremum is, indeed, minimum. Note that, whatever the approach, the characteristic f_0 is calculated on the basis of the same relationship.

The distribution $p(x)$ is often assumed to be normal, even if the actual distribution is not. With such an assumption, the evaluation of the f_0 and K values (as functions of the mean $\langle x \rangle$ and variance D_x) can be substantially simplified.

Example 13.5 Find the characteristic f_0 and the coefficient K in the linearized relationship

$$Z = f_0 + K\dot{X}$$

statistically equivalent to the nonlinear relationship

$$Y = f(X)$$

for a relay with a level of saturation A. The input process $X(t)$ is normal.

Solution Using the relay equation

$$y = f(x) = A \operatorname{sign} x$$

and the distribution

$$p(x) = \frac{1}{\sqrt{2\pi D_x}} \exp\left[-\frac{(x - \langle x \rangle)^2}{2D_x}\right]$$

for the input process $X(t)$, we have

$$\langle y \rangle = \int_{-\infty}^{\infty} f(x)p(x)\, dx = -A \int_{-\infty}^{0} p(x)\, dx + A \int_{0}^{\infty} p(x)\, dx$$

$$= A\Phi\left(\frac{\langle x \rangle}{\sqrt{2D_x}}\right)$$

$$D_y = \int_{-\infty}^{\infty} f^2(x)p(x)\, dx - \langle y \rangle^2 = A^2 \int_{-\infty}^{\infty} p(x)\, dx - \langle y \rangle^2$$

$$= A^2\left[1 - \Phi^2\left(\frac{\langle x \rangle}{\sqrt{2D_x}}\right)\right]$$

where $\Phi(\alpha)$ is the Laplace function. Then we obtain

$$f_0(\langle x \rangle, D_x) = A\Phi\left(\frac{\langle x \rangle}{\sqrt{2D_x}}\right)$$

The K value, depending on the approach, is either

$$K(\langle x \rangle, D_x) = \frac{A}{\sqrt{D_x}}\left[1 - \Phi^2\left(\frac{\langle x \rangle}{\sqrt{2D_x}}\right)\right]$$

or

$$K(\langle x \rangle, D_x) = \frac{\partial}{\partial \langle x \rangle}\left(\frac{2A}{\sqrt{\pi}}\int_0^{\langle x \rangle/2D_x} e^{-t^2}\,dt\right) = \frac{2A}{\sqrt{2\pi D_x}}e^{-\langle x \rangle^2/(2D_x)}$$

Example 13.6 Linearize Eq. (13.25) by linearizing the function $Y = \alpha T^2$. The process $T(t)$ is normal with zero mean.

Solution Since $\dot{T}(t) = 0$, the coefficient K does not have to be determined. The characteristic f_0 can be calculated as

$$f_0 = \int_{-\infty}^{\infty} \alpha T^2 \frac{1}{\sqrt{2\pi D_T}} e^{-T^2/(2D_T)}\,dT$$

$$= \frac{2\alpha}{\sqrt{2\pi D_T}} \int_0^{\infty} T^2 e^{-T^2/(2D_T)}\,dT = \frac{2\alpha D_T}{\sqrt{2\pi D_T}} \int_0^{\infty} e^{-T^2/(2D_T)}\,dT$$

$$= \frac{2\alpha}{\sqrt{\pi}} D_T \int_0^{\infty} e^{-t^2}\,dt = \alpha D_T$$

so that the frequency of the linearized system is

$$\sigma^2 = \omega_0^2 + \alpha D_T$$

[compare with $\sigma^2 = \omega_0^2 + \frac{3}{2}\alpha D_T$ in Eq. (13.28)].

13.2.3. Fokker–Planck equation for an elastic system

The behavior of many elastic systems subjected to random dynamic loading can be characterized by the following equations:

$$\ddot{z}_k + 2r_k \dot{z}_k + f_k(z_1, z_2, \ldots, z_n) = \frac{Q_k(t)}{M_k}, \qquad k = 1, 2, \ldots, n \quad (13.31)$$

Here z_k is the principal coordinate, r_k is the coefficient of damping, $Q_k(t)$ is the excitation force, M_k is the mass of the kth mode, and $f_k(z_1,$

$z_2, \ldots, z_n)$ is the restoring force. Equation (13.3) provides an example of such equations. Treating the functions z_k and $\dot{z}_k$ as independent generalized coordinates, we have

$$A_{z_k} = \lim_{\Delta t \to 0} \frac{\langle z_{k\tau} - z_k \rangle}{\Delta t} = \lim_{\Delta t \to 0} \frac{\langle \Delta z_k \rangle}{\Delta t} = \dot{z}_k$$

$$A_{\dot{z}_k} = \lim_{\Delta t \to 0} \frac{\langle \dot{z}_{k\tau} - \dot{z}_k \rangle}{\Delta t} = \lim_{\Delta t \to 0} \frac{\langle \Delta \dot{z}_k \rangle}{\Delta t}$$

$$= \lim_{\Delta t \to 0} \left\{ \frac{1}{\Delta t} \left\langle \left[-2r_k \dot{z}_k - f_k(z_1, z_2, \ldots, z_n) \right] \Delta t + \int_t^{t+\Delta t} \frac{Q_{k(t)}\, dt}{M_k} \right\rangle \right\}$$

$$= -2r_k \dot{z}_k - f_k(z_1, z_2, \ldots, z_n)$$

$$B_{z_k z_k} = \lim_{\Delta t \to 0} \frac{\langle (z_{k\tau} - z_k)^2 \rangle}{\Delta t} = \lim_{\Delta t \to 0} \frac{\langle (\Delta z_k)^2 \rangle}{\Delta t} = \lim_{\Delta t \to 0} (\Delta t \dot{z}_k^2) = 0$$

$$B_{z_k \dot{z}_l} = B_{\dot{z}_l z_k} = \lim_{\Delta t \to 0} \frac{\langle (z_{k\tau} - z_k)(\dot{z}_{l\tau} - \dot{z}_l) \rangle}{\Delta t} = \lim_{\Delta t \to 0} (\Delta t \dot{z}_k \langle \Delta \dot{z}_l \rangle) = 0$$

$$B_{\dot{z}_k \dot{z}_l} = \lim_{\Delta t \to 0} \frac{\langle (\dot{z}_{k\tau} - \dot{z}_k)(\dot{z}_{l\tau} - \dot{z}_l) \rangle}{\Delta t}$$

$$= \lim_{\Delta t \to 0} \left\{ \frac{1}{\Delta t} \left\langle \left[-2r_k \dot{z}_k - f_k(z_1, z_2, \ldots, z_n) \Delta t + \int_t^{t+\Delta t} \frac{Q_k(t)\, dt}{M_k} \right] \right. \right.$$

$$\left. \left. \times \left[-2r_l \dot{z}_l - f_l(z_1, z_2, \ldots, z_n) \Delta t + \int_t^{t+\Delta t} \frac{Q_l(t)\, dt}{M_l} \right] \right\rangle \right\}$$

$$= \frac{\pi}{M_k M_l} S_{Q_k Q_l}$$

After substituting the above relationships into Eq. (11.43), we obtain

$$\frac{\partial}{\partial t} p(z_1, z_2, \ldots, z_n; \dot{z}_1, \dot{z}_2, \ldots, \dot{z}_n)$$

$$= -\sum_{k=1}^n \frac{\partial}{\partial z_k}(\dot{z}_k p) + \sum_{k=1}^n \frac{\partial}{\partial \dot{z}_k}\{[2r_k \dot{z}_k + f_k(z_1, z_2, \ldots, z_n)]p\}$$

$$+ \frac{1}{2} \sum_{k=1}^n \sum_{l=1}^n \frac{\partial^2}{\partial \dot{z}_k \partial \dot{z}_l}\left(\frac{\pi S_{Q_k Q_l}}{M_k M_l} p\right)$$

$$= -\sum_{k=1}^n \dot{z}_k \frac{\partial p}{\partial z_k} + \sum_{k=1}^n 2r_k\left(p + \dot{z}_k \frac{\partial p}{\partial \dot{z}_k}\right) + \sum_{k=1}^n \frac{\partial}{\partial \dot{z}_k}[f_k(z_1, z_2, \ldots, z_n)p]$$

$$+ \frac{\pi}{2} \sum_{k=1}^n \sum_{l=1}^n \frac{S_{Q_k Q_l}}{M_k M_l} \frac{\partial^2 p}{\partial \dot{z}_k \partial \dot{z}_l} \tag{13.32}$$

This is the Fokker–Planck equation for a nonlinear elastic system. For a stationary process, $p = p(z_1, z_2, \ldots, z_n; \dot{z}_1, \dot{z}_2, \ldots, \dot{z}_n)$, this equation can be simplified:

$$-\sum_{k=1}^{n} \dot{z}_k \frac{\partial p}{\partial z_k} + \sum_{k=1}^{n} 2r_k\left(p + \dot{z}_k \frac{\partial p}{\partial \dot{z}_k}\right) + \sum_{k=1}^{n} \frac{\partial}{\partial \dot{z}_k}[f_k(z_1, z_2, \ldots, z_n)p]$$

$$+ \frac{\pi}{2} \sum_{k=1}^{n} \sum_{l=1}^{n} \frac{S_{Q_k Q_l}}{M_k M_l} \frac{\partial^2 p}{\partial \dot{z}_k \partial \dot{z}_l} = 0 \quad (13.33)$$

13.2.4. Fokker–Planck equation for a one-degree-of-freedom nonlinear system

Comparing Eq. (13.23) with Eq. (13.31) and considering the change in the notation for the principal coordinate, we conclude that the restoring force function is expressed as

$$f_k(z_1, z_2, \ldots, z_n) = \lambda^2 z + \alpha z^3$$

and the generalized excitation force is

$$Q_k(t) = M_k q(t)$$

Then, considering that the probability $p = p(z, \dot{z})$ is time-independent, Eq. (13.33) yields

$$-\dot{z} \frac{\partial p}{\partial z} + 2r\left(p + \dot{z} \frac{\partial p}{\partial \dot{z}}\right) + \frac{\partial}{\partial \dot{z}}[(\lambda^2 z + \alpha z^3)p] + \frac{\pi}{2} \frac{S_Q}{M^2} \frac{\partial^2 p}{\partial \dot{z}^2} = 0$$

This equation can be rewritten as

$$\frac{\partial}{\partial \dot{z}}\left[(\lambda^2 z + \alpha z^3)p + \frac{\pi S_Q}{4M^2 r} \frac{\partial p}{\partial z}\right] - \left(\frac{\partial}{\partial z} - 2r \frac{\partial}{\partial \dot{z}}\right)\left(\dot{z}p + \frac{\pi S_Q}{4M^2 r} \frac{\partial p}{\partial \dot{z}}\right) = 0$$

and is fulfilled if the following relationships take place:

$$(\lambda^2 z + \alpha z^3)p + \frac{\pi S_Q}{4M^2 r} \frac{\partial p}{\partial z} = 0 \quad (13.34)$$

$$\dot{z}p + \frac{\pi S_Q}{4M^2 r} \frac{\partial p}{\partial \dot{z}} = 0 \quad (13.35)$$

Equations (13.34) and (13.35) can be easily integrated:

$$p_1(z) = \frac{1}{C_1} \exp\left[-\frac{2M^2 r \lambda^2}{\pi S_Q}\left(z^2 + \frac{\alpha}{2\lambda^2} z^4\right)\right] \quad (13.36)$$

$$p_2(\dot{z}) = \frac{1}{C_2} \exp\left(-\frac{2M^2 r}{\pi S_Q} \dot{z}^2\right) \quad (13.37)$$

The constants C_1 and C_2 of integration could be obtained from the conditions of normalization for the functions $p_1(z)$ and $p_2(\dot{z})$:

$$C_1 = \int_{-\infty}^{\infty} \exp\left[-\frac{2M^2 r\lambda^2}{\pi S_Q}\left(z^2 + \frac{\alpha}{2\lambda^2}z^4\right)\right] dz$$

$$= \frac{\lambda}{\sqrt{2\alpha}} \exp\left(\frac{M^2 r\lambda^4}{2\pi S_Q \alpha}\right) K_{1/4}\left(\frac{M^2 r\lambda^4}{2\pi S_Q \alpha}\right) \qquad (13.38)$$

$$C_2 = \int_{-\infty}^{\infty} \exp\left(-\frac{M^2 r}{\pi S_Q}\dot{z}^2\right) d\dot{z} = \pi\sqrt{\frac{S_Q}{M^2 r}} \qquad (13.39)$$

where $K_\nu(z)$ is the Kelvin function. In these equations the following formula was used:

$$\int_0^\infty e^{-\mu x^4 - 2\nu x^2} dx = \frac{1}{4}\sqrt{\frac{2\nu}{\mu}} \exp\left(\frac{\nu^2}{2\mu}\right) K_{1/4}\left(\frac{\nu^2}{2\mu}\right) \qquad (13.40)$$

(see, for instance, I. S. Gradshteyn and I. M. Ryzhik, "Table of Integrals, Series, and Products," Academic Press, 1980). Clearly, the integral in the formula (13.38) can always be evaluated numerically.

Let us determine the probability that the process $z(t)$ will exceed a level z^*. The two-dimensional probability density function for the process $z(t)$ and its time derivative $\dot{z}(t)$ is

$$p(z, \dot{z}) = p_1(z)p_2(\dot{z}) = \frac{1}{C_1 C_2} \exp\left[-\frac{Mr^2}{\pi S_Q}\left(\lambda^2 z^2 + \frac{\alpha}{2}z^4 + \dot{z}^2\right)\right] \qquad (13.41)$$

Then

$$p\{z > z^*\} = \frac{N_{z^*}}{N_0} = \frac{\int_0^\infty \dot{z} p(z^*, \dot{z}) \, d\dot{z}}{\int_0^\infty \dot{z} p(0, \dot{z}) \, d\dot{z}} = \frac{p_1(z^*)}{p_2(0)}$$

$$= \exp\left[-\frac{M^2 r}{\pi S_Q}\left(\lambda^2 z^{*2} + \frac{\alpha}{2}z^{*4}\right)\right] \qquad (13.42)$$

Dynamic systems for which one can find an exact solution to the Fokker–Planck equation are rarely encountered, and therefore each system that lends itself to such a solution is often used as an etalon for the assessment of the accuracy of various approximate solutions.

Example 13.7 Using the formula (13.42), calculate the probabilities of exceeding a level z^* for the case

$$\frac{\pi S_Q}{M^2 r \omega^2} = 1 \text{ cm}^2, \qquad \frac{\alpha}{\omega^2} = 8 \text{ cm}^{-3}$$

Compare the obtained results with the results based on the linear approach ($\alpha = 0$) and the method of statistical linearization.

Solution The calculated probabilities are shown in Table 13.1. The table gives an idea of the difference in results obtained using the method of statistical linearization and the method based on the Fokker–Planck equation.

Example 13.8 (Svetlitsky, 1976) An aircraft of mass M is moving horizontally under the action of the thrust $F = F_0 + \Delta F$, where F_0 is the nominal thrust and ΔF is its random component. The equation of motion of the aircraft can be written as

$$M\dot{v} + R(v) = F_0 + \Delta F$$

where v is its speed and $R(v)$ is the drag force. Assuming that the random component ΔF of the thrust can be approximated by a "white noise" with zero mean and has a correlation function

$$K_{\Delta F} = \sigma_{\Delta F}^2 \, \delta(t - t')$$

determine the steady-state distribution of the probability $p(v_1)$ of the random component v_1 of the aircraft's speed $v = v_0 + v_1$.

Solution Expanding the drag force function $R(v)$ in the Taylor series we have

$$R(v) = R(v_0) + v_1 \left.\frac{\partial R}{\partial v}\right|_{v=v_0} + \frac{v_1^3}{6} \left.\frac{\partial^3 F}{\partial v^3}\right|_{v=v_0}$$

This expansion reflects a natural assumption that the drag $R(v)$ is an odd function, and therefore only the odd terms should be considered in the

TABLE 13.1 Probabilities of exceeding the given level: predictions based on different approaches

z^*, cm	Linear system ($\alpha = 0$)	Method of statistical linearization	Method of Fokker–Planck equation
0	1	1	1
0.2	0.961	0.887	0.955
0.5	0.779	0.472	0.606
1.0	0.368	0.050	0.0067
1.5	0.105	0.001	0

Taylor series. Then the equation of motion of the aircraft can be written as

$$\dot{v}_1 + R_1(v_1) = \varepsilon(t) \tag{13.43}$$

where
$$R_1 = \frac{1}{M}\left(v_1 \left.\frac{\partial R}{\partial v}\right|_{v=v_0} + \frac{v_1^3}{6} \left.\frac{\partial^3 R}{\partial v^3}\right|_{v=v_0}\right)$$

and the function

$$\varepsilon(t) = \frac{\Delta F}{M}$$

can be viewed as a random excitation.

The integration of Eq. (13.43) from t to $t + \Delta t$ yields

$$\Delta v_1 = -R_1 \Delta t + \int_t^{t+\Delta t} \varepsilon(t)\, dt$$

The coefficients A and B in the Fokker–Planck equation can be determined, using this formula, as follows:

$$A = \lim_{\Delta t \to 0} \frac{1}{\Delta t} \langle \Delta v_1 \rangle = -R_1(v_1)$$

$$B = \lim_{\Delta t \to 0} \frac{1}{\Delta t} \langle \Delta v_1(t)\, \Delta v_1(t') \rangle = \sigma^2 = \frac{\sigma_{\Delta F}^2}{M^2}$$

Then the Fokker–Planck equation for the probability $p = p(v_1)$ can be written as

$$\frac{\sigma^2}{2} \frac{d^2 p}{\partial v_1^2} + \frac{d}{dv_1}[R_1(v_1)p] = 0$$

or, after integration,

$$\frac{\sigma^2}{2} \frac{dp}{dv_1} + R_1(v_1)p = C_1$$

where C_1 is the constant of integration. This constant can be found, based on the fact that the probability p of very large fluctuations of the speed v_1 is small, and so is its derivative dp/dv_1. Therefore, one can put $C_1 = 0$ and the solution to the obtained equation is

$$p(v_1) = C \exp\left[-\frac{2}{\sigma^2} \int_0^{v_1} R_1(v_1)\, dv_1\right] \tag{13.44}$$

where C is a new constant of integration. This constant can be found from the condition of normalization

$$\int_{-\infty}^{\infty} p(v_1)\, dv_1 = 1$$

If, for instance,
$$R_1 = \alpha_1 v_1 + \alpha_2 v_1^3$$
then the formula (13.44) yields
$$p(v_1) = C \exp\left[-\frac{2}{\sigma^2}\left(\frac{\alpha_1}{2} v_1^2 + \frac{\alpha_2}{4} v_1^4\right)\right]$$
and the variance of the speed v_1 is
$$D_{v_1} = C \int_{-\infty}^{\infty} v_1^2 \exp\left[-\frac{2}{\sigma^2}\left(\frac{\alpha_1}{2} v_1^2 + \frac{\alpha_2}{4} v_1^4\right)\right] dv_1$$

The constant C, obtained from the normalization condition, is as follows:
$$C = \frac{1}{\int_{-\infty}^{\infty} \exp\left[-\frac{2}{\sigma^2}\left(\frac{\alpha_1}{2} v_1^2 + \frac{\alpha_2}{4} v_1^4\right)\right] dv_1}$$

The integral in this formula can be evaluated numerically or, using Eq. (13.43), can be expressed as
$$\int_{-\infty}^{\infty} \exp\left[-\frac{2}{\sigma^2}\left(\frac{\alpha_1}{2} v_1^2 + \frac{\alpha_2}{4} v_1^4\right)\right] dv_1 = \frac{1}{4}\sqrt{\frac{2\alpha_1}{\alpha_2}} \exp\left(\frac{\alpha_1^2}{2\sigma^2 \alpha_2}\right) K_{1/4}\left(\frac{\alpha_1^2}{2\sigma^2 \alpha_2}\right)$$
where $K_v(z)$ is the Kelvin function (see earlier in Sec. 13.2.4).

Example 13.9 (Svetlitsky, 1976) A very light piston whose mass is so small that the inertia force need not be considered is subjected to a constant force F_0 and, in addition, to a random force ΔF, so that the motion of the piston can be described by the equation
$$R\dot{x} = F_0 + \Delta F(t)$$
where R is the coefficient of viscous damping. The random force ΔF can be approximated by a "white noise" $n(t)$ with zero mean and a correlation function $K_{\Delta F} = \sigma^2 \delta(\tau)$. At the initial moment of time $t = 0$, the piston is located in the middle of a cylinder whose total length is $2l$. What is the probability that the piston will not hit the end walls of the cylinder during the time t?

Solution Using the same approach as in the solution to the previous problem, we find that the coefficients of the Fokker–Planck equation are
$$A = \frac{F_0}{R}, \quad B = \frac{\sigma^2}{R^2}$$
so that this equation can be written as
$$\frac{\partial p}{\partial t} + \frac{F_0}{R}\frac{\partial p}{\partial x} - \frac{\sigma^2}{2R^2}\frac{\partial^2 p}{\partial x^2} = 0 \quad (13.45)$$

Seeking the solution to this equation in the form

$$p(x, t) = X(x)T(t)$$

we have

$$X(x)\dot{T}(t) + \frac{F_0}{R} X'(x)T(t) - \frac{\sigma^2}{2R^2} X''(x)T(t) = 0$$

or

$$\frac{\dot{T}(t)}{T(t)} = -\frac{F_0}{R} \frac{X'(x)}{X(x)} + \frac{\sigma^2}{2R^2} \frac{X''(x)}{X(x)} = \lambda^2 = \text{constant}$$

Since the left-hand part of this equation is x independent and the right part is t independent, each of these parts must be equal to the same constant value λ^2. Thus, we have the following two equations: the equation

$$\dot{T}(t) + \lambda^2 T(t) = 0$$

for the time function (principle coordinate) $T(t)$ and the equation

$$X''(x) - \frac{2F_0 R}{\sigma^2} X'(x) + \frac{2R^2\lambda^2}{\sigma^2} X(x) = 0$$

for the location function $X(x)$. These equations have the following solutions:

and
$$T(t) = C_1 e^{-\lambda^2 t}$$
$$X(x) = e^{(F_0 R/\sigma^2)x}(C_1 \cos \lambda_1 x + C_2 \sin \lambda_1 x)$$

where C_1 and C_2 are the constants of integration and

$$\lambda_1 = \sqrt{\frac{2R^2\lambda^2}{\sigma^2} - \frac{F_0^2 R^2}{\sigma^4}} = \frac{R}{\sigma}\sqrt{2\lambda^2 - \frac{F_0^2}{\sigma^2}}$$

is the frequency of the piston's oscillations.

Clearly, the piston will never hit the end walls of the cylinder if the probability density function $p(x, t)$ is zero at $x = \pm l$. This leads to the following equations for the constants C_1 and C_2:

$$C_1 \cos \lambda_1 l + C_2 \sin \lambda_1 l = 0$$
$$C_1 \cos \lambda_1 l - C_2 \sin \lambda_1 l = 0$$

This system of homogeneous equations yields $C_2 = 0$ and $\cos \lambda_1 l = 0$, so that

$$\lambda_1 l = \frac{2k+1}{2}\pi$$

or

$$\lambda_1 = \frac{2k+1}{2}\frac{\pi}{l}$$

430   Chapter Thirteen

With this λ_1 value, we obtain

$$\lambda_k^2 = \frac{\sigma^2}{2R^2}\left[\left(\frac{2k+1}{2}\right)^2 \frac{\pi^2}{l^2} + \frac{R^2 F_0^2}{\sigma^4}\right]$$

Thus, the solution to the governing Fokker–Planck equation (13.45) can be written as follows:

$$p(x, t) = e^{(RF_0/\sigma^2)x} \sum_{k=0}^{\infty}\left(C_k \cos\frac{2k+1}{2}\frac{\pi x}{l} e^{-\lambda_k^2 t}\right)$$

At the initial moment of time $t = 0$, the position of the piston is defined, and therefore the initial probability density function is a delta function: $p(x, 0) = \delta(x)$. Hence,

$$\sum_{k=0}^{\infty}\left(C_k \cos\frac{2k+1}{2}\frac{\pi x}{l}\right) = \delta(x) e^{-(RF_0/\sigma^2)x}$$

In order to obtain the expression for the coefficients C_k, we multiply both parts of this equation by $\cos[(2k+1)/2](\pi x/l)$ and integrate over the cylinder length, i.e., from $-l$ to l. This results in the equation

$$C_k \int_{-l}^{l} \cos^2\frac{2k+1}{2}\frac{\pi x}{l}\,dx = \int_{-l}^{l} \delta(x) e^{-(RF_0/\sigma^2)x} \cos\frac{2k+1}{2}\frac{\pi x}{l}\,dx$$

which yields

$$C_k = \frac{1}{l}$$

Hence, the solution to the Fokker–Planck equation is

$$p(x, t) = \frac{1}{l} e^{(RF_0/\sigma^2)x} \sum_{k=0}^{\infty} e^{-\lambda_k^2 t} \cos\frac{2k+1}{2}\frac{\pi x}{l}$$

The probability that the piston will remain within the cylinder during the time t, i.e., will not hit its end walls, can be found as

$$P(t) = \frac{1}{l}\int_{-l}^{l} e^{(RF_0/\sigma^2)x} \sum_{k=0}^{\infty}\left(e^{-\lambda_k^2 t}\cos\frac{2k+1}{2}\frac{\pi x}{l}\right) dx$$

$$= \frac{2}{l}\cos\frac{RF_0}{\sigma^2} l \sum_{k=0}^{\infty} e^{-\lambda_k^2 t}(-1)^{k+2} \frac{\frac{2k+1}{2}\frac{\pi}{l}}{\left(\frac{RF_0}{\sigma^2}\right)^2 + \left(\frac{2k+1}{2}\frac{\pi}{l}\right)^2}$$

The series in the obtained solution converges well and can be used for practical calculations.

Example 13.10 Using the solution obtained in the previous example, determine the mean time required for the piston to reach the end walls of the cylinder.

Solution In the previous example we obtained an expression for the probability $P(t)$ that the piston does not hit the end walls during the time t. By differentiating the obtained expression with respect to this time, one can obtain the probability density function $f(t) = dP(t)/dt$. The mean time of reaching the boundaries (walls) can then be calculated as

$$\langle t \rangle = \int_0^\infty t f(t) \, dt = -tP \Big|_{t=0}^\infty + \int_0^\infty P(t) \, dt = \int_0^\infty P(t) \, dt$$

$$= \frac{2}{l} \cosh \frac{RF_0}{\sigma^2} l \sum_{k=0}^\infty e^{-\lambda_k^2 t} \frac{(-1)^{k+2}}{\lambda_k^2} \frac{\dfrac{2k+1}{2} \dfrac{\pi}{l}}{\left(\dfrac{RF_0}{\sigma^2}\right)^2 + \left(\dfrac{2k+1}{2} \dfrac{\pi}{l}\right)^2}$$

This series also converges well and can be used for practical computations.

13.3. Vibrations Caused by Periodic Impulses: Stochastic Instability

13.3.1. Dynamic response of a one-degree-of-freedom system to a train of instantaneous periodic impulses

Vibrations caused by periodic impulses have numerous applications in engineering and physics. Examples are: functioning of diesel engines, rolling mills, gear teeth, punch presses, pneumatic tools (such as hammers, miner's hacks and machines, chisels, etc.); performance of clock mechanisms; machine gun and rapid-shooting artillery fire; vortex separation due to unfavorable shapes of civil, aircraft, space and ocean structures (aero- and hydrodynamic flutter); ship slamming in regular seas; working of synchrocyclotrons and other accelerators of charged particles; heart beating, walking, running, jumping, etc.

Shock-excited vibrations caused by periodic instantaneous impulses can be described by the following equation:

$$\ddot{T} + 2r\dot{T} + \omega_0^2 T + \alpha T^3 = \frac{P}{M} \sum_{k=0}^n \delta(t - kT_s), \qquad n, 1, 2, \ldots \quad (13.46)$$

where $T(t)$, $\dot{T}(t)$ and $\ddot{T}(t)$ are the displacement, the velocity, and the acceleration, respectively, ω_0 is the frequency of undamped linear vibrations, r is the coefficient of damping, α is the parameter of non-linearity, P is the magnitude of the impulses, M is the mass of the system, T_s is the time between the impulses (the period of the impulse train), t is time, and $\delta(t)$ is Dirac's delta function (see Example 7.11, Sec. 7.3). The assumed type of nonlinearity (Duffing oscillator) is widespread in engineering and physics.

The fundamental result of the theory in question was obtained by G. Duffing, who derived the following formula for the dimensionless steady-state amplitude of undamped linear vibrations:

$$A = \tfrac{1}{2}|\cosec \theta_0| \tag{13.47}$$

Here A is the ratio of the amplitude of vibrations caused by periodic impulses to the amplitude $\bar{A}_1 = P/(M\omega_0)$ of vibrations due to a single impulse, and the phase angle θ_0 is found as

$$\theta_0 = \tfrac{1}{2}\omega_0 T_s \tag{13.48}$$

Duffing's derivation was based on the conditions of periodicity:

$$T(0) = T(T_s), \qquad \dot{T}(0) = \dot{T}(T_s) + \frac{P}{M} \tag{13.49}$$

These conditions mean that the displacement $T(t)$ does not change as a consequence of the application of the impulse, while the velocity $\dot{T}(t)$ experiences a step P/M.

The following two practically important questions arise in connection with Duffing's solution:

1. Can the most severe conditions of the system's performance (say, maximum amplitudes or accelerations) occur during the transient period of vibrations?
2. Are the obtained solutions stable? If they are not, how does such an instability affect the system's response and how shouldone describe the system's behavior in the state of unstable vibrations?

13.3.2. Linear vibrations

In the linear case ($\alpha = 0$) the solution to Eq. (13.46) can be found by using the Duhamel integral (7.16):

$$T(t) = \int_0^t Q(\xi) h(t - \xi)\, d\xi \tag{13.50}$$

Here the excitation force $Q(\xi)$ and the impulse response function $h(t-\xi)$ are expressed by the formulas:

$$Q(t) = P \sum_{k=0}^{n} \delta(t - kT_s)$$

$$h(t - \xi) = \frac{1}{M\omega} e^{-r(t-\xi)} \sin \omega(t - \xi)$$
(13.51)

where $\omega\sqrt{\omega_0^2 - r^2}$ is the frequency of damped vibrations. After substituting Eqs (13.51) into Eq. (13.50) we obtain

$$T(t) = \frac{P}{M\omega} \int_0^t e^{-r(t-\xi)} \sum_{k=0}^{n} \delta(\xi - kT_s) \sin \omega(t - \xi) \, d\xi$$

$$= \frac{P}{M\omega} \sum_{k=0}^{n} \int_0^t e^{-r(t-\xi)} \delta(\xi - kT_s) \sin \omega(t - \xi) \, d\xi$$

Since, for any function $f(\xi)$,

$$\int_0^t \delta(\xi - a) f(\xi) \, d\xi = f(a)$$

(see Example 7.11, Sec. 7.3), Eq. (13.50) results in the following formula for the displacement:

$$T(t) = \frac{P}{M\omega} \sum_{k=0}^{n} e^{-r(t-kT_s)} \sin \omega(t - kT_s)$$
(13.52)

This equation provides a solution in "global" time, counted from the moment of application of the first ($n = 0$) impulse. In order to obtain a solution in "local" time, counted from the moment of application of the nth impulse, we substitute in Eq. (13.52) the value $t + nT_s$ for t:

$$T(t) = \frac{P}{M\omega} \sum_{k=0}^{n} e^{-r[t+(n-k)T_s]} \sin \omega[t + (n-k)T_s]$$

$$= \frac{P}{M\omega} \sum_{i=0}^{n} e^{-r(t+iT_s)} \sin \omega(t + iT_s)$$

$$= \frac{P}{M\omega} e^{-rt} \sum_{k=0}^{n} (C_n \sin \omega t + D_n \cos \omega t)$$

$$= \frac{P}{M\omega} e^{-rt} A_n \cos (\omega t + \varepsilon_n)$$
(13.53)

Here the coefficients C_n and D_n are expressed as

$$C_n = \sum_{k=0}^{n} e^{-krT_s} \cos k\omega T_s$$

$$= \frac{e^{rT_s} - \cos \omega T_s - e^{-nrT_s} \cos (n+1)\omega T_s + e^{-(n+1)rT_s} \cos n\omega T_s}{2(\cosh rT_s - \cos \omega T_s)}$$
(13.54)

$$D_n = \sum_{n=0}^{n} e^{-krT_s} \sin k\omega T_s$$

$$= \frac{\sin \omega T_s - e^{-nrT_s} \sin (n+1)\omega T_s + e^{-(n+1)rT_s} \sin n\omega T_s}{2(\cosh rT_s - \cos \omega T_s)}$$

The amplitude and the phase angle can be calculated by the formulas:

$$A_n = \sqrt{C_n^2 + D_n^2}$$

$$= \sqrt{e^{-rT_s} \frac{\cosh (n+1)rT_s - \cos (n+1)\omega T_s}{\cosh rT_s - \cos \omega T_s}}$$

$$= \sqrt{\frac{1 + e^{-2(n+1)rT_s} - 2e^{-(n+1)rT_s} \cos (n+1)\omega T_s}{1 + e^{-2rT_s} - 2e^{-rT_s} \cos \omega T_s}}$$
(13.55)

$$\varepsilon_n = -\arctan \frac{C_n}{D_n}$$

$$= -\arctan \frac{e^{rT_s} - \cos \omega T_s - e^{-nrT_s} \cos (n+1)\omega T_s + e^{-(n+1)rT_s} \cos n\omega T_s}{\sin \omega T_s - e^{-nrT_s} \sin (n+1)\omega T_s + e^{-(n+1)rT_s} \sin \omega T_s}$$
(13.56)

In the special case of undamped vibrations ($t = 0$), we have

$$A_n = |\operatorname{cosec} \theta_0 \sin (n+1)\theta_0|, \qquad \varepsilon_n = -\frac{\pi}{2} + n\theta_0, \qquad n = 0, 1, 2, \ldots$$
(13.57)

These equations show that, with an increase in n, the amplitude of nonresonant undamped vibrations does not approach any finite limit at all, but changes periodically with the "frequency" θ_0 within the range (0, cosec θ_0). When $\theta_0 = m\pi$, $m = 1, 2, 3, \ldots$, an impulse resonance takes place, and the dimensionless amplitude becomes $A_n = n$. Hence, in the resonance condition, the amplitude of the undamped vibrations increases linearly, without any limit, with an n-fold increase after the nth impulse.

In another special case, when $n \to \infty$, we obtain

$$A_n = A_\infty = (1 + e^{-2rT_s} - 2e^{-rT_s} \cos \omega T_s)^{-1/2}$$

$$\varepsilon_n = \varepsilon_\infty = -\arctan \frac{e^{rT_s} - \cos \omega T_s}{\sin \omega T_s} \quad (13.58)$$

It is evident from these equations that damped vibrations, unlike undamped ones, achieve a steady-state condition in the time domain. When damping is small and has an effect only on the final steady-state character of the vibrations, but not on the magnitude of the amplitude and the phase angle, these may be obtained by simply putting $rT_s = 0$ in the solutions (13.58):

$$A_\infty \simeq \tfrac{1}{2} |\text{cosec } \theta_0|, \quad \varepsilon_\infty \simeq -\theta_0 \quad (13.59)$$

Hence, the amplitude of the steady-state nonresonant slightly damped vibrations turns out to be half the maximum amplitude in the undamped case. The first formula in (13.59) coincides with the Duffing formula (13.47).

In the third special case, when $\theta = \tfrac{1}{2}\omega T_s = m\pi$, $m = 1, 2, 3, \ldots$ (impulse resonance in a system with damping), we obtain

$$A_n = \frac{1 - e^{-(n+1)rT_s}}{1 - e^{-rT_s}}, \quad \varepsilon_n = -\frac{\pi}{2} \quad (13.60)$$

For steady-state vibrations ($n \to \infty$),

$$A_\infty = \frac{1}{1 - e^{-rT_s}} \quad (13.61)$$

Minimum amplitudes occur when $\theta = [(2m + 1)/2]\pi$, $m = 1, 2, 3, \ldots$. In this case,

$$A_n = \frac{1 + e^{-(n+1)rT_s}}{1 + e^{-rT_s}}, \quad \varepsilon_n = -\frac{\pi}{2} \quad (13.62)$$

The minimum steady-state amplitude ($n \to \infty$) is

$$A_\infty = \frac{1}{1 + e^{-rT_s}} \quad (13.63)$$

Some numerical results are plotted in Fig. 13.2. As evident from this figure, the steady-state amplitudes at resonance conditions always exceed the transient ones, while the transient amplitudes at nonresonance conditions may considerably exceed the steady-state ones. This is important to keep in mind when choosing the most

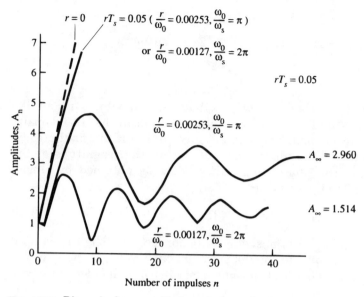

Figure 13.2 Dimensionless amplitudes of damped vibrations due to periodic impulses

severe conditions for engineering evaluations, whether deterministic or probabilistic. In the case of small damping, the ratio of the maximum transient amplitude to the steady-state amplitude, obtained on the basis of the formulas (13.57) and (13.59), is

$$\frac{A_{n,\,\max}}{A_\infty} = 2|\sin(n+1)\theta_0| \qquad (13.64)$$

Hence, the maximum transient nonresonant amplitude can be twice as large as the steady-state amplitude.

13.3.3. Nonlinear steady-state undamped vibrations

The (exact) steady-state solution to the nonlinear equation

$$\ddot{T} + \omega_0^2 T + \alpha T^3 = \frac{P}{M} \sum_{k=0}^{\infty} \delta(t - kT_s) \qquad (13.65)$$

can be sought in the form

$$T = A\,\frac{P}{M\omega_0}\,\mathrm{cn}\,(\alpha t + \varepsilon_s,\,k_s) = A\,\frac{P}{M\omega_0}\,\mathrm{cn}\,u \qquad (13.66)$$

where cn u is an elliptic cosine, k_s is the modulus of an elliptic function, ε is the initial phase angle, and σ is the parameter of the nonlinear frequency. Using the formulas

$$(\text{cn } u)' = -\text{sn } u \text{ dn } u, \quad (\text{sn } u)' = \text{cn } u \text{ dn } u, \quad (\text{dn } u)' = -k_s^2 \text{ sn } u \text{ cn } u$$

for differentiating elliptic functions, we obtain

$$\dot{T} = -A \frac{P}{M\omega_0} \sigma \text{ sn } u \text{ dn } u \tag{13.67}$$

$$\ddot{T} = -A \frac{P}{M\omega_0} \sigma^2 \text{ cn } u(1 - 2k_s^2 \text{ sn}^2 u) \tag{13.68}$$

where sn u is an elliptic sine and dn $u = \sqrt{1 - k_s^2 \text{ sn}^2 u}$ is the function of delta amplitude. After substituting Eqs (13.66) and (13.68) in Eq. (13.65), we obtain

$$\sigma^2 = \omega_0^2 + \alpha A^2 \left(\frac{P}{M\omega_0}\right)^2 = \omega_0^2(1 - \bar{\alpha}A^2) \tag{13.69}$$

$$k_s^2 = \frac{\alpha A^2}{2\sigma^2} \left(\frac{P}{M\omega_0}\right)^2 = \frac{1}{2[1 + 1/(\bar{\alpha}A^2)]} \tag{13.70}$$

where

$$\bar{\alpha} = \alpha \left(\frac{P}{M\omega_0^2}\right)^2 \tag{13.71}$$

is the dimensionless parameter of nonlinearity.

Introducing Eqs (13.66) and (13.67) in the periodicity conditions (13.49), we have

$$\text{cn }(\varepsilon, k_s) = \text{cn }(\sigma T_s + \varepsilon, k_s) \tag{13.72}$$

$$-A\sigma \text{ sn }(\varepsilon, k_s) \text{ dn }(\varepsilon, k_s) = -A\sigma \text{ sn }(\sigma T_s + \varepsilon, k_s) \text{ dn }(\sigma T_s + \varepsilon, k_s) + \omega_0 \tag{13.73}$$

Equation (13.72) yields

$$\text{sn }(\varepsilon, k_s) = \pm \text{sn }(\sigma T_s + \varepsilon, k_s), \quad \text{dn }(\varepsilon, k_s) = \mp \text{dn }(\sigma T_s + \varepsilon, k_s) \tag{13.74}$$

Since it is known from the theory of elliptic functions that

$$\left. \begin{array}{l} \text{sn }(u + v) = \dfrac{\text{sn } u \text{ cn } v \text{ dn } v + \text{sn } v \text{ cn } u \text{ dn } u}{1 - k_s^2 \text{ sn}^2 u \text{ sn}^2 v} \\[2ex] \text{cn }(u + v) = \dfrac{\text{cn } u \text{ cn } v - \text{sn } v \text{ dn } u \text{ dn } v}{1 - k_s^2 \text{ sn}^2 u \text{ sn}^2 v} \end{array} \right\}$$

we find, by putting $u = \varepsilon$ and $v = \sigma T_s + \varepsilon$, that

$$\operatorname{sn}(\sigma T_s + 2\varepsilon, k_s) = 0, \qquad \operatorname{cn}(\sigma T_s + 2\varepsilon, k_s) = 1 \qquad (13.75)$$

Then we obtain the following equation for the amplitude of the elliptic function:

$$\phi = \operatorname{am}(\sigma T_s + 2\varepsilon, k_s) = 2\pi i, \qquad i = 0, \pm 1, \pm 2, \ldots \qquad (13.76)$$

and therefore

$$\sigma T_s + 2\varepsilon = \int_0^{2\pi i} \frac{d\varepsilon}{\sqrt{1 - k_s^2 \sin^2 \zeta}} = 4iK(k_s) \qquad (13.77)$$

where

$$K(k_s) = \int_0^{\pi/2} \frac{d\xi}{\sqrt{1 - k_s^2 \sin^2 \zeta}} \qquad (13.78)$$

is a complete elliptic integral of the first kind. From Eq. (13.77) we obtain the following formula for the initial phase angle:

$$\varepsilon = 2iK(k_s) - \tfrac{1}{2}\sigma T_s, \qquad i = 0, \pm 1, \pm 2, \ldots \qquad (13.79)$$

In order to find the amplitude A of vibrations, one can substitute the formulas (13.74) into (13.73). This yields

$$A = \frac{\omega_0}{2\sigma F} \qquad (13.80)$$

where the function F is expressed as

$$F = F(\varepsilon, k_s) = \operatorname{sn}(\varepsilon, k_s) \operatorname{dn}(\varepsilon, k_s) \qquad (13.81)$$

Introducing the σ value from Eq. (13.69) into Eq. (13.80) and solving the obtained equation for the dimensionless amplitude A, we obtain

$$A = \sqrt{\frac{\sqrt{1 + \bar{\alpha}/F^2} - 1}{2\bar{\alpha}}} \qquad (13.82)$$

Since the period of the elliptic cosine $\operatorname{cn}(t, k_s)$ is $4K(k_s)$, the period of the function $\operatorname{cn}(\sigma t + \varepsilon, k_s)$, i.e., the period of the vibrations, is $(4/\sigma)K(k_s)$, and the corresponding frequency is

$$p = \frac{\pi \sigma}{2K(k_s)} \qquad (13.83)$$

In the case of linear vibrations ($\alpha = 0$), we obtain $k_s = 0$, $K(k_s) = \pi/\alpha$, $p = \sigma = \omega_0$, $\varepsilon = i\pi - \tfrac{1}{2}\omega_0 T_s$, $\operatorname{sn}(\varepsilon, 0) = \sin \varepsilon = -\sin \omega_0 T_s/2$, and $\operatorname{dn}(\varepsilon, 0) = 1$, so that $F = -\sin \omega_0 T_s/2 = -\sin \theta_0$ and the formula

(13.82) yields

$$A = \lim_{\bar{\alpha} \to 0} \sqrt{\frac{\sqrt{(1 - \bar{\alpha}/F^2)} - 1}{2\bar{\alpha}}} = \frac{1}{2|F|} = \frac{1}{2} |\cosec \theta_0| \qquad (13.84)$$

i.e., leads to the Duffing formula (13.47).

In the case of a single impulse, the initial conditions

$$T(0) = 0, \qquad T_s(0) = \frac{P}{M} \qquad (13.85)$$

should be applied. These yield

$$\cn(\varepsilon, k_s) = 0, \qquad -A_1 \sigma \sn(\varepsilon, k_s) \dn(\varepsilon, k_s) = \omega_0 \qquad (13.86)$$

so that

$$A_1 \sigma \sqrt{1 - k_s^2} = \omega_0 \qquad (13.87)$$

Considering Eq. (13.69) and solving the obtained equation for the amplitude A_1, we obtain

$$A_1 = \sqrt{\frac{\sqrt{(1 + 2\bar{\alpha})} - 1}{\bar{\alpha}}} \qquad (13.88)$$

The function (13.81) is in this case

$$F = \frac{1}{2} \sqrt{\frac{1}{2}\left(1 + \frac{1}{\sqrt{(1 + 2\bar{\alpha})}}\right)} \qquad (13.89)$$

The amplitude A of periodically excited vibrations has its minimum when $F = F_{\max}$. From Eq. (13.81) we find

$$\sn^2 u = \frac{1}{2k_s^2} \left(1 - \sqrt{1 - 4k_s^2 F^2}\right) \qquad (13.90)$$

Since an elliptic sine cannot be greater than unity ($\sn^2 u \leq 1$), the following relationship must be fulfilled: $F^2 \leq 1 - k_s^2$. Therefore the maximum F value can be calculated as

$$F_{\max} = \sqrt{1 - k_s^2} \qquad (13.91)$$

From the formulas (13.82) and (13.91), we find that the minimum value of the dimensionless amplitude A is

$$A_{\min}^2 = \frac{1}{\bar{\alpha}}\left(\sqrt{1 + \frac{\bar{\alpha}}{2}} - 1\right) \qquad (13.92)$$

and that the maximum value of the function F is expressed by the formula:

$$F^2_{max} = \frac{1 + \sqrt{1 + \bar{\alpha}/2}}{2\sqrt{1 + \bar{\alpha}/2}} \quad (13.93)$$

The calculated amplitudes of nonlinear shock-excited vibrations of a plate subjected to periodic lateral impulsive loads are plotted in Fig. 13.3. This figure indicates that a linear approach can be misleading when evaluating the response of a plate to periodic shock loads it can overestimate.

Based on the results obtained for a linear system, one can expect that the steady-state solution obtained for an undamped nonlinear system can be applied for the prediction of its response in the case of small damping as well, keeping in mind that such damping is sufficient to lead, sooner or later, to steady-state conditions, but need not be accounted for when evaluating the system's response.

13.3.4. Stochastic instability

Some deterministic systems reveal stochastic properties, i.e., they oscillate in a quasirandom manner in spite of the nonrandom nature of the excitation. Quasirandom vibrations of deterministic systems subjected to deterministic excitation could be caused, in particular, by the stochastic instability of the phase angle. The physical nature

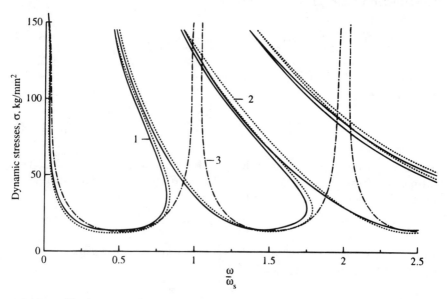

Figure 13.3 Maximum total stresses: (1) simply supported plate; (2) clamped plate; (3) linear vibrations (simply supported plate)

of such an instability in the case of periodic instantaneous impulses can be explained as follows.

The phase angle of linear vibrations occurring immediately after the nth impulse is expressed by the second formula in (13.57) as follows:

$$\varepsilon_n = -\frac{\pi}{2} + n\theta_0 = -\frac{\pi}{2} + n\pi\frac{\omega_0}{\omega_s} \qquad (13.94)$$

Let us assume that, in an approximate analysis, this formula can be applied to nonlinear vibrations as well, and that the frequency of nonlinear vibrations, expressed by the formula (13.83), can be evaluated as

$$p \simeq \sigma = \omega_0\sqrt{1 + \bar{\alpha}A^2} \simeq \omega_0(1 + \tfrac{1}{2}\bar{\alpha}A^2)$$

Then we have

$$\varepsilon_n \approx -\frac{\pi}{2} + n\pi\left(\eta + \frac{1}{2}\frac{\bar{\alpha}}{\omega_0^2}\eta A_n^2\right) \qquad (13.95)$$

where $\eta = \omega_0/\omega_s$ is the ratio of the linear frequency of the undamped vibrations to the frequency of the train of the impulses. Due to inevitable random disturbances in the frequency of vibrations and in the magnitude of the applied impulses, the values η and A_n^2 are never known with precision. Therefore these values should be treated as random variables, though, perhaps, with very small variances D_η and $D_{A_n^2}$. Using the expression (13.95), we obtain the following formula for the variance of the phase angle ε_n:

$$D_{\varepsilon_n} = n^2\pi^2\left[D_\eta + \left(\frac{1}{2}\frac{\bar{\alpha}}{\omega_0^2}\right)^2(\langle\eta\rangle^2 D_{A_n^2} + \langle A_n^2\rangle^2 D_\eta + D_\eta D_{A_n^2})\right] \qquad (13.96)$$

Here $\langle\eta\rangle$ and $\langle A_n^2\rangle$ are the mean values of the variables η and A_n^2. As evident from this formula, the uncertainty in the phase angle, measured by the magnitude of its variance, increases in the time domain with an increase in the n number squared. Therefore at the moment of time when a new impulse is applied, the information about the initial phase angle for the preceding period is partially or completely lost. Because of this, the amplitude caused by this impulse and to a great extent affected by the phase angle at the moment of the impact cannot be evaluated with certainty and should be treated as a random variable. Such a situation is especially strongly pronounced in nonlinear systems, where both variances D_η and $D_{A_n^2}$ contribute to the uncertainty in the variance D_{ε_n} of the phase angle.

A simulation procedure based on the approximate solution

$$T = \frac{P}{M\omega} e^{-n} A_n \cos \varepsilon_n, \quad n = 0, 1, 2, \ldots \quad (13.97)$$

to Eq. (13.46) has been carried out based on the following equations for the amplitude A_n and the phase angle ε_n:

$$\left.\begin{aligned}
A_{n+1}^2 &= \frac{1}{2\bar{\alpha}} \left(\left\{ \left[\frac{\omega^2}{\omega_0^2} + \bar{\alpha}(A_n e^{-rT_s} \cos \varepsilon_{n+1}^+)^2 \right]^2 \right. \right. \\
&\quad + \bar{\alpha} \left[2 e^{-rT_s}(A_n p_{n+1}^- \sin \varepsilon_{n+1}^+ - e^{-rT_s}) \right]^2 \right\}^{1/2} \\
&\quad - \left[\frac{\omega^2}{\omega_0^2} - \bar{\alpha}(A_n e^{-rT_s} \cos \varepsilon_{n+1}^+)^2 \right] \right) \\
\varepsilon_{n+1}^+ &= \arctan \left[\frac{p_{n+1}^-}{p_{n+1}^+} \left(\tan \varepsilon_{n+1}^- - \frac{e^{rT_s}}{A_n^+ p_{n+1}^- \cos p_{n+1}^-} \right) \right]
\end{aligned} \right\} \quad (13.98)$$

where the following notation is used:

$$\left.\begin{aligned}
p_{n+1}^- &= \frac{\sigma_{n+1}^-}{\omega_0} = \sqrt{\frac{\omega^2}{\omega_0^2} + \bar{\alpha}(A_n e^{-rT_s})^2} \\
p_{n+1}^+ &= \frac{\sigma_{n+1}^+}{\omega_0} = \sqrt{\frac{\omega^2}{\omega_0^2} + \bar{\alpha} A_{n+1}^2}
\end{aligned} \right\} \quad (13.99)$$

$$\varepsilon_{n+1}^- = \varepsilon_n^+ + p_{n+1}^- T_s \quad (13.100)$$

The initial values of the amplitude A_n and the phase angle ε_n^+, required to begin the calculations using the formulas (13.98), can be found by putting $A_{-1} = 0$ and $\varepsilon_{-1} = 0$ in the formulas (13.99) and (13.100). This results in the equations:

$$A_0^2 = \frac{1}{2\bar{\alpha}} \left(\sqrt{\frac{\omega^4}{\omega_0^4} + 4\bar{\alpha}} - \frac{\omega^2}{\omega_0^2} \right), \quad \varepsilon_0^+ = -\frac{\pi}{2} \quad (13.101)$$

Then

$$p_0^+ = \frac{1}{2} \sqrt{\frac{\omega^4}{\omega_0^4} + 4\bar{\alpha}} \quad (13.102)$$

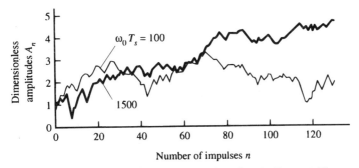

Figure 13.4 Nondimensional amplitudes of stochastically unstable nonlinear vibrations caused by periodic impulses

An example of the calculated response is shown in Fig. 13.4 and proves the quasistochastic nature of the oscillations. Thus, it is natural to use a probabilistic approach to describe the behavior of the system in stochastically unstable conditions.

When using such an approach, we find it more convenient to proceed from energy considerations. The velocities of the oscillations before and after the nth impulse are related as

$$T_i^+ = T_i^- + \frac{P}{M} \qquad (13.103)$$

where T_i^- is the velocity before and T_i^+ is the velocity after the sequential impulse is applied. Since the kinetic energy is

$$K = \tfrac{1}{2} M \dot{T}^2$$

we have

$$K^+ = K^- + H_0 + PT_t^- \qquad (13.104)$$

where $H_0 = P^2/(2M)$ is the energy due to a single impulse, or the energy absorbed by the system after the first shock ($n = 0$). An instantaneous impulse does not change the potential energy, and therefore Eq. (13.104) holds for the total energy as well:

$$H^+ = H^- + H_0 + PT_i^- \qquad (13.105)$$

We assume

$$\dot{T} \approx \dot{T}_{\max} \cos \varepsilon \qquad (13.106)$$

where

$$\dot{T}_{max} = \sqrt{\frac{2H}{M}} \qquad (13.107)$$

is the maximum velocity and ε is the phase angle. Then the relationship (13.105) can be rewritten as

$$H^+ = H^- + H_0 + 2\sqrt{H_0 H^-} \cos \varepsilon^- \qquad (13.108)$$

Assuming, for the sake of simplicity, that the energy of free damped vibrations fades away exponentially, one obtains the following formula for the total energy at the moment of time immediately following the $(n + 1)$th shock:

$$H_{n+1} = H_n e^{-2rT_s} + H_0 + 2e^{-rT_s}\sqrt{H_0 H_n} \cos \varepsilon_{n+1}^-, \qquad n = 0, 1, 2, \ldots \qquad (13.109)$$

Here ε_{n+1}^- is the phase angle prior to the application of the $(n + 1)$th shock. Introducing a dimensionless energy $h_n = H_n/H_0$ and using the notation $\tau = e^{-rT_s}$, we obtain the following formula for the dimensionless energy h:

$$h_{n+1} = 1 + \tau^2 h_n + 2\tau\sqrt{h_n} \cos \varepsilon_{n+1}^-, \qquad n = 0, 1, 2, \ldots \qquad (13.110)$$

If one wishes to obtain a deterministic solution for the evolution of this energy in time, it would be necessary to introduce an additional relationship for the change in the phase angle. Such a relationship can be obtained from the condition that the shock loading does not change the system's position. This would lead to the condition $T_n^+ = T_n^-$. In the present case, however, when trying to find the probabilistic and not the deterministic characteristics of the oscillations, one can use, examining an extreme case, the "*stochastic phase approximation*" as an additional condition. This approximation presumes that when the standard deviation $\sqrt{D_{\varepsilon_n}}$ of the phase angle at the moment of an impact exceeds 2π, the phase angle can be viewed as an "absolutely random," i.e., absolutely uncertain (unpredictable). Assuming the phase angle ε_{n+1}^- to be a uniformly distributed random variable and treating the dimensionless energy h_{n+1} value as a deterministic function of the random variable ε_{n+1}^-, we obtain the following question for the probability transition function:

$$Q(h_{n+1}, h_n) = \frac{1}{\pi}\left[4\tau^2 h_n - (h_{n+1} - \tau^2 h_n - 1)^2\right]^{-1/2}, \qquad n = 0, 1, 2, \ldots \qquad (13.111)$$

The evolution of the probability density distribution function can be computed based on the Smoluchowski equation (11.40):

$$f_{n+1}(h_{n+1}) = \int_{\alpha_n^-}^{\alpha_n^+} f_n(h_n) Q(h_{n+1}, h_n) \, dh_n, \qquad n = 0, 1, 2, \ldots \quad (13.112)$$

The initial probability density function defines a nonrandom value and is expressed by a delta function $f_0(h_0) = \delta(h_0)$. Note that in the nondissipative case ($\tau = 1$) the function (13.111) is symmetric with respect to its arguments, and the kernel of the integral equation (13.112) is also symmetric. This means that the random process of undamped shock-excited vibrations is reversible in the time domain.

In this analysis we assume that nonlinearity is small and affects only the very fact of the development of stochasticity, but does not result in any essential change in the extreme h_n values. Therefore the limits of integration in Eq. (13.112) can be evaluated on the basis of the formulas (13.60) and (13.62), so that

$$\alpha_n^\pm = h_n \simeq \left(\frac{1 \mp \tau_{n+1}}{1 \mp \tau} \right)^2, \qquad n = 0, 1, 2, \ldots \quad (13.113)$$

These limits define the widest range of the possible energy changes. It may occur, however, that large values of h_{n+1} are not possible with small values of $\alpha_n^- = h_n^{\min}$. On the other hand, small h_{n+1} values may not be possible for large enough values of $\alpha_n^+ = h_n^{\max}$. Such α_n^+ values may take place, for instance, in a hypothetical case when all the preceding impulses occurred at the moments of time, corresponding to the extreme conditions $\cos \bar{\varepsilon} = \pm 1$. In actual calculations, the necessity to narrow the interval of integration, compared to the formula (13.113) prediction, will be indicated by the negative value of the expression in the brackets in (13.111). Such narrowed limits can be found by simply requiring that this expression is positive. This yields

$$\alpha_n^\pm = h_n = \left(\frac{\sqrt{h_{n+1}} \pm 1}{\tau} \right)^2 \quad (13.114)$$

The governing equation (13.112) does not lend itself to a simple analytical solution. It has been solved numerically, assuming the following initial distribution:

$$f_1(h_1) = Q(h_1, h_0) = Q(h_1, 1) = \frac{1}{\pi} [4\tau^2 - (h_1 - \tau^2 - 1)^2]^{-1/2} \quad (13.115)$$

and the initial limits for the dimensionless energy:

$$\alpha_i^\pm = (1 \pm \tau)^2 \tag{13.116}$$

Some computed probability density functions are plotted in Fig. 13.5 for the case $rT_s = 0.2$. In the accompanying table the magnitude of the extreme values $h_n^{\max}$ and $h_n^{\min}$ are indicated, as well as the mean value $\langle h_n \rangle$ of the dimensionless energy. As one can see from this figure, the initial delta-like probability density function for the dimensionless energy "spreads" in due course, and, for damped vibrations, approaches a steady-state condition.

One important result of the performed analysis is that for large enough n values, the energy distribution density function is strongly skewed to the left: low values of the energy are much more likely than large ones. Therefore the mean value of the dimensionless vibration energy $\langle h_n \rangle$ is considerably closer to $h_n^{\min}$ than to $h_n^{\max}$.

13.3.5. Vibrations of an elongated plate due to periodic impulses

As an application of the analysis carried out in the previous section, examine shock-excited vibrations of an elongated rectangular plate,

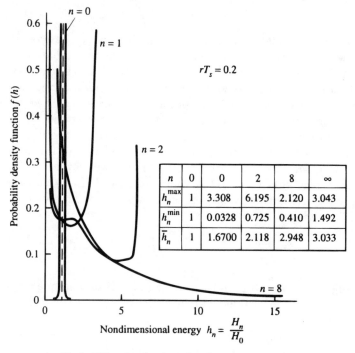

Figure 13.5 Probability density function for nondimensional energy of stochastically unstable vibrations caused by periodic impulses

subjected to periodic shock loads, uniformly distributed over the plate's surface. Such loads can occur, for instance, in the ship's bottom forward because of the ship's slamming. We proceed from the following formula for the total energy of an elongated rectangular plate supported by a nondeformable contour and subjected to periodic shock loads:

$$H = \frac{1}{2} \rho h_p \int_{-a/2}^{a/2} \left(\frac{\partial w}{\partial t}\right)^2 dx + \frac{1}{2} D \int_{-a/2}^{a/2} \frac{\partial^2 w}{\partial x^2} dx + \frac{1}{2} h_p \sigma_x^0 \int_{-a/2}^{a/2} \left(\frac{\partial x}{\partial x}\right)^2 dx \tag{13.117}$$

Here

$$\sigma_x^0 = \frac{E}{2a} \int_{-a/2}^{a/2} w_x^2 \, dx \tag{13.118}$$

is the "membrane" (in-plane) stress, $w = w(x, t)$ is the deflection function, $D = Eh_p^3/[12(1 - v^2)]$ is the plate's flexural rigidity, h_p is its thickness, E is Young's modulus of the material, v is Poisson's ratio, a is the plate's width, and ρ is the material's density. The origin of the rectangular coordinates xyz is at the center of the plate in its midplane.

We restrict our analysis to the first mode of vibrations only. Then the deflection function $w(x, t)$ can be sought in the form

$$w(x, t) = T(t)X(x) \tag{13.119}$$

where

$$X(x) = \cos \frac{2\gamma}{a} x + C \cosh \frac{2\gamma}{a} x \tag{13.120}$$

is the vibration mode and $T(t)$ is the principal coordinate. For a simply supported plate, $C = 0$ and $\gamma = \pi/2$. For a clamped plate, $C = 0.1329$ and $\gamma = 2.365$. Substituting (13.119) into (13.117), we obtain

$$H = \tfrac{1}{2} M(\dot{T}^2 + \omega_0^2 T^2 + \tfrac{1}{2}\alpha T^4) \tag{13.121}$$

where

$$\left.\begin{array}{l} M = \rho h_p \displaystyle\int_{-a/2}^{a/2} X^2(x)\, dx, \quad \omega_0^2 = \dfrac{D}{M} \displaystyle\int_{-a/2}^{a/2} [X''(x)]^2 \, dx \\[2ex] \alpha = \dfrac{Eh_0}{2a^3} \dfrac{\kappa^2}{M}, \quad \kappa = a \displaystyle\int_{-a/2}^{a/2} [X'(x)]^2 \, dx \end{array}\right\} \tag{13.122}$$

For a simply supported plate,

$$M = \frac{1}{2}\rho h_p a, \qquad \omega_0 = \left(\frac{\pi}{a}\right)^2 \sqrt{\frac{D}{\rho h_p}}$$
$$\alpha = 3(1-\nu^2)\frac{\omega^2}{h_p^2}, \qquad \kappa = \frac{\pi^2}{2} = 4.93 \tag{13.123}$$

For a clamped plate,

$$M = \frac{1+C^2}{2}\rho h_p a, \qquad \omega_0 = \left(\frac{2\gamma}{a}\right)^2 \sqrt{\frac{D}{\rho h_p}}$$
$$\alpha = 0.9231(1-\nu^2)\frac{\omega^2}{h_p^2}, \qquad \kappa = 6.26 \tag{13.124}$$

By using the Hamilton form of the equations of motion, the following equation for the principle coordinate $T(t)$ can be obtained from (13.121):

$$\ddot{T} + \omega_0^2 T + \alpha T^3 = 0 \tag{13.125}$$

This equation is identical to Eq. (13.65) and therefore all the results obtained in the Sec. 13.3.3 are valid for the case in question.

The total maximum stress due to the combined action of the bending and the membrane forces is

$$\sigma = c_1 \frac{E}{1-\nu^2}\left(\frac{h_p}{a}\right)^2 \frac{\bar{A}_1}{h_p} A\left(1 + c_2 \frac{\bar{A}_1}{h_p} A\right) \tag{13.126}$$

where the linear term is due to the bending stress and the term proportional to the A^2 reflects the effect of the membrane stress. The following notation is used in the formula (13.126):

$$c_1 = \frac{\pi^2}{9}$$
$$c_2 = 0.5$$
$$\bar{A}_1 = \frac{P}{M\omega} = \frac{4}{\pi^3}\frac{qa^2}{\sqrt{D\rho h_p}} = 0.1290\frac{qa^2}{\sqrt{D\rho h_p}} \tag{13.127}$$
$$\frac{\bar{A}_1}{h_p} = \frac{\sqrt{\bar{\alpha}}}{3(1-\nu^2)} = 0.5773\frac{\sqrt{\bar{\alpha}}}{1-\nu^2}$$

in the case of a simply supported plate ($x = 0$), and

$$\left.\begin{aligned} c_1 &= 18.1 \\[6pt] c_2 &= 0.1731 \\[6pt] \bar{A}_1 &= \frac{P}{M\omega} = \frac{\sin\gamma}{(1+C^2)\gamma^3}\frac{qa^2}{\sqrt{D\rho h_p}} = 0.0526\frac{qa^2}{\sqrt{D\rho h_p}} \\[6pt] \frac{\bar{A}_1}{h_p} &= \frac{2\gamma^2}{\kappa}\sqrt{\frac{1+C^2}{3(1-\nu^2)}}\;\bar{\alpha} = 1.0408\sqrt{\frac{\bar{\alpha}}{1-\nu^2}} \end{aligned}\right\} \quad (13.128)$$

in the case of a clamped plate ($x = \pm a/2$). The q value in the above formulas is the impact load per unit plate area, so that the generalized impulse P can be evaluated as

$$P = \int_{-a/2}^{a/2} qX(x)\,dx = \frac{qa}{\gamma}(\sin\gamma + C\sinh\gamma) \qquad (13.129)$$

It is equal to $P = (2/\pi)qa = 0.637qa$ for a simply supported plate and to $P = 0.3qa$ for a clamped plate. The $\bar{\alpha}$ value in the formulas (13.127) and (13.128) is defined by formula (13.71).

The amplitudes and maximum stresses for an elongated 5 mm thick and 0.5 m wide steel plate ($E = 2 \times 10^{10}$ kg/m², $\rho = 800$ kg s²/m⁴, $\nu = 0.3$) subjected to periodic impulsive loading $q = 10$ kg s/m² are plotted in Figs 13.3 and 13.6. The cases where the plate is simply supported or clamped on a nondeformable contour are considered.

The probability density distribution functions for the random amplitudes and stresses can be obtained on the basis of the distribution density function $f(h)$ for the dimensionless energy. This energy is related to the dimensionless amplitude A of vibrations as follows:

$$h = A^2(1 + \tfrac{1}{2}\bar{\alpha}A^2) \qquad (13.130)$$

The probability density functions for the amplitude A and the total stress

$$\sigma = \frac{E}{1-\nu^2}\left(\frac{h_p}{a}\right)^2 c_0 \sqrt{\frac{\bar{\alpha}}{1-\nu}}\, A\left(1 + c_2\sqrt{\frac{\bar{\alpha}}{1-\nu}}\,A\right) \qquad (13.131)$$

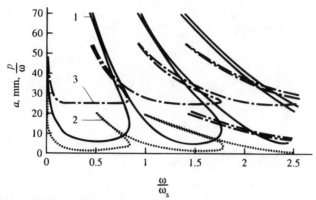

Figure 13.6 Amplitudes and frequencies of a rectangular plate subjected to periodic lateral impulsive loads

1. Amplitude of a simply supported plate
2. Amplitude of a clamped plate
3. Nonlinear/linear frequency ratio in a simply supported plate

can be calculated as

$$f_A(A) = f_h(h) \frac{dh}{dA} = 2A(1 + \bar{\alpha}A) f_h[h(A)] \tag{13.132}$$

$$f_\sigma(\sigma) = f_A(A) \frac{dA}{d\sigma}$$

$$= \frac{2(1-\nu)^2}{c_0 E} \left(\frac{a}{h_p}\right)^2 \sqrt{\frac{1-\nu}{\bar{\alpha}}} \frac{A(1-\bar{\alpha}A)}{1 + c_2\sqrt{[\bar{\alpha}/(1-\nu)]}A} f_h[h(A)] \tag{13.133}$$

Here, the dimensionless amplitude A is expressed according to (13.131) as

$$A = \frac{1}{2c_2}\sqrt{\frac{1-\nu}{\bar{\alpha}}}\left[\sqrt{1 + \frac{c_2}{c_0}\left(\frac{2a}{h_p}\right)^2 (1-\nu^2)\frac{\sigma}{E}} - 1\right] \tag{13.134}$$

In these equations, $c_0 = \pi^2/(2\sqrt{3}) = 2.85$ for a simply supported plate and $c_0 = 18.8$ for a clamped plate.

Further Reading (see Bibliography): 6, 11, 14, 15, 17, 23, 24, 36, 45, 49, 52, 67, 69, 70, 78, 79, 89, 91, 92, 93, 108, 123, 126, 132, 146.

Chapter 14

Geometric Tolerance

"To err is human, to forgive is divine, but to include errors in your design is statistical."
LESLIE KISCH

"Be grateful for luck, but do not depend on it."
WILLIAM FEATHER

"God and Devil are, in effect, similar beings. Only the first one is with a 'plus", and the second one is with a 'minus'."
UNKNOWN PHYSICIST

14.1. Random Geometry

Actual geometry of mechanical elements is never the same as the design ("nominal") geometry. Here are some commonly used terms to describe sizes, allowances, and tolerances for fits between cylindrical parts, such as bearings, shrink and drive fits, etc.:

Allowance. Minimum clearance (positive allowance) or maximum interference (negative allowance) between mating parts.

Tolerance. Total permissible variation of size (dimension).

Limits of size. Applicable maximum and minimum sizes.

Clearance fit. Limits of size are so prescribed that a clearance always results when mating parts are assembled.

Interference fit. Limits of size are so prescribed that an interference always results on assembled parts.

Transition fit. This may have either a clearance an interference or assembly.

Basic size. One from which limits of size are derived by the application of allowances and tolerances.

Unilateral tolerance. In this case a variation in size is permitted in only one direction from the basic size.

Fits are divided into the following general categories:

1. *Running and sliding fits,* which are intended to provide similar running performance with suitable lubrication allowance throughout the range of sizes.
2. *Locational clearance fits,* which are intended for normally stationary parts that can, however, be freely assembled or dismantled.
3. *Transition fits,* applied in those cases when accuracy of location is important, but a small amount of either clearance or interference is allowed.
4. *Location interference fits,* used where accuracy of location is of primary importance and for parts requiring rigidity and alignment with no special requirements for bore pressure.
5. *Force (shrink) fits,* which are characterized by approximately constant bore pressures throughout the range of sizes.

Running and sliding fits are further subdivided into close-sliding fits intended for accurate location of parts that must assemble without perceptible play; sliding fits intended for parts that are supposed to move and turn easily but are not intended to run freely (in large size such parts may seize under small temperature changes); precision-running fits intended for precision work at slow speeds and light journal pressures, but are not suitable where appreciable temperature differences are encountered; close-running fits intended for accurate machinery with moderate surface speeds and journal pressures, where accurate location and minimum play is desired; medium-running fits intended for higher running speeds or heavy journal pressures, or both; free-running fits intended for use where accuracy is not essential, or where large temperature variations are likely to be present, or both; and loose-running fits intended for use with materials, such as cold-rolled shafting or tubing, made to commercial tolerances. Location clearance fits are subdivided into various classes which run from snug fits for parts requiring accuracy of location, through medium clearance fits (spigots), to the looser fastener fits where freedom of assembly is of primary importance. Force (shrink) fits are divided into light-drive fits intended for applications requiring light assembly pressures (thin sections, long fits, cast-iron external members); medium-drive fits suitable for ordinary steel parts or for shrink fits or light sections; heavy-drive fits used for heavier steel parts or shrink fits in medium sections; force fits suitable for parts that can be highly stressed.

To achieve the required degree of accuracy, when machining and processing the machine parts intended for a specific application it is necessary to add tolerance figures to the basic size (dimension), i.e., to indicate the amount of variation permitted in the part or the total variation permissible in a given dimension. Clearly, if the given part must fit properly with some other part, both must be given tolerances in keeping with the desired allowance and the available manufacturing processes. Cost of manufacturing is also important, since, typically, a decrease in tolerance results in an increase in the cost of the part. In addition, if a large number of elements is produced, they will all differ in a random manner from each other.

Although the traditional engineering practice treats the geometric tolerances as deterministic values, it is obvious that the actual dimensions of the real parts produced are random variables. At first glance this should lead to a more complicated procedure for various inspections, as well as for establishing the accuracy of the particular manufacturing technology. The truth is, however, that application of a probabilistic approach enables one to quantitatively estimate the effect of the dimensional variability on the performance of mechanical elements and to keep this variability within allowable limits. Consideration of the probabilistic nature of the machine parts geometry enables one to increase the speed and reduce the cost of quality control of large populations of parts. Statistical rules of such a control enable an inspector to decide, based on an evaluation of just several samples, whether the entire population should be rejected or could be accepted. Such rules enable the inspector not only to estimate the unknown parameter of the distribution of a random dimension (say, its mean or standard deviation) but also to determine the region in which this parameter will be supposedly found with the given probability. It should be emphasized that since such a region is constructed based on the observed data, it can change from one sampling to another and is therefore itself a random variable (domain). Thus, one should view the problem of finding a certain parameter within a particular region as the problem of establishing the probability of the event that this region (interval) contains the actual value of the parameter in question. Actually, the problem that is typically encountered by a designer is to construct a statistical rule that would enable determination, for the given (observed) collection of data (measurements), of a region (*confidence intervals, tolerance limits, control limits*) containing the actual value of the parameter in question with the required (desired) probability $1 - \alpha$, where α is a small number. Clearly, in a good design the deviation of the parts dimensions from the nominal value of the given dimension should be such that only a small allowable fraction α of the manufactured parts would fall outside the specified limits for the required fit. Note that a simplified version of the problem in question, when the role of the sample size was not accounted for, has been examined in Sec. 10.9.

14.2. Confidence Intervals

In accordance with the central limit theorem (see Sec. 5.3), the confidence intervals can be constructed when estimating normally distributed characteristics of parts dimensions, on the basis of the following considerations.

Let a population ("stream," "universe") of the items of interest (say, measured for a specific dimension X) be sampled by taking $N(j = 1, 2, \ldots, N)$ samples (groups of items) each of the size n ($i = 1, 2, \ldots, n$). Let the dimension in question be a normally distributed random variable with the probability density function

$$f(x; \theta) = \frac{1}{\sqrt{2\pi}\,\sigma} \exp\left[-\frac{(x-\theta)^2}{2\sigma}\right]$$

where θ is a so-far unknown mean value of the random variable X and σ is the standard deviation of this variable. This deviation is assumed to be known, or appropriately chosen. Situations when the standard deviation is unknown will be examined later.

Let, as a result of observations (measurements), a collection $x_1, x_2, \ldots, X_n$ of measured values (realizations) of the random variable X be obtained for each sample, and the mean

$$\theta^* = \langle x \rangle = \frac{1}{n}\sum_{i=1}^{n} x_i$$

and variance

$$\sigma^2 = \frac{1}{n}\sum_{i=1}^{n}(x_i - \langle x \rangle)^2$$

of this collection be evaluated. In the treatment that follows, it is taken into account that the mean θ^* can vary considerably from sample to sample and therefore should be viewed itself as a random variable.

Since the variable θ^* is a linear function of a collection of normally distributed random variables, it is also a normal distributed variable. Its mean value θ (the "mean of the mean"), if the sizes of the samples are large, should not be different from the mean value of the basic variable X. As far as the variance D_{θ^*} of the variable θ^* is concerned, it can be evaluated as

$$D_{\theta^*} = \frac{1}{n^2}(D_{x_1} + D_{x_2} + \cdots + D_{x_n}) = \frac{1}{n^2}(n\sigma^2) = \frac{\sigma^2}{n}$$

Examine now an auxiliary random variable $Z = \sqrt{n}(\theta^* - \theta)$. Since θ^* is normally distributed, the variable Z is also normally distributed and, since the mean value of θ^* is equal to θ, has zero mean and the variance $D_z = nD_{\theta^*} = \sigma^2$. Hence, the probability density function of the random variable Z is

$$f_z(z) = \frac{1}{\sqrt{2\pi}\,\sigma} e^{-z^2/2\sigma^2}$$

Let us require that the probability that the variable Z falls within the interval $(-k\sigma, k\sigma)$, where k is a given number, is equal to $1 - \alpha$, where α is a small number. Then we have

$$P(-k\sigma < z < k\sigma) = P\left(\theta^* - \frac{k\sigma}{\sqrt{n}} < \theta < \theta^* + \frac{k\sigma}{\sqrt{n}}\right)$$

$$= \int_{-k\sigma}^{k\sigma} f_z(z)\,dz = \frac{1}{\sqrt{2\pi}\,\sigma} \int_{-k\sigma}^{k\sigma} e^{-z^2/(2\sigma^2)}\,dz$$

$$= \Phi\left(\frac{k}{\sqrt{2}}\right) = 1 - \alpha \qquad (14.1)$$

where

$$\Phi\left(\frac{k}{\sqrt{2}}\right) = \frac{2}{\sqrt{\pi}} \int_0^{k/\sqrt{(2)}} e^{-t^2}\,dt$$

is the Laplace function. Thus, the interval

$$\left(\theta^* - \frac{k\sigma}{\sqrt{n}},\; \theta^* + \frac{k\sigma}{\sqrt{n}}\right)$$

is the confidence interval with the confidence level $1 - \alpha$. The formula (14.1) provides an estimate for the mean value of the measured dimension (or any other characteristic of interest) for the given standard deviation. This formula enables one to evaluate the number k of standard deviations that results in the probability $1 - \alpha$ that the mean value $\theta = \langle x \rangle$ of the random variable X is found within the obtained confidence interval.

The confidence interval for the estimation of the variance θ of a random variable X, when its mean value a is known, can be obtained, using similar reasoning, as

$$\left(\frac{n\theta^*}{a_2}, \frac{n\theta^*}{a_1}\right)$$

The numbers a_1 and a_2 in the confidence limits should be chosen, using the tables for the χ^2 distribution with n degrees of freedom (see Appendix E), in such a way that the probability that the random variable $n\theta^*/\theta$ falls within the interval (a_1, a_2) be $1 - \alpha$:

$$P\left(a_1 < \frac{n\theta^*}{\theta} < a_2\right) = P\left(\frac{n\theta^*}{a^2} < \theta < \frac{n\theta^*}{a_1}\right) = 1 - \alpha \quad (14.2)$$

If the mean value of the random variable X is unknown, the confidence interval for the variance θ can be constructed in the same fashion as in the case when this value is known, i.e., on the basis of the formula

$$P\left(a_1 < \frac{n\theta^*}{\theta} < a_2\right) = 1 - \alpha \quad (14.3)$$

where the random variable $(n-1)\theta^*/\theta$ has a χ^2 distribution with $n-1$ degrees of freedom.

The confidence interval for the mean value θ, if the variance is unknown, can be obtained as

$$(\theta^* - k\sigma, \theta^* + k\sigma)$$

and the probability that the θ value falls within this interval is

$$P(-k\sigma < \theta^* - \theta < k\sigma) = P(\theta^* - k\sigma < \theta < \theta^* + k\sigma) = 1 - \alpha \quad (14.4)$$

The variable $(\theta^* - \theta)/\sigma$, in which the σ value can be determined from the formula

$$\sigma^2 = \frac{1}{n}\sum_{i=1}^{n}(x_i - \theta^*)^2$$

follows the Student distribution with $n - 1$ degrees of freedom (see Sec. 3.3.12).

Example 14.1 (Haugen, 1980) Measurements are made of the diameters of nine shafts, and the obtained diameters (in inches) are: 4.330, 4.287, 4.450, 4.295, 4.340, 4.407, 4.295, 4.388, 4.356. Find the tolerance limits, for which not more than 10 percent will fall outside this limit, with the confidence of 99 percent.

Solution Using the table for tolerance factors for normal distributions from Appendix B, we find that for $n = 9$, $\gamma = 0.99$, and $\alpha = 0.1$, the k value is $k = 3.822$. The mean and the standard deviation of the sample are $\langle d \rangle = 4.350$ in and $\sqrt{D_d} = 0.056$ in. Then the tolerance limits are $4.350 - 3.822 \times 0.056 = 4.140$ in and $4.350 + 3.822 \times 0.056 = 4.57$ in.

Example 14.2 (Haugen, 1980) According to specification, a manufacturer was required to produce cylinders having a bore with a lower limit on the diameter than can be assured, with probability $\gamma = 95$ percent and that $(1 - \alpha) = 99$ percent of his production be above this limit. A sample of 30 units are taken, and the sample mean and sample standard deviations were $\langle d \rangle = 9.87$ mm and $\sqrt{D_d} = 0.059$ mm, respectively. Determine the required lower tolerance limit.

Solution From the table for odd-sided tolerance limit factors for normal distribution (Appendix B), for $n = 30$, $\gamma = 0.95$, and $\alpha = 0.01$ we find that $k = 3.064$. Then the required lower tolerance limit can be found as

$$\langle d \rangle - k\sqrt{D_d} = 9.87 - 3.064 \times 0.059 = 9.69 \text{ mm}$$

Further Reading (see Bibliography): 2, 3, 64, 65.

Chapter

15

Random Loads and Responses in Some Engineering Systems

> *"I do not get satisfaction from the formulas, until I have a feeling for the numerical values of the results."*
> WILLIAM THOMSON (LORD KELVIN)
>
> *"By asking impossible obtain the best possible."*
> ITALIAN SAYING
>
> *"It is easy to see, it is hard to foresee."*
> BENJAMIN FRANKLIN

15.1. Cars

In this chapter we will briefly summarize some probabilistic methods used to evaluate loads and responses in particular engineering systems. It can be seen that the same methods can be successfully employed for the analysis of quite different engineering objects.

15.1.1. Loads

The mean sources of random loading on surface-going vehicles are: road unevenness; excentricity in, and the nonuniformity of the rotation of, the wheels; and disbalance of various rotating parts, such as wheels, rotating parts within the engine, and the transmission.

It is usually assumed that the "microprofile" of the road is a stationary random function of the traveled distance. In addition, it is assumed that the ordinates of this function are normally distributed and that the microprofile of the road changes in the vertical direction only. The spectrum of loading due to the random road roughness $X(s)$

can be written as a function of the "road frequency:"

$$\omega_1 = \frac{2\pi}{S_e}$$

where S_e is the effective length ("period") of the road profile, as follows:

$$S_x(\omega_1) = \frac{2}{\pi} \int_0^\infty K_x(s) \cos \omega_1 s \, ds \qquad (15.1)$$

Here

$$K_x(s) = \int_0^\infty S_x(\omega_1) \cos \omega_1 s \, d\omega_1 \qquad (15.2)$$

is the correlation function of the road microprofile $X(s)$. While the spectrum $S_x(\omega_1)$ characterizes the frequency of occurrence of the given lengths of the road irregularities (the prevailing lengths in the random profile of the road's longitudinal cross-section), the correlation function $K_x(s)$ gives an idea of how strongly these irregularities are pronounced, i.e., how "fast" the microprofile $X(s)$ changes with the change in the traveled distance s.

The correlation function of the road microprofile can be written, similarly to many other cases encountered in engineering problems, as

$$K_x(s) = D_x e^{-\alpha|s|} \qquad (15.3)$$

where

$$D_x = K_x(0) = \int_0^\infty S_x(\omega_1) \, d\omega_1 \qquad (14.4)$$

is the variance of the ordinates of the road microprofile and α is the parameter characterizing the "smoothness" of the road. Then the spectrum of the road profile can be determined from the formula (15.3) as follows:

$$S_x(\omega_1) = \frac{2\alpha D_x}{\pi(\omega_1^2 + \alpha^2)} \qquad (15.5)$$

It should be emphasized that, while the road engineer is interested in the actual microprofile of the road under design or construction, the automobile designer is interested primarily in a "generalized" road spectrum. This spectrum is supposed to reflect not only the actual "averaged" profiles of numerous roads the automobile will be operated on during its lifetime, but should also be able to account for the important features of the car design itself: its type and speed, the projected lifetime, type of suspension, etc. Such "design" spectra are often taken in the form:

$$S_x^*(\omega_1) = \frac{S_x(\omega_1)}{D_x} = a\frac{v^2 + b}{v^4 + c} \tag{15.6}$$

where $v = v\omega_1$, v is the nominal car speed, and the parameters a, b, and c consider the type of road surface and other factors that might affect the design.

When the excitation spectrum is chosen, a car operated on an uneven road can be treated as a linear dynamic system experiencing random loadings $Q_1(t)$ and $Q_2(t)$ transmitted to the car body through the front and the rear wheels. The sought responses are the displacements and the accelerations of the car frame, stresses in vulnerable structural elements of the car, etc. These responses can be calculated in accordance with the general methods and formulas discussed in Chapter 8. In order to use these formulas, one has to know the expressions for the response functions $\Phi(i\omega_1)$. These expressions can be obtained either theoretically or experimentally. If an experimental approach is taken, the automobile should be tested on a test stand, on which it is subjected to harmonic excitation, or on a specially constructed road with an artificially constructed wavy surface.

15.1.2. Responses

Examine a simplest case of random vibrations of an automobile in its longitudinal plane when moving on a road characterized by a spectral function of the type (15.5). In the analysis that follows it is assumed, for the sake of simplicity, that the road microprofile is the same under the left and the right wheels, so that the car is subjected to the longitudinal vibrations only and does not experience vibrations in the transverse plane.

The equations of motion of the car body (Fig. 15.1) can be written as follows:

$$\left.\begin{array}{r}M_1\ddot{z}_1 + M_2\ddot{z}_2 = P_1 + P_2 \\ I\dfrac{\ddot{z}_1 - \ddot{z}_2}{L} = P_1 a - P_2 a\end{array}\right\} \tag{15.7}$$

where $M_1 = M(b/L)$ and $M_2 = M(a/L)$ are the effective masses of the fore- and the afterbody of the car, M is the mass of the car, L is the distance between the wheel axes (car base), a and b are the distances

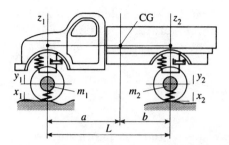

Figure 15.1 A car on an uneven road.

of the car's center of gravity from the front and the rear wheel axes, respectively, I is the moment of inertia of the car's mass, P_1 and P_2 are the forces of interaction between the car body and its front and rear suspensions, and z_1 and z_2 are the vertical displacements of the points of the car body located above the front and the rear wheels, respectively.

The moment of inertia, I, of the car can be computed, for many contemporary cars and trucks, by a simplified formula $I = Mab$. With this formula, the equations (15.7) become uncoupled and can be simplified as follows:

$$P_1 = M_1 \ddot{z}_1, \qquad P_2 = M_2 \ddot{z}_2 \tag{15.8}$$

These formulas indicate that, in an approximate analysis, the vibrations of the fore- and the afterbody of the car can be evaluated independently.

Examine the vibrations of the car's forebody. These vibrations can be described by the equation:

$$m\ddot{y} = -P + K_t(x - y) \tag{15.9}$$

where m is the mass of the bridge, y is its displacement, K_t is the spring constant of the tires, and x is the height of the road unevenness under the wheels. On the other hand, the force P of interaction between the car's body and the suspension can be calculated as

$$P = K_s(z - y) + c(\dot{z} - \dot{y}) \tag{15.10}$$

where K_s is the spring constant of the suspension, c is the damping coefficient of the shock absorber, and z is the displacement of the point of the car's body located above the front wheel. Introducing the first formula in Eq. (15.8) into Eqs (15.13) and (15.14), we obtain the following relationships for the displacements x, y, and z:

$$\left. \begin{array}{l} K_t x - m\ddot{y} - K_t y - M_1 \ddot{z} = 0 \\ c\dot{y} + K_s y - M_1 \ddot{z} - c\dot{z} - K_s z = 0 \end{array} \right\} \tag{15.11}$$

In order to obtain the expression for the response function for the displacement z, one should consider that the external excitation $X(t)$ due to the road microprofile be described by a sinusoide. In this case the relationship between the steady-state acceleration $\ddot{z}$ and the displacement z is

$$\ddot{z} = -\omega^2 z$$

where ω is the vibration frequency. Similarly, the relationship between the acceleration $\ddot{y}$ and the corresponding displacement y can be written as

$$\ddot{y} = -\omega^2 y$$

Then the equations (15.11) yield

$$\left.\begin{aligned}K_t x + (m\omega^2 - K_t)y + M_1\omega^2 z &= 0 \\ c\dot{y} + K_s y + (M_1\omega^2 - K_s)z - c\dot{z} &= 0\end{aligned}\right\} \quad (15.12)$$

Solving the first equation in (15.12) for the displacement y and introducing the obtained expression into the second equation, we obtain the following equation for the displacements $z(t)$ and $x(t)$ and their velocities $\dot{z}(t)$ and $\dot{x}(t)$:

$$c[K_t - (M_1 + m)\omega^2]\dot{z} + [K_t K_s - (M_1 K_t + M_1 K_s + mK_s)\omega^2 + M_1 m\omega^4]z \\ - cK_t \dot{x} - K_t K_s x = 0 \quad (15.13)$$

The velocities $\dot{z}$ and $\dot{x}$ of harmonic oscillations are related to the displacements z and x as $\dot{z} = i\omega z$ and $\dot{x} = i\omega x$, where $i = \sqrt{-1}$ is the "imaginary unity." This enables us to obtain from (15.13) the following expression for the complex frequency characteristic:

$$\Phi(i\omega) = \frac{z}{x}$$

$$= \frac{K_t(K_s + ic\omega)}{M_1 m\omega^4 - [M_1(K_t + K_s) + mK_s]\omega^2 + K_t K_s - ic[(M_1 + m)\omega^3 - K_t\omega]}$$

$$(15.14)$$

With the obtained expression for the complex frequency characteristic, one can determine the spectrum of the "output process" of the displacements z by the formula

$$S_z(\omega) = |\Phi(i\omega)|^2 S_0(\omega) = a^2(\omega) S_0(\omega) \qquad (15.15)$$

Here $a(\omega) = |\Phi(i\omega)|$ is the response function of the system and $S_0(\omega)$ is the spectrum of the "input" random process $X(t)$. Since the spectrum $S_x(\omega_1)$ of the random microprofile of the road is obtained as a function of the "space" frequency ω_1, one should calculate first the spectral function $S_0(\omega)$ from the experimentally obtained spectrum $S_x(\omega_1)$. This can be done on the basis of the formula:

$$S_0(\omega) = \frac{1}{v} S_x\left(\frac{\omega}{v}\right) \qquad (15.16)$$

where v is the car speed.

Example 15.1 The effective mass of the car forebody is $M_1 = 82$ kgf $\times$ s² $\times$ m^{-1}, the mass of the front bridge is $m = 10.2$ kgf $\times$ s² $\times$ m^{-1}, the spring constant of the suspension is $K_s = 50$ kgf cm^{-1}, spring constant of the front tires is $K_t = 330$ kgf cm^{-1}, damping coefficient of the shock absorber is $c = 3.2$ kgf $\times$ s $\times$ cm^{-1}, the spectrum parameter that characterizes the road "smoothness" is $\alpha = 0.133$ m^{-1}, the variance of the road microprofile $X(t)$ is $D_x = 2.0$ cm², and the car speed is $v = 15$ m s^{-1}. Evalu-

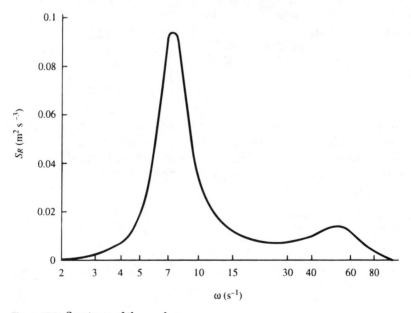

Figure 15.2 Spectrum of the road unevenness.

ate the standard deviation of the vertical acceleration of the car forebody. Compare the obtained result with the maximum acceleration of 2.5 m s^{-2} of a human body during normal walk.

Solution The "time" spectral function $S_0(\omega)$ can be found from the "space" spectral function $S_x(\omega_1)$ as follows:

$$S_0(\omega) = \frac{1}{v} S_x\left(\frac{\omega}{v}\right) = D_x \frac{\alpha_1}{\alpha_1^2 + \omega^2}$$

where $\alpha_1 = \alpha v$. For $v = 15$ m s^{-1} we have: $\alpha_1 = \alpha v = 2.0$ s^{-1}. The spectral function of the acceleration of the car's forebody is

$$S_a(\omega) = S_0(\omega) |\Phi_a(i\omega)|^2 = \omega^4 S_0(\omega) |\Phi(i\omega)|^2 = \omega^4 S_0(\omega) a^2(\omega)$$

The calculated spectral function of the induced accelerations is shown in Fig. 15.2. The variance of the acceleration computed, based on the calculated spectral function, is $D_a = 1.05$ m^2 s^{-4} and the standard deviation is $\sigma_a = \sqrt{D_a} = 1.02$ m s^{-2}. This value is less than half of the maximum acceleration of a human body during normal walk.

15.2. Ships

15.2.1. Loads

Loads experienced by ocean-going ships and other marine vehicles can be broken down into the following three major categories:

1. Loads associated with the unbalanced main and auxiliary engines and mechanisms, with various defects in the propelling and shafts geometry and mounting, and other "inner" imperfections in the vessel's machinery.
2. Hydrodynamic loads due to the operation of the propellers.
3. Wave-induced loads.

The consequences of the loads of the first category are typically addressed by imposing adequate requirements for the accuracy in manufacturing and mounting of the main power plant, auxiliary mechanisms, shafts systems, and propellers. The analyses are carried out on the basis of a probabilistic approach, with consideration of various uncertainties envolved, so as to make sure that the amplitudes and accelerations of the vibrating parts and the induced dynamic stresses will stay within the allowable limits.

The loads of the second category are due to

1. The change in the propeller blades' thrust and drag during one propeller revolution, because of the nonuniform distribution of the flow, as well as the screening effect of the ship's hull (these loads are transmitted to the hull through the bearings).

2. Direct hydrodynamic loading experiencing by the ship's afterbody from the operating propeller.

The frequency of the hydrodynamic loads due to the operation of the propeller is equal to the frequency of the shaft rotation times the number of propeller blades. These loads consist of a regular (nonrandom) and irregular (random) components. The random components are due to sea waves, ship motion, and wake turbulence. Although the application of spectral methods for the prediction of the dynamic responses of ship structures to random excitation is, in principle, possible, it is considered too complicated and is not widely used so far. Naval architects prefer to resort to full-scale and model experiments to analyze these responses and to ensure that they are kept within the allowable limits.

As to the loads of the third category, i.e., wave-induced loads, they can be subdivided into the following two types:

1. Loads that are related linearly to the wave parameters (these are loads causing ship motion, hull bending and torsion, and the "springing" type of wave-induced vibrations).
2. Loads due to various nonlinear effects (ship "slamming," wave-induced vibrations of the "whipping" type, shipping of green water, wave impacts on the ship's sides, etc.).

The "springing" type of ship hull vibration is not affected by the "nonstraightwallness" of the ship sides, but is affected by the "rectilinearness" of the waterlines. The wave-induced vibrations of the "whipping" type are due primarily to the "nonstraightwallness" of the ship's hull at the ship's ends, especially at her forebody. Both the "springing" and the "whipping" types of hull vibrations are forced vibrations of a continuous fashion, although with different frequencies: the frequency of the "springing" vibration is equal to the frequency of the wave encounter, while the dominating "whipping" frequency is twice as high. This is due to the fact that the excitation forces causing the "springing" type of vibration are proportional to the vertical velocities of the ship's hull, while the forces causing "whipping" are proportional to the vertical velocities squared. As far as the "slamming" type of vibration is concerned, it can be defined as shock-excited free vibrations caused by a severe hydrodynamic impact at the ship's bottom forward during her resonance longitudinal motion at ballast condition in rough head seas. This type of vibration, although less strongly pronounced, can be due also to the impact of the bow flare and occur in the full load conditions. Whatever the nature of the hydrodynamic loading, the continuous forced vibration of the "whipping" type and the shock-excited free vibration of the "slamming" type are due to nonlinear effects.

The probabilistic methods and, particularly, the spectral theory of random processes are widely applied for the evaluation of the parameters of the ship motion in waves, as well as for the evaluation of the responses of the ship hull structures to wave loads. The irregular sea waves are treated in oceanography, ocean engineering, and naval architecture as a result of superposition of a large number of regular sinusoidal waves with different frequencies and different directions of propagation, and with random amplitudes and phase angles (Fig. 15.3). Since the probabilistic characteristics of the sea waves change, as a rule, substantially slower than the processes of ship motion and wave-induced bending (which are of the primary interest to ship navigators and naval architects), the sea waves are typically treated in ocean engineering as a stationary (and ergodic) random process.

Although the actual sea waves are three dimensional, in practical ship design problems two-dimensional spectra are most often used.

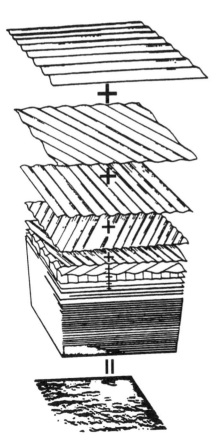

Figure 15.3 A sum of many simple sine waves makes a sea.

Similarly to what has been said in the previous section about the spectra used in the design of surface-going vehicles, the "design" spectra employed in naval architecture do not have to be necessarily the spectra of the actual sea state, whether short or long term, but are supposed, first of all, to adequately reflect various requirements for the ship's seaworthiness and strength.

The ordinates of the two-dimensional irregular waves can be represented as an infinite series:

$$\zeta_w(\xi, t) = \sum_{j=1}^{\infty} r_j \cos(k_j \xi + \omega_j t + \varepsilon_j) = \sum_{j=1}^{\infty} \rho_j e^{i(k_j \xi + \omega_j t)}$$

where $k_j = \omega_j^2/g = 2\pi/\lambda_j$ are the wave numbers, λ_j are the lengths of the components of the irregular waves, $\rho_j = r_j e^{i\varepsilon_j}$ are the complex wave amplitudes, and $i = \sqrt{-1}$ is the "imaginary unity." The amplitudes ρ_j of the component waves are assumed to be statistically independent (uncorrelated), so that the variance D_j of the ordinates of the jth wave can be calculated, assuming $j = l$, as follows:

$$D_j = \frac{1}{T} \int_0^T \rho_j(t) \rho_l(t+\tau) \, dt = \frac{1}{T} \int_0^T \rho_j^2(t) \, dt$$

(for $j \neq l$ this integral is equal to zero). The correlation function of the process $\zeta_w(\xi, t)$ is

$$K_w(\tau) = \frac{1}{T} \int_0^T \zeta_w(\xi, t) \zeta_w(\xi, t+\tau) \, dt$$

$$= \frac{1}{T} \int_0^T \sum_j \rho_j e^{i(k_j \xi + \omega_j t)} \sum_l \rho_l e^{i[k_l \xi + \omega_l(t+\tau)]} \, dt = \sum_{j=1}^{\infty} D_j e^{i\omega_j \tau}$$

or, in the real form,

$$K_w(\tau) = \sum_{j=1}^{\infty} D_j \cos \omega_j \tau$$

The coordinate ξ plays the same role with respect to the wave numbers k_j (the "frequencies" of the wave shape) as the time t plays with respect to the actual frequencies ω_j. The random process $\zeta_w(\xi, t)$, which possesses the ergodic property, is stationary both "in time" and "in space." Therefore, by analogy with the last formula, one can write

$$K_w(\eta) = \sum_{j=1}^{\infty} D_j \cos k_j \eta, \qquad \eta = \xi_2 - \xi_1$$

Random Loads and Responses 469

TABLE 15.1 Wind and its Effect on Sea

Beaufort number	Wind speed, knots	Wind description	Effects at sea
0	0–0.9	Calm	Sea like a mirror
1	1–3	Light air	Scale-like ripples form, but without foam crests
2	4–6	Light breeze	Small wavelets, short but more pronounced. Crests have a glassy appearance and do not break
3	7–10	Gentle breeze	Large wavelets. Crests begin to break. Foam has glassy appearance. Perhaps scattered white horses
4	11–16	Moderate breeze	Small waves, becoming longer. Fairly frequent white horses
5	17–21	Fresh breeze	Moderate waves, taking a more pronounced long form. Many white horses are formed. Chance of some spray
6	22–27	Strong breeze	Large waves begin to form. While foam crests are more extensive everywhere. Some spray
7	28–33	Moderate breeze	Sea heaps up and white foam from breaking waves begins to be blown in streaks along the direction of the wind. Spindrift begins
8	34–40	Fresh gale	Moderately high waves of greater lengths. Edges of crests break into spindrift. Foam is blown in well-marked streaks along the direction of the wind
9	41–47	Strong gale	High waves. Dense streaks of foam along the direction of the wind. Sea begins to roll. Spray may affect visibility
10	48–55	Whole gale and/or storm	Very high waves with long overhanging crests. The resulting foam in great patches is blown in dense white streaks along the direction of the wind. On the whole, the surface of the sea takes a white appearance. The rolling of the sea becomes heavy and shock like. Visibility is affected
11	56–63	Storm and/or violent storm	Exceptionally high waves. Small- and medium-sized vessels might for a long time be lost to view behind the waves. The sea is completely covered with long white patches of foam lying along the direction of the wind. Everywhere, the edges of the wave crests are blown into froth. Visibility seriously affected
12	64 or higher	Hurricane and typhoon	The air is filled with foam and spray. Sea is completely white with driving spray. Visibility is very seriously affected

As can be seen from the above formulas, the variance of the wave ordinates can be computed as

$$D_w = K_w(0) = \sum_{j=1}^{\infty} D_j$$

It has been established that, for the given wave conditions (see state), the distribution of the wave ordinates follows the normal law with zero mean:

$$f_\zeta(\zeta_w) = \frac{1}{\sqrt{2\pi D_w}} e^{-\zeta_w^2/(2D_w)}$$

Then the wave amplitudes, for sufficiently narrow-band sea spectra, are distributed in accordance with the Rayleigh law:

$$f_r(r) = \frac{r}{D_w} e^{-r^2/(2D_w)}$$

The mean, $\langle r \rangle$, and the variance, D_r, of the amplitudes of irregular sea waves are related to the variance, D_w, of the wave ordinates by the formulas:

$$\langle r \rangle = \sqrt{\frac{\pi}{2} D_w}, \qquad D_r = \frac{4 + \pi}{2} D_w$$

where the variance D_w of the wave ordinates is related to the significant wave height $h_{1/3}$ by the formula (see Sec. 3.3.5)

$$D_w = 0.455 \left(\frac{h_{1/3}}{2}\right)^2$$

The currently accepted wind scale and the description of the corresponding state of the sea are shown in Table 15.1. The numerical values of the parameters D_w, D_r, and $\langle r \rangle$, calculated from the above formulas, are shown in the Table 15.2. As can be seen from the table data, the Rayleigh distribution underestimates the actual wave heights for relatively calm seas and overestimates them for rough seas. A better description of the sea state, when the wave spectrum is not narrow banded, can be achieved by using the Rice distribution for the wave heights (see Sec. 3.3.6).

The sea wave spectra used in naval architecture characterize the distribution of the wave energy between regular components of various frequencies of the most severe, i.e., fully developed, sea conditions. With an increase in the significant wave heights the maxima of

TABLE 15.2 Wind and Its Effect on Wave Parameters

Wave parameters	Wind					
	Calm	Light	Weak	Moderate		

Wave parameters	Calm	Light	Weak	Moderate	Waves	Brisk	Storm
$h_{1/3}$ (m)	0	0–0.14	0.14–0.42	0.42–0.70	0.70–1.12	1.12–1.96	1.96–3.36	3.36–4.76	4.76–6.16	>6.16
D_w (m²)	0	0–0.002	0.002–0.02	0.02–0.056	0.056–0.143	0.143–0.435	0.435–1.29	1.29–2.6	2.6–4.6	>4.3
D_r (m²)	0	0–0.008	0.008–0.07	0.07–0.325	0.325–0.510	0.510–1.56	1.56–4.60	4.60–6.65	6.65–15.5	>15.5
$\langle r \rangle$ (m)	0	0–0.06	0.06–0.175	0.175–0.290	0.29–0.47	0.47–0.82	0.82–1.4	1.4–2.0	2.0–2.6	>2.6
$\langle r \rangle$ (m)†	0.17	0.19	0.23	0.29	0.38	0.50	1.04	1.21	1.28	>1.28
Actual ω_{max} (s⁻¹)‡	—	—	—	1.2	0.8	0.6	0.4	0.3	0.3	—

† Mean values of the half-heights of waves observed by Roll in North Atlantic.
‡ Wave frequencies, corresponding to the maximums of the spectral densities of wave ordinates.

the spectral densities shift toward lower frequencies. This is due to the fact that, when the wave energy increases, longer waves are more stable and are able to conserve the wave energy better.

There are numerous sea wave spectra suggested for the purpose of ship design. Although the majority of such spectra are intended for the North Atlantic region, there are many other spectra designed for different ocean areas and different ship types and applications (Great Lakes wave spectra, Mediterranean wave spectra, North Sea wave spectra, Caspian Sea wave spectra, etc.). For open sea conditions, the 15th International Towing Tank Conference (ITTC) recommended the use of the following International Ship and Offshore Structures Congress (ISOSC) spectral formulation for fully developed seas (intended for navigation in the North Atlantic):

$$S_w(\omega) = 0.11 h_{1/3}^2 \frac{\langle \tau \rangle}{2\pi} \left(\frac{\omega \langle \tau \rangle}{2\pi} \right)^{-5} \exp\left[-0.44 \left(\frac{\omega \langle \tau \rangle}{2\pi} \right)^{-4} \right]$$

Here $h_{1/3}$ is the significant wave height and $\langle \tau \rangle$ is the mean wave period. This period can be evaluated by the formula

$$\langle \tau \rangle \equiv 2\pi \frac{m_0}{m_1}$$

where

$$m_k = \int_0^\infty \omega^k S(\omega) \, d\omega$$

are the moments of the wave spectrum. In practice, an approximate formula

$$\langle \tau \rangle = 4.55 h_{1/3}^{0.4}$$

can be used to assess the mean wave period.

The ITTC spectrum is a modified Pierson–Moskowitz spectrum obtained experimentally for the developed sea state in the North Atlantic:

$$S_w(\omega) = 16.2 \times 10^{-3} \frac{g^2}{\omega^2} \exp\left[-0.74 \left(\frac{g}{\omega v_w} \right)^4 \right]$$

In this formula, g is the gravitational acceleration and v_w is the wind speed.

In the case of limited fetch (the size of the area of the wave function), the 17th ITTC recommended that the following Joint North Sea Wave Project (JONSWAP) wave spectrum is used:

$$S_w(\omega) = 155 \frac{h_{1/3}^2}{\langle\tau\rangle^4 \omega^5} \exp\left(-\frac{944}{\langle\tau\rangle^4 \omega^4}\right)(3.3)^Y$$

Here

$$Y = \exp\left[-\left(\frac{0.191\omega\langle\tau\rangle - 1}{\sigma\sqrt{2}}\right)^2\right]$$

is the "peak enhancement" factor and the parameter σ is equal to $\sigma = 0.07$ for $\omega \le 5.24/\langle\tau\rangle$ and to $\sigma = 0.09$ for $\omega > 5.24/\langle\tau\rangle$. The JONSWAP project originated in 1967 as a cooperative experiment on sea waves between a number of institutes in the Netherlands, Germany, the United States and the United Kingdom. The two basic aims of the study were to measure the growth of waves under limited fetch conditions and the attenuation of waves coming into shallow water. The main purpose in studying wave spectra under limited fetch conditions was to determine the form of the source function of the energy balance for the wave spectrum during wave growth. Knowledge of this function is an essential requirement in numerical wave models. It was found that the main features of the observed source function could be accounted for by nonlinear wave-wave interactions. The energy transfer due to these interactions was found to control the overall balance within the wave spectrum to the extent that measured wave spectra (with narrow peaks, "overshoot" factors, and fetch-dependent high-frequency tails) could all be explained in terms of this nonlinear transfer process. The JONSWAP spectrum represents a generalization of the Pierson–Moskowitz spectrum through the inclusion of fetch as an additional parameter to the wind speed. It provides a good description of sea conditions in the North Sea and other fetch-limited water areas. Typical irregular sea spectra are shown in Fig. 15.4.

The ITTC, Pierson–Moskowitz, and JONSWAP spectra are "short-term" spectra: they assume that the significant wave height $h_{1/3}$ and the mean wave period $\langle\tau\rangle$ do not change during the time considered. Various "long-term" spectra consider the variability of these parameters for a ship navigating during her lifetime in different ocean areas. These spectra can be constructed by investigating the joint probability ("frequency of occurrence") of the significant wave height and the mean wave period. It has been suggested, for instance,

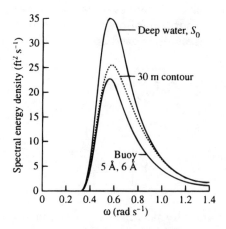

Figure 15.4 Deepwater spectrum and computed spectra at two points in a port approach channel.

that the long-term distribution of significant wave heights and mean periods follows a Weibull, or log-normal, distribution.

15.2.2. Responses

Whatever the spectrum, the ship's responses (such as, for example, the parameters of her motion or the wave-induced bending moments and stresses) can be determined by treating the ship in irregular seas as a dynamic system, which transforms the input process of sea waves into the responses (output processes) of interest. If, for instance, wave-induced bending moments are considered, one should first determine, experimentally (in a model basin) or theoretically (based on the theory of ship motions), the response functions—the ratios of the amplitudes of the bending moments when sailing in regular waves to the radius (half the height) of these waves. The response function should be evaluated not only for different wave frequencies ω but for different ship speeds, v, as well:

$$a_M(\omega; \text{Fr}) = \frac{M_a(\omega, \text{Fr})}{r}$$

The Froude number $\text{Fr} = v/\sqrt{gL}$ (L is the ship's length) is a similitude criterion for ships of different lengths and speeds. This number should be calculated with consideration of the actual sea conditions, i.e., with taking into account the loss of the ship's speed in waves, as well as, sometimes, the deliberate speed reduction by navigators to avoid shipping of green water on deck, or ship slamming, to improve ship's course keeping, her steerability, etc.

After the response function $a_M(\omega; \text{Fr})$ is determined, the long-term variance, D_M, of the bending moments can be computed from the chosen long-term wave spectrum, $S_w(\omega)$, by the formula:

$$D_M = \int_0^\infty S_w(\omega) a_M^2(\omega; \text{Fr})\, d\omega$$

Here, as has been mentioned, the Froude number, Fr, should be evaluated for the accepted ship's speed, typical for the sea conditions, i.e., conditions that correspond to the spectrum $S_w(\omega)$.

In order to determine the long-term probabilistic characteristics of the bending moments, one has to assume a certain law of the long-term probability distribution for these moments. This law can be the Weibull law, the log-normal law, or a law based on extreme value distributions. Then one can evaluate the probability that the long-term bending moment will exceed the given (allowable) level M^* and establish the section modulus of the ship's hull in such a way that the probability that this level is exceeded during the ship's lifetime is sufficiently low (say, as low as $P = 10^{-7}$ to 10^{-8}).

Another approach can be based on the ship's response to the short-term wave spectra $S_j(\omega)$ for the jth wave conditions, characterized by the given significant wave height $h_{1/3}(h_j)$ and the mean wave period $\langle \tau \rangle (\tau_j)$. If the ship is operated in the jth wave conditions for her entire lifetime, the variance, D_j, of the wave bending moments induced by such waves would be

$$D_j = \int_0^\infty S_j(\omega) a_M^2(\omega; \text{Fr}_j)\, d\omega$$

If one assumes that the distribution of the random amplitudes of the wave bending moments is rayleighian, then the probability that a certain level M^* will be exceeded, when sailing in the jth sea condition, can be calculated as

$$P_j = P(M_a > M^*) = e^{-(M^*)^2/(2D_j)}$$

The ship encounters different wave conditions during her lifetime, and the probability q_j of encountering the jth wave condition is, of course, different for different conditions, as well as for different ships, ocean areas, ship routes, seasons, etc. If an ocean-going vessel is intended for navigation in the North Atlantic region, which is one of the most severe and, at the same time, the most actively navigated region of the world ocean, the probability q of encountering the given wave condition, characterized by the significant wave height $h_{1/3}$, can be taken from Table 15.3.

The resulting probability P that the level M^* of the bending moment is exceeded can be calculated by the following "*total probability*" formula:

$$P = \sum_j p_j q_j = \sum_j q_j e^{-(M^*)^2/(2D_j)} = e^{-(M^*)^2/(2D_M)}$$

This formula can be applied, for the given (allowable) P value, to different M^* levels. The design value M^* of the wave bending moment (i.e., the value that will be used to establish the adequate section modulus of the ship's hull) can be accepted as a level that could be exceeded with a sufficiently low allowable probability P. The equivalent variance D_M can be found as

$$D_M = -\frac{(M^*)^2}{2 \ln P}$$

Example 15.2 Establish the required section modulus of the ship's hull on the basis of a probabilistic approach. Apply this approach to the evaluation of the wave-induced bending moments only. The yield strength σ_Y of the hull material is known and cannot be exceeded during the ship's lifetime.

Solution The bearing capacity of a ship hull subjected to overall bending is $C = S\sigma_Y$, where S is the hull's section modulus and σ_Y is the yield strength of the hull material. The load acting on the hull can be represented as a sum of the bending moment, M_s, in still water, and the additional wave-induced bending moment M_w:

$$D = M_s + M_w$$

In accordance with an approach described in Chapter 10, the function of nonfailure (safety margin) can be evaluated as

$$\psi = C - D = S\sigma_Y - (M_s + M_w)$$

For the sake of simplicity, let us assume that the mean value of the yield stress is a deterministic quantity, i.e., equal to its nominal value ($\langle \sigma_Y \rangle = \sigma_Y$), and that the variance of this stress is zero ($D_{\sigma_Y} = 0$). In addition, let us assume that the bending moment M_s in still water is also a deterministic value (and therefore $\langle m_s \rangle = M_s$ and $D_{M_s} = 0$). Finally, we assume that for a straight-walled ship one can put the mean value of the wave bending moment equal to zero ($\langle m_w \rangle = 0$). With these assumptions, we have the

TABLE 15.3 Probabilities of Encountering Given Wave Conditions

$h_{1/3}$ (m)	1.68	3.36	5.04	6.72	8.40	10.07
q	0.1500	0.0501	0.0092	0.876×10^{-3}	0.437×10^{-4}	0.115×10^{-5}

following expressions for the mean and the variance of the function of nonfailure:

$$\langle \psi \rangle = S\sigma_Y - M_s, \qquad D_\psi = D_M$$

Then the safety index can be expressed as

$$\delta_R = \frac{\langle \psi \rangle}{\sqrt{2D_\psi}} = \frac{S\sigma_Y - M_s}{\sqrt{2D_M}}$$

Solving this equation for the section modulus S, we conclude that the strength condition for the overall strength of the ship's hull can be written as follows:

$$S \geq \frac{M_s + \delta_R \sqrt{2D_M}}{\sigma_Y}$$

If a deterministic approach were used (when the wave bending moment is treated as a nonrandom value) one would have to assume $D_M = \frac{1}{2}M_w^2$. Then the obtained strength condition would be written as

$$S \geq \frac{M_s + \delta_R M_w}{\sigma_Y}$$

Comparing the last two formulas, one can see that, with the taken assumptions, the role of the probabilistic approach is reduced to the evaluation of the variance of the wave bending moments. The safety index δ_R plays the role of a safety factor with respect to the wave bending moments. Whatever approach is used to determine the design variance D_M of the wave bending moment, one can write the formula for the probability, that the design value M_w of this moment is exceeded, as follows:

$$P = e^{-M_w^2/(2D_M)}$$

so that the standard deviation of the wave bending moment is

$$\sqrt{D_M} = \frac{M_w}{\sqrt{-2 \ln P}}$$

and the strength condition for the ship's hull can be written as

$$S \geq \frac{M_s + nM_w}{\sigma_Y}$$

The magnitude of the factor $n = \delta_R/\sqrt{-\ln P}$ can be established on the basis of the accumulated experience for the existing ocean-going ships. Analysis of the building rules of different classification societies has shown that the probability level of $P = 10^{-8}$ can be taken as a standard one. The obtained condition can be used for a comparative assessment of the ultimate strength of various ship hulls, as well as for the extrapolation of the accumulated experience in the construction and operation of ocean-going ships on new vessels, whose material and/or type are not considered by the existing rules of ship construction.

15.3. Aircrafts

15.3.1. Loads

The emergence of jet aviation in the postwar period imposed elevated requirements on the reliability of aircraft structures. This, as well as the recognition and rapid advancement of probabilistic and statistical methods, resulted in a wide application of these methods in flight mechanics and in the analysis and design of aircraft structures.

The aircraft, its power plant, and the board equipment (both general and auxiliary) is a complex system. Design, manufacturing, and testing of this system and its numerous subsystems, as well as processing and analysis of various test data, are carried out in avionic engineering on the basis of probabilistic and statistical methods. The reliability of all the aircraft apparatus and flight safety is based on probabilistic reliability engineering, not to mention all the radio, electronic, telemetric, and other "high-technology" systems, equipment, and devices. In the majority of cases, the necessity in application of probabilistic methods is due to the random nature of the processes under investigation, such as, for instance, the loading of aircraft structures during its flight in a turbulent atmosphere, or during its motion on runways or backtracks. In other cases, although a problem can be solved in the first approximation as a deterministic one, the application of probabilistic methods enables one to consider the effects of various random disturbances and measurement errors.

Typical loads acting on an aircraft structure can be broken down into two major categories:

1. Static loads:
    (a) Loads due to the execution of a maneuver
    (b) Loads due to the operation of various outboard systems
    (c) Loads associated with the static stability of the plane (e.g., reversals, divergence, etc.)
    (d) Thermally induced loads
2. Dynamic loads:
    (a) Acoustic loads (pressure pulsations in the boundary layer, loads due to flow separation from badly streamlined structures,

loads due to the jets from the engines or from the propellers, etc.)

(b) Loads causing forced vibration of the aircraft structures (these are due to the amospheric turbulence, buffeting at high angles of attack, or during transonic flight, excitations from the engines, etc.)

(c) Loads associated with the dynamic aeroelastic stability (flutter, self-excited structural vibrations in the transonic zone, etc.)

We limit our discussion of the application of probabilistic methods in avionic engineering to the field of the statistical description of atmospheric turbulence and the dynamic loading on aircraft structures.

The motion of air in the Earth's atmosphere has a strong influence on the flight dynamics and structural response of an aircraft, its structures, and equipment. In the aircraft design, one distinguishes between the small-scale motions of atmospheric turbulence and large-scale air motions in the atmosphere (wind). In the description that follows we will address small-scale motions (atmospheric turbulence) only.

From the standpoint of an avionic engineer or an aircraft navigator, the turbulent atmosphere can be viewed as a three-dimensional time-independent random field. If the correlation theory of random media is applied, the turbulent atmosphere is treated as an isotropic field characterized by the longitudinal and the transverse correlation functions $K_r(r)$ and $K_n(r)$ for the random speeds of the atmospheric pulsations. These functions can be written as

$$K_r(r) = \sigma_t^2 e^{-|r|/L}, \qquad K_n(r) = \sigma_t^2 \left(1 - \frac{r}{2L}\right) e^{-|r|/L}$$

where σ_t is the intensity of the atmospheric turbulence (the standard deviations of the air speeds) and L is its scale. The atmospheric turbulence is considered low when $\sigma_t < 0.5$ m s^{-1}. It is considered high when $\sigma_t > 2.5$ m s^{-1}. The length L characterizes the correlations between the numerous low-level air gusts in different points of the turbulence field. This length is found to be approximately equal to the flight altitude when flying over a more or less flat surface ($L \approx H$) and to be twice as large as the flight altitude ($L \approx 2H$) when flying over high hills and mountains. At high altitudes, the scale of the turbulence is strongly dependent on the climate and meteorological conditions in the given geographical region.

If the aircraft's speed V can be assumed constant within the distances r comparable with the turbulence correlation length L, then, by putting $r = V\tau$, one obtains

$$K_r(\tau) = \sigma_t^2 e^{-|\tau|/T}, \qquad K_n(\tau) = \sigma_t^2\left(1 - \frac{\tau}{2\tau}\right)e^{-|\tau|/T}$$

Here τ is the time required for the aircraft to move from one point of the space to another, and $T = L/V$ (correlation time) is the time required to cover a distance equal to the turbulence scale L.

The spectral densities corresponding to the longitudinal and the transverse correlation functions of the atmospheric turbulence can be computed as

$$S_r(\omega) = \frac{1}{2\pi}\int_{-\infty}^{\infty} K_r(r)e^{-i\omega r}\, dr = \frac{\sigma_t^2 L}{\pi(1 + L^2\omega^2)}$$

$$S_n(\omega) = \frac{1}{2\pi}\int_{-\infty}^{\infty} K_n(r)e^{-i\omega r}\, dr = \frac{\sigma_t^2 L(1 + 3L^2\omega^2)}{\pi(1 + L^2\omega^2)^2}$$

Putting $L = TV$ into these formulas, we obtain

$$S_r(\omega) = \frac{\sigma_t^2 T}{\pi(1 + T^2\omega_1^2)}, \qquad S_n(\omega) = \frac{\sigma_t^2 T}{\pi}\frac{1 + 3T^2\omega_1^2}{(1 + T^2\omega_1^2)^2}$$

where the "time-related" frequency ω_1 is associated with the "space-related" frequency ω by the formula $\omega_1 = \omega V$.

The dynamic loading on aircraft structures is typically broken down into vibrational loading of a general nature and acoustic loading, although there is no stringent dividing line between these two types of loading. The reason why acoustic loading and sound-excited vibrations are singled out from the general dynamic loadings and mechanical (nonacoustic) vibrations is due to the wish to separate the behaviors and dynamic processes that are thought to be primarily responsible for the service life of thin-walled structures of an aircraft from the loadings that are thought to be primarily responsible for the ultimate (short-term) strength of the aircraft structures. In addition, methods for analyzing thin-walled structures with stiffeners are based on the theory of thin shells and are often different from methods used to analyze the behavior of other avionic structures and structural elements. The major sources of acoustic loads are: pressure pulsations due to aircraft buffeting (up to 180 dB); noise due to the screened jets (up to 180 dB), compressors and blowers (up to 160 dB); propellers (up to 155 dB); and pressure pulsations in the turbulent boundary layer (up to 145 dB). Some structural elements are prone to acoustic fatigue caused by acoustic load levels as low as 130 dB. This might cause serious problems, since the number of cycles of variable pressures that is accumulated by aircraft plating during one hour of flight is, on average, as high as 2×10^6.

Although thin-walled aircraft structures indeed are operated typically in a highly intense acoustic environment, acoustic vibrations can also be caused by nonacoustic sources, i.e., are not necessarily associated with the propagation of sound in the surrounding environment or in the structure itself. If the pressure pulsations (often accompanied by the separation of the turbulent flow from badly streamlined structures) are localized in zones close to the aircraft surface, such pulsations can result in extensive vibrations of thin-walled structural elements, although they will radiate very little acoustic waves. On the other hand, the jets produced by the working engines generate almost purely acoustic excitations in areas remote from the aircraft surface, which are, however, often too weak to cause appreciable structural vibrations. However, even if the excitation is purely acoustic and the vibrations are significant, these vibrations, although referred to as acoustic ones, are primarily mechanical, i.e., are affected to only a small extent by the initiation and propagation of sound in the structure itself. The aircraft plating and its fasteners (stringers, frames, compensation units, etc.) are the most susceptible to acoustic loading and vibrations. The frequencies of acoustic vibrations of these elements are typically within 50 and 2500 Hz.

15.3.2. Responses

The major responses of aircraft structures to random excitation are buffeting, flutter, and continuous resonant vibrations. Buffeting (aerodynamic shaking due to flow separation), flutter (continuous vibrations caused by flow separation), and resonant vibrations of the aircraft structures have been studied extensively, with application of probabilistic approaches, since the late 1930s. Examples are: the evaluation of the probability of encounter with the given type and level of loading, both in the air and on the ground; new phenomena associated with the flight conditions at speeds close to and exceeding the speed of sound in air; application of "smart" materials and structures; evaluation of the performance of the system "elastic-aircraft automatic-control system," acoustic strength of aircraft structures, etc.

The general (nonacoustic) vibrations can be due to the interaction of the aircraft structure and the air flow (vibrations caused by flow separation belong to this category) or to the operation of the main power plants and continuously operated auxiliary machinery, or can be caused by short-term intensive loading or periodic shocks. Flow separation from badly streamlined structures results in aerodynamic pressure pulsations, and, in addition, creates vortices which can cause vibrations in relatively remote areas. For instance, vortices separated from the wings or the fuselage can cause vibration of the aircraft's tail. Flow separation is usually of a random nature. Vibrations due to the working engine are due to excitations transmitted directly through the engine foundation. These vibrations occupy a very wide range of frequencies: low frequencies (between 3 and 40 Hz) due to various pumps; mid-range frequencies (between 40 and 70 Hz) due to vibrations of gas turbines and shafts; high frequencies (up to hundreds of Hz), typical of vibrational combustion and variable forces due to the static and dynamic disbalance of rotors.

Although buffeting and flutter can occur as a result of various types of flow separation or as a result of action of the turbulent wake, transonic buffeting is of primary importance. In addition to elastic vibrations, severe buffeting can interfere with the aircraft motion as a solid body and can affect its stability and maneuverability. Buffeting with overloadings from ± 0.1 to ± 0.2 g is tolerable, overloadings from ± 0.2 to ± 0.6 g are uncomfortable, and overloadings from ± 0.6 to 1.0 g are unbearable, even if they last for only a few secods. Tail buffeting is less dangerous than wing buffeting, but can be severe if the tail is located inadequately with respect to the wing.

Evaluation of the response of an aircraft structure to dynamic loading, whether vibrational or acoustic, is carried out on the basis of the correlation or, more often, spectral theory of the transformation of random excitations by dynamic systems. The computations are often simplified, assuming that the excitation spectrum is constant within the transfer function of the system, especially in the areas of the resonance peaks.

During flight tests, it is important to choose, in an optimal way, the test "sample." The probabilistic formulation of this problem can be based on the evaluation of the minimum of the penalty function which considers losses of two types: (1) direct losses (or the risk of the losses) due to the given aircraft "sample" (its cost, as well as the cost of the tests; losses due to the delay in other tests which might have been conducted on this aircraft, etc.) and (2) indirect losses due to insufficient completeness on insufficient reliability of the preliminary information, obtained from calculations and ground tests about the aircraft strength. The direct losses decrease with an increase in the sequential number of the manufactured aircraft of the given series: flight tests on an aircraft with well-developed main systems, well-investigated flight and technical characteristics, and well-understood characteristics of stability and control require smaller expenses. The indirect losses, however, increase with an increase in the sequential number of manufactured aircraft, since the later the structural shortcomings are detected, the higher is the cost of correcting them. It has been found, at least in the American practice, that the total cost of the direct and indirect losses can be minimized if the third or the fourth aircraft is selected for flight tests.

15.4. Earthquakes

15.4.1. Loading

It is commonly recognized that various earthquake problems should be approached on the basis of probabilistic methds. In effect, one of the major problems of the design of structures that would be able to withstand earthquake loadings—choosing the design level of such loadings—has always been approached from the viewpoint of the intuitive assessment of the "most probable" strength of the expected earthquake, using the accumulated experience for the given seismic region, as well as common sense. The probabilistic nature of seismic forces has two basic sources. The first is that each particular earthquake is a random event and the second is due to the fact that the location of the epicenter, the depth of the earthquake cell (center), the amount of the released energy, the earthquake duration, and other factors are random variables, and the earthquake-induced ground motion and the dynamic response of the civil engineering structures are random processes.

In contemporary earthquake engineering, the accumulated statistics of the oscillations of the ground during intensive earthquakes enable one to carry out more or less reliable probabilistic assessment of the strength of structures subjected to earthquake loading. The probabilistic approach to the prediction of the structural responses to earthquake excitations is based on an assumption that the ground accelerations in each actual earthquake make up a random process whose probabilistic characteristics can be established using the measured data from past earthquakes. The objective is to determine the probabilistic characteristics of the "output" processes of displacements, accelerations, and stresses from the probabilistic characteristics of the "input" process of the ground accelerations. In such a formulation, the basic problem of earthquake engineering has a lot in common with similar problems in other areas of engineering. At the same time, an earthquake engineer, unlike a specialist in, say, probabilistic telecommunications, automatic control, or mechanical design, is interested in the probabilistic characteristics of the maximum values of extremely rare events. In addition, an earthquake engineer needs to know the complete probabilistic description of the structural responses, i.e., the probability distributions of the stresses, displacements, etc. A knowledge of the means, variances, and correlation functions is often not sufficient to predict the earthquake responses and consequences. At the same time, the earthquake excitations are different in different geographical areas and at different times, and in the majority of cases are, essentially, nonstationary.

The following major probabilistic approaches are used in earthquake engineering:

1. Approaches that treat seismic excitation as a nonstationary random process
2. Approaches that treat seismic loading as a stationary random process
3. Approaches that treat seismic loading as a markovian process and
4. Approaches that treat seismic excitation as a series of uncorrelated shock impulses

In accordance with the approach that views earthquake loading as a nonstationary process, the ground acceleration during an earthquake is represented as

$$W_0(t) = \sum_{k=1}^{\infty} A_k(t) Z_k(t)$$

where $A_k(t)$ are deterministic functions of time and $Z_k(t)$ are stationary random processes. In an approximate analysis one can ignore the change in the frequency content of the ground oscillations during earthquakes and consider only the first term in this series:

$$W_0(t) = A(t) Z(t)$$

A real accelerogram of a strong earthquake, as well as the functions $A(t)$ and $Z(t)$, are shown in Fig. 15.5. One assumes that the functions $A(t)$ and $Z(t)$ depend on a finite number of random parameters q_v, $v = 1, 2, \ldots, m$, characterizing the earthquake: the depth of the site of origin, its energy, the distance from the epicenter, the characteristics of the Earth's crust along the propagation of seismic waves, the engineering and geological situation at the given location, etc. Assuming that the first r parameters q_v describe the shape of the envelope and the remaining $m-r$ parameters describe the properties of the function $Z(t)$, one can write

$$W_0(t) = A(q_1, q_2, \ldots, q_r, t) Z(q_{t+1}, \ldots, q_m, t)$$

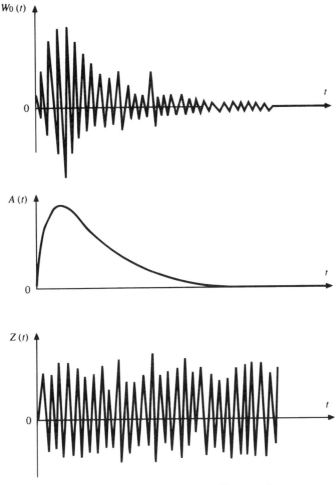

Figure 15.5 The actual ground acceleration, $W_0(t)$, can be represented as a sum of the envelope, $A(t)$, and high-frequency stationary random process, $Z(t)$.

Particularly, the envelope function can be accepted as

$$A(t) = A_0 e^{-ct}$$

or, for greater flexibility,

$$A(t) = A_0 e^{-c_0 t} - A_1 e^{-c_1 t}$$

For a probabilistic description of the ground acceleration, one should know a multivariate probability distribution function $P = P(q_1, q_2, \ldots, q_m)$ for the parameters q_v. It is usually assumed that this function can be established on the basis of seismological and other data. Then, for each typical combination of the parameters q_v, one can find, on the basis of processing the corresponding "representative" accelerograms, the correlation function and the spectral density of the random process $Z(t)$, and then calculate the function $W_0(t)$. The mean value $\langle w_0 \rangle$ of the ground acceleration can usually be assumed to be equal to zero. The probabilistic description of the seismic loading in the case of many terms in the series for the acceleration function can be obtained in a similar way. After the correlation function of the ground acceleration has been determined, the probabilistic characteristics of the dynamic response of the structure can be established on the basis of the general methods of the theory of random processes.

15.4.2. Responses

As an example, examine a system with n degrees of freedom, subjected to seismic loading. This can be, say, a multistorey building experiencing horizontal lateral vibration. The building can be idealized, from the standpoint of its response to an earthquake excitation, as a vertical cantilever beam carrying n concentrated masses on the floor levels. The solution to the differential equations of seismic vibrations can be written, using the method of principal coordinates for a system with discrete masses and assuming zero initial conditions of the induced displacements, as follows:

$$y_k(t) = \sum_{i=1}^{n} X_{ik} T_i(t), \quad k = 1, 2, \ldots, n$$

Here $y_k(t)$ is the lateral displacement of the kth concentrated mass M_k, X_{ik} is the ordinate of the ith mode of free lateral vibrations of the beam at the point of location of the kth mass, and $T_i(t)$ is the principal coordinate of the ith mode. This coordinate can be evaluated, using the Duhamel integral, as

$$T_i(t) = -\frac{D_i}{\omega_i} \int_0^t W_0(\tau) e^{-r\omega_i(t-\tau)} \sin \omega_i(t-\tau) \, d\tau$$

where ω_i is the frequency of free vibrations of the ith mode, the factor D_i is expressed as

$$D_i = \frac{\sum_{v=1}^{n} M_v X_{iv}}{\sum_{v=1}^{n} M_v X_{iv}^2}$$

$W_0(t)$ is the lateral component of the seismic acceleration of the ground (cantilever base), r is the coefficient of viscous damping, and t is time. The bivariate correlation functions $K_{ij}(t_1, t_2)$ of the principal coordinates $T_i(t)$ can be found from the correlation function $K_0(t_1, t_2)$ of the acceleration of the building's foundation using the formula

$$K_{ij}(t_1, t_2) = \langle T_i(t_1)T_j(t_2) \rangle$$
$$= \int_0^{t_1} \int_0^{t_2} D_i D_j K_0(\tau_1, \tau_2) h_i(t_1 - \tau_1) h_j(t_2 - \tau_2) \, d\tau_1 \, d\tau_2$$

where

$$h_i(t) = \frac{1}{\omega_i} e^{-r\omega_i t} \sin \omega_i t$$

is the impulse transfer function (the response to a unit impulse) of the ith mode. The variances of the induced displacements $y_k(t)$ can be evaluated as

$$D_{y_k} = \langle y_k^2(t) \rangle = \sum_{i=1}^n \sum_{j=1}^n K_{ij}(t, t) X_{ik} X_{jk}$$

The existing investigations indicate that for small enough values r of viscous damping and the nondense frequency spectrum, the bivariate correlations of the vibration modes play an insignificant role, so that only the autocorrelation functions $K_{ii}(t, t)$ can be considered. The main difficulties in the practical implementation of the described approach are associated with the evaluation of the correlation function K_0, characterizing the probabilistic properties of the seismic excitation.

Substantial simplifications can be achieved if the ground acceleration can be treated as a stationary random process. Such a treatment can sometimes be justified by the fact that on some accelerograms of strong earthquakes, or at least on some portions of these accelerograms, the ordinates of the excitation envelope and the frequency content of the loading change rather slowly. At the same time, seismic forces obtained on the basis of an assumption that the seismic loading can be treated as a stationary excitation are deemed to be higher than those obtained using an approach that considers the fact that the earthquake loading is a nonstationary process. Therefore such a simplification can be justified in engineering practice. The correlation function of the steady-state ground accelerations can be computed as

$$K_0(\tau) = K_0(t_1, t_2) = K_0(t_2 - t_1) \simeq \frac{1}{T} \int_0^T W_0(t) W_0(t + \tau)\, dt$$

where $W_0(t)$ is the recorded random process of the ground acceleration and T is the recording time, which should be chosen long enough.

The correlation functions in earthquake engineering can be satisfactorily approximated by the equation

$$K_0(\tau) = D_0 e^{-\alpha|\tau|} \cos \beta \tau$$

where $D_0 = K_0(0)$ is the variance of the ground accelerations. The parameters α and $\bar{\omega}_0$ characterize the correlation time and the prevailing frequency of the ground oscillations. The measured values of these parameters for various earthquakes are $\alpha = 6$ to 9 and $\beta = 14$ to 20. With the above formula for the correlation function, the spectral density of the ground accelerations is (see Example 8.1, in Sec. 8.4)

$$S_0(\omega) = \frac{2}{\pi} \int_0^\infty K_0(\tau) \cos \omega \tau\, d\tau = \frac{2\alpha D_0}{\pi} \frac{\omega^2 + b^2}{\omega^4 + 2a\omega^2 + b^4}$$

where $a = \alpha^2 - \beta^2$ and $b^2 = \alpha^2 + \beta^2$. The function $S_0(\omega)$ has its maximum at $\omega = \sqrt{b(2\beta - b)}$. Calculations indicate that the peak values of the spectrum $S_0(\omega)$ occur for periods $\tau_e = 2\pi/\omega = 0.30 - 0.45$ s.

The dynamic response of a structure to a seismic excitation is, as a rule, non-steady-state, even though the excitation itself might be stationary. Moreover, since the non-steady-state seismic vibrations taking place at the first period of the excitation are characterized by the highest induced stresses, the dynamic response to seismic excitations must be evaluated with consideration of the non-steady-state conditions. The transfer function for non-steady-state conditions can be evaluated by the formula

$$|\Phi(i\omega, t)|^2 = |\Phi_0(i\omega)|^2$$
$$\times \{1 + e^{-2r\omega_0 t} - 2e^{-r\omega_0 t}[\cos(\omega - \omega_0)t - r \sin(\omega - \omega_0)t]\}$$
$$- re^{-r\omega_0 t} \frac{\omega_0^4 - \omega^2 \omega_0^2 - \omega\omega_0(\omega_0^2 - \omega^2)}{M^2(\omega^4 - 2\omega^2 \omega_0^2 + \omega_0^4)^2} \sin(\omega - \omega_0)t$$

Here ω_0 is the frequency of free vibrations of the system and

$$\Phi_0(i\omega) = \frac{1}{M(\omega_0^2 - \omega^2 + 2ri\omega_0^2)}$$

is the transfer function in steady-state conditions. Calculations indicate that the maximum values of the dynamic factor during non-steady-state oscillations exceed the steady-state dynamic factor by 35 to 40 percent. If the total duration of an earthquake is about 15 to 25 s, the duration of its non-steady-state portion is about 10 s.

Further Reading (see Bibliography): 6, 10, 14, 17, 29, 36, 48, 51, 52, 57, 68, 75, 77, 78, 80, 81, 90, 95, 99, 112, 114, 115, 119, 120, 121, 128, 139, 140, 142, 146.

Chapter

16

Processing of Random Data

"All life is an experiment. The more experiments you make the better."
RALPH WALDO EMERSON

"There is no such thing as failed experiment. There are only experiments with unpredictable outcomes."
UNKNOWN ENGINEER

"... But of all these principles, least squares is the most simple: by the others we would be led into the most complicated calculations."
KARL FRIEDRICH GAUSS

16.1. Evaluation of the Moments of Random Variables

Approximate values of the moments of random variables obtained on the basis of processing of measured data are called *estimates* of these variables. We will use notations with a tilde for an estimate of the given random variable: $\langle \tilde{x} \rangle$ for $\langle x \rangle$, $\tilde{D}_x$ for D_x, $\tilde{K}_{xy}$ for K_{xy}, etc. The set of values (realizations) $x_1, x_2, \ldots, x_n$ of a random variable X, obtained as a result of n experiments, is known as *sample of size n*. It is supposed that the experiments are carried out in identical conditions and independently. When the size of the sample increases without any limit ($n \to \infty$), a *consistent estimate* converges in probability to the estimated value (parameter). If the mean value of the estimate coincides with the sought parameter for any sample size, the estimate is called the *unbiased estimate*:

$$\langle \tilde{x} \rangle = \frac{1}{n} \sum_{i=1}^{n} x_i = C + \frac{1}{n} \sum_{i=1}^{n} (x_i - C) \qquad (16.1)$$

where C is an arbitrary number, introduced for the sake of convenience of the calculations ("false zero"). If the mean value is unknown, the unbiased estimate of the variance is

$$\tilde{D}_x = \frac{1}{n-1} \sum_{i=1}^{n} (x_i - \langle \tilde{x} \rangle)^2 = \frac{1}{n-1} \sum_{i=1}^{n} (x_i - C)^2 - \frac{n}{n-1} (\langle \tilde{x} \rangle - C)^2 \quad (16.2)$$

If the unknown random variable follows the normal distribution, the unbiased estimate of the standard deviation is

$$\tilde{\sigma}_x = k_n \left[\frac{1}{n-1} \sum_{i=1}^{n} (x_i - \langle \tilde{x} \rangle)^2 \right]^{1/2} \quad (16.3)$$

where

$$k_n = \sqrt{\frac{n-1}{2}} \frac{\Gamma[(n-1)/2]}{\Gamma(n/2)} \quad (16.4)$$

and $\Gamma(\alpha)$ is the gamma function defined by the formula (3.38). The tabulated values of the parameter k_n are given in Table 16.1. If the mean value is known, the unbiased estimate of the variance is

$$\tilde{D}_x = \frac{1}{n} \sum_{i=1}^{n} (x_i - \langle x \rangle)^2 \quad (16.5)$$

If the sought random value is normally distributed, the unbiased estimate for the standard deviation can be computed as follows:

$$\tilde{\sigma}_x = k_{n+1} \left[\frac{1}{n} \sum_{i=1}^{n} (x_i - \langle x \rangle)^2 \right]^{1/2} \quad (16.6)$$

If $x_1, y_1, \ldots, x_n, y_n$ are the values of the random variables X and Y, and are obtained as a result of n independent experiments carried out in identical conditions, the unbiased estimate of the correlation

TABLE 16.1 Factor k_n in the Formula for the Unbiased Estimate of the Standard Deviation

n	3	4	5	6	7	10	12	15
k_n	1.1284	1.0853	1.0640	1.0506	1.0423	1.0280	1.0230	1.0181
n	20	25	30	35	40	45	50	
k_n	1.0134	1.0640	1.0104	1.0087	1.0072	1.0064	1.0056	1.0051

moment of these variables can be evaluated by the formula

$$\tilde{K}_{xy} = \frac{1}{n-1} \sum_{i=1}^{n} (x_i - \langle \tilde{x} \rangle)(y_i - \langle \tilde{y} \rangle) \qquad (16.7)$$

if the means of the random variables X and Y are unavailable, or by the formula

$$\tilde{K}_{xy} = \frac{1}{n} \sum_{i=1}^{n} (x_i - \langle x \rangle)(y_i - \langle y \rangle) \qquad (16.8)$$

if these means are known. The estimate of the correlation coefficient r_{xy} is

$$\tilde{r}_{xy} = \frac{\tilde{K}_{xy}}{\tilde{\sigma}_x \tilde{\sigma}_y} \qquad (16.9)$$

Example 16.1 In order to determine the resolution of a measuring device, whose systematic error is practically equal to zero, five independent measurements were carried out. The results of these measurements are shown in Table 16.2.

Determine the unbiased estimate of the variance of the errors of the measuring device if (a) the magnitude of the measured value is known and is equal to 2800 m; (b) the magnitude of the measured value is unknown.

Solution When the magnitude of the measured value is known, the unbiased estimate of the variance is

$$\tilde{D}_x = \frac{1}{n} \sum_{i=1}^{n} (x_i - \langle x \rangle)^2 = \frac{6439}{5} = 1287.8 \text{ m}^2$$

When the magnitude of the measured value is unknown, the estimate of this value is

$$\langle \tilde{x} \rangle = \frac{1}{n} \sum_{i=1}^{n} x_i = 2809 \text{ m}$$

Therefore the unbiased estimate of the variance in this case is

$$\tilde{D}_x = \frac{1}{n-1} \sum_{i=1}^{n} (x_i - \langle \tilde{x} \rangle)^2 = \frac{6034}{4} = 1508.5 \text{ m}^2$$

TABLE 16.2 Measurements Results

Measurement number	1	2	3	4	5
x_i (m)	2781	2836	2807	2763	2858

16.2. Fiducial Probabilities and Confidence Intervals

The *confidence interval* is an interval that covers, with the given *fiducial (confidence) probability* α, the parameter θ under estimation. The width 2ε of a symmetrical confidence interval is determined by the condition

$$P\{|\theta - \tilde{\theta}| \le \varepsilon\} = \alpha \qquad (16.10)$$

where $\tilde{\theta}$ is the estimate of the parameter θ and the probability in the left part of this condition is determined by the distribution law for the estimate $\tilde{\theta}$. The limits $\tilde{\theta} - \varepsilon$ and $\tilde{\theta} + \varepsilon$ of the confidence interval are called *confidence limits*. Fiducial (confidence) probabilities and confidence intervals are used to obtain an idea of how accurate the statistical estimate is and of how close the estimate is to the characteristics of the theoretical distribution. In Sec. 14.2 it was shown, in application to geometric tolerances, how confidence intervals can be obtained for normally distributed random variables, with consideration of the variability of the parameters of the distribution. In this section we will provide, without derivations, the basic formulas for the evaluation of fiducial probabilities and confidence intervals, for the normal and the exponential laws of distribution.

If $(x_1, x_2, \ldots, x_n)$ is the sampling from a normal general population, the fiducial probability α for the mean value is determined by the formula

$$\alpha = P\{|\tilde{x} - \langle x \rangle| \le \varepsilon\} = \Phi\left(\frac{\varepsilon\sqrt{n}}{\sigma_x}\right) \qquad (16.11)$$

if the standard deviation σ_x is available. If the standard deviation is unavailable, the fiducial probability α should be calculated by the formula

$$\alpha = \int_{-t_\alpha}^{t_\alpha} S_n(t)\, dt \qquad (16.12)$$

where the integrand represents the Student distribuion (see Sec. 3.3.12 and Appendix E) expressed by the formula (3.55). The limits of integration in the formula (16.12) are

$$t_\alpha = \frac{\varepsilon\sqrt{n}}{\tilde{\sigma}_x} = \frac{\varepsilon\sqrt{n}}{[1/(n-1)\sum_{i=1}^{n}(x_i - \langle \tilde{x} \rangle)^2]^{1/2}} \qquad (16.13)$$

The values t_α can be found from Appendix E as a function of the number k, which is related to the number n of the degrees of freedom as $k = n - 1$, and the fiducial probability α.

The fiducial probability α for the standard deviation σ_x can be calculated as

$$\alpha = P\{|\tilde{\sigma}_x - \sigma_x| \le \varepsilon\} = P\left\{\frac{\sqrt{k}}{1+q} < \chi < \frac{\sqrt{k}}{1-q}\right\} = \int_{\sqrt{k}/(1+q)}^{\sqrt{k}/(1-q)} P_k(\chi)\, d\chi \qquad (16.14)$$

where the function $P_k(\chi) = P(\chi^2, k)$ (χ^2-distribution or Pearson distribution) is expressed by the formula (3.36). The values of the integral $\int_{\sqrt{k}/(1+q)}^{\sqrt{k}/(1-q)} P_k(\chi)\, d\chi$ are tabulated (see Appendix D). The confidence interval $(\gamma_1 \tilde{\sigma}_x, \gamma_2 \tilde{\sigma}_x)$ for the σ_x value, for which the probability of being found outside the left limit $\gamma_1 \tilde{\sigma}_x$ and the right limit $\gamma_2 \tilde{\sigma}_x$ is the same and equal to $(1-\alpha)/2$, is determined by the formula

$$P\left\{\frac{\sigma_x}{\tilde{\sigma}_x} \le \gamma_1\right\} = P\left\{\frac{\sigma_x}{\tilde{\sigma}_x} \ge \gamma_2\right\} = P\left\{\chi^2 \ge \frac{n-1}{\gamma_2^2}\right\} = P\left\{\chi^2 \ge \frac{n-1}{\gamma_1^2}\right\} = \frac{1-\alpha}{2} \qquad (16.15)$$

In order to evaluate the factors γ_1 and γ_2 for the given fiducial probability α and the number $k = n - 1$ one can use the tables in Appendix D.

For the exponential law of the distribution of the random variable X, the confidence interval $(v_1 \langle \tilde{x} \rangle, v_2 \langle \tilde{x} \rangle)$ for the mean value $\langle x \rangle$ can be found from the equation

$$P\left\{\frac{\langle x \rangle}{\langle \tilde{x} \rangle} \le v_1\right\} = P\left\{\frac{\langle x \rangle}{\langle \tilde{x} \rangle} \ge v_2\right\} = P\left\{\chi^2 \le \frac{2n}{v_2} = \chi^2_{1-\delta}\right\}$$

$$= P\left\{\chi^2 \ge \frac{2n}{v_1} = \chi^2_\delta\right\} = \frac{1-\alpha}{2} = \delta \qquad (16.16)$$

From this relationship we have $v_1 = 2n/\chi^2_\delta$, $v_2 = 2n/\chi^2_{1-\delta}$. The values χ^2_δ and $\chi^2_{1-\delta}$ are determined from Appendix D for the probabilities, which are equal to δ and $1-\delta$, respectively, and the number of the degrees of freedom is $k = 2n$.

When the sample size is large ($n > 15$), the limits of the confidence interval for the $\langle x \rangle$ value can be evaluated by the following approximate formulas:

$$\frac{4\sum_{i=1}^n x_i}{(\sqrt{4n-1}+\varepsilon_0)^2},\quad \frac{4\sum_{i=1}^n x_i}{(\sqrt{4n-1}-\varepsilon_0)^2} \qquad (16.17)$$

where ε_0 should be found from the equation $\alpha = \Phi(\varepsilon_0)$.

Example 16.2 The mean value of the distance to the land mark, obtained on the basis of four independent measurements, is equal to 2250 m. The mean value of the error of the measuring device is $E = 40$ m. Evaluate the con-

fidence interval for the measured distance with the probability 95 percent.
Solution The probability that the true value of the distance $\langle x \rangle$ will be covered by the interval $(\langle \tilde{x} \rangle - \varepsilon, \langle \tilde{x} \rangle + \varepsilon)$ with random limits, provided that the mean E is available, can be evaluated as

$$P\{|\langle \tilde{x} \rangle - \langle x \rangle| \le \varepsilon\} = \frac{\rho}{\sqrt{\pi} E_1} \int_{-\varepsilon}^{\varepsilon} \exp\left(-\rho^2 \frac{z^2}{E_1^2}\right) dz = \hat{\Phi}\left(\frac{\varepsilon}{E_1}\right)$$

where $E_1 = E/\sqrt{n}$ is the mean deviation of the random variable $\langle \tilde{x} \rangle = \frac{1}{n} \sum_{i=1}^{n} x_i$. Solving the equation $\hat{\Phi}(\varepsilon\sqrt{n}/E) = 0.95$, which can be done on the basis of Appendix B, we find that $\varepsilon\sqrt{n}/E = 2.91$, so that $\varepsilon = (2.91/\sqrt{n})E = (2.91 \times 40)/2 = 58.2$ m. Then the sought upper and lower limits of the confidence interval are $2250 + 58.2 = 2308.2$ m and $2250 - 58.2 = 2191.8$ m, respectively.

Example 16.3 The standard deviation of the error of the altimeter is $\sigma_x = 15$ m. The measurement errors follow normal probability distributions, and there are no systematic errors in the measurements. It is required that the error in the evaluation of the mean altitude $\langle \tilde{x} \rangle$ exceeds -30 m with the probability 99 percent. How many altimeters should one have on board the aircraft?
Solution The conditions of the problem can be written as

$$P\{-30 < \langle \tilde{x} \rangle - \langle x \rangle < \infty\} = 0.99$$

The random variable

$$Z = \langle \tilde{x} \rangle - \langle x \rangle = \frac{1}{n} \sum_{i=1}^{n} x_i - \langle x \rangle$$

is a linear function of normally distributed random variables, and therefore its distribution is normal as well. The parameters of this distribution are

$$\langle z \rangle = \left\langle \frac{1}{n} \sum_{i=1}^{n} x_i - \langle x \rangle \right\rangle = 0, \qquad D_z = \frac{\sigma_x^2}{n}$$

Then we have

$$P(-30 < Z < \infty) = \frac{1}{\sigma_z \sqrt{2\pi}} \int_{-30}^{\infty} e^{-z^2/(2\sigma_z^2)} dz = \frac{1}{2}\left[1 + \Phi\left(\frac{30}{\sigma_z}\right)\right] = 0.99$$

Solving the equation

$$\Phi = \left(\frac{30\sqrt{n}}{\sigma_x}\right) = 0.98$$

which can be done by using Appendix B tables, we find

$$\frac{30\sqrt{n}}{\sigma_x} = 2.33$$

so that

$$n = \left(\frac{2.33\sigma_x}{30}\right)^2 = 1.36$$

Hence, there should be at least two altimeters on board the aircraft.

Example 16.4 During the control testing of 16 electric bulbs, the estimates of the mean and the standard deviation of their service time were evaluated. These values turned out to be $\langle \tilde{x} \rangle = 3000$ hours and $\tilde{\sigma}_x = 20$ hours, respectively. The service time of the bulbs follows the normal law. Determine the confidence interval for the mean value and the standard deviation for the fiducial probability $P = 0.9$.

Solution The limits of the confidence interval for the mean value can be found from the equation

$$\alpha = \int_{-t_\alpha}^{t_\alpha} S_n(t) \, dt$$

Using the tabulated data for the t_α value with $k = n - 1$ and $\alpha = 0.9$ (see Appendix E) we find

$$t_\alpha = \frac{\varepsilon \sqrt{n}}{\tilde{\sigma}_x} = 1.753$$

so that

$$\varepsilon = \frac{1.753 \tilde{\sigma}_x}{\sqrt{n}} = 8.765 \text{ hours}$$

The upper and the lower limits of the confidence interval for the $\langle x \rangle$ value are $3000 + 8.765 = 3008.765$ hours and $3000 - 8.765 = 2991.235$ hours, respectively. Using the tabulated data for the values γ_1 and γ_2 with $k = 15$ and $\alpha = 0.9$ (see Appendix E), we have $\gamma_1 = 0.775$ and $\gamma_2 = 1.437$. Thus, for the fiducial probability 0.9, the values of the standard deviation σ_x, compatible with the observed data, fall within the limits from $0.775 \tilde{\sigma}_x = 15.50$ hours to $1.437 \tilde{\sigma}_x = 28.74$ hours.

Example 16.5 Using the conditions of the previous example, evaluate the probability that the absolute value of the error, with which the mean value $\langle x \rangle$ is determined, will not exceed 10 hours, and that the error, with which the standard deviation σ_x is determined, will not exceed 2 hours.

Solution The probability of the event

$$-10 \text{ hours} < \langle \tilde{x} \rangle - \langle x \rangle < 10 \text{ hours}$$

is determined by the Student distribution

$$\alpha = P\{|\langle x \rangle - \langle \tilde{x} \rangle| < \varepsilon\} = \int_{-t_\alpha}^{t_\alpha} S_n(t) \, dt$$

Using the tabulated data (see Appendix E) for the t_α value, with $t_\alpha = \varepsilon \sqrt{n}/\tilde{\sigma}_x = 10 \sqrt{16}/20 = 2$ and the number of the degrees of freedom $k = n - 1 = 15$, we find $\alpha = 0.93$. Using the χ distribution, one can determine the probability of the unequality

$$-2 \text{ hours} < \tilde{\sigma}_x - \sigma_x < 2 \text{ hours}$$

as follows:

$$\alpha = P\{|\tilde{\sigma}_x - \sigma_x| < \varepsilon\} = \int_{\sqrt{k/(1+q)}}^{\sqrt{k/(1-q)}} P_k(\chi) \, d\chi$$

TABLE 16.3 Calculated Characteristics of the Confidence Interval

n	χ_δ^2	$\chi_{1-\delta}^2$	v_1	v_2
3	12.60	1.63	0.48	3.68
5	18.30	3.94	0.55	2.54
10	31.40	10.90	0.64	1.83

With $q = \varepsilon/\tilde{\sigma}_x = 2/20 = 0.1$ and the number of the degrees of freedom $k = n = 15$, using the tabulated data, from Appendix E, we find $\alpha = 0.41$.

Example 16.6 The random variable T follows the exponential law of the probability distribution

$$f(t) = \frac{1}{\langle t \rangle} e^{-t/\langle t \rangle}$$

The estimate of the parameter $\langle t \rangle$ is determined by the formula $\langle \hat{t} \rangle = (1/n) \sum_{i=1}^{n} t_i$. Find the limits of the confidence interval for the $\langle t \rangle$ value as functions of the estimate $\langle \hat{t} \rangle$, so that the condition

$$P(v_1 \langle \hat{t} \rangle > \langle t \rangle) = P(v_2 \langle \hat{t} \rangle < \langle t \rangle) = \frac{1-\alpha}{2}$$

is fulfilled with $\alpha = 0.9$, for $n = 3, 5, 10, 20, 30,$ and 40.

Solution In accordance with the condition of the problem, we have

$$P(v_1 \langle \hat{t} \rangle > \langle t \rangle) = P(v_2 \langle \hat{t} \rangle < \langle t \rangle) = \frac{1-\alpha}{2} = \delta$$

This condition can be rewritten as

$$P\left(\frac{2n\langle \hat{t} \rangle}{\langle t \rangle} > \frac{2n}{v_1} = \chi_\delta^2\right) = P\left(\frac{2n\langle \hat{t} \rangle}{\langle t \rangle} < \frac{2n}{v_2} = \chi_{1-\delta}^2\right) = \delta$$

The random variable $U = 2n\langle \hat{t} \rangle / \langle t \rangle$ is distributed in accordance with Pierson's law (χ^2-distribution) with $2n$ degrees of freedom, and for $n > 15$ the random variable $Z = \sqrt{2U}$ has a close-to-normal distribution, with

TABLE 16.4 Calculated Characteristics of the Confidence Interval

n	ε_0	v_1	v_2
20	1.65	0.72	1.53
30	1.65	0.76	1.40
40	1.65	0.79	1.33

$\langle z \rangle = \sqrt{2n-1}$ and $\sigma_z = 1$. Therefore, for $n < 15$, we have

$$v_1 = \frac{2n}{\chi_\delta^2}, \quad v_2 = \frac{2n}{\chi_{\delta-1}^2}$$

Using Appendix D data, we find χ_δ^2 and $\chi_{1-\delta}^2$ for the number of degrees of freedom $2n$ and the probabilities δ and $1-\delta$, and calculate the v_1 and v_2 values. This can be done in the form of Table 16.3.

For $n > 15$, in accordance with the formulas (16.17), we have

$$v_1 = \frac{4n}{(\sqrt{4n-1}+\varepsilon_0)^2}, \quad v_2 = \frac{4n}{(\sqrt{4n-1}-\varepsilon_0)^2}$$

The calculated v_1 and v_2 values are given in Table 16.4. The ε_0 is determined for $\alpha = 0.9$ from the table in Appendix B.

16.3. Least-Square Method

The least-square method can be used to determine the estimates for the parameters of the functional relationships between the variables whose values are obtained experimentally. The type of functional relationship is supposed to be known.

If one obtained $n+1$ couples (x_i, y_i) of values for the argument x_i and the function y_i, then the parameters of the approximating function are being chosen in such a way that the sum

$$S = \sum_{i=0}^{n} [y_i - F(x_i)]^2 \quad (16.18)$$

be minimum. If a polynomial

$$F(x) = Q_m(x) = a_0 + a_1 x + \cdots + a_m x^m \quad (m \leq n) \quad (16.19)$$

is taken as an approximating function, then the estimates $\tilde{a}_k$ of its coefficients can be found from the following system of $m+1$ "normal" equations:

$$\sum_{i=0}^{m} s_{k+j} a_j \equiv s_k a_0 + s_{k+1} a_1 + \cdots + s_{k+m} a_m = v_k, \quad k = 0, 1, \ldots, m$$

$$(16.20)$$

where

$$s_k = \sum_{i=0}^{n} x_i^k, \quad k = 0, 1, \ldots, 2m \quad (16.21)$$

and

$$v_k = \sum_{i=0}^{n} y_i x_i^k, \quad k = 0, 1, \ldots, m \quad (16.22)$$

If the x_i values are known precisely and the y_i values are independent and equally accurate, then the estimate for the variance $\tilde{\sigma}^2$ of the y_i value can be evaluated by the formula

$$\tilde{\sigma}^2 = \frac{1}{n-m} S_{\min} \qquad (16.23)$$

where $S_{\min}$ is the S value calculated on the basis of an assumption that the coefficients of the polynomial $F(x) = Q_m(x)$ are substituted with their estimates, which are found from the above system of "normal" equations. For the normally distributed random variable y_i, this method leads to the most accurate evaluation of the approximating function $F(x)$.

The estimates $\tilde{\sigma}^2_{a_k}$ of the variances of the coefficients $\tilde{a}_k$ and the correlation moments $\tilde{K}_{a_k a_j}$ can be found as

$$\tilde{\sigma}_{a_k} = M_{kk}\tilde{\sigma}^2, \qquad \tilde{K}_{a_k a_j} = M_{kj}\tilde{\sigma}^2 \qquad (16.24)$$

where $M_{kj} = \Delta_{kj}/\Delta$, $\Delta = |d_{kj}|$ is the determinant of the system of the $m+1$ "normal" equation, $d_{kj} = s_{k+j}$ $(j, k = 0, 1, \ldots, m)$, and Δ_{kj} is the cofactor of the element d_{kj} is the cofactor of the element d_{kj} in the determinant Δ.

If the function $y(x)$ is linear $(m = 1)$,

$$y = a_0 + a_1 x$$

then

$$\tilde{a}_0 = \frac{s_2 v_0 - s_1 v_1}{s_2 s_0 - s_1^2}, \qquad \tilde{a}_1 = \frac{s_0 v_1 - s_1 v_0}{s_2 s_0 - s_1^2}$$

$$\tilde{\sigma}^2_{a_0} = \frac{s_2}{s_2 s_0 - s_1^2} \frac{S_{\min}}{n-1}, \qquad \tilde{\sigma}^2_{a_1} = \frac{s_0}{s_2 s_0 - s_1^2} \frac{S_{\min}}{n-1}$$

$$\tilde{K}_{a_0 a_1} = \frac{s_1}{s_2 s_0 - s_1^2} \frac{S_{\min}}{n-1}$$

The confidence intervals for the coefficients a_k are

$$\tilde{a}_k - \gamma \tilde{\sigma}_{a_k} < a_k < \tilde{a}_k + \gamma \tilde{\sigma}_{a_k}$$

where the γ values can be obtained from the tables for the Student distribution (Appendix E), with the given probability α and the number $k = n - m$ of the degrees of freedom.

Example 16.7 When investigating the temperature effect on the rate ω of a chronometer, the data given in Table 16.5 were obtained. The calculated

TABLE 16.5 Measured Temperature Effect on the Rate of a Chronometer

t_i (°C)	5.0	9.6	16.0	19.6	24.4	29.8	34.4
ω_i	2.60	2.01	1.34	1.08	0.94	1.06	1.25

values $\tilde{\omega}$ of the rate ω satisfy the relationship

$$\tilde{\omega} = a_0 + a_1(t - 15) + a_2(t - 15)^2$$

For the fiducial probability $\alpha = 0.9$, determine the estimates for the coefficients a_k, the estimates of the square deviations $\tilde{\sigma}$ (single measurement) and $\tilde{\sigma}_{a_k}$ for the coefficients a_k, and establish the confidence intervals for the coefficients a_k and for the mean square deviation σ which characterizes the accuracy of each particular measurement.

Solution We form the "normal" equations for the evaluation of the coefficients a_k and M_{kk}. In order to reduce the magnitudes of the coefficients in these equations, we introduce a new variable

$$x = \frac{t}{15} - 1$$

and seek the approximating function as

$$y = a'_0 + a'_1 x + a'_2 x^2$$

The calculated coefficients of the "normal" equations are

$$s_0 = 7, \quad s_1 = 2.254, \quad s_2 = 3.712, \quad s_3 = 3.056, \quad s_4 = 4.122$$

$$v_0 = 10.28, \quad v_1 = 1.215, \quad v_2 = 5.017$$

and the "normal" equations can be written as

$$\left.\begin{array}{r}7\tilde{a}'_0 + 2.254\tilde{a}'_1 + 3.712\tilde{a}'_2 = v_0 \\ 2.254\tilde{a}'_0 + 3.712\tilde{a}'_1 + 3.056\tilde{a}'_2 = v_1 \\ 3.712\tilde{a}'_0 + 3.056\tilde{a}'_1 + 4.122\tilde{a}'_2 = v_2\end{array}\right\}$$

From these equations we find

$$\tilde{a}'_0 = 0.2869 v_0 + 0.0986 v_1 - 0.3314 v_2$$
$$\tilde{a}'_1 = 0.0986 v_0 + 0.7248 v_1 - 0.6260 v_2$$
$$\tilde{a}'_2 = -0.3314 v_0 - 0.6260 v_1 + 1.0051 v_2$$

Introducing the v_k values into these equations, we obtain

$$\tilde{a}'_0 = 1.404, \quad \tilde{a}'_1 = -1.246, \quad \tilde{a}'_2 = 0.874\,11$$

The values M_{kk} are coefficients in front of v_k values in each of the obtained relationships for $\tilde{a}'_k$. Hence,

$$M_{00} = 0.2869, \quad M_{11} = 0.7248, \quad M_{22} = 1.0051$$

The calculated S_{min} value which is needed to compute the estimates of the variance of the particular measurement y_i and the variances of the coefficients $\tilde{a}_k$ is $S_{min} = 0.005\,223$. The calculations of the S_{min} values are carried out in Table 16.6.

Then we find

$$\tilde{\sigma}^2 = \frac{S_{min}}{6-2} = 0.001\,306, \qquad \tilde{\sigma} = 0.036\,14$$

$$\tilde{\sigma}^2_{\tilde{a}_0'} = M_{00}\sigma^2 = 0.000\,374\,6, \qquad \tilde{\sigma}^2_{\tilde{a}_1'} = M_{11}\sigma^2 = 0.000\,946\,4,$$

$$\tilde{\sigma}^2_{\tilde{a}_2'} = M_{22}\sigma^2 = 0.001\,312$$

Hence,

$$\tilde{\sigma}_{\tilde{a}_0'} = 0.019\,36, \qquad \tilde{\sigma}_{\tilde{a}_1'} = 0.030\,76, \qquad \tilde{\sigma}_{\tilde{a}_2'} = 0.036\,23$$

Returning to the argument t, we have

$$\tilde{\omega} = \tilde{a}_0 + \tilde{a}_1(t-15) + \tilde{a}_2(t-15)^2$$

where

$$\tilde{a}_0 = \tilde{a}_0' = 1.404, \qquad \tilde{a}_1 = \frac{\tilde{a}_1'}{15} = -0.083\,06, \qquad \tilde{a}_2 = \frac{\tilde{a}_2'}{15^2} = 0.003\,885$$

The corresponding estimates of the mean square deviations are

$$\tilde{\sigma}_{a_0} = \tilde{\sigma}_{\tilde{a}_0'} = 0.019\,36, \qquad \tilde{\sigma}_{a_1} = \frac{\tilde{\sigma}_{\tilde{a}_1'}}{15} = 0.001\,291, \qquad \tilde{\sigma}_{a_2} = \frac{\tilde{\sigma}_{\tilde{a}_2'}}{15^2} = 0.000\,161\,0$$

The confidence intervals for the coefficients a_k, for $\alpha = 0.90$ and the number of the degrees of freedom $k = n - m = 4$, can be found from the value $\gamma = 2.132$, obtained from the table for the Student distribution

TABLE 16.6 Calculated S_{min} Value Affecting the Estimate of the Variance of the i-th Measurement

i	$a_0' + a_1'x_i$	$a_2'x_i^2$	$\tilde{\omega}_i$	ε_i	ε_i^2
0	2.2352	0.3889	2.624	−0.024	0.000576
1	1.8527	0.1133	1.966	0.044	0.001936
2	1.3207	0.0039	1.325	0.015	0.000225
3	1.0217	0.0823	1.104	−0.024	0.000576
4	0.6230	0.3436	0.967	−0.027	0.000729
5	0.1745	0.8515	1.026	0.034	0.001156
6	−0.2067	1.4613	1.255	−0.005	0.000025
				$S_{min} =$	0.005223

(Appendix E). The confidence intervals for a_k are

$$\tilde{a}_k - \gamma\tilde{\sigma}_{a_k} < a_k < \tilde{a}_k + \gamma\tilde{\sigma}_{a_k}$$

and therefore

$$1.363 < a_0 < 1.446$$

$$0.08031 < a_1 < 0.08581$$

$$0.003542 < a_2 < 0.004228$$

The confidence interval for the mean square deviation σ, which characterizes the accuracy of a particular measurement, is

$$\gamma_1 \tilde{\sigma} < \sigma < \gamma_2 \tilde{\sigma}$$

The values γ_1 and γ_2 can be determined from the tabulated data (see Appendix E) for $k = 4$ and $\alpha = 0.90$. Then we have $\gamma_1 = 0.649$, $\gamma_2 = 2.370$, and therefore

$$0.02345 < \sigma < 0.08565$$

16.4. Monte Carlo Simulation

Monte Carlo simulation is a powerful numerical method for solving mathematical problems, not necessarily of a probabilistic nature, by modeling random variables. This method was created by J. von Neumann and S. Ulam in 1949. The name of this method stems from the name of the city of Monte Carlo, Monaco, famous for its casinos, and is due to the fact that roulette is one of the simplest mechanical devices for generating random numbers.

Let us start with a simple example, where the area of a plane figure of an arbitrary shape is to be computed. This can be done by placing the figure inside a figure of a simple shape (say, a square or a circle) and by putting N randomly (and evenly) distributed points inside the outer figure, both inside and outside the figure with the complicated boundary. The area of the figure with the complicated boundary can be approximately calculated as the ratio of the number n of points fallen within this figure to the total number N of points times the area of the outer figure (of a simple configuration). The larger the number N the more accurate is the calculated area of the inner figure. In effect, such a Monte Carlo simulation is rarely used to calculate the areas of plane figures: there are other methods suitable for that. Although these methods might be substantially more complicated than the described procedure, they ensure a substantially better accuracy. However, the Monte Carlo simulation also enables one to evaluate in a similar, very simple fashion, three- and multidimensional "volumes" in a multidimensional space. In some cases the Monte Carlo method is the only numerical method that can be used to solve the problem.

Monte Carlo simulation has, as follows from the examined example, two major features. First, its algorithm is very simple. As a rule, one develops a computer program for carrying out a single random test (in the examined example this test was the selection of a random point within the outer figure and checking whether or not this point belongs to the inner figure as well). This test is repeated N times. Each test is independent from the others, and the results of all the conducted tests are then averaged. Therefore Monte Carlo simulation is sometimes called the *method of statistical testing*. The second important feature is its low accuracy. It can be shown that the computation error in Monte Carlo testing is proportional to the $\sqrt{D/N}$ value, where D is a certain constant and N is the number of tests. This indicates that in order to reduce the error by ten times (i.e., in order to obtain one more decimal in the result), one has to increase the number N (i.e., the work volume) by a hundred times. Because of this one cannot achieve very high accuracy. For this reason, Monte Carlo simulation can be effectively used in those problems in which an accuracy of about 5 to 10 percent is acceptable.

Monte Carlo simulation enables one to model any process that is affected by random factors. Moreover, for many mathematical and applied problems not associated with any random phenomena, one can "artificially" create a probabilistic model that would enable these problems to be solved. In some cases, it might be advisable to replace the actual random process with an "artificial" model and to use this model to facilitate the solution to the basic problem.

In order to use Monte Carlo simulation, one has to be able to generate random numbers. There are three major ways of obtaining random numbers: tables of random numbers, generators of random numbers, and methods of pseudorandom numbers.

A table of random numbers can be obtained, for instance, in the following way. Let one write the numbers 0, 1, 2, ..., 9 on ten identical cards, put these cards in a hat, mix them up, and pull out one card at a time, each time putting it back and mixing up the cards again. The figures obtained in such a way are written down in table form. This table can be introduced in a computer and taken out when necessary. The RAND Corporation table of random numbers contains one million figures. These were obtained by using a special "electronic roulette."

The generators of random numbers typically use noises in transistors. If the level of the noise exceeds the given threshold an even number of times during the time Δt, one writes down "zero." If the level of the noise exceeds the given threshold an odd number of times, one writes down "one."

Since the "quality" of the utilized random numbers is checked by special tests, it does not matter how these numbers were obtained, as long as they satisfy the accepted system of tests. One can even try to calculate them on the basis of a formula. It must be, of course, a rather sophisticated formula. Numbers obtained on the basis of a formula and simulating the magnitude of a random variable γ are called pseudorandom numbers. The first algorithm for obtained pseudorandom numbers was suggested by von Neumann and is called the *method of the middle of squares*.

Let us clarify this method by using the following example. Let us take a four-digit number $\gamma_0 = 0.9876$, square it ($\gamma_0^2 = 0.97535376$), and take the four numbers of it in the middle: $\gamma_1 = 0.6353$. Then we square the γ_1 number ($\gamma_1^2 = 0.28654609$) and take again the four middle numbers. Then we get $\gamma_2 = 0.6546$, and so on. There are many other algorithms for generating pseudorandom numbers. The merits of such an approach are obvious: only several simple operations are needed to obtain one number, so that the speed of generating random numbers is of the same order of magnitude as the speed of the computer itself. In addition, this method requires very little memory of the computer and every number can be easily reproduced. The majority of the Monte Carlo simulations is currently performed by using pseudorandom numbers.

When solving various problems one has to model different random variables. In the early stages of the use of Monte Carlo simulations, every random variable was generated on the basis of its own "roulette." For instance, in order to find the realizations of a random variable with the distribution

X:	x_1	x_2	x_3	x_4
	0.5	0.25	0.125	0.125

one used a roulette with sectors whose areas were related as $4:2:1:1$. This, however, turned out to be unnecessary: the realization of any random variable can be obtained by transforming the values of a single "standard" random variable. Typically, the role of such a random variable is played by a random variable γ, uniformly distributed within the segment (0, 1). The process of obtaining the realizations of a random variable ζ by transforming one or several values of γ is known as drawing of a random variable ζ.

Assume that one has to obtain the values of a discrete random variable ζ with the distribution

$$\zeta: \begin{array}{|c|c|c|c|} \hline x_1 & x_2 & \cdots & x_n \\ \hline p_1 & p_2 & \cdots & p_n \\ \end{array}$$

To do this, we consider the interval $0 < y < 1$ and break it down into n segments whose lengths are $p_1, p_2, \ldots, p_n$. The coordinates of the points of separation of these segments are $y = p_1$, $y = p_1 + p_2$, ..., $y = p_1 + p_2 + \cdots + p_{n-1}$. Let us enumerate the obtained segments as 1, 2, ..., n. Each time we have "to carry out an experiment" and to draw a value ζ, we will choose the value γ and will build a point $y = \gamma$. If this point falls within a segment with a number i, we can consider that $\zeta = x_i$ in this experiment. Indeed, since the random variable γ is uniformly distributed on the interval (0, 1), the probability that γ will fall within a certain interval is equal to the length of its interval:

$$P(0 < \gamma < p_1) = p_1$$
$$P(p_1 < \gamma < p_1 + p_2) = p_2$$
$$\ldots\ldots\ldots\ldots\ldots\ldots\ldots\ldots\ldots$$
$$P(p_1 + p_2 + \cdots + p_{n-1} < \gamma < 1) = p_n$$

In accordance with our procedure, $\zeta = x_i$ when

$$p_1 + p_2 + \cdots + p_{i-1} < \gamma < p_1 + p_2 + \cdots + p_i$$

and the probability of this event is p_i. This procedure can be easily computerized.

It can be shown that, in the case of a continuous random variable ζ, its values can be determined for the selected value γ from the equation

$$\int_a^\zeta p(x)\, dx = \gamma$$

If, for instance, the variable ζ is uniformly distributed within the interval (a, b), then its values can be found from the equation

$$\int_a^\zeta \frac{dx}{b-a} = \gamma$$

and therefore

$$\zeta = a + \gamma(b-a)$$

Monte Carlo simulation is widely and effectively used in many areas of engineering and applied science: queueing theory, nuclear physics, reliability engineering, for the evaluation of definite integrals, especially multidimensional ones, etc. Here we will limit ourselves just to two examples from the field of reliability engineering.

Example 16.8 The electrical device consists of several resistances R_i, capacitances, C_k, etc. Let the quality of the device be determined by the value of one output parameter U, which can be evaluated from the known values of the parameters of all the items:

$$U = f(R_1, R_2, \ldots, C_1, C_2, \ldots)$$

If, for instance, U is the voltage on the working part of the electric network, then, using Ohm's laws, one can find the U value. Using the Monte Carlo method, determine the effect of the deviation of the parameters of the elements from their nominal values on the magnitude U.

Solution In order to use the Monte Carlo simulation, one has to know the probabilistic characteristics of the elements $R_1, R_2, \ldots, C_1, C_2, \ldots$ and to be able to compute the parameter U for the given probabilistic characteristics of the elements. The probabilistic characteristics of the elements can be found experimentally. The calculation procedure is as follows. For each element one draws the value of each parameter and then calculates the U value. Repeating this test N times, one obtains the values $U_1, U_2, \ldots, U_N$, and then calculates the mean

$$\langle u \rangle \simeq \frac{1}{N} \sum_{j=1}^{N} u_j$$

and the variance

$$D_u \simeq \frac{1}{N-1} \left(\sum_{j=1}^{N} u_j^2 - \langle u \rangle^2 \right)$$

Example 16.9 Using the Monte Carlo method, assess the mean time-to-failure of a product for the given mean time-to-failure of each element.

Solution If one knows the probability density functions for each of the elements of the product, then the mean time-to-failure of the product can be calculated in the same fashion as it has been done in the previous example. Indeed, for each element one can draw the random value of a mean time-to-failure for this element, and then, depending on whether the failure of a single element will lead to the failure of the entire product (consequent connection of elements) or not (parallel, redundant, connection of elements), evaluate the value of the mean time-to-failure for the entire product. Repeating this test many times, one can find

$$\langle t \rangle \simeq \frac{1}{N} \sum_{i=1}^{N} \langle t_i \rangle$$

where t_i is the value of the mean time-to-failure in a ith test. This and the previous examples indicate that the substance of the Monte Carlo simulation is quite simple. One has to know the probabilistic characteristics (parameters) of all the elements of the product and be able to calculate the quantity of interest as a function of these characteristics. Then the randomness of these characteristics can be accounted for on the basis of the Monte Carlo simulation.

16.5. Statistical Methods of Quality Control

Statistical control enables one to establish the quality of products on the basis of testing of the product samples, with guaranteed probabilities α of rejecting a sound batch ("vendor's risk") and β of accepting a faulty batch ("customer's risk"). The batch is considered to be a good one if a certain parameter that characterizes the quality of the batch does not exceed the given value and is considered a bad one if this parameter exceeds another given value. For instance, a parameter that characterizes the quality of the batch can be the number of faulty items in the batch, or the mean value of a certain parameter, or, if the homogeneity of the product is important, the standard deviation of a certain parameter. The evaluation of the sample size and the criteria for the acceptance or rejection of the product for the given probabilities α and β is known as the design of the sampling plan. Three major sampling methods are used: single sampling, double sampling, and chain sampling (Wald's method).

16.5.1. Single sampling

When single sampling is used, one determines the sample size n_0 and the acceptance number v. If the value of the controlled parameter in the sampling is below v, the batch is accepted, otherwise it is rejected. If the number of faulty items in a sample of the size n_0 is controlled, the probabilities α and β can be found as follows:

$$\left.\begin{aligned}\alpha &= P(M > v \,|\, L = l_0) = 1 - \sum_{m=0}^{v} \frac{C(l_0, m)C(N - l_0, n_0 - m)}{C(N, n_0)} \\ \beta &= P(M \leq v \,|\, L = l_1) = \sum_{m=0}^{v} \frac{C(l_1, m)C(N - l_1, n_0 - m)}{C(N, n_0)}\end{aligned}\right\} \quad (16.25)$$

Here v is the acceptance number, l_0 is the lower boundary of the number l of the faulty items in the batch, l_1 is the upper boundary of the number of the faulty items ($l_1 > l_0$), L is the total number of the faulty items in the batch, N is the batch size, and n_0 is the sample size. For small n_0 values ($n_0 \leq 0.1N$), the binomial distribution can be used:

$$\left.\begin{aligned}\alpha &= 1 - \sum_{m=0}^{v} C(n_0, m) p_0^m (1 - p_0)^{n_0 - m} = 1 - P(p_0, n_0, v) \\ \beta &= \sum_{m=0}^{v} C(n_0, m) p_1^m (1 - p_1)^{n_0 - m} = P(p_1, n_0, v)\end{aligned}\right\} \quad (16.26)$$

Here $p_0 = l_0/N$, $p_1 = l_1/N$, and the $P(p, n, d)$ values can be calculated from the Appendix A data. If, in addition, the probabilities p_0 and p_1 are small (smaller than, say, 10 percent), then the Poisson distribution can be applied:

$$\alpha = \sum_{m=\nu+1}^{\infty} \frac{a_0^m}{m!} e^{-a_0} = 1 - P(\chi^2 \geq \chi_{q_0}^2)$$
$$\beta = 1 - \sum_{m=\nu+1}^{\infty} \frac{a_1^m}{m!} e^{-a_1} = P(\chi^2 \geq \chi_{q_1}^2)$$
(16.27)

$$\alpha = \sum_{m=\nu+1}^{\infty} \frac{a_0^m}{m!} e^{-a_0} = 1 - P(\chi^2 \geq \chi_{q_0}^2)$$
$$\beta = 1 - \sum_{m=\nu+1}^{\infty} \frac{a_1^m}{m!} e^{-a_1} = P(\chi^2 \geq \chi_{q_1}^2)$$
(16.27)

where

$$a_0 = n_0 p_0, \qquad a_1 = n_0 p_1, \qquad \chi_{q_0}^2 = 2a_0, \qquad \chi_{q_1}^2 = 2a_1$$

The values $\sum_{m=\nu+1}^{\infty} (a^m/m!)e^{-a}$ are given in Appendix A, and the probabilities $P(\chi^2 > \chi_q^2)$ can be found from Appendix D for the number of the degrees of freedom $k = 2(\nu + 1)$.

If the mean value

$$\langle x \rangle = \frac{1}{n_0} \sum_{i=1}^{n} x_i \qquad (16.28)$$

of a parameter in a sample is controlled and the value x_i of this parameter for one item follows the normal law with the known standard deviation σ, then

$$\alpha = \frac{1}{2}\left[1 - \Phi\left(\frac{\nu - \xi_0}{\sigma/\sqrt{n_0}}\right)\right], \quad \beta = \frac{1}{2}\left[1 - \Phi\left(\frac{\xi_1 - \nu}{\sigma/\sqrt{n_0}}\right)\right] \quad (16.29)$$

If $\langle x \rangle$ is not smaller than ν and $\xi_0 > \xi_1$, the batch is accepted. If $\langle x \rangle$ is smaller than ν the batch is rejected and the sign "minus" in front of the Laplace function in the above formulas should be changed to "plus."

If the parameter under control is exponentially distributed:

$$f(x) = \lambda e^{-\lambda x} \qquad (16.30)$$

then

$$\alpha = 1 - P(\chi^2 \geq \chi_{q_0}^2), \qquad \beta = P(\chi^2 > \chi_{q_1}^2) \qquad (16.31)$$

where

$$\chi_{q_0}^2 = 2n_0 \lambda_0 v, \qquad \chi_{q_1}^2 = 2n_0 \lambda_1 v \qquad (16.32)$$

and the probability $P(\chi^2 \geq \chi_q^2)$ can be found from the Appendix D table for the number of degrees of freedom $k = 2n_0$. Here λ_0 and λ_1 are the lower and the upper boundaries of the parameter λ. For $n_0 > 15$ one can use the following approximate formula:

$$P(\chi^2 \geq \chi_q^2) = \frac{1}{2}\left[1 - \Phi\left(\frac{\chi_q^2 - 2n_0}{2\sqrt{n_0}}\right)\right] \qquad (16.33)$$

If the homogeneity of the product is controlled and the parameter that characterizes the quality of the product is normally distributed, then

$$\alpha = 1 - P(\tilde{\sigma} \leq q_0 \sigma_0), \qquad \beta = P(\tilde{\sigma} \leq q_1 \sigma_1) \qquad (16.34)$$

where $q_0 = v/\sigma_0$, $q_1 = v/\sigma_1$, and

$$\tilde{\sigma}^2 = \frac{1}{n_0} \sum_{i=1}^{n} (x_i - \langle x \rangle)^2 \qquad (16.35)$$

if the mean value $\langle x \rangle$ of the parameter is known, or

$$\tilde{\sigma}^2 = \frac{1}{n_0 - 1} \sum_{i=1}^{n} \left(x_i - \frac{1}{n_0} \sum_{j=1}^{n} x_j\right)^2 \qquad (16.36)$$

if the $\langle x \rangle$ value is unavailable. The probabilities $P(\tilde{\sigma} \leq q\sigma)$ are computed using the χ distribution with (see Section 3.3.9) $k = n_0$ degrees of freedom if the $\langle x \rangle$ value is known and with $k = n_0 - 1$ if the $\langle x \rangle$ value is unavailable.

16.5.2. Double sampling

When double sampling is used, one determines the sizes n_1 and n_2 of the first and the second samples and the acceptance numbers v_1, v_2, and v_3 (typically, $v_1 < [n_1/(n_1 + n_2)]v_3 < v_2$). If in the first sample the controlled parameter does not exceed v_1, the batch is accepted if it is greater than v_2, the batch is rejected; in all other cases the second sampling is taken. If the value of the controlled parameter determined from the sample of the size $n_1 + n_2$ does not exceed v_3, the batch is accepted; otherwise it is rejected.

The formula for the probabilities α and β if the number of faulty items in the sample is controlled are as follows:

$$\left. \begin{aligned} \alpha &= 1 - \sum_{m_1=0}^{v_2} \frac{C(l_0, m_1) C(N - l_0, n_1 - m_1)}{C(N, n_1)} \\ &+ \sum_{m_1=v_1+1}^{v_2} \left\{ \frac{C(l_0, m_1) C(N - l_0, n_1 - m_1)}{C(N, n_1)} \right. \\ &\times \left. \left[1 - \sum_{m_2=0}^{v_3 - m_1} \frac{C(l_0 - m_1, m_2) C(N - l_0 - n_1 + m_1, n_2 - m_2)}{C(N - n_1, n_2)} \right] \right\} \\ \beta &= \sum_{m_1=0}^{v_1} \frac{C(l_1, m_1) C(N - l_1, n_1 - m_1)}{C(N, n_1)} \\ &+ \sum_{m_1=v_1+1}^{v_2} \left[\frac{C(l_1, m_1) C(N - l_1, n_1 - m_1)}{C(N, n_1)} \right. \\ &\times \left. \sum_{m_2=0}^{v_3 - m_1} \frac{C(l_1 - n_1, m_2) C(N - l_1 - n_1 + m_1, n_2 - m_2)}{C(N - n_1, n_2)} \right] \end{aligned} \right\} \quad (16.37)$$

As in the case of single sampling, these formulas can be simplified for certain relationships between the numbers n_1, n_2, N, l_0, and l_1.

The cases where the mean value of the parameter is controlled or where the homogeneity of the product is important are not examined.

16.5.3. Chain sampling

When chain sampling (Wald's method) for the variable size n of the sample and the random value of the controlled parameter in the sample is used, the likelihood coefficient γ is calculated. The quality control is performed until the γ value will not get out of the "corridor" between the values $B = \beta/(1 - \alpha)$ and $A = (1 - \beta)/\alpha$. If $\gamma \leq B$, the batch is accepted. If $\gamma \geq A$, the batch is rejected. If $B < \gamma < A$, testing should be continued. This method was examined in detail in Section 10.11.2.2 in connection with technical diagnostics problems.

If the number m of faulty items in the sample is controlled, the coefficient $\gamma = \gamma(n, m)$ is computed as

$$\gamma = \frac{C(l_1, m) C(N - l_1, n - m)}{C(l_0, m) C(N - l_0, n - m)} \quad (16.38)$$

When $n \leq 0.1N$, a simplified formula, based on the binomial law, can be applied:

$$\gamma = \frac{p_1^m (1-p_1)^{n-m}}{p_0^m (1-p_0)^{n-m}} \tag{16.39}$$

where $p_0 = l_0/N$, $p_1 = l_1/N$. The batch is accepted if $m \leq h_1 + nh_3$ and is rejected if $m \geq h_2 + nh_3$ (Fig. 16.1). Testing is continued if $h_1 + nh_3 < m < h_2 + nh_3$. Here

$$\left. \begin{array}{l} h_1 = \dfrac{\log B}{\log p_1/p_0 + \log (1-p_0)/(1-p_1)} \\[2mm] h_2 = \dfrac{\log A}{\log p_1/p_0 + \log (1-p_0)/(1-p_1)} \\[2mm] h_3 = \dfrac{\log (1-p_0)/(1-p_1)}{\log p_1/p_0 + \log (1-p_0)/(1-p_1)} \end{array} \right\} \tag{16.40}$$

Area I in Fig. 16.1 is the acceptance area, area III is the rejection area, and area II gives the n and m values, which indicate that testing should be continued.

We do not examine the cases where the mean value of the quality parameter is controlled or where the homogeneity of the product is important.

Example 16.10 A batch of $N = 40$ items is accepted if the number of faulty items in the batch does not exceed $l_0 = 8$. If the number of the faulty items is higher than $l_1 = 20$, the batch is rejected. Using the single sampling method, determine the probabilities α and β for the sample size $n_0 = 10$ if the acceptance number is $v = 3$.

Solution The sought probabilities are

$$\alpha = 1 - \sum_{m=0}^{3} \frac{C(8, m)C(32, 10-m)}{C(40, 10)} = 0.089$$

$$\beta = \sum_{m=0}^{3} \frac{C(20, m)C(20, 10-m)}{C(40, 10)} = 0.136$$

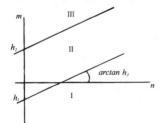

Figure 16.1 Quality Control using Chain Sampling (Wald's method)

Example 16.11 Using the condition of the previous example and the double sampling method, determine the probabilities α and β for the case $n_1 = n_2 = 5$, $v_1 = 0$, $v_2 = 2$, $v_3 = 3$.

Solution Using the formulas

$$\alpha = 1 - \sum_{m=0}^{2} \frac{C(8, m)C(32, 5 - m)}{C(40, 5)} + \sum_{m_1 = 1}^{2} \left\{ \frac{C(8, m_1)C(32, 5 - m_1)}{C(40, 5)} \right.$$
$$\left. \times \left[1 - \sum_{m_2 = 0}^{3 - m_1} \frac{C(8 - m_1, m_2)C(27 + m_1, 5 - m_2)}{C(35, 5)} \right] \right\}$$

$$\beta = \frac{C(20, 0)C(20, 5)}{C(40, 5)} + \sum_{m_1 = 1}^{2} \left[\frac{C(20, m_1)C(20, 5 - m_1)}{C(40, 5)} \right.$$
$$\left. \times \sum_{m_2 = 0}^{3 - m_1} \frac{C(20 - m_1, m_2)C(15 + m_1, 5 - m_2)}{C(35, 5)} \right]$$

we obtain $\alpha = 0.105$, $\beta = 0.134$.

Example 16.12 Using the condition and the results obtained in Example 16.10, design the chain sampling plan and determine $n_{\min}$ for the batches with $L = 0$ and $L = N$.

Solution With $\alpha = 0.089$ and $\beta = 0.136$, we have

$$B = \frac{\beta}{1 + \alpha} = 0.149, \quad \log B = -0.826$$

$$A = \frac{1 - \beta}{\alpha} = 9.710, \quad \log A = 0.987$$

In order to determine $n_{\min}$ in the case when all the items in the batch are sound, we compute the chain values $\log \gamma(n; 0)$ as follows:

$$\log \gamma(1; 0) = \log (N - l_1)! + \log (N - l_0 + 1)!$$
$$- \log (N - l_0)! - \log (N - l_1 + 1)!$$
$$\log \gamma(n + 1; 0) = \log \gamma(n; 0) - \log (N - l_0 - n)! + \log (N - l_1 - n)!$$

Then we have

$\log \gamma(1; 0) = 0.7959;$ $\log \gamma(2; 0) = 0.5833$

$\log \gamma(3; 0) = 0.3614;$ $\log \gamma(4; 0) = 0.1295$

$\log \gamma(5; 0) = -0.1136;$ $\log \gamma(6; 0) = -0.3688$

$\log \gamma(7; 0) = -0.6377;$ $\log \gamma(8; 0) = -0.9217$

Since the condition $\log \gamma(n; 0) < \log B$ is fulfilled only beginning with $n = 8$, then $n_{\min} = 8$.

For a batch consisting the faulty items only, $n = m$. Then we find: $\log \gamma(1; 1) = 0.3979$. For further n values we use the formula

$$\log \gamma(n + 1; m + 1) = \log \gamma(n; m) + \log (l_1 - m) - \log (l_0 - m)$$

Then we obtain

$$\log \gamma(2; 2) = 0.8316, \quad \log \gamma(3; 3) = 1.3087 > \log A = 0.9870$$

Hence, in this case, $n_{\min} = 3$.

Example 16.13 Using single sampling, determine the sample size n_0 and the acceptance number v for a batch size $N = 20$. The batch is considered acceptable if the number of defections does not exceed $l_0 = 4$ pieces, and must be rejected when the number of defectives exceeds $l_1 = 10$. The vendor's risk is $\alpha = 0.10$ and the customer's risk is $\beta = 0.15$.

Solution The formulas (16.25) result in the following equations for the unknowns n_0 and v:

$$S_\alpha(n_0, v) = \sum_{m=0}^{v} \frac{C(4, m)C(10, n_0 - m)}{C(20, n_0)} = 1 - \alpha = 0.90$$

$$S_\beta(n_0, v) = \sum_{m=0}^{v} \frac{C(10, m)C(10, n_0 - m)}{C(20, n_0)} = \beta = 0.15$$

These equations can be rewritten as follows:

$$S_\alpha(n_0, v) = \frac{1}{C(20, n_0)} [C(16, n_0) + 4C(10, n_0 - 1) + 6C(16, n_0 - 2)$$
$$+ 4C(16, n_0 - 3) + C(16, n_0 - 4)]$$

$$S_\beta(n_0, v) = \frac{1}{C(20, n_0)} [C(10, n_0) + 10C(10, n_0 - 1) + 45C(10, n_0 - 2)$$
$$+ 120C(10, n_0 - 3) + 210C(10, n_0 - 4) + 252C(10, n_0 - 5)$$
$$+ 210C(10, n_0 - 6) + \cdots + C(10, n_0 - 10)]$$

Then we have

$$S_\alpha(n_0, 1) = \frac{1}{C(20, n_0)} [C(16, n_0) + 4C(16, n_0 - 1)]$$

and therefore

$$S_\alpha(3, 1) = \frac{1}{1140} (560 + 4 \times 120) = 0.9123 > 0.90$$

$$S_\alpha(4, 1) = \frac{1}{4845} (1820 + 4 \times 560) = 0.8380 < 0.90$$

Similarly, since

$$S_\alpha(n_0, 2) = \frac{1}{C(20, n_0)} [C(16, n_0) + 4C(16, n_0 - 1) + 6C(16, n_0 - 2)]$$

then

$$S_\alpha(6, 2) = \frac{1}{38\,760} (8008 + 4 \times 4368 + 6 \times 1820) = 0.9391 > 0.90$$

$$S_\alpha(7, 2) = \frac{1}{77\,520} (11\,440 + 4 \times 8008 + 6 \times 4368) = 0.8989 < 0.90$$

The equation for the customer's risk yields

$$S_\beta(n_0, 1) = \frac{1}{C(20, n_0)} [C(10, n_0) + 10C(10, n_0 - 1)]$$

so that

$$S_\beta(5, 1) = \frac{1}{15\,504} (252 + 10 \times 210) = 0.1517 > 0.15$$

$$S_\beta(6, 1) = \frac{1}{38\,760} (210 + 10 \times 252) = 0.0704 < 0.17$$

and since

$$S_\beta(n_0, 2) = \frac{1}{C(20, n_0)} [C(10, n_0) + 10C(10, n_0 - 1) + 45C(10, n_0 - 2)]$$

then

$$S_\beta(7, 2) = \frac{1}{77\,520} (120 + 10 \times 210 + 45 \times 252) = 0.1761 > 0.15$$

$$S_\beta(8, 2) = \frac{1}{125\,970} (45 + 10 \times 120 \times 45 \times 210) = 0.0849 < 0.15$$

Since the n_0 and ν values must be expressed by integers, we conclude that the only combination of these values that approximately satisfies the conditions $1 - \alpha = 0.90$ and $\beta = 0.15$ is $n_0 = 7$, $\nu = 2$. Hence, the desired accuracy in meeting the requirements for the risks α and β can be achieved if 7 pieces out of a 20 piece batch are selected for testing. If the number of defectives in this sample does not exceed 2 pieces, the probability that the number of defectives in the entire batch does not exceed 4 pieces is 90 percent, and therefore this batch can be accepted. If the number of defectives is larger than 2, the batch should be rejected.

Example 16.14 A batch of the size $N > 1000$ is undergoing quality control. The acceptable quality level is such that if the percentage defective is $p \leq p_0 = 0.02$, the batch can be accepted, and if $p \geq p_1 = 0.10$, the batch should be rejected. Determine the sample size n_0 and the acceptance number ν for the vendor's risk $\alpha = 0.10$ and the customer's risk $\beta = 0.15$.

Solution The sample size n_0 is expected to be considerably smaller than the batch size N. This enables one to use the binomial distribution to compute the numbers n_0 and ν. From the formulas (16.26) we find

$$S_\alpha(n_0, v) = 1 - P(p_0, n_0, v) = \sum_{m=0}^{v} C(n_0, m)(0.02)^m(0.98)^{n_0-m} = 0.90$$

$$S_\beta(n_0, v) = P(p_1, n_0, v) = \sum_{m=0}^{v} C(n_0, m)(0.10)^m(0.90)^{n_0-m} = 0.15$$

These equations can be written as

$$S_\alpha(n_0, v) = (0.98)^{n_0} + n_0(0.02)(0.98)^{n_0-1} + \frac{n_0(n_0-1)}{2}(0.02)^2(0.98)^{n_0-2} + \cdots$$

$$S_\beta(n_0, v) = (0.90)^{n_0} + n_0(0.10)(0.90)^{n_0-1} + \frac{n_0(n_0-1)}{2}(0.10)^2(0.90)^{n_0-2} + \cdots$$

Then we have

$$S_\alpha(n_0, 1) = (0.98)^{n_0} + n_0(0.02)(0.98)^{n_0-1}$$

$$S_\alpha(26, 1) = (0.98)^{26} + 26(0.02)(0.98)^{25} = 0.9052 > 0.90$$

$$S_\alpha(27, 1) = (0.98)^{27} + 27(0.02)(0.98)^{26} = 0.8990 < 0.90$$

$$S_\alpha(n_0, 2) = (0.98)^{n_0} + n_0(0.02)(0.98)^{n_0-1} + \frac{n_0(n_0-1)}{2}(0.02)^2(0.98)^{n_0-2}$$

$$S_\alpha(55, 2) = (0.98)^{55} + 55(0.02)(0.98)^{54} + \frac{55 \times 54}{2}(0.02)^2(0.98)^{53} = 0.9023 > 0.90$$

$$S_\alpha(56, 2) = (0.98)^{56} + 56(0.02)(0.98)^{55} + \frac{56 \times 55}{2}(0.02)^2(0.98)^{54} = 0.8982 < 0.90$$

$$S_\beta(n_0, 1) = (0.90)^{n_0} + n_0(0.10)(0.90)^{n_0-1}$$

$$S_\beta(32, 1) = (0.90)^{32} + 32(0.10)(0.90)^{31} = 0.1564 > 0.15$$

$$S_\beta(33, 1) = (0.90)^{33} + 33(0.10)(0.90)^{32} = 0.1442 < 0.15$$

$$S_\beta(n_0, 2) = (0.90)^{n_0} + n_0(0.1)(0.90)^{n_0-1} + \frac{n_0(n_0-1)}{2}(0.10)^2(0.90)^{n_0-2}$$

$$S_\beta(45, 2) = (0.90)^{45} + 45(0.10)(0.90)^{44} + \frac{45 \times 44}{2}(0.10)^2(0.90)^{43} = 0.1590 > 0.15$$

$$S_\beta(46, 2) = (0.90)^{46} + 46(0.10)(0.90)^{45} + \frac{46 \times 45}{2}(0.10)^2(0.90)^{44} = 0.1484 < 0.15$$

These data indicate that the acceptance number v can be assumed to be equal to 2 and that the sample size should be around $n_0 = 50$. In this case the formulas (16.26) yield

$$\alpha = 1 - \sum_{m=0}^{2} C(50, m)(0.02)^m(0.98)^{50-m}$$

$$= 1 - (0.98)^{50} - 50(0.02)(0.98)^{49} - \frac{50 \times 49}{2}(0.02)^2(0.98)^{48}$$

$$= 1 - 0.3642 - 0.3716 - 0.1858 = 0.080$$

$$\beta = \sum_{m=0}^{2} C(50, m)(0.10)^m(0.90)^{50-m}$$

$$= (0.90)^{50} + 50(0.10)(0.90)^{49} + \frac{50 \times 49}{2}(0.10)^2(0.90)^{48}$$

$$= 0.0052 + 0.0286 + 0.0779 = 0.112$$

Example 16.15 Solve the problem in the previous example using Poisson's distribution.

Solution Applying the formulas (16.27), we have

$$S_\alpha(n_0, v) = 1 - P(\chi^2 \geq \chi^2_{q_0}) = \sum_{m=v+1}^{\infty} \frac{a_0^m}{m!} e^{-a_0} = 0.90$$

$$S_\beta(n_0, v) = 1 - P(\chi^2 \geq \chi^2_{q_1}) = \sum_{m=v+1}^{\infty} \frac{a_1^m}{m!} e^{-a_1} = 0.15$$

Using the Appendix A data, we find that the equation $S_\alpha = 0.90$ is fulfilled in the following cases:

$$v = 1, \quad a_0 = 0.5309 \ (n_0 = 26.5)$$
$$v = 2, \quad a_0 = 1.0811 \ (n_0 = 54.0)$$
$$v = 3, \quad a_0 = 1.6539 \ (n_0 = 82.7)$$

The equation $S_\beta = 0.15$ takes place for

$$v = 1, \quad a_1 = 3.4569 \ (n_0 = 34.6)$$
$$v = 2, \quad a_1 = 4.7766 \ (n_0 = 47.8)$$
$$v = 3, \quad a_1 = 6.0174 \ (n_0 = 60.2)$$

Since the sample size n_0 and the acceptance number v must be expressed by integers, one should assume $n_0 = 50$, $v = 2$. This result agrees well with the results obtained in Example 16.14. The calculations, however, are much simpler.

16.6. Delphi Method

Useful information can often be obtained by processing opinions of a certain number of experts about a parameter of interest, instead of conducting actual experiments or processing actual experimental data. The experts do not interact and express their opinions independently, typically by answering sets of questions from previously developed questionnaires. The estimates of a particular expert are then treated as independent samples of a random variable, and statistical methods of processing random data are used to process the expert information. The *Delphi method* is the best known and the most widely used procedure of this type.

The Delphi method, as is any other statistical method of processing expert information, is based on an assumption that the deviations of expert estimates from the "true value" of the parameter of interest are due to random causes. The task is to restore such a "true value" with a minimum error. The method enables one to assess the meaningfulness of the estimates, the degree of the agreement of experts opinions, etc. Clearly, the degree of agreement of the experts opinions can be used as a suitable criterion of the quality of the resultant estimates. Let us describe a possible modification of the Delphi method.

Let a parameter X be estimated by N experts. The kth expert, whose competence ("weight") is α_k, estimated the value X as x_k. Then the estimate a of the value X can be found as

$$a = \phi(x_1, \ldots, x_N) = \frac{\sum_{k=1}^{N} x_k \alpha_k}{\sum_{k=1}^{N} \alpha_k} \qquad (16.41)$$

and the degree of disagreement between the expert opinions can be evaluated as

$$D = \frac{\sum_{k=1}^{N} (x_k - a)^2 \alpha_k}{\sum_{k=1}^{N} \alpha_k} \qquad (16.42)$$

where the a value is defined by the formula (16.41). The statistical significance of the obtained results can be defined as an interval in which the value of interest can be found with the given (required, desired) probability $1 - P$, where P is the probability of error:

$$\langle a \rangle - \Delta \leq a \leq \langle a \rangle + \Delta \qquad (16.43)$$

Here $\langle a \rangle$ is the mean value of the parameter a and Δ is the deviation of the actual (random) value of this parameter from its mean value. If the a value has a normal distribution with the mean $\langle a \rangle$ and the variance D, then

$$\Delta = \frac{t\sigma}{\sqrt{N}} \qquad (16.44)$$

where the t value has Student's distribution with $N - 1$ degrees of freedom.

Example 16.16 The expert's estimates of a certain parameter are: $x_1 = 33$, $x_2 = 35$, $x_3 = 32.2$, $x_4 = 34$, $x_5 = 38$, $x_6 = 34$, $x_7 = 37$, $x_8 = 40$, $x_9 = 36$, $x_{10} = 35.5$. Assuming that all the experts have the same "weights" $\alpha_k = 1$, determine the confidence interval for the random variable X, with the probability of error $P = 0.05$.

Solution From (16.41) we find that $a = 35.5$. Then the formula (16.42) yields $D = 4.90$, so that $\sigma = \sqrt{D} = 2.2136$. From the tables for Student's distribution (Appendix G), for the number of the degrees of freedom $N - 1 = 9$ and the probability of error $P = 0.05$, we find $t = 2.262$. Then the formula (16.44) yields $\Delta = 1.583$. Finally, using (16.43), we obtain (33.917; 33.083) with the probability 0.95. Hence, the parameter in question can be found, with the probability 0.95, within the range between 33.917 and 33.083.

The Delphi method can also be applied for the selection of suitable criteria (indices) that best characterize the given item or process, i.e., can be used in decision making on the basis of processing the generalized experience of experts. A possible modification of the method, aimed at the selection of statistically significant quantitative parameters that are able to adequately characterize an item or a process, can be described as follows. This modification can be used, for instance, to determine the most likely failure modes in a reliability analysis (see also Sec. 10.7).

The choice of adequate criteria (parameters) is based on the questionnaire suggested to n experts, each of which is requested to describe the features characterizing, in their opinion, the given item, process, product, or a class of items, or a failure mode in the case of a reliability evaluation. Then the experts are requested to answer another questionnaire in which, based on the processing of the first questionnaire, the adequate parameters are listed in order of their significance, i.e., all the parameters are "ranked." Let the total number of the parameters be m. If, in the experts' opinion, some parameters have the same significance, they are assigned the same sequential number. Then one can develop a ranking matrix for the ranks (significance numbers) x_{ij}, where i refers to an expert ($i = 1, 2, \ldots, n$) and j refers to a criterion ($j = 1, 2, \ldots, m$). The sums of all the columns and all the lines in this matrix should satisfy the condition:

$$\sum_{j=1}^{m} \sum_{i=1}^{n} x_{ij} = \sum_{i=1}^{n} \sum_{j=1}^{m} x_{ij} \qquad (16.45)$$

Using the ranking matrix, one assesses the degree of correlation of the experts' opinions. The discrepancies in the experts' opinions can be due both to different qualifications of different experts and to different opinions of different experts, because of the insufficient knowledge about the item. The *concordance coefficient* can be selected in such a way that it is equal to unity ($C = 1$) when all the experts' opinions coincide for each of the criteria and is equal to zero ($C = 0$) if all the opinions are different. If there are no equal elements (ranks) x_{ij} in the ranking matrix, the concordance coefficient can be calculated by the formula:

$$C = \frac{12S}{n^2 m(m^2 - 1)} \qquad (16.46)$$

where n is the number of experts, m is the number of parameters, and S is the correlation factor, which can be evaluated as

$$S = \sum_{j=1}^{m} \left(\sum_{i=1}^{n} x_{ij} - \frac{1}{m} \sum_{j=1}^{m} \sum_{i=1}^{n} x_{ij} \right) \tag{16.47}$$

If some of the elements/ranks in the ranking matrix are the same, then the formula for the concordance coefficient can be modified as follows:

$$C = \frac{12S}{n^2 m(m^2 - 1) - n \sum_{i=1}^{n} T_i} \tag{16.48}$$

where

$$T_i = \sum_{j=1}^{m} t_j (t_j^2 - 1) \tag{16.49}$$

and t is the number of ranks of the jth type in each of the lines of the ranking matrix. If the calculated C value is small (say, smaller than 0.1), one should conclude that the experts are chosen wrongly and/or there is not sufficient knowledge about the object (item, process). If the C value is too large (say, larger than 0.9), one should conclude that the analysis is carried out too formally, too superficially, without an in-depth study of the object. In all cases, when the calculated C value is too small or too large, the process of questioning the experts should be repeated. The significance of the deviation of the concordance coefficient from zero can be checked on the basis of the following Fischer criterion:

$$Z = \frac{1}{2} \ln \left[(n-1) \frac{C}{1-C} \right] \tag{16.50}$$

The calculated Z value is compared with a value Z_α determined for a low level $\alpha = 0.01$ to 0.05, and the degrees of freedom v_1 and v_2 which are calculated as

$$v_1 = m - 1 - \frac{2}{n}, \quad v_2 = n - 1$$

The Z_α values for $\alpha = 0.05$ (5 percent significance level) are given in Table 16.7 as a function of the numbers v_1 and v_2. If $Z < Z_\alpha$, one can conclude, with the probability $P \geq 1 - \alpha$, that there is no agreement among the experts. In this case, one should conduct new analysis or

TABLE 16.7 Fischer Criterion for 5 per cent Significance Level

v_2	v_1									
	1	2	3	4	5	6	8	12	24	∞
1	2.452	2.648	2.687	2.707	2.719	2.728	2.738	2.748	2.759	2.769
2	1.459	1.472	1.476	1.479	1.480	1.481	1.482	1.483	1.484	1.485
3	1.578	1.128	1.114	1.106	1.099	1.095	1.090	1.084	1.078	1.072
4	1.021	0.969	0.943	0.927	0.917	0.909	0.899	0.888	0.877	0.864
5	0.944	878	844	824	810	800	786	771	755	737
6	895	819	780	756	739	727	711	693	673	650
7	861	778	735	708	690	677	658	637	613	586
8	836	748	701	672	652	638	618	594	568	537
9	816	724	676	645	624	608	586	561	532	498
10	801	706	655	623	601	584	561	535	504	466
11	789	691	639	606	582	565	541	513	480	439
12	779	679	625	591	567	549	523	494	459	416
13	770	668	613	578	554	535	509	478	442	396
14	763	659	604	568	542	523	496	465	426	378
15	757	652	595	558	533	513	485	453	414	363
16	751	645	588	550	524	504	476	443	402	349
17	747	639	581	543	517	496	468	434	392	337
18	742	634	575	537	510	489	460	426	383	325
19	739	630	570	532	504	483	454	418	374	315
20	735	626	565	526	499	478	447	412	367	306
21	732	622	561	522	494	472	442	406	360	297
22	729	618	557	518	489	468	437	400	354	289
23	727	615	554	514	485	464	432	395	348	282
24	725	612	551	511	482	460	428	390	342	275
25	722	610	548	507	478	456	424	386	338	268
26	720	608	545	504	475	453	421	382	333	262
27	719	605	543	502	472	450	418	379	329	257
28	717	603	540	499	470	447	415	375	325	252
29	715	601	538	497	467	444	412	372	321	247
30	714	599	536	495	465	442	409	369	318	242
60	693	574	507	463	431	406	370	326	265	164
∞	673	549	478	432	397	371	331	280	208	000

substitute the experts with new ones, who would agree better. This can be done, for instance, by excluding one of the experts from the team and by evaluating the coefficient C_1 for the remaining experts. If $C_1 > C$, this expert should be excluded from the team. If $C_1 < C$, the expert remains in the team. Such evaluations should be carried out for each expert. As a result of these evaluations, the degree of agreement of experts remaining in the team increases.

At the next step, one assesses the difference in the roles of the different criteria and the significance of the effects of different criteria. This can be done on the basis of the calculated variances. The significance of the rank x_{ij} is due to the following three independent components (inputs):

1. The component (input) which is due to the given expert.
2. The component (input) which is due to the given criterion (feature).
3. The remainder, which can be treated as a normally distributed random variable with zero mean and nonzero variance.

Hence, the total variance of the given criterion can be represented as a sum of the three components:

$$\sum_{i=1}^{n} \sum_{j=1}^{m} (x_{ij} - \langle x_{ij} \rangle)^2 = \sum_{i=1}^{n} \sum_{j=1}^{m} (\langle x_i \rangle - \langle x_{ij} \rangle)^2 + \sum_{i=1}^{n} \sum_{j=1}^{m} (\langle x_j \rangle - \langle x_{ij} \rangle)^2$$
$$+ \sum_{i=1}^{n} \sum_{j=1}^{m} (x_{ij} - \langle x_i \rangle - \langle x_j \rangle + \langle x_{ij} \rangle)^2 \quad (16.51)$$

Here

$$\langle x_{ij} \rangle = \frac{1}{mn} \sum_{i=1}^{n} \sum_{j=1}^{m} x_{ij} \quad (16.52)$$

is the total mean rank,

$$\langle x_j \rangle = \frac{1}{m} \sum_{j=1}^{m} x_{ij} \quad (16.53)$$

is the mean rank for the ith expert, and

$$\langle x_i \rangle = \frac{1}{n} \sum_{i=1}^{n} x_{ij} \quad (16.54)$$

is the mean rank for the jth expert. The assessment of the difference in the role of different criteria are carried out by comparing the variances:

$$D_1 = \frac{1}{m-1} \sum_{i=1}^{n} \sum_{j=1}^{m} (\langle x_j \rangle - \langle x_{ij} \rangle)^2$$
$$= \frac{1}{m-1} \left(n \sum_{j=1}^{m} \langle x_j \rangle^2 - mn \langle x_{ij} \rangle^2 \right) \quad (16.55)$$

between the different criteria, with the remaining variance being

$$D_r = \frac{1}{(m-1)(n-1)} \sum_{i=1}^{n} \sum_{j=1}^{m} (x_{ij} - \langle x_i \rangle - \langle x_j \rangle + \langle x_{ij} \rangle)^2$$
$$= \frac{1}{(m-1)(n-1)} \left(\sum_{i=1}^{n} \sum_{j=1}^{m} x_{ij}^2 - m \sum_{i=1}^{n} \langle x_i \rangle^2 - n \sum_{j=1}^{m} \langle x_j \rangle^2 + mn \langle x_{ij} \rangle^2 \right)$$
$$(16.56)$$

The significance of the difference in the variances D_1 and D_r can be evaluated on the basis of the Fischer criterion:

$$Z = \frac{1}{2} \ln \left(\frac{D_1}{D_r}\right) \qquad (16.57)$$

for the degrees of freedom v_1 and v_2 defined as $v_1 = m - 1$ and $v_2 = (m - 1)(n - 1)$. If the Z value calculated on the basis of this formula is not smaller than the value Z_α calculated for a low enough α value (say, $\alpha = 0.05$), then the difference in the variances D_1 and D_r is significant. This means that the difference in the roles of different criteria is substantial, and the influence of the chosen criteria is substantial as well. If the calculated Z value is smaller than the Z_α value, then the difference in the variances D_1 and D_r is insignificant, i.e., the difference in the roles of different criteria is small and the influence of the chosen criteria is not essential. If this is the case, one should broaden the number of the criteria and start a new questioning process.

When conducting statistical analysis, one assesses the significance in the distribution of the criteria, in order to establish the structure of the effects of different criteria. The criteria for which the difference in the probability distributions is insignificant can be put in the same group. The assessment of the significance in the distributions for different criteria can be substituted by the assessment of the difference in the mean values of the ranks are small, belong to the same group, and are not different from the standpoint of their roles (degrees of influence).

Examine how the difference in the mean values of the ranks can be assessed. The ranks for each criterion form random samples. Each of these samples is characterized by its mean $\langle x_j \rangle$ and its variance

$$D_j = \frac{1}{n-1} \sum_{j=1}^{n} (x_{ij} - \langle x_j \rangle)^2, \qquad j = 1, 2, \ldots, n \qquad (16.58)$$

The significance of the difference in the mean value of different ranks for different criteria can be evaluated on the basis of the Student distribution tables, by using a sequential comparison of the mean ranks for different criteria:

$$t_{\langle x_k \rangle - \langle x_l \rangle} = \frac{\langle x_k \rangle - \langle x_l \rangle}{\sqrt{D_{\langle x_k \rangle - \langle x_l \rangle}}} \qquad (16.59)$$

TABLE 16.8 t_α values in the Student distribution

v	5%†	1%‡	v	5%	1%	v	5%	1%	v	5%	1%
1	6.31	63.7	10	1.81	3.17	19	1.73	2.86	28	1.70	2.76
2	2.92	9.92	11	1.80	3.11	20	1.73	2.85	29	1.70	2.76
3	2.35	5.84	12	1.78	3.05	21	1.72	2.83	30	1.70	2.75
4	2.13	4.60	13	1.77	3.01	22	1.72	2.82	40	1.68	2.70
5	2.01	4.03	14	1.76	2.98	23	1.71	2.81	60	1.67	2.66
6	1.94	3.71	15	1.75	2.95	24	1.71	2.80	120	1.66	2.62
7	1.89	3.50	16	1.75	2.92	25	1.71	2.79			
8	1.86	3.36	17	1.74	2.90	26	1.71	2.78			
9	1.83	2.25	18	1.73	2.88	27	1.71	2.77			

† One-sided restrictions.
‡ Two-sided restrictions.

where $k, l = 1, 2, \ldots, m$ and $k \neq l$. The variance $D_{\langle x_k \rangle - \langle x_l \rangle}$ can be calculated as

$$D_{\langle x_k \rangle - \langle x_l \rangle} = D_{\langle x_k \rangle} + D_{\langle x_l \rangle} \qquad (16.60)$$

and

$$D_{\langle x_j \rangle} = \frac{1}{n} D_j, \qquad j = 1, 2, \ldots, m \qquad (16.61)$$

The corresponding number of the degrees of freedom is

$$v = 2(n - 1)$$

If the table for the Student distribution (Appendix E) results in the absolute value $|t_{\langle x_k \rangle - \langle x_l \rangle}|$ which exceeds the value t_α determined on the basis of the Student distribution table, then one can conclude, with the probability $P \geq 1 - \alpha$, that the difference in the mean values $\langle x_k \rangle$ and $\langle x_l \rangle$ is insignificant and that the criteria under examination are not different, from the standpoint of their significance (influence), as well, and can be put into the same group. The t_α values for $\alpha = 0.05$ and $\alpha = 0.01$ are given in Table 16.8.

Example 16.17 An analyst is interested in the assessment of a research project using different criteria (say, novelty, technical risk, commercial risk, cost of fulfillment and implementation, etc). The analyst invited $n = 12$ experts who suggested $m = 8$ suitable criteria to evaluate the project. The obtained ranking matrix is shown in Table 16.9.

Solution The condition (16.45) is fulfilled:

$$\sum_{j=1}^{8} \sum_{i=1}^{12} x_{ij} = \sum_{i=1}^{12} \sum_{j=1}^{8} x_{ij} = 387$$

Hence, the data in Table 16.9 is consistent. The calculated sums of the columns indicate that criterion 3 ($\sum_i x_{i3} = 21$) is the most important one and criterion 8 ($\sum_i x_{i8} = 80$) is the least important.

TABLE 16.9 Ranking Matrix

Experts	x_1	x_2	x_3	x_4	x_5	x_6	x_7	x_8	$\sum_{j=1}^{8} x_{ij}$
1	3	4	2	5	1	6	8	8	37
2	4	6	2	5	6	7	1	7	38
3	2	5	1	6	3	4	7	7	35
4	3	4	1	3	2	5	6	8	32
5	1	2	3	3	5	2	6	4	26
6	2	3	1	4	1	5	6	8	30
7	6	4	3	1	2	2	5	7	30
8	4	5	2	1	3	2	6	8	31
9	3	2	1	6	4	3	5	6	30
10	5	8	3	2	4	7	1	6	36
11	2	4	1	8	3	2	5	6	31
12	2	3	1	7	2	7	4	5	31
$\sum_{i=1}^{12} x_{ij}$	37	50	21	51	36	52	60	80	387

Since in the matrix in Table 16.9 some of the ranks are equal, we use the formula (16.48) to evaluate the concordance coefficient. Using the formula (16.49), we obtain

$$T_1 = 2^3 - 2 = 6$$

$$T_2 = (2^3 - 2) + (2^3 - 2) = 12$$

$$T_3 = 2^3 - 2 = 6$$

$$T_4 = T_6 = T_7 = T_8 = T_9 = T_{11} = T_{12} = 6$$

$$T_5 = T_{12} = 12$$

$$T_{10} = 0$$

The S value, in accordance with the formula (16.47), is

$$S = \left(37 - \frac{1}{8} \times 387\right)^2 + \left(50 - \frac{1}{8} \times 387\right)^2 + \cdots + \left(80 - \frac{1}{8} \times 387\right)^2 = 2190$$

The concordance coefficient can be found on the basis of the formula (16.48) as

$$C = \frac{12 \times 2190}{12^2(8^3 - 8) - (\frac{1}{2} \times 8 + 1 + 1)} = 0.370$$

Determine how far this coefficient is from zero. The formula (16.50) yields

$$Z = \frac{1}{2} \ln \frac{(12 - 1)0.370}{0.370} = 0.930$$

The degrees of freedom are

$$\nu_1 = (8-1) - \frac{2}{12} \simeq 7, \qquad \nu_2 = (12-1)\nu_1 \simeq 75$$

For $\alpha = 0.05$, with $\nu_1 = 7$ and $\nu_2 = 75$, we find $Z_\alpha = 0.35$. This value is smaller than $Z = 0.930$. Therefore one can state, with the probability $P \geq 0.95$, that there is a nonrandom agreement between the experts that the selected criteria reflect well the quality of the research project.

Let us assess now the differences in the roles of different criteria on the project in question, and the significance of these roles. This can be established on the basis of the formulas (16.55) and (16.56). The mean values of the ranks of the different criteria are

$\langle x_1 \rangle = 3.08, \qquad \langle x_2 \rangle = 4.16, \qquad \langle x_3 \rangle = 1.75, \qquad \langle x_4 \rangle = 4.25$

$\langle x_5 \rangle = 3.00, \qquad \langle x_6 \rangle = 4.33, \qquad \langle x_7 \rangle = 5.00, \qquad \langle x_8 \rangle = 6.67$

The mean values of the ranks, as determined by the experts, are

$\langle x_1 \rangle = 4.63, \qquad \langle x_2 \rangle = 4.75, \qquad \langle x_3 \rangle = 4.38, \qquad \langle x_4 \rangle = 4.00$

$\langle x_5 \rangle = 3.25, \qquad \langle x_6 \rangle = 3.75, \qquad \langle x_7 \rangle = 3.75, \qquad \langle x_8 \rangle = 3.88$

$\langle x_9 \rangle = 3.75, \qquad \langle x_{10} \rangle = 4.50, \qquad \langle x_{11} \rangle = 3.88, \qquad \langle x_{12} \rangle = 3.88$

The total mean rank is $\langle x_{ij} \rangle = 4.03$. Then the formulas (16.55) and (16.56) yield $D_1 = 26.1$, $D_r = 3.1$.

The significance of the difference in the variances D_1 and D_r can be checked, using the formula (16.57). We have

$$Z = \frac{1}{2} \ln \frac{26.1}{3.1} = 1.06$$

With $\nu_1 = 7$ and $\nu_2 = 75$, for $\alpha = 0.05$, we find $Z_\alpha = 0.35$. Since $Z > Z_\alpha$, we conclude, with the probability $P \geq 0.95$, that the difference in the influence of the criteria under investigation is statistically significant, and the chosen criteria are suitable to evaluate the project.

Determine now the structure of the influence of the role of the chosen criteria, i.e., the influence of each of the criteria on the general evaluation of the project. Different groups of the governing criteria can be found based on the assessment of the difference in the mean values of the ranks of different criteria. The standard deviations of the criteria can be found, using the formula (16.58), as follows:

$\sqrt{D_1} = 1.44, \qquad \sqrt{D_2} = 1.69, \qquad \sqrt{D_3} = 0.87, \qquad \sqrt{D_4} = 2.30$

$\sqrt{D_5} = 1.54, \qquad \sqrt{D_6} = 2.10, \qquad \sqrt{D_7} = 2.13, \qquad \sqrt{D_8} = 1.30$

The highest mean value has been given by the experts to the criterion x_8. This value is $\langle x_8 \rangle = 6.67$ and the corresponding standard deviation is $\sqrt{D_8} = 1.30$. Then follows the criterion x_7 with $\langle x_7 \rangle = 5.00$ and $\sqrt{D_7} = 2.13$. The formulas (16.60) and (16.61) yield

$$D_{\langle x_8 \rangle - \langle x_7 \rangle} = \frac{1.3^2 + 2.13^2}{12} = 0.5184, \qquad \sqrt{D_{\langle 8 \rangle - \langle 7 \rangle}} = 0.72$$

Using the formulas (16.59) and (16.62), we have

$$t_{\langle x_8 \rangle - \langle x_7 \rangle} = \frac{6.67 - 5.00}{0.72} = 2.32$$

$$v = 2(12 - 1) = 22$$

For $\alpha = 0.05$ and $v = 22$, Table 16.5 yields $t_\alpha = 1.72$.

Since $t = 2.32 > t_\alpha = 1.72$, then one can state, with the probability $P \geq 0.95$, that the difference in the mean values of criteria 8 and 7 is statistically significant, and therefore these criteria should be put, based on their role, in different groups. The other t values can be calculated in a similar way:

$$t_{\langle x_7 \rangle - \langle x_6 \rangle} = 0.77, \quad t_{\langle x_6 \rangle - \langle x_5 \rangle} = 1.76, \quad t_{\langle x_2 \rangle - \langle x_1 \rangle} = 1.68$$

$$t_{\langle x_6 \rangle - \langle x_4 \rangle} = 0.09, \quad t_{\langle x_5 \rangle - \langle x_1 \rangle} = 0.13$$

$$t_{\langle x_6 \rangle - \langle x_2 \rangle} = 0.22, \quad t_{\langle x_5 \rangle - \langle x_3 \rangle} = 2.45$$

Comparing the calculated $t_{\langle x_k \rangle - \langle x_l \rangle}$ values with $t_{0.05} = 1.72$, one can choose the following groups of criteria:

1. x_6, x_4, x_2, x_1
2. x_5, x_1
3. x_3

As a result of our calculations, we found that the criterion x_1 belongs to two groups. The comparison of the values $t_{\langle x_5 \rangle - \langle x_1 \rangle} = 0.13$ and $t_{\langle x_2 \rangle - \langle x_1 \rangle} = 1.68$ shows, however, that the criterion x_1 should belong to the same group as the criterion x_5. The final groups are as follows:

1. x_8
2. x_7, x_6, x_4, x_2
3. x_5, x_1
4. x_3

The criterion x_3 is the most influential, followed by the criteria $x_5, x_1, x_2, x_4, x_6, x_7, x_8$.

Bibliography

1. Abramson, N. M.: "Information Theory and Coding," McGraw-Hill, New York, 1963.
2. Ang, A. H.-S., and Tang, W.-H.: "Probability Concepts in Engineering Planning and Design," vol. 1, "Basic Principles," John Wiley, New York, 1975.
3. Ang, A. H.-S., and Tang, W.-H.: "Probability Concepts in Engineering Planning and Design," vol. 2, "Design, Risk, and Reliability," John Wiley, New York, 1984.
4. Ash, R.: "Information Theory," Interscience, New York, 1965.
5. Asmussen, S.: "Applied Probability and Queues," John Wiley, New York, 1987.
6. Augusti, G., Barrata, A., and Casciati, F.: "Probabilistic Methods in Structural Engineering," Chapman and Hall, London, 1983.
7. Ayer, A. J.: "Probability and Evidence," Columbia University Press, New York, 1972.
8. Barleir, R. E., and Proschan, F.: "Statistical Theory of Reliability and Life Testing," Holt, Rinehart, and Winston, New York, 1975.
9. Beckmann, P.: "Probability in Communication Engineering," Harcourt Brace & World, New York, 1967.
10. Benjamin, J. R., and Cornell, C. A.: "Probability, Statistics, and Decision for Civil Engineers," McGraw-Hill, New York, 1970.
11. Beranek, L. L. (ed.): "Noise and Vibration Control," McGraw-Hill, New York, 1971.
12. Bharucha-Reid, A. T.: "Elements of the Theory of Markov Processes and Their Applications," McGraw-Hill, New York, 1960.
13. Bhat, U. N.: "Elements of Applied Stochastic Processes," John Wiley, New York, 1972.
14. Bolotin, V. V.: "Statistical Methods in Structural Mechanics," Holden Day, San Francisco, 1969.
15. Bolotin, V. V.: "Random Vibrations of Elastic Systems," Martinus Nijhoff, The Hague, 1984.
16. Bolotin, V. V.: "Prediction of Service Life for Machines and Structures," ASME Press, New York, 1989.
17. Brebhia, C. A., and Walker, S.: "Dynamic Analysis of Offshore Structures," Newnes Butterworth, London, 1979.
18. Brook, R. H. W.: "Reliability Concepts in Engineering Manufacture," Butterworths, London, 1972.
19. Brown, R. G.: "Introduction of Random Signal Analysis and Kalman Filtering," John Wiley, New York, 1983.
20. Bury, K. V.: "Statistical Models in Applied Science," John Wiley, New York, 1976.
21. Childers, D. G. (ed.): "Modern Spectrum Analysis," John Wiley, New York, 1978.
22. Clarke, A. B., and Disney, R. L.: "Probability and Random Processes for Engineers and Scientists," John Wiley, New York, 1970.
23. Clarkson, B. L. (ed.): "Stochastic Problems in Dynamics," Pitman, London, 1977.
24. Clough, R. W., and Penzien, J.: "Dynamics of Structures," McGraw-Hill, New York, 1975.

25. Collins, J. A.: "Failure of Materials in Mechanical Design: Analysis, Prediction, and Prevention," John Wiley, New York, 1981.
26. Cooper, G. R., and McGillem, C. D.: "Probabilistic Methods of Signal and System Analysis," 2nd edn., Holt, Rinehart, and Winston, New York, 1986.
27. Cox, D. R., and Miller, H. D.: "The Theory of Stochastic Processes," John Wiley, New York, 1965.
28. Cox, D. R., and Smith, W. L.: "Queues," Chapman and Hall, London, 1961.
29. Crandall, S. H., and Mark, W. D.: "Random Vibrations in Mechanical Systems," Academic Press, New York, 1963.
30. Davenport, W. B.: "Probability and Random Processes: An Introduction for Applied Scientists and Engineers, McGraw-Hill, New York, 1970.
31. David, F. N.: "Games, Gods and Gambling: The Origins and History of Probability and Statistical Ideas from the Earliest Times to the Newtonian Era," Hafner Publishing Co., New York, 1962.
32. Devore, J. L.: "Reliability and Statistics for Engineering and Sciences," 3rd ed., Wadsworth & Brooks/Cole, Pacific Grave, CA, 1991.
33. Ditlevsen, O.: "Uncertainty Modeling," McGraw-Hill, New York, 1981.
34. Doob, J. L.: "Stochastic Processes," John Wiley, New York, 1953.
35. Dubins, L., and Savage, L.: "How to Gamble if You Must," McGraw-Hill, New York, 1965.
36. Eggwartz, S., and Lind, N. (eds.): "Probabilistic Methods in Mechanics and Structures," Springer, Berlin, 1985.
37. Elishakoff, G.: "Probabilistic Methods in the Theory of Structures," John Wiley, New York, 1983.
38. Evans, W. H.: "Probability and Its Applications for Engineers," Marcel Dekker, New York, 1992.
39. Feinstein, A.: "Foundations of Information Theory," McGraw-Hill, New York, 1958.
40. Feller, W.: "An Introduction to Probability Theory and Its Applications," vol. 1, 3rd edn, 1968, vol. 2, 2nd edn, 1971, John Wiley, New York.
41. Ferry Borges, J., and Castanheta, M.: "Structural Safety," National Civil Engineering Laboratory, Lisbon, Portugal, 1971.
42. Franks, L. E.: "Signal Theory," Prentice-Hall, Englewood Cliffs, New Jersey, 1979.
43. Freedman, D.: "Markov Chains," Holden Day, San Francisco, 1971.
44. Fry, T. C.: "Probability and Its Engineering Uses," 2nd edn, Van Nostrand, Princeton, New Jersey, 1965.
45. Fryba, L.: "Vibration of Solids and Structures Under Moving Loads," Noordhoff International, Groningen, 1972.
46. Fuchs, H. O., and Stephens, R. J.: "Metal Fatigue in Engineering," John Wiley, New York, 1980.
47. Furman, T. T.: "Approximate Methods in Engineering Design," Academic Press, New York, 1981.
48. Ghiocel, D., and Lungu, D.: "Wind, Snow, and Temperature Effects in Structures Based on Probability," Abacus Press, Tunbridge Wells, 1975.
49. Gleik, J.: "Chaos: Making a New Science," Viking, New York, 1987.
50. Gnedenko, B. V., Belyayev, Y. K., and Solovyev, A. D.: "Mathematical Methods of Reliability Theory," Academic Press, New York, 1969.
51. Goda, Y.: "Random Seas and Design of Maritime Structures," University of Tokyo Press, Tokyo.
52. Gould, P. L., and Abu-Sitta, S. H.: "Dynamic Response of Structures of Wind and Earthquakes," Pentech Press, London, 1980.
53. Gray, R. M., and Davisson, L. D.: "Random Processes: A Mathematical Approach for Engineers," Prentice-Hall, Englewood Cliffs, New Jersey, 1986.
54. Grosch, D. L.: "A Primer on Reliability Theory," John Wiley, New York, 1989.
55. Gumbel, E. J.: "Statistical Theory of Extreme Values and Some Practical Applications," National Bureau of Standards, Washington, D.C., 1954.
56. Gumbel, E. J.: "Statistics of Extremes," Columbia University Press, New York, 1958.

57. Gupta, A., and Singh, R. P.: "Fatigue Behavior of Offshore Structures," Springer-Verlag, New York, 1986.
58. Hahn, G. J., and Shapiro, S. S.: "Statistical Models in Engineering," John Wiley, New York, 1967.
59. Hald, A.: "Statistical Theory with Engineering Applications," John Wiley, New York, 1952.
60. Hammersley, J. M., and Handscomb, D.: "Monte Carlo Methods," John Wiley, New York, 1964.
61. Hamming, R. W.: "Digital Filters," 3rd edn, Prentice-Hall, Englewood Cliffs, New Jersey, 1989.
62. Hamming, R. W.: "The Art of Probability for Scientists and Engineers," Addison-Wesley, Redwood City, California, 1991.
63. Harr, M. E.: "Reliability-Based Design in Civil Engineering," McGraw-Hill, New York, 1987.
64. Haugen, E. B.: "Probabilistic Approaches to Design," John Wiley, New York, 1968.
65. Haugen, E. B.: "Probabilistic Mechanical Design," John Wiley, New York, 1980.
66. Helstrom, C. W.: "Probability and Stochastic Processes for Engineers," Macmillan, New York, 1991.
67. Howard, R. A.: "Dynamic Probabilistic Systems," John Wiley, New York, 1971.
68. Hsu, T. H.: "Applied Offshore Structure Engineering," Gulf, Houston, Texas, 1984.
69. Hurty, W. C., and Rubinstein, M. F.: "Dynamics of Structures," Prentice-Hall, Englewood Cliffs, New Jersey, 19??.
70. Ibrahim, R. A.: "Parametric Random Vibration," John Wiley, New York, 1985.
71. Kapur, K. C. and Lamberson, L. R., "Reliability in Engineering Design," John Wiley, New York, 1977.
72. Khinchin, A. I.: "Mathematical Foundations of Statistical Mechanics," Dover, New York, 1949.
73. Khinchin, A. I.: "Mathematical Foundations of Information Theory," Dover, New York, 1957.
74. Khinchin, A.: "Works on the Theory of Queues," John Wiley, New York, 1961.
75. Kinsman, B.: "Wind Waves, Thin Generation and Propagation in the Ocean Surface," Prentice-Hall, New York, 1965.
76. Kirenson, G.: "Durability and Reliability in Engineering Design," Hayden Book Co., New York, 1971.
77. Kree, P., and Soize, C.: "Mathematics of Random Phenomena—Random Vibration of Mechanical Structures," Reisel Publishing Co., Dordrecht, 1983.
78. Larson, H. J., and Shubert, B. O.: "Probabilistic Models in Engineering Sciences," vol. 1, "Random Variables and Stochastic Processes," vol. 2, "Random Noise, Signals, and Dynamic Systems," John Wiley, New York, 1979.
79. Lin, Y.-K.: "Probabilistic Theory of Structural Dynamics," McGraw-Hill, New York, 1967 (2nd edn, 1976).
80. Lomnitz, C., and Rosenblueth, E. (eds.): "Seismic Risk and Engineering Decisions," Elsevier, New York, 1977.
81. Lyon, R. H.: "Statistical Energy Analysis of Dynamical Systems: Theory and Application," MIT Press, Boston, 1975.
82. McCuen, R. H.: "Statistical Methods for Engineers," Prentice-Hall, Englewood Cliffs, New Jersey, 1985.
83. Mann, N. R., Schafer, R. E., and Singpurwalla, N. D.: "Methods for Statistical Analysis of Reliability of Life Data," John Wiley, New York, 1974.
84. Martz, H. F., and Waller, R. A.: "Bayesian Reliability Analysis," John Wiley, New York, 1982.
85. Medhi, J.: "Stochastic Processes," John Wiley, New York, 1981.
86. Middleton, D.: "An Introduction of Statistical Communication Theory," McGraw-Hill, New York, 1960.
87. Moan, T., and Shinozuka, M.: "Structural Safety and Reliability," Elsevier, Amsterdam, 1981.

88. Montgomery, D. C. and Runger, G. C., "Applied Statistics and Probability for Engineers," John Wiley, New York, 1994.
89. Newland, D. E.: "An Introduction to Random Vibrations and Spectral Analysis," Longman, Essex, 1984.
90. Newmark, N. M., and Rosenblueth, E.: "Fundamentals and Earthquake Engineering," Prentice-Hall, Englewood Cliffs, New Jersey, 1971.
91. Nigam, N. C.: "Introduction to Random Vibrations," MIT Press, Cambridge, Massachusetts, 1983.
92. Nigam, N. C., and Narayaman, S.: "Applications of Random Vibrations," Springer-Verlag, Narasa Publishing House, New Delhi, 1994.
93. Norton, M. P.: "Fundamentals of Noise and Vibration Analysis for Engineers," Cambridge University Press, London, 1989.
94. Ochi, M. K.: "Applied Probability and Stochastic Processes in Engineering and Physical Sciences," John Wiley, New York, 1990.
95. Packeja, H. B. (ed.): "The Dynamics of Vehicles," Swets & Zeitlinger, Amsterdam, 1976.
96. Papoulis, A.: "The Fourier Integral and Its Applications," McGraw-Hill, New York, 1962.
97. Papoulis, A.: "Probability, Random Variables, and Stochastic Processes," 3rd edn, McGraw-Hill, New York, 1991.
98. Parzen, E.: "Modern Probability Theory and Its Applications," John Wiley, New York, 1960.
99. Patel, M. H.: "Dynamics of Offshore Structures," Butterworth, London, 1989.
100. Peck, D., and Trapp, O. D.: "Accelerated Testing Handbook," Technology Associates and Bell Telephone Laboratories, Portola, California, 1980.
101. Peebles, P. Z.: "Probability, Random Variables, and Random Signal Principles," McGraw-Hill, New York, 1980.
102. Priestley, M.: "Spectra Analysis and Time Series," 2 vols., Academic Press, London, 1981.
103. Provan, J. W. (ed.): "Probabilistic Fracture Mechanics and Reliability," Martinus Nijhoff, The Netherlands, 1986.
104. Pugachev, V. S.: "Probability Theory and Mathematical Statistics for Engineers," Pergamon Press, Oxford, 1984.
105. Raiffa, M.: "Decision Analysis: Introductory Lectures on Choices under Uncertainty," Addison-Wesley, Reading, Massachusetts, 1968.
106. Rao, S. S.: "Reliability-Based Design," McGraw-Hill, New York, 1992.
107. Roberts, N. H.: "Mathematical Methods in Reliability Engineering," McGraw-Hill, New York, 1964.
108. Robson, J. D.: "An Introduction to Random Vibrations," Edinburgh University Press, Edinburgh, 1963.
109. Ross, S.: "Applied Probability Models with Optimization Applications," Holden Day, San Francisco, 1970.
110. Ross, S.: "Introduction to Probability Models," Academic Press, New York, 1972.
111. Ross, S.: "Introduction to Probability and Statistics for Engineers and Scientists," John Wiley York, 1987.
112. Sachs, P.: "Wind Forces in Engineering," Pergamon Press, Oxford, 1972.
113. Sachs, P.: "Wind Forces in Engineering," Pergamon Press, Oxford, 1978.
114. Sandler, G. H.: "System Reliability Engineering," Prentice-Hall, Englewood Cliffs, New Jersey, 1963.
115. Sarpkaya, T., and Isaacson, M.: "Mechanics of Wave Forces on Offshore Structures," Van Nostrand Reinhold, New York, 1981.
116. Schaeffer, R. L. and McClare, J. T., "Probability and Statistics for Engineers," Duxbury Press, Belmont, CA, 1995.
117. Schwartz, M., and Shaw, L.: "Signal Processing," McGraw-Hill, New York, 1975.
118. Shooman, M. L.: "Probabilistic Reliability: An Engineering Approach," McGraw-Hill, New York, 1968.
119. Simiu, E., and Scanlan, R. H.: "Wind Effects on Structures," John Wiley, New York, 1986.

120. Slibar, A., and Springer, H. (eds.): "The Dynamics of Vehicles," Swets and Zeitlinger, Amsterdam, 1978.
121. Smith, G. N.: "Probability and Statistics in Civil Engineering," Nikols, New York, 1986.
122. Soong, T. T.: "Probabilistic Modeling and Analysis in Science and Engineering," John Wiley, New York, 1992.
123. Spanos, P., and Robson, J. B.: "Random Vibration and Statistical Linearization," John Wiley, New York, 1990.
124. Stark, H., and Woods, J. W.: "Probability, Random Processes, and Estimation Theory for Engineers," Prentice-Hall, Englewood Cliffs, New Jersey, 1986.
125. Sveshnikov, A. A.: "Problems in Probability Theory, Mathematical Statistics and Theory of Random Functions," Dover, New York, 1968.
126. Svetlitsky, V. A.: "Random Vibrations of Mechanical Systems (in Russian)," Machinostrayenie, Moscow, 1976.
127. Takacs, K.: "An Introduction to the Theory of Queues," Oxford University Press, New York, 1962.
128. Thoft-Christensen, P., and Baker, M. J.: "Structural Reliability Theory and Its Applications," Springer-Verlag, Berlin, 1982.
129. Thomas, J. B.: "An Introduction to Statistical Communication Theory," John Wiley, New York, 1969.
130. Thomas, J. B.: An Introduction to Applied Probability and Random Processes," R. E. Krieger Publishing Co., Huntington, New York, 1981.
131. Thomas, J. B.: "An Introduction to Communication Theory and Systems," Springer, New York, 1988.
132. Thomson, W. T.: "Theory of Vibration with Application," Prentice-Hall, Englewood Cliffs, New Jersey, 1975.
133. Tobias, P. A., and Trindade, D. C.: "Applied Reliability," Van Nostrand Reinhold, New York, 1986.
134. Tribus, M.: "Rational Descriptions, Decisions, and Designs," Pergamon Press, New York, 1969.
135. Ventzel, E. S., and Ovcharov, L. A.: "The Theory of Random Processes and Its Engineering Applications (in Russian)," Nauka, Moscow, 1991.
136. Volkov, S. D.: "Statistical Strength Theory," Gordon and Breach, New York, 1962.
137. von Mises, R.: "Probability, Statistics, and Truth," 2nd version English edn, Macmillan Publishing Co., New York, 1957.
138. Walpole, R. E. and Myers, R. H.: "Probability and Statistics for Engineers and Scientists," 5-th ed., Macmillan, New York, 1993.
139. Wiegel, R. L.: "Oceanographic Engineering," Prentice-Hall, New York, 1963.
140. Wiegel, R. L.: "Earthquake Engineering," Prentice-Hall, Englewood Cliffs, New Jersey, 1970.
141. Wiener, N.: "Extrapolation, Interpolation, and Smoothing of Stationary Time Series with Engineering Applications," John Wiley, New York, 1960.
142. Willermeit, H. P. (ed.): "The Dynamics of Vehicles," Swets & Zeitlinger, Lisse, 1980.
143. Wolff, R. W.: "Stochastic Modeling and the Theory of Queues," Prentice-Hall, Englewood Cliffs, New Jersey, 1989.
144. Wong, E., and Hajek, B.: "Stochastic Process in Engineering Systems," Springer-Verlag, New York, 1985.
145. Yaglom, A. M.: "Correlation Theory of Stationary and Related Random Functions," 2 vols., Springer-Verlag, New York, 1987.
146. Yang, C.-Y.: "Random Vibrations of Structures," John Wiley, New York, 1986.
147. Yakobori, T.: "The Strength, Fracture, and Fatigue of Materials," Noordhoff, Groningen, 1965.

Appendix A

Poisson Distribution Tables

TABLE A.1 Poisson's distribution $P(m, a) = (a^m/m!)e^{-a}$

	a								
m	0.1	0.2	0.3	0.4	0.5	0.6	0.7	0.8	0.9
0	0.9048	0.8187	0.7408	0.6703	0.6065	0.5488	0.4966	0.4493	0.4066
1	0.0905	0.1638	0.2222	0.2681	0.3033	0.3293	0.3476	0.3595	0.3659
2	0.0045	0.0164	0.0333	0.0536	0.0758	0.0988	0.1217	0.1438	0.1647
3	0.0002	0.0019	0.0033	0.0072	0.0126	0.0198	0.0284	0.0383	0.0494
4		0.0001	0.0002	0.0007	0.0016	0.0030	0.0050	0.0077	0.0111
5				0.0001	0.0002	0.0004	0.0007	0.0012	0.0020
6							0.0001	0.0002	0.0003

	a									
m	1	2	3	4	5	6	7	8	9	10
0	0.3679	0.1353	0.0498	0.0183	0.0067	0.0025	0.0009	0.0003	0.0001	0.0000
1	0.3679	0.2707	0.1494	0.0733	0.0337	0.0149	0.0064	0.0027	0.0011	0.0005
2	0.1839	0.2707	0.2240	0.1465	0.0842	0.0446	0.0223	0.0107	0.0050	0.0023
3	0.0613	0.1804	0.2240	0.1954	0.1404	0.0892	0.0521	0.0286	0.0150	0.0076
4	0.0153	0.0902	0.1680	0.1954	0.1755	0.1339	0.0912	0.0572	0.0337	0.0189
5	0.0031	0.0361	0.1008	0.1563	0.1755	0.1606	0.1277	0.0916	0.0607	0.0378
6	0.0005	0.0120	0.0504	0.1042	0.1462	0.1606	0.1490	0.1221	0.0911	0.0631
7	0.0001	0.0037	0.0216	0.0595	0.1044	0.1377	0.1490	0.1396	0.1171	0.0901
8		0.0009	0.0081	0.0298	0.0653	0.1033	0.1304	0.1396	0.1318	0.1126
9		0.0002	0.0027	0.0132	0.0363	0.0688	0.1014	0.1241	0.1318	0.1251
10			0.0008	0.0053	0.0181	0.0413	0.0710	0.0993	0.1186	0.1251
11			0.0002	0.0019	0.0082	0.0225	0.0452	0.0722	0.0970	0.1137
12			0.0001	0.0006	0.0034	0.0126	0.0263	0.0481	0.0728	0.0948
13				0.0002	0.0013	0.0052	0.0142	0.0296	0.0504	0.0729
14				0.0001	0.0005	0.0022	0.0071	0.0169	0.0324	0.0521
15					0.0002	0.0009	0.0033	0.0090	0.0194	0.0347
16						0.0003	0.0014	0.0045	0.0109	0.0217
17						0.0001	0.0006	0.0021	0.0058	0.0128
18							0.0002	0.0009	0.0029	0.0071
19							0.0001	0.0004	0.0014	0.2237
20								0.0002	0.0006	0.0019
21								0.0001	0.0003	0.0009
22									0.0001	0.0004
23										0.0002
24										0.0001

TABLE A.2 Probabilities† $\bar{R}(m, a) = 1 - R(m, a) = 1 - \sum_{k=0}^{m} (a^k/k!)e^{-a}$

m	$a = 0.1$	$a = 0.2$	$a = 0.3$	$a = 0.4$	$a = 0.5$
0	9.5163^{-2}	1.8127^{-1}	2.5918^{-1}	3.2968^{-1}	8.9347^{-1}
1	4.6788^{-3}	1.7523^{-2}	3.6936^{-2}	6.1552^{-2}	9.0204^{-2}
2	1.5465^{-4}	1.1485^{-3}	3.5995^{-3}	7.9263^{-3}	1.4388
3	3.8468^{-6}	5.6840^{-5}	2.6581^{-4}	7.7625^{-4}	1.7516^{-3}
4		2.2592^{-6}	1.5785^{-5}	6.1243^{-5}	1.7212^{-4}
5				4.0427^{-6}	1.4165^{-5}
6					1.0024^{-6}

m	$a = 0.6$	$a = 0.7$	$a = 0.8$	$a = 0.9$
0	4.5119^{-1}	5.0341^{-1}	5.5067^{-1}	5.9343^{-1}
1	1.2190	1.5580	1.9121	2.2752
2	2.3115^{-2}	3.4142^{-2}	4.7423^{-2}	6.2857^{-2}
3	3.3581^{-3}	5.7535^{-3}	9.0799^{-3}	1.3459
4	3.9449^{-4}	7.8554^{-4}	1.4113	2.3441^{-3}
5	3.8856^{-5}	9.0026^{-5}	1.8434^{-4}	3.4349^{-4}
6	3.2931^{-6}	8.8836^{-6}	2.0747^{-5}	4.3401^{-5}
7			2.0502^{-6}	4.8172^{-6}

m	$a = 1$	$a = 2$	$a = 3$	$a = 4$	$a = 5$
0	6.3212^{-1}	8.6466^{-1}	9.5021^{-1}	9.8168^{-1}	9.9326^{-1}
1	2.6424	5.9399	8.0085	9.0842	9.5957
2	8.0301^{-2}	3.2332	5.7681	7.6190	8.7535
3	1.8988	1.4288	3.5277	5.6653	7.3497
4	3.6598^{-3}	5.2653^{-2}	1.8474	3.7116	5.5951
5	5.9418^{-4}	1.6564	8.3918^{-2}	2.1487	3.8404
6	8.3241^{-5}	4.5338^{-3}	3.3509	1.1067	2.3782
7	1.0219	1.0967	1.1905	5.1134^{-2}	1.3337
8	1.1252^{-2}	2.3745^{-4}	3.8030^{-3}	2.1363	6.8094^{-2}
9		4.6498^{-5}	1.1025	8.1322^{-3}	3.1828
10		8.3082^{-6}	2.9234^{-4}	2.8398	1.3695
11		1.3646	7.1387^{-5}	9.1523^{-4}	5.4531^{-3}
12			1.6149	2.7372	2.0189
13			3.4019^{-6}	7.6328^{-5}	6.9799^{-4}
14				1.9932	2.2625
15				4.8926^{-6}	6.9008^{-5}
16				1.1328	1.9869
17					5.4163^{-6}
18					1.4017

† $P(m, a) = (a^m/m!)e^{-a}$ can be found in terms of $\bar{R}(m, a)$ as follows:
$P(m, a) = \bar{R}(m - 1, a) - \bar{R}(m, a)$ $(m > 0)$, $P(0, a) = 1 - \bar{R}(0, a)$.

TABLE A.2 (Continued)

m	$a=6$	$a=7$	$a=8$	$a=9$	$a=10$
0	9.9752^{-1}	9.9909^{-1}	9.9966^{-1}	9.9988^{-1}	9.9995^{-1}
1	9.8265	9.9270	9.9698	9.9877	9.9950
2	9.3803	9.7036	9.8625	9.9377	9.9723
3	8.4880	9.1823	9.5762	9.7877	9.8966
4	7.1494	8.2701	9.0037	9.4504	9.7075
5	5.5432	6.9929	8.0876	8.8431	9.3291
6	3.9370	5.5029	6.8663	7.9322	8.6986
7	2.5602	4.0129	5.4704	6.7610	7.7978
8	1.5276	2.7091	4.0745	5.4435	6.6718
9	8.3924^{-2}	1.6950^{-1}	2.8338^{-1}	4.1259^{-1}	5.4207^{-1}
10	4.2621	9.8521^{-2}	1.8411	2.9401	4.1696
11	2.0092	5.3350	1.1192	1.9699	3.0322
12	8.8275^{-3}	2.7000	6.3797^{-2}	1.2423	2.0844
13	3.6285	1.2811	3.4181	7.3851^{-2}	1.3554
14	1.4004	5.7172^{-3}	1.7257	4.1466	8.3458^{-2}
15	5.0910^{-4}	2.4066	8.2310^{-3}	2.2036	4.8740
16	1.7488	9.5818^{-4}	3.7180	1.1106	2.7042
17	5.6917^{-5}	3.6178	1.5943	5.3196^{-3}	1.4278
18	1.7597	1.2985	6.5037^{-4}	2.4264	7.1865^{-3}
19	5.1802^{-5}	4.4402^{-5}	2.5294	1.0560	3.4543
20	1.4551	1.4495	9.3969^{-5}	4.3925^{-4}	1.5883
21		4.5263^{-6}	3.3407	1.7495	6.9965^{-4}
22		1.3543	1.1385	6.6828^{-5}	2.9574
23			3.7255^{-6}	2.4519	1.2012
24			1.1722	8.6531^{-6}	4.6949^{-5}
25				2.9414	1.7680
26					6.4229^{-6}
27					2.2535

m	$a=11$	$a=12$	$a=13$	$a=14$	$a=15$
1	9.9998^{-1}	9.9999^{-1}			
1	9.9980	9.9991	9.9997^{-1}	9.9999^{-1}	
2	9.9879	9.9948	9.9978	9.9991	9.9996^{-1}
3	9.9508	9.9771	0.9895	9.9953	9.9979
4	9.8490	9.9240	9.9626	9.9819	9.9914
5	9.6248	9.7966	9.8927	9.9447	9.9721
6	9.2139	9.5418	9.7411	9.8577	9.9237
7	8.5681	9.1050	9.4597	9.6838	9.8200
8	7.6801	8.4497	9.0024	9.3794	9.6255
9	6.5949	7.5761	8.3419	8.9060	9.3015
10	5.4011	6.5277	7.4832	8.2432	8.8154
11	4.2073	5.3840	6.4684	7.3996	8.1525
12	3.1130	4.2403	5.3690	6.4154	7.3239
13	2.1871	3.1846	4.2696	5.3555	6.3678
14	1.4596	2.2798	3.2487	4.2956	5.3435
15	9.2604^{-2}	1.5558	2.3639	3.3064	4.3191
16	5.5924	1.0129	1.6451	2.4408	3.3588
17	3.2191	6.2966^{-2}	1.0954	1.7280	2.5114
18	1.7687	3.7416	6.9833^{-2}	1.1736	1.8053

TABLE A.2 (Continued)

m	$a=11$	$a=12$	$a=13$	$a=14$	$a=15$
19	9.2895^{-3}	2.1280	4.2669	7.6505^{-2}	1.2478
20	4.6711	1.1598	2.5012	4.7908	8.2972^{-3}
21	2.2519	6.0651^{-3}	1.4081	2.8844	5.3106
22	1.0423	3.0474	7.6225^{-3}	1.6712	3.2744
23	4.6386^{-4}	1.4729	3.9718	9.3276^{-3}	1.9465
24	1.9871	6.8563^{-4}	1.9943	5.0199	1.1165
25	8.2050^{-5}	3.0776	9.6603^{-4}	2.6076	6.1849^{-3}
26	3.2693	1.3335	4.5190	1.3087	3.3119
27	1.2584	5.5836^{-5}	2.0435	6.3513^{-4}	1.7158
28	4.6847^{-6}	2.2616	8.9416^{-5}	2.9837	8.6072^{-4}
29	1.6882	8.8701^{-6}	3.7894	1.3580	4.1843
30		3.3716	1.5568	5.9928^{-5}	1.9731
31		1.2432	6.2052^{-6}	2.5665	9.0312^{-5}
32			2.4017	1.0675	4.0155
33				4.3154^{-6}	1.7356
34				1.6968	7.2978^{-6}
35					2.9871
36					1.1910

m	$a=16$	$a=17$	$a=18$	$a=19$	$a=20$
0					
1					
2	9.9998^{-1}	9.9999^{-1}			
3	9.9991	9.9996	9.9998^{-1}	9.9999^{-1}	
4	9.9960	9.9982	9.9992	9.9996	9.9998^{-1}
5	9.9862	9.9933	9.9968	9.9985	9.9993
6	9.9599	9.9794	9.9896	9.9948	9.9974
7	9.9000	9.9457	9.9711	9.9849	9.9922
8	9.7801	9.8740	9.9294	9.9613	9.9791
9	9.5670	9.7388	9.8462	9.9114	9.9500
10	9.2260	9.5088	9.6963	9.8168	9.8919
11	8.7301	9.1533	9.4511	9.6533	9.7861
12	8.0688	8.6498	9.0833	9.3944	9.6099
13	7.2545	7.9913	8.5740	9.0160	9.3387
14	6.3247	7.1917	7.9192	8.5025	8.9514
15	5.3326	6.2855	7.1335	7.8521	8.4349
16	4.3404	5.3226	6.2495	7.0797	7.7893
17	3.4066	4.3598	5.3135	6.2164	7.0297
18	2.5765	3.4504	4.3776	5.3052	6.1858
19	1.8775	2.6368	3.4908	4.3939	5.2974
20	1.3183	1.9452	2.6928	3.5283	4.4091
21	8.9227^{-2}	1.3853	2.0088	2.7450	3.5630
22	5.8241	9.5272^{-2}	1.4491	2.0687	2.7939
23	3.6686	6.3296	1.0111	1.5098	2.1251
24	2.2315	4.0646	6.8260^{-2}	1.0675	1.5677
25	1.3119	2.5245	4.4608	7.3126^{-2}	1.1218
26	7.4589^{-3}	1.5174	2.8234	4.8557	7.7887^{-2}
27	4.4051	8.8335^{-3}	1.7318	3.1268	5.2481

TABLE A.2 (Continued)

m	a = 16	a = 17	a = 18	a = 19	a = 20
28	2.1886	4.9838	1.0300	1.9536	3.4334
29	1.1312	2.7272	5.9443^{-3}	1.1850	2.1818
30	5.6726^{-4}	1.4484	3.3308	6.9819^{-3}	1.3475
31	2.7620	7.4708^{-4}	1.8133	3.9982	8.0918^{-3}
32	1.3067	3.7453	9.5975^{-4}	2.2267	4.7274
33	6.0108^{-5}	1.8260	4.9416	1.2067	2.6884
34	2.6903	8.6644^{-5}	2.4767	6.3674^{-4}	1.4890
35	1.1724	4.0035	1.2090	3.2732	8.0366^{-4}
36	4.9772^{-6}	1.8025	5.7519^{-5}	1.6401	4.2290
37	2.0599	7.9123^{-6}	2.6684	8.0154^{-5}	2.1708
38		3.3882	1.2078	3.8224	1.0875
39		1.4162	5.3365^{-6}	1.7797	5.3202^{-5}
40			2.3030	8.0940^{-6}	2.5426
41				3.5975	1.1877
42				1.5634	5.4252^{-6}
43					2.4243
44					1.0603

Example We must find the probability that an event A will occur no more than twice if $a = 7$.

We have

$$R(2, 7) = 1 - \bar{R}(2, 7) = 1 - 9.7036^{-1} = 1 - 0.97036 = 0.02964$$

Note
1. If a number in the table has no exponent, then the exponent of the previous number in the column is the exponent. For example, $\bar{R}(33, 19) = 1.2067 \times 10^{-3}$.
2. For $a > 20$ the probability $R(m, a)$ can be approximated as

$$R(m, a) \approx \Phi\left(\frac{m + 0.5 - a}{\sqrt{a}}\right) + 0.5,$$

where $\Phi(x)$ is the error function (Appendix B).

Appendix B

Error (Laplace) Function Tables

TABLE B.1 Standard cumulative normal distribution $R = (2\pi)^{-1/2} \int_z^\infty e^{-z^2/2}\,dz$†

z	0.00	0.01	0.02	0.03	0.04	0.05	0.06	0.07	0.08	0.09
0.0	0.5000	0.5040	0.5080	0.5120	0.5160	0.5199	0.5239	0.5279	0.5319	0.5359
0.1	0.5398	0.5438	0.5478	0.5517	0.5557	0.5596	0.5636	0.5675	0.5714	0.5753
0.2	0.5793	0.5832	0.5871	0.5910	0.5948	0.5987	0.6026	0.6064	0.6103	0.6141
0.3	0.6179	0.6217	0.6255	0.6293	0.6331	0.6368	0.6406	0.6443	0.6480	0.6517
0.4	0.6554	0.6591	0.6628	0.6664	0.6700	0.6736	0.6772	0.6808	0.6844	0.6879
0.5	0.6915	0.6950	0.6985	0.7019	0.7054	0.7088	0.7123	0.7157	0.7190	0.7224
0.6	0.7257	0.7291	0.7324	0.7357	0.7389	0.7422	0.7454	0.7486	0.7517	0.7549
0.7	0.7580	0.7611	0.7642	0.7673	0.7703	0.7734	0.7764	0.7794	0.7823	0.7852
0.8	0.7881	0.7910	0.7939	0.7967	0.7995	0.8023	0.8051	0.8078	0.8106	0.8133
0.9	0.8159	0.8186	0.8212	0.8238	0.8264	0.8289	0.8315	0.8340	0.8365	0.8389
1.0	0.8413	0.8438	0.8461	0.8485	0.8508	0.8531	0.8554	0.8577	0.8599	0.8621
1.1	0.8643	0.8665	0.8686	0.8708	0.8729	0.8749	0.8770	0.8790	0.8810	0.8830
1.2	0.8849	0.8869	0.8888	0.8907	0.8925	0.8944	0.8962	0.8980	0.8997	0.90147
1.3	0.90320	0.90490	0.90658	0.90824	0.90988	0.91149	0.91309	0.91466	0.91621	0.91774
1.4	0.91924	0.92073	0.92220	0.92364	0.92507	0.92647	0.92785	0.92922	0.93056	0.93189
1.5	0.93319	0.93448	0.93574	0.93699	0.93822	0.93943	0.94062	0.94179	0.94295	0.94408
1.6	0.94520	0.94630	0.94738	0.94845	0.94950	0.95053	0.95154	0.95254	0.95352	0.95449
1.7	0.95543	0.95637	0.95728	0.95818	0.95907	0.95994	0.96080	0.96164	0.96246	0.96327
1.8	0.96407	0.96485	0.96562	0.96638	0.96712	0.96784	0.96856	0.96926	0.96995	0.97062
1.9	0.97128	0.97193	0.97257	0.97320	0.97381	0.97441	0.97500	0.97558	0.97615	0.97670
2.0	0.97725	0.97778	0.97831	0.97882	0.97932	0.97982	0.98030	0.98077	0.98124	0.98169
2.1	0.98214	0.98257	0.98300	0.98341	0.98382	0.98422	0.98461	0.98500	0.98537	0.98574
2.2	0.98610	0.98645	0.98679	0.98713	0.98745	0.98778	0.98809	0.98840	0.98870	0.98899
2.3	0.98928	0.98956	0.98983	0.9²0097	0.9²0358	0.9²0613	0.9²0863	0.9²1106	0.9²1344	0.9²1576
2.4	0.9²1802	0.9²2024	0.9²2240	0.9²2451	0.9²2656	0.9²2857	0.9²3053	0.9²3244	0.9²3431	0.9²3613
2.5	0.9²3790	0.9²3963	0.9²4132	0.9²4297	0.9²4457	0.9²4614	0.9²4766	0.9²4915	0.9²5060	0.9²5201

† $\Phi(\alpha) = 1 - 2R(\alpha\sqrt{2})$, where $\Phi(\alpha) = \dfrac{2}{\sqrt{\pi}} \int_0^\alpha e^{-t^2}\,dt$

B. Error (Laplace) Function Tables

2.6	0.9^25339	0.9^55473	0.9^25604	0.9^25731	0.9^25855	0.9^25975	0.9^26093	0.9^26207	0.9^26319	0.9^26427
2.7	0.9^26533	0.9^56636	0.9^26736	0.9^26833	0.9^26928	0.9^27020	0.9^27110	0.9^27197	0.9^27282	0.9^27365
2.8	0.9^27445	0.9^57523	0.9^27599	0.9^27673	0.9^27744	0.9^27814	0.9^27882	0.9^27948	0.9^28012	0.9^28074
2.9	0.9^28134	0.9^58193	0.9^28250	0.9^28305	0.9^28359	0.9^28411	0.9^28462	0.9^28511	0.9^28559	0.9^28605
3.0	0.9^28650	0.9^58694	0.9^28736	0.9^28777	0.9^28817	0.9^28856	0.9^28893	0.9^28930	0.9^28965	0.9^28999
3.1	0.9^30324	0.9^30646	0.9^30957	0.9^31260	0.9^31553	0.9^31836	0.9^32112	0.9^32378	0.9^32636	0.9^32886
3.2	0.9^33129	0.9^33363	0.9^33590	0.9^33810	0.9^34024	0.9^34230	0.9^34429	0.9^34623	0.9^34810	0.9^34991
3.3	0.9^35166	0.9^35335	0.9^35499	0.9^35658	0.9^35811	0.9^35959	0.9^36103	0.9^36242	0.9^36376	0.9^36505
3.4	0.9^36631	0.9^36752	0.9^36869	0.9^36982	0.9^37091	0.9^37197	0.9^37299	0.9^37398	0.9^37493	0.9^37585
3.5	0.9^37674	0.9^37759	0.9^37842	0.9^37922	0.9^37999	0.9^38074	0.9^38146	0.9^38215	0.9^38282	0.9^38347
3.6	0.9^38409	0.9^38469	0.9^38527	0.9^38583	0.9^38637	0.9^38689	0.9^38739	0.9^38787	0.9^38834	0.9^38879
3.7	0.9^38922	0.9^38964	0.9^40039	0.9^40426	0.9^40799	0.9^41158	0.9^41504	0.9^41838	0.9^42159	0.9^42468
3.8	0.9^42765	0.9^43052	0.9^43327	0.9^43593	0.9^43848	0.9^44094	0.9^44331	0.9^44558	0.9^44777	0.9^44988
3.9	0.9^45190	0.9^45385	0.9^45573	0.9^45753	0.9^45926	0.9^46092	0.9^46253	0.9^46406	0.9^46554	0.9^46696
4.0	0.9^46833	0.9^46964	0.9^47090	0.9^47211	0.9^47327	0.9^47439	0.9^47546	0.9^47649	0.9^47748	0.9^47843
4.1	0.9^47934	0.9^48022	0.9^48106	0.9^48186	0.9^48263	0.9^48338	0.9^48409	0.9^48477	0.9^48542	0.9^48605
4.2	0.9^48665	0.9^48723	0.9^48778	0.9^48832	0.9^48882	0.9^48931	0.9^48978	0.9^50226	0.9^50655	0.9^51066
4.3	0.9^51460	0.9^51837	0.9^52199	0.9^52545	0.9^52876	0.9^53193	0.9^53497	0.9^53788	0.9^54066	0.9^54332
4.4	0.9^54587	0.9^54831	0.9^55065	0.9^55288	0.9^55502	0.9^55706	0.9^55902	0.9^56089	0.9^56268	0.9^56439
4.5	0.9^56602	0.9^56759	0.9^56908	0.9^57051	0.9^57187	0.9^57318	0.9^57442	0.9^57561	0.9^57675	0.9^57784
4.6	0.9^57888	0.9^57987	0.9^58081	0.9^58172	0.9^58258	0.9^58340	0.9^58419	0.9^58494	0.9^58566	0.9^58634
4.7	0.9^58699	0.9^58761	0.9^58821	0.9^58877	0.9^58931	0.9^58983	0.9^60320	0.9^60789	0.9^61235	0.9^61661
4.8	0.9^62067	0.9^62453	0.9^62822	0.9^63173	0.9^63508	0.9^63827	0.9^64131	0.9^64420	0.9^64696	0.9^64958
4.9	0.9^65208	0.9^65446	0.9^65673	0.9^65889	0.9^66094	0.9^66289	0.9^66475	0.9^66652	0.9^66821	0.9^66981

SOURCE: A. Hald, "Statistical Tables and Formulas," John Wiley, New York, 1952, Table II.

TABLE B.2 Standard cumulative normal distribution function

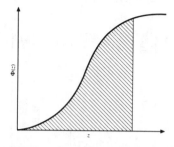

$$\Phi(z) = \frac{1}{\sqrt{2\pi}} \int_{-\infty}^{z} \exp\left(\frac{-x^2}{2}\right) dx$$

for $0.00 \leq z \leq 4.00$
$1 - \Phi(z) = \Phi(-z)$

z	0.00	0.01	0.02	0.03	0.04	0.05	0.06	0.07	0.08	0.09
0.0	0.5000	0.5040	0.5080	0.5120	0.5160	0.5199	0.5239	0.5279	0.5319	0.5359
0.1	0.5398	0.5438	0.5478	0.5517	0.5557	0.5596	0.5636	0.5675	0.5714	0.5753
0.2	0.5793	0.5832	0.5871	0.5910	0.5948	0.5987	0.6026	0.6064	0.6103	0.6141
0.3	0.6179	0.6217	0.6255	0.6293	0.6331	0.6368	0.6406	0.6443	0.6480	0.6517
0.4	0.6554	0.6591	0.6628	0.6664	0.6700	0.6736	0.6772	0.6808	0.6844	0.6879
0.5	0.6915	0.6985	0.6985	0.7019	0.7054	0.7088	0.7123	0.7157	0.7190	0.7224
0.6	0.7257	0.7291	0.7324	0.7357	0.7389	0.7422	0.7454	0.7486	0.7517	0.7549
0.7	0.7580	0.7611	0.7642	0.7673	0.7703	0.7734	0.7764	0.7794	0.7823	0.7852
0.8	0.7881	0.7910	0.7939	0.7967	0.7995	0.8023	0.8051	0.8078	0.8106	0.8133
0.9	0.8159	0.8186	0.8212	0.8238	0.8264	0.8289	0.8315	0.8340	0.8365	0.8389
1.0	0.8413	0.8438	0.8461	0.8485	0.8508	0.8531	0.8554	0.8577	0.8599	0.8621
1.1	0.8643	0.8665	0.8686	0.8708	0.8729	0.8749	0.8770	0.8790	0.8810	0.8830
1.2	0.8849	0.8869	0.8888	0.8907	0.8925	0.8944	0.8962	0.8980	0.8997	0.9015
1.3	0.9032	0.9049	0.9066	0.9082	0.9099	0.9115	0.9131	0.9147	0.9162	0.9177
1.4	0.9192	0.9207	0.9222	0.9236	0.9251	0.9265	0.9279	0.9292	0.9306	0.9319
1.5	0.9332	0.9345	0.9357	0.9370	0.9382	0.9394	0.9406	0.9418	0.9430	0.9440
1.6	0.9452	0.9463	0.9474	0.9485	0.9495	0.9505	0.9515	0.9525	0.9535	0.9545
1.7	0.9554	0.9564	0.9573	0.9582	0.9591	0.9599	0.9608	0.9616	0.9625	0.9633
1.8	0.9641	0.9649	0.9656	0.9664	0.9671	0.9678	0.9686	0.9693	0.9700	0.9706
1.9	0.9713	0.9719	0.9726	0.9732	0.9738	0.9744	0.9750	0.9756	0.9762	0.9767
2.0	0.9773	0.9778	0.9783	0.9788	0.9793	0.9798	0.9803	0.9808	0.9812	0.9817
2.1	0.9821	0.9826	0.9830	0.9834	0.9838	0.9842	0.9846	0.9850	0.9854	0.9857
2.2	0.9861	0.9865	0.9868	0.9871	0.9875	0.9878	0.9881	0.9884	0.9887	0.9890
2.3	0.9893	0.9896	0.9898	0.9^2010	0.9^2061	0.9^2035	0.9^2086	0.9^2111	0.9^2134	0.9^2158
2.4	0.9^2180	0.9^2202	0.9^2224	0.9^2245	0.9^2266	0.9^2286	0.9^2305	0.9^2324	0.9^2343	0.9^2361
2.5	0.9^2379	0.9^2396	0.9^2413	0.9^2430	0.9^2446	0.9^2461	0.9^2477	0.9^2492	0.9^2506	0.9^2520
2.6	0.9^2534	0.9^2547	0.9^2560	0.9^2573	0.9^2586	0.9^2598	0.9^2609	0.9^2621	0.9^2632	0.9^2643
2.7	0.9^2653	0.9^2664	0.9^2674	0.9^2683	0.9^2693	0.9^2702	0.9^2711	0.9^2720	0.9^2728	0.9^2737
2.8	0.9^2745	0.9^2752	0.9^2760	0.9^2767	0.9^2774	0.9^2781	0.9^2788	0.9^2795	0.9^2801	0.9^2807
2.9	0.9^2813	0.9^2819	0.9^2825	0.9^2831	0.9^2836	0.9^2841	0.9^2846	0.9^2851	0.9^2856	0.9^2861
3.0	0.9^2865	0.9^2869	0.9^2874	0.9^2878	0.9^2882	0.9^2886	0.9^2889	0.9^2893	0.9^2897	0.9^2900
3.1	0.9^3032	0.9^3065	0.9^3096	0.9^3126	0.9^3155	0.9^3184	0.9^3211	0.9^3238	0.9^3264	0.9^3289
3.2	0.9^3313	0.9^3336	0.9^3359	0.9^3381	0.9^3402	0.9^3423	0.9^3443	0.9^3462	0.9^3481	0.9^3499
3.3	0.9^3517	0.9^3534	0.9^3550	0.9^3566	0.9^3581	0.9^3596	0.9^3610	0.9^3624	0.9^3638	0.9^3651
3.4	0.9^3663	0.9^3675	0.9^3687	0.9^3698	0.9^3709	0.9^3720	0.9^3730	0.9^3740	0.9^3749	0.9^3759
3.5	0.9^3767	0.9^3776	0.9^3784	0.9^3792	0.9^3800	0.9^3807	0.9^3815	0.9^3822	0.9^3822	0.9^3835
3.6	0.9^3841	0.9^3847	0.9^3853	0.9^3858	0.9^3864	0.9^3869	0.9^3874	0.9^3879	0.9^3883	0.9^3888
3.7	0.9^3892	0.9^3896	0.9^4004	0.9^4043	0.9^4116	0.9^4116	0.9^4150	0.9^4184	0.9^4216	0.9^4257
3.8	0.9^4277	0.9^4305	0.9^4333	0.9^4359	0.9^4385	0.9^4409	0.9^4433	0.9^4456	0.9^4478	0.9^4499
3.9	0.9^4519	0.9^4539	0.9^4557	0.9^4575	0.9^4593	0.9^4609	0.9^4625	0.9^4641	0.9^4655	0.9^4670

B. Error (Laplace) Function Tables

TABLE B.3 Values of the function $\Phi_M(x) = \dfrac{2}{\sqrt{\pi}} \displaystyle\int_0^x e^{-t^2}\, dt$

x	0	1	2	3	4	5	6	7	8	9
0.0	0.000 00	0.011 28	0.022 56	0.033 84	0.045 11	0.056 37	0.067 62	0.078 86	0.090 08	0.101 28
0.1	0.112 46	0.123 62	0.134 76	0.145 87	0.156 95	0.168 00	0.179 01	0.189 99	0.200 94	0.211 84
0.2	0.222 70	0.233 52	0.244 30	0.255 02	0.265 70	0.276 33	0.286 90	0.297 42	0.307 88	0.318 28
0.3	0.328 63	0.338 91	0.349 13	0.359 28	0.369 36	0.379 38	0.389 33	0.399 21	0.409 01	0.418 74
0.4	0.428 39	0.437 97	0.447 47	0.456 89	0.466 22	0.475 48	0.484 66	0.493 74	0.502 75	0.511 67
0.5	0.520 50	0.529 24	0.537 90	0.546 46	0.554 94	0.563 32	0.571 62	0.579 82	0.587 92	0.595 94
0.6	0.603 86	0.611 68	0.619 41	0.627 05	0.634 59	0.642 03	0.649 38	0.656 63	0.663 78	0.670 84
0.7	0.677 80	0.684 67	0.691 43	0.698 10	0.704 68	0.711 16	0.717 54	0.723 82	0.730 01	0.736 10
0.8	0.742 10	0.748 00	0.753 81	0.759 52	0.765 14	0.770 67	0.776 10	0.781 44	0.786 69	0.791 84
0.9	0.796 91	0.801 88	0.806 77	0.811 56	0.816 27	0.820 89	0.825 42	0.829 87	0.834 23	0.838 51
1.0	0.842 70	0.846 81	0.850 84	0.854 78	0.858 65	0.862 44	0.866 14	0.869 77	0.873 33	0.876 80
1.1	0.880 20	0.883 53	0.886 79	0.889 97	0.893 08	0.896 12	0.899 10	0.902 00	0.904 84	0.907 61
1.2	0.901 31	0.912 96	0.915 53	0.918 05	0.920 50	0.922 90	0.925 24	0.927 51	0.929 73	0.931 90
1.3	0.934 01	0.936 06	0.938 06	0.940 02	0.941 91	0.943 76	0.945 56	0.947 31	0.949 02	0.950 67
1.4	0.952 28	0.953 85	0.955 38	0.956 86	0.958 30	0.959 70	0.961 05	0.962 37	0.963 65	0.964 90
1.5	0.966 10	0.967 28	0.968 41	0.969 52	0.970 59	0.971 62	0.972 63	0.973 60	0.974 55	0.975 46
1.6	0.976 35	0.977 21	0.978 04	0.978 84	0.979 62	0.980 38	0.981 10	0.981 81	0.982 49	0.983 15
1.7	0.983 79	0.984 41	0.985 00	0.985 58	0.986 14	0.986 67	0.987 19	0.987 69	0.988 17	0.988 64
1.8	0.989 09	0.989 52	0.989 94	0.990 35	0.990 74	0.991 11	0.991 47	0.991 82	0.992 16	0.992 48
1.9	0.992 79	0.993 09	0.993 38	0.993 66	0.993 92	0.994 18	0.994 43	0.994 66	0.994 89	0.995 11
2.0	0.995 32	0.995 52	0.995 72	0.995 91	0.996 09	0.996 26	0.996 42	0.996 58	0.996 73	0.996 88
2.1	0.997 02	0.997 16	0.997 28	0.997 41	0.997 52	0.997 64	0.997 75	0.997 85	0.997 95	0.998 05
2.2	0.998 14	0.998 22	0.998 31	0.998 39	0.998 46	0.998 54	0.998 61	0.998 67	0.998 74	0.998 80
2.3	0.998 86	0.998 91	0.998 97	0.999 02	0.999 06	0.999 11	0.999 16	0.999 20	0.999 24	0.999 28
2.4	0.999 31	0.999 35	0.999 38	0.999 41	0.999 44	0.999 47	0.999 50	0.999 52	0.999 55	0.999 57
2.5	0.999 59	0.999 61	0.999 63	0.999 65	0.999 67	0.999 69	0.999 71	0.999 72	0.999 74	0.999 75
2.6	0.999 76	0.999 78	0.999 79	0.999 80	0.999 81	0.999 82	0.999 83	0.999 84	0.999 85	0.999 86
2.7	0.999 87	0.999 87	0.999 88	0.999 89	0.999 89	0.999 90	0.999 91	0.999 91	0.999 92	0.999 92
2.8	0.999 92	0.999 93	0.999 93	0.999 94	0.999 94	0.999 94	0.999 95	0.999 95	0.999 95	0.999 96
2.9	0.999 96	0.999 96	0.999 96	0.999 97	0.999 97	0.999 97	0.999 97	0.999 97	0.999 98	0.999 98
3.0	0.999 98	0.999 98	0.999 98	0.999 98	0.999 98	0.999 98	0.999 98	0.999 99	0.999 99	0.999 99

TABLE B.4 Values of the function $\Phi_o(x) = \dfrac{1}{\sqrt{2\pi}} \displaystyle\int_o^x e^{-u^2/2} \, du$

	Hundredths									
x	0	1	2	3	4	5	6	7	8	9
0.0	0.0000	0.0040	0.0080	0.0120	0.0160	0.0200	0.0239	0.0279	0.0319	0.0359
0.1	0.0398	0.0438	0.0478	0.0517	0.0557	0.0596	0.0636	0.0675	0.0714	0.0753
0.2	0.0793	0.0832	0.0871	0.0910	0.0948	0.0987	0.1026	0.1064	0.1103	0.1141
0.3	0.1179	0.1217	0.1255	0.1293	0.1331	0.1368	0.0406	0.0443	0.0480	0.0517
0.4	0.0554	0.0591	0.0628	0.0664	0.0700	0.0736	0.0772	0.0808	0.0844	0.0879
0.5	0.0915	0.0950	0.0985	0.2019	0.2054	0.2088	0.2123	0.2157	0.2190	0.2224
0.6	0.2257	0.2291	0.2324	0.0357	0.0389	0.0422	0.0454	0.0486	0.0517	0.0549
0.7	0.0580	0.0611	0.0642	0.0673	0.0703	0.0734	0.0764	0.0794	0.0823	0.0852
0.8	0.0881	0.0910	0.0939	0.0967	0.0995	0.3023	0.3051	0.3078	0.3106	0.3133
0.9	0.3159	0.3186	0.3212	0.3238	0.3264	0.0289	0.0315	0.0340	0.0365	0.0389
1.0	0.0413	0.0437	0.0461	0.0485	0.0508	0.0581	0.0554	0.0577	0.0599	0.0621
1.1	0.0643	0.0665	0.0686	0.0708	0.0729	0.0749	0.0770	0.0790	0.0810	0.0830
1.2	0.0849	0.0869	0.0888	0.0907	0.0925	0.0944	0.0962	0.0980	0.0997	0.4015
1.3	0.4032	0.4049	0.4066	0.4082	0.4099	0.4115	0.4131	0.4147	0.4162	0.4177
1.4	0.0192	0.0207	0.0222	0.0236	0.0251	0.0265	0.0279	0.0292	0.0306	0.0319
1.5	0.0332	0.0345	0.0357	0.0370	0.0382	0.0394	0.0406	0.0418	0.0429	0.0441
1.6	0.0452	0.0463	0.0474	0.0484	0.0495	0.0505	0.0515	0.0525	0.0535	0.0545
1.7	0.0554	0.0564	0.0573	0.0582	0.0591	0.0599	0.0608	0.0616	0.0625	0.0633
1.8	0.0641	0.0649	0.0656	0.0664	0.0671	0.0678	0.0686	0.0693	0.0699	0.0706
1.9	0.0713	0.0719	0.0726	0.0732	0.0738	0.0744	0.0750	0.0756	0.0761	0.0767
2.0	0.0772	0.0778	0.0783	0.0788	0.0793	0.0798	0.0803	0.0808	0.0812	0.0817
2.1	0.0821	0.0826	0.0830	0.0834	0.0838	0.0842	0.0846	0.0850	0.0854	0.0857
2.2	0.0861	0.0864	0.0868	0.0871	0.0875	0.0878	0.0881	0.0884	0.0887	0.0890
2.3	0.0893	0.0896	0.0898	0.0901	0.0904	0.0906	0.0909	0.0911	0.0913	0.0916
2.4	0.0918	0.0920	0.0922	0.0925	0.0927	0.0929	0.0931	0.0932	0.0934	0.0936
2.5	0.0938	0.0940	0.0941	0.0943	0.0945	0.0946	0.0948	0.0949	0.0951	0.0952
2.6	0.0953	0.0955	0.0956	0.0957	0.0959	0.0960	0.0961	0.0962	0.0963	0.0964
2.7	0.0965	0.0966	0.0967	0.0968	0.0969	0.0970	0.0971	0.0972	0.0973	0.0974
2.8	0.0974	0.0975	0.0976	0.0977	0.0977	0.0978	0.0979	0.0979	0.0980	0.0981
2.9	0.0981	0.0982	0.0982	0.0983	0.0984	0.0984	0.0985	0.0985	0.0985	0.0986
3.0	0.0987	0.0987	0.0987	0.0988	0.0988	0.0989	0.0989	0.0989	0.0990	0.0990

TABLE B.5 Values of the error function $\Phi(x) = \dfrac{1}{\sqrt{2\pi}} \displaystyle\int_0^x e^{-z^2/2}\, dz$

x	0	1	2	3	4	5	6	7	8	9
0.0	0.000 00	0.003 99	0.007 98	0.011 97	0.015 95	0.019 94	0.023 92	0.027 90	0.031 88	0.035 86
0.1	0.039 83	0.043 80	0.047 76	0.051 72	0.055 67	0.059 62	0.063 56	0.067 49	0.071 42	0.075 35
0.2	0.079 26	0.083 17	0.087 06	0.090 95	0.094 83	0.098 71	0.102 57	0.106 42	0.110 26	0.114 09
0.3	0.117 91	0.121 72	0.125 52	0.129 30	0.133 07	0.136 83	0.140 58	0.144 31	0.148 03	0.151 73
0.4	0.155 42	0.159 10	0.162 76	0.166 40	0.170 03	0.173 64	0.177 24	0.180 82	0.184 39	0.187 93
0.5	0.191 46	0.194 97	0.198 47	0.201 94	0.205 40	0.208 84	0.212 26	0.215 66	0.219 04	0.222 40
0.6	0.225 75	0.229 07	0.232 37	0.235 65	0.238 91	0.242 15	0.245 37	0.248 57	0.251 75	0.254 90
0.7	0.258 04	0.261 15	0.264 24	0.267 30	0.270 35	0.273 37	0.276 37	0.279 35	0.282 30	0.285 24
0.8	0.288 14	0.291 03	0.293 89	0.296 73	0.299 55	0.302 34	0.305 11	0.307 85	0.310 57	0.313 27
0.9	0.315 94	0.318 59	0.321 21	0.323 81	0.326 39	0.328 94	0.331 47	0.333 98	0.336 46	0.338 91
1.0	0.341 34	0.343 75	0.346 14	0.348 50	0.350 83	0.353 14	0.355 43	0.357 69	0.359 93	0.362 14
1.1	0.364 33	0.366 50	0.368 64	0.368 76	0.372 86	0.374 93	0.376 98	0.379 00	0.381 00	0.382 98
1.2	0.384 93	0.386 86	0.388 77	0.390 65	0.392 51	0.394 35	0.396 17	0.397 96	0.399 73	0.401 47
1.3	0.403 20	0.404 90	0.406 58	0.408 24	0.409 88	0.411 49	0.413 09	0.414 66	0.416 21	0.417 74
1.4	0.419 24	0.420 73	0.422 20	0.423 64	0.425 07	0.426 47	0.427 86	0.429 22	0.430 56	0.431 89
1.5	0.433 19	0.434 48	0.435 74	0.436 99	0.438 22	0.439 43	0.440 62	0.441 79	0.442 95	0.444 08
1.6	0.445 20	0.446 30	0.447 38	0.448 45	0.449 50	0.450 53	0.451 54	0.452 54	0.453 52	0.454 49
1.7	0.455 43	0.456 37	0.457 28	0.458 18	0.459 07	0.459 94	0.460 80	0.461 64	0.462 46	0.463 27
1.8	0.464 07	0.464 85	0.465 62	0.466 38	0.467 12	0.467 84	0.468 56	0.469 26	0.469 95	0.470 62
1.9	0.471 28	0.471 93	0.472 57	0.473 20	0.473 81	0.474 41	0.475 00	0.475 58	0.476 15	0.476 70
2.0	0.477 25	0.477 78	0.478 31	0.478 82	0.479 32	0.479 82	0.480 30	0.480 77	0.481 24	0.481 69
2.1	0.482 14	0.482 57	0.483 00	0.483 41	0.483 82	0.484 22	0.484 61	0.485 00	0.485 37	0.485 74
2.2	0.486 10	0.486 45	0.486 79	0.487 13	0.487 45	0.487 78	0.488 09	0.488 40	0.488 70	0.488 99
2.3	0.489 28	0.489 56	0.489 83	0.490 10	0.490 36	0.490 61	0.490 86	0.491 11	0.491 34	0.491 58
2.4	0.491 80	0.492 02	0.492 24	0.492 45	0.492 66	0.492 86	0.493 05	0.493 24	0.493 43	0.493 61
2.5	0.493 79	0.493 96	0.494 13	0.494 30	0.494 46	0.494 61	0.494 77	0.494 92	0.495 06	0.495 20
2.6	0.495 34	0.495 47	0.495 60	0.495 73	0.495 85	0.495 98	0.496 09	0.496 21	0.496 32	0.496 43
2.7	0.496 53	0.496 64	0.496 74	0.496 83	0.496 93	0.497 02	0.497 11	0.497 20	0.497 28	0.497 36
2.8	0.497 44	0.497 52	0.497 60	0.497 67	0.497 74	0.497 81	0.497 88	0.497 95	0.498 01	0.498 07
2.9	0.498 13	0.498 19	0.498 25	0.498 31	0.498 36	0.498 41	0.498 46	0.498 51	0.498 56	0.498 61

x	Φ(x)	x	Φ(x)	x	Φ(x)	x	Φ(x)	x	Φ(x)
3.0	0.498 65	3.1	0.499 03	3.2	0.499 31	3.3	0.499 52	3.4	0.499 66
3.5	0.499 77	3.6	0.499 84	3.7	0.499 89	3.8	0.499 93	3.9	0.499 95
4.0	0.499 968								
4.5	0.499 997								
5.0	0.499 999 97								

TABLE B.6 Values of the function $F(x) = \dfrac{1}{\sqrt{2\pi}} \displaystyle\int_{-\infty}^{x} e^{-t^2/2}\, dt$

x	0	1	2	3	4	5	6	7	8	9
0.0	0.500 00	0.503 99	0.507 98	0.511 97	0.515 95	0.519 94	0.523 92	0.527 90	0.531 88	0.535 86
0.1	0.539 83	0.543 80	0.547 76	0.551 72	0.555 67	0.559 62	0.563 56	0.567 49	0.571 42	0.575 35
0.2	0.579 26	0.583 17	0.587 06	0.590 95	0.594 83	0.598 71	0.602 57	0.606 42	0.610 26	0.614 09
0.3	0.617 91	0.621 72	0.625 52	0.629 30	0.633 07	0.636 83	0.640 58	0.644 31	0.648 03	0.651 73
0.4	0.655 42	0.659 10	0.662 76	0.666 40	0.670 03	0.673 64	0.677 24	0.680 82	0.684 39	0.687 93
0.5	0.691 46	0.694 97	0.698 47	0.701 94	0.705 40	0.708 84	0.712 26	0.715 66	0.719 04	0.722 40
0.6	0.725 75	0.729 07	0.732 37	0.735 65	0.738 91	0.742 15	0.745 37	0.748 57	0.751 75	0.754 90
0.7	0.758 04	0.761 15	0.764 24	0.767 30	0.770 35	0.773 37	0.776 37	0.779 35	0.782 30	0.785 24
0.8	0.788 14	0.791 03	0.793 89	0.796 73	0.799 55	0.802 34	0.805 11	0.807 85	0.810 57	0.813 27
0.9	0.815 94	0.818 59	0.821 21	0.823 81	0.826 39	0.828 94	0.831 47	0.833 98	0.836 46	0.838 91
1.0	0.841 34	0.843 75	0.846 14	0.848 50	0.850 83	0.853 14	0.855 43	0.857 69	0.859 93	0.862 14
1.1	0.864 33	0.866 50	0.868 64	0.870 76	0.872 86	0.874 93	0.876 98	0.879 00	0.881 00	0.882 98
1.2	0.884 93	0.886 86	0.888 77	0.890 65	0.892 51	0.894 35	0.896 17	0.897 96	0.899 73	0.901 47
1.3	0.903 20	0.904 90	0.906 58	0.908 24	0.909 88	0.911 49	0.913 08	0.914 66	0.916 21	0.917 74
1.4	0.919 24	0.920 73	0.922 20	0.923 64	0.925 07	0.926 47	0.927 86	0.929 22	0.930 56	0.931 89
1.5	0.933 19	0.934 48	0.935 74	0.936 99	0.938 22	0.939 43	0.940 62	0.941 79	0.942 95	0.944 08
1.6	0.945 20	0.946 30	0.947 38	0.948 45	0.949 50	0.950 53	0.951 54	0.952 54	0.953 52	0.954 49
1.7	0.955 43	0.956 37	0.957 28	0.958 18	0.959 07	0.959 94	0.960 80	0.961 64	0.962 46	0.963 27
1.8	0.964 07	0.964 85	0.965 62	0.966 38	0.967 12	0.967 84	0.968 56	0.969 26	0.969 95	0.970 62
1.9	0.971 28	0.971 93	0.972 57	0.973 20	0.973 81	0.974 41	0.975 00	0.975 58	0.976 15	0.976 70
2.0	0.977 25	0.977 78	0.978 31	0.978 82	0.979 32	0.979 82	0.980 30	0.980 77	0.981 24	0.981 69
2.1	0.982 14	0.982 57	0.983 00	0.983 41	0.983 82	0.984 22	0.984 61	0.985 00	0.985 37	0.985 74
2.2	0.986 10	0.986 45	0.986 79	0.987 13	0.987 45	0.987 78	0.988 09	0.988 40	0.988 70	0.988 99
2.3	0.989 28	0.989 56	0.989 83	0.990 10	0.990 36	0.990 61	0.990 86	0.991 11	0.991 34	0.991 58
2.4	0.991 80	0.992 02	0.992 24	0.992 45	0.992 66	0.992 86	0.993 05	0.993 24	0.993 43	0.993 61
2.5	0.993 79	0.993 96	0.994 13	0.994 30	0.994 46	0.994 61	0.994 77	0.994 92	0.995 06	0.995 20
2.6	0.995 34	0.995 47	0.995 60	0.995 73	0.995 85	0.995 98	0.996 09	0.996 21	0.996 32	0.996 43
2.7	0.996 53	0.996 64	0.996 74	0.996 83	0.996 93	0.997 02	0.997 11	0.997 20	0.997 28	0.997 36
2.8	0.997 44	0.997 52	0.997 60	0.997 67	0.997 74	0.997 81	0.997 88	0.997 95	0.998 01	0.998 07
2.9	0.998 13	0.998 19	0.998 25	0.998 31	0.998 36	0.998 41	0.998 46	0.998 51	0.998 56	0.998 61
3.0	0.998 65	0.998 69	0.998 74	0.998 78	0.998 82	0.998 86	0.998 89	0.998 93	0.998 96	0.999 00
3.1	0.999 03	0.999 06	0.999 10	0.999 13	0.999 16	0.999 18	0.999 21	0.999 24	0.999 26	0.999 29
3.2	0.999 31	0.999 34	0.999 36	0.999 38	0.999 40	0.999 42	0.999 44	0.999 46	0.999 48	0.999 50
3.3	0.999 52	0.999 53	0.999 55	0.999 57	0.999 58	0.999 60	0.999 61	0.999 62	0.999 64	0.999 65
3.4	0.999 66	0.999 68	0.999 69	0.999 70	0.999 71	0.999 72	0.999 73	0.999 74	0.999 75	0.999 76
3.5	0.999 77	0.999 78	0.999 78	0.999 79	0.999 80	0.999 81	0.999 81	0.999 82	0.999 83	0.999 83
3.6	0.999 84	0.999 85	0.999 85	0.999 86	0.999 86	0.999 87	0.999 87	0.999 88	0.999 88	0.999 89
3.7	0.999 89	0.999 90	0.999 90	0.999 90	0.999 91	0.999 91	0.999 92	0.999 92	0.999 92	0.999 92
3.8	0.999 93	0.999 93	0.999 93	0.999 94	0.999 94	0.999 94	0.999 94	0.999 95	0.999 95	0.999 95
3.9	0.999 95	0.999 95	0.999 96	0.999 96	0.999 96	0.999 96	0.999 96	0.999 96	0.999 97	0.999 97
4.0	0.999 97	0.999 98	0.999 99	0.999 99	0.999 99	—	—	—	—	—

B. Error (Laplace) Function Tables

TABLE B.7 Areas under the normal curve from K_α to ∞

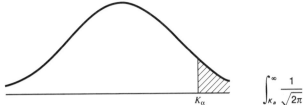

$$\int_{K_a}^{\infty} \frac{1}{\sqrt{2\pi}} e^{-x^2/2}\, dx = \alpha$$

K_a	0.00	0.01	0.02	0.03	0.04	0.05	0.06	0.07	0.08	0.09
0.0	0.5000	0.4960	0.4920	0.4880	0.4840	0.4801	0.4761	0.4721	0.4681	0.4641
0.1	0.4602	0.4562	0.4522	0.4483	0.4443	0.4404	0.4364	0.4325	0.4286	0.4247
0.2	0.4207	0.4168	0.4129	0.4090	0.4052	0.4013	0.3974	0.3936	0.3897	0.3859
0.3	0.3821	0.3783	0.3745	0.3707	0.3669	0.3632	0.3594	0.3557	0.3520	0.3483
0.4	0.3446	0.3409	0.3372	0.3336	0.3300	0.3264	0.3238	0.3192	0.3156	0.3121
0.5	0.3085	0.3050	0.3015	0.2981	0.2946	0.2912	0.2877	0.2843	0.2810	0.2776
0.6	0.2743	0.2709	0.2676	0.2643	0.2611	0.2578	0.2546	0.2514	0.2483	0.2451
0.7	0.2420	0.2389	0.2358	0.2327	0.2296	0.2266	0.2236	0.2206	0.2177	0.2148
0.8	0.2119	0.2090	0.2061	0.2033	0.2005	0.1977	0.1949	0.1922	0.1894	0.1867
0.9	0.1841	0.1814	0.1788	0.1762	0.1736	0.1711	0.1685	0.1660	0.1635	0.1611
1.0	0.1587	0.1562	0.1539	0.1515	0.1492	0.1469	0.1446	0.1423	0.1401	0.1379
1.1	0.1357	0.1335	0.1314	0.1292	0.1271	0.1251	0.1230	0.1210	0.1190	0.1170
1.2	0.1151	0.1131	0.1112	0.1093	0.1075	0.1056	0.1038	0.1020	0.1003	0.0985
1.3	0.0968	0.0951	0.0934	0.0918	0.0901	0.0885	0.0869	0.0853	0.0838	0.0823
1.4	0.0808	0.0793	0.0778	0.0764	0.0749	0.0735	0.0721	0.0708	0.0694	0.0681
1.5	0.0668	0.0655	0.0643	0.0630	0.0618	0.0606	0.0594	0.0582	0.0571	0.0559
1.6	0.0548	0.0537	0.0526	0.0516	0.0505	0.0495	0.0485	0.0475	0.0465	0.0455
1.7	0.0446	0.0436	0.0427	0.0418	0.0409	0.0401	0.0392	0.0384	0.0375	0.0367
1.8	0.0359	0.0351	0.0344	0.0336	0.0329	0.0322	0.0314	0.0307	0.0301	0.0294
1.9	0.0287	0.0281	0.0274	0.0268	0.0262	0.0256	0.0250	0.0244	0.0239	0.0233
2.0	0.0228	0.0222	0.0217	0.0212	0.0207	0.0202	0.0197	0.0192	0.0188	0.0183
2.1	0.0179	0.0174	0.0170	0.0166	0.0162	0.0158	0.0154	0.0150	0.0146	0.0143
2.2	0.0139	0.0136	0.0132	0.0129	0.0125	0.0122	0.0119	0.0116	0.0113	0.0110
2.3	0.0107	0.0104	0.0102	0.00990	0.00964	0.00939	0.00914	0.00889	0.00866	0.00842
2.4	0.00820	0.00798	0.00776	0.00755	0.00734	0.00714	0.00695	0.00676	0.00657	0.00639
2.5	0.00621	0.00604	0.00587	0.00570	0.00554	0.00539	0.00523	0.00508	0.00494	0.00480
2.6	0.00466	0.00453	0.00440	0.00427	0.00415	0.00402	0.00391	0.00379	0.00368	0.00357
2.7	0.00347	0.00336	0.00326	0.00317	0.00307	0.00298	0.00289	0.00280	0.00272	0.00264
2.8	0.00256	0.00248	0.00240	0.00233	0.00226	0.00219	0.00212	0.00205	0.00199	0.00193
2.9	0.00187	0.00181	0.00175	0.00169	0.00164	0.00159	0.00154	0.00149	0.00144	0.00139

K_a	0.0	0.1	0.2	0.3	0.4	0.5	0.6	0.7	0.8	0.9
3	0.00135	$0.0^3 968$	$0.0^3 687$	$0.0^3 483$	$0.0^2 337$	$0.0^3 233$	$0.0^3 159$	$0.0^3 108$	$0.0^4 723$	$0.0^4 481$
4	$0.0^4 317$	$0.0^4 207$	$0.0^4 133$	$0.0^5 854$	$0.0^5 541$	$0.0^5 340$	$0.0^6 211$	$0.0^6 130$	$0.0^6 793$	$0.0^6 479$
5	$0.0^6 287$	$0.0^6 170$	$0.0^7 996$	$0.0^7 579$	$0.0^7 333$	$0.0^7 190$	$0.0^7 107$	$0.0^8 599$	$0.0^5 332$	$0.0^8 182$
6	$0.0^9 987$	$0.0^9 530$	$0.0^9 282$	$0.0^9 149$	$0.0^{10} 777$	$0.0^{10} 402$	$0.0^{10} 206$	$0.0^{10} 104$	$0.0^{11} 523$	$0.0^{11} 260$

SOURCE: From Frederick E. Croxton, "Elementary Statistics with Applications in Medicine," Prentice-Hall, Englewood Cliffs, New Jersey, 1953, p. 323.

TABLE B.8 Tolerance factors for normal distributions†

						α				
			$\gamma = 0.75$					$\gamma = 0.90$		
n	0.25	0.10	0.05	0.01	0.001	0.25	0.10	0.05	0.01	0.001
2	4.498	6.301	7.414	9.531	11.920	11.407	15.978	18.800	24.167	30.227
3	2.501	3.358	4.187	5.431	6.844	4.132	5.847	6.919	8.974	11.309
4	2.035	2.892	3.431	4.471	5.657	2.932	4.166	4.943	6.440	8.149
5	1.825	2.599	3.088	4.033	5.117	2.454	3.494	4.152	5.423	6.879
6	1.704	2.429	2.889	3.779	4.802	2.196	3.131	3.723	4.870	6.188
7	1.624	2.318	2.757	3.611	4.593	2.034	2.902	3.452	4.521	5.750
8	1.568	2.238	2.663	3.491	4.444	1.921	2.743	3.264	4.278	5.446
9	1.525	2.178	2.593	3.400	4.330	1.839	2.626	3.125	4.098	5.220
10	1.492	2.131	2.537	3.328	4.241	1.775	2.535	3.018	3.959	5.046
11	1.465	2.093	2.493	3.271	4.169	1.724	2.463	2.933	3.849	4.906
12	1.443	2.062	2.456	3.223	4.110	1.683	2.404	2.863	3.758	4.792
13	1.425	2.036	2.424	3.183	4.059	1.648	2.355	2.805	3.682	4.697
14	1.409	2.013	2.398	3.148	4.016	1.619	2.314	2.756	3.618	4.615
15	1.395	1.994	2.375	3.118	3.979	1.594	2.278	2.713	3.562	4.545
16	1.383	1.977	2.355	3.092	3.946	1.572	2.246	2.676	3.514	4.484
17	1.372	1.962	2.337	3.069	3.917	1.552	2.219	2.643	3.471	4.430
18	1.363	1.948	2.321	3.048	3.891	1.535	2.194	2.614	3.433	4.382
19	1.355	1.936	2.307	3.030	3.867	1.520	2.172	2.588	3.399	4.339
20	1.347	1.925	2.294	3.013	3.846	1.506	2.152	2.564	3.368	4.300
21	1.340	1.915	2.282	2.998	3.827	1.493	2.135	2.543	3.340	4.264
22	1.334	1.906	2.271	2.984	3.809	1.482	2.118	2.524	3.315	4.232
23	1.328	1.898	2.261	2.971	3.793	1.471	2.103	2.506	3.292	4.203
24	1.322	1.891	2.252	2.959	3.778	1.462	2.089	2.489	3.270	4.176
25	1.317	1.883	2.244	2.948	3.764	1.453	2.077	2.474	3.251	4.151
26	1.313	1.877	2.236	2.938	3.751	1.444	2.065	2.460	3.232	4.127
27	1.309	1.871	2.229	2.929	3.740	1.437	2.054	2.447	3.215	4.106
28	1.305	1.865	2.222	2.920	3.728	1.430	2.044	2.435	3.199	4.085
29	1.301	1.860	2.216	2.911	3.718	1.423	2.034	2.424	3.184	4.066
30	1.297	1.855	2.210	2.904	3.708	1.417	2.025	2.413	3.170	4.049
31	1.294	1.850	2.204	2.896	3.699	1.411	2.017	2.403	3.157	4.032
32	1.291	1.846	2.199	2.890	3.690	1.405	2.009	2.393	3.145	4.106
33	1.288	1.842	2.194	2.883	3.682	1.400	2.001	2.385	3.133	4.001
34	1.285	1.838	2.189	2.877	3.674	1.395	1.994	2.376	3.122	3.987
35	1.283	1.834	2.185	2.871	3.667	1.390	1.988	2.368	3.112	3.974
36	1.280	1.830	2.181	2.866	3.660	1.386	1.981	2.361	3.102	3.961
37	1.278	1.827	2.177	2.860	3.653	1.381	1.975	2.353	3.092	3.949
38	1.272	1.824	2.173	2.855	3.647	1.377	1.969	2.346	3.083	3.938
39	1.273	1.821	2.169	2.850	3.641	1.374	1.964	2.340	3.075	3.927
40	1.271	1.818	2.166	2.846	3.635	1.370	1.959	2.334	3.066	3.917
41	1.269	1.815	2.162	2.841	3.629	1.366	1.954	2.328	3.059	3.907
42	1.267	1.812	2.159	2.837	3.624	1.363	1.949	2.322	3.051	3.897
43	1.266	1.810	2.156	2.833	3.619	1.360	1.944	2.316	3.044	3.888
44	1.264	1.807	2.153	2.829	3.614	1.357	1.940	2.311	3.037	3.879
45	1.262	1.805	2.150	2.826	3.609	1.354	1.935	2.306	3.030	3.871

TABLE B.8 (Continued)

					α				
		$\gamma = 0.75$					$\gamma = 0.90$		
n	0.25	0.10	0.05	0.01	0.001	0.25	0.10	0.05	0.01	0.001
46	1.261	1.802	2.148	2.822	3.605	1.351	1.931	2.301	3.024	3.863
47	1.259	1.800	2.145	2.819	3.600	1.348	1.927	2.297	3.018	3.855
48	1.258	1.798	2.143	2.815	3.596	1.345	1.924	2.292	3.012	3.847
49	1.256	1.796	2.140	2.812	3.592	1.343	1.920	2.288	3.006	3.840
50	1.255	1.794	2.138	2.809	3.588	1.340	1.916	2.284	3.001	3.833

					α				
		$\gamma = 0.95$					$\gamma = 0.99$		
n	0.25	0.10	0.05	0.01	0.001	0.25	0.10	0.05	0.01	0.001
2	22.858	32.019	37.674	48.430	60.573	114.363	160.193	188.491	242.300	303.054
3	5.922	8.380	9.916	12.861	16.208	13.378	18.930	22.401	29.055	36.616
4	3.779	5.369	6.370	8.299	10.502	6.614	9.398	11.150	14.527	18.383
5	3.002	4.275	5.079	6.634	8.415	4.643	6.612	7.855	10.260	13.015
6	2.604	3.712	4.414	5.775	7.337	3.743	5.337	6.345	8.301	10.548
7	2.361	3.369	4.007	5.248	6.676	3.233	4.613	5.488	7.187	9.142
8	2.197	3.136	3.732	4.891	6.226	2.905	4.147	4.936	6.468	8.234
9	2.078	2.967	3.532	4.631	5.899	2.677	3.822	4.550	5.966	7.600
10	1.987	2.839	3.379	4.433	5.649	2.508	3.582	4.265	5.594	7.129
11	1.916	2.737	3.259	4.277	5.452	2.378	3.397	4.045	5.308	6.766
12	1.858	2.655	3.162	4.150	5.291	2.274	3.250	3.870	5.079	6.477
13	1.810	2.587	3.081	4.044	5.158	2.190	3.130	3.727	4.893	6.240
14	1.770	2.529	3.012	3.955	5.045	2.120	3.029	3.608	4.737	6.043
15	1.735	2.480	2.954	3.878	4.949	2.060	2.945	3.507	4.605	5.876

					α				
		$\gamma = 0.75$					$\gamma = 0.90$		
n	0.25	0.10	0.05	0.01	0.001	0.25	0.10	0.05	0.01	0.001
16	1.705	2.437	2.903	3.812	4.865	2.009	2.872	3.421	4.492	5.732
17	1.679	2.400	2.858	3.754	4.791	1.965	2.808	3.345	4.393	5.607
18	1.655	2.366	2.819	3.702	4.725	1.926	2.753	3.279	4.307	5.497
19	1.635	2.337	2.784	3.656	4.667	1.891	2.703	3.221	4.230	5.399
20	1.616	2.310	2.752	3.615	4.614	1.860	2.659	3.168	4.161	5.312
21	1.599	2.286	2.723	3.577	4.567	1.833	2.620	3.121	4.100	5.234
22	1.584	2.264	2.697	3.543	4.523	1.808	2.584	3.078	4.044	5.163
23	1.570	2.244	2.673	3.512	4.484	1.785	2.551	3.040	3.993	5.098
24	1.557	2.225	2.651	3.483	4.447	1.764	2.522	3.004	3.947	5.039
25	1.545	2.208	2.631	3.457	4.413	1.745	2.494	2.972	3.904	4.985
26	1.534	2.193	2.612	3.432	4.382	1.727	2.469	2.941	3.865	4.935
27	1.523	2.178	2.595	3.409	4.353	1.711	2.446	2.914	3.828	4.888
28	1.514	2.164	2.579	3.388	4.326	1.695	2.424	2.888	3.794	4.845
29	1.505	2.152	2.554	3.368	4.301	1.681	2.404	2.864	3.763	4.805
30	1.497	2.140	2.549	3.350	4.278	1.668	2.385	2.841	3.733	4.768

TABLE B.8 (Continued)

	$\gamma = 0.95$					$\gamma = 0.99$				
n	0.25	0.10	0.05	0.01	0.001	0.25	0.10	0.05	0.01	0.001
31	1.489	2.129	2.536	3.332	4.256	1.656	2.367	2.280	3.706	4.732
32	1.481	2.118	2.524	3.316	4.235	1.664	2.351	2.801	3.680	4.699
33	1.475	2.108	2.512	3.300	4.215	1.633	2.335	2.782	3.655	4.668
34	1.468	2.099	2.501	3.286	4.197	1.623	2.320	2.764	3.632	4.639
35	1.462	2.090	2.490	3.272	4.179	1.613	2.306	2.748	3.611	4.611
36	1.455	2.081	2.479	3.258	4.161	1.604	2.293	2.732	3.590	4.585
37	1.450	2.073	2.470	3.246	4.146	1.595	2.281	2.717	3.571	4.560
38	1.446	2.068	2.464	3.237	4.134	1.587	2.269	2.703	3.552	4.537
39	1.441	2.060	2.455	3.226	4.120	1.579	2.257	2.690	3.534	4.514
40	1.435	2.052	2.445	3.213	4.104	1.571	2.247	2.677	3.518	4.493
41	1.430	2.045	2.437	3.202	4.090	1.564	2.236	2.665	3.502	4.472
42	1.426	2.039	2.429	3.192	4.077	1.557	2.227	2.653	3.486	4.453
43	1.422	2.033	2.422	3.183	4.065	1.551	2.217	2.642	3.472	4.434
44	1.418	2.027	2.415	3.173	2.053	1.545	2.208	2.631	3.458	4.416
45	1.414	2.021	2.408	3.165	4.042	1.539	2.200	2.621	3.444	4.399
46	1.410	2.016	2.402	3.156	4.031	1.533	2.192	2.611	3.431	4.383
47	1.406	2.011	2.396	3.148	4.021	1.527	2.184	2.602	3.419	4.367
48	1.403	2.006	2.390	3.140	4.011	1.522	2.176	2.593	3.407	4.352
49	1.399	2.001	2.384	3.133	4.002	1.517	2.169	2.584	3.396	4.337
50	1.396	1.969	2.379	3.126	3.993	1.512	2.162	2.576	3.385	4.323

† Factors K such that the probability is γ that at least a proportion $1 - \alpha$ of the distribution will be included between $\bar{x} \pm K\alpha$, where $\bar{x}$ and α are estimates of the mean and the standard deviation computed from a sample of n.
SOURCE: C. Eisenhart, M. W. Hastay, and W. A. Wallis, "Techniques of Statistical Analysis," McGraw-Hill, New York, 1947, Chap. 2.

TABLE B.9 One-sided tolerance limit factors k for normal distribution, 0.95 confidence, and $n - 1$ degrees of freedom

n	0.75000	0.90000	0.95000	0.97500	0.99000	0.99900	0.99990	0.99999
2	11.763	20.581	26.260	31.257	37.094	49.276	59.304	68.010
3	3.806	6.155	7.656	8.986	10.553	13.857	16.598	18.986
4	2.618	4.162	5.144	6.015	7.042	9.214	11.019	12.593
5	2.150	3.407	4.203	4.909	5.741	7.502	8.966	10.243
6	1.895	3.006	3.708	4.329	5.062	6.612	7.901	9.025
7	1.732	2.755	3.399	3.970	4.642	6.063	7.244	8.275
8	1.618	2.582	3.187	3.723	4.354	5.688	6.796	7.763
9	1.532	2.454	3.031	3.542	4.143	5.413	6.469	7.390
10	1.465	2.355	2.911	3.402	3.981	5.203	6.219	7.105
11	1.411	2.275	2.815	3.292	3.852	5.036	6.020	6.878
12	1.366	2.210	2.736	3.201	3.747	4.900	5.858	6.694
13	1.328	2.155	2.671	3.125	3.659	4.787	5.723	6.540
14	1.296	2.109	2.614	3.060	3.585	4.690	5.609	6.409
15	1.268	2.068	2.566	3.005	3.520	4.607	5.510	6.297
16	1.243	2.033	2.524	2.956	3.464	4.535	5.424	6.199
17	1.220	2.002	2.486	2.913	3.414	4.471	5.348	6.113
18	1.201	1.974	2.453	2.875	3.370	4.415	5.281	6.037
19	1.183	1.949	2.423	2.841	3.331	4.364	5.221	5.968
20	1.166	1.926	2.396	2.810	3.295	4.318	5.167	5.906
21	1.152	1.905	2.371	2.781	3.263	4.277	5.118	5.850
22	1.138	1.886	2.349	2.756	3.233	4.239	5.073	5.799
23	1.125	1.869	2.328	2.732	3.206	4.204	5.031	5.752
24	1.114	1.853	2.309	2.710	3.181	4.172	4.994	5.709
25	1.103	1.838	2.292	2.690	3.158	4.142	4.959	5.670
26	1.093	1.824	2.275	2.672	3.136	4.115	4.926	5.633
27	1.083	1.811	2.260	2.654	3.116	4.089	4.896	5.598
28	1.075	1.799	2.246	2.638	3.098	4.066	4.868	5.566
29	1.066	1.788	2.232	2.623	3.080	4.043	4.841	5.536
30	1.058	1.777	2.220	2.608	3.064	4.022	4.816	5.508
31	1.051	1.767	2.208	2.595	3.048	4.002	4.793	5.481
32	1.044	1.758	2.197	2.582	3.034	3.984	4.771	5.456
33	1.037	1.749	2.186	2.570	3.020	3.966	4.750	5.433
34	1.031	1.740	2.176	2.559	3.007	3.950	4.730	5.410
35	1.025	1.732	2.167	2.548	2.995	3.934	4.712	5.389
36	1.019	1.725	2.158	2.538	2.983	3.919	4.694	5.369
37	1.014	1.717	2.149	2.528	2.972	3.904	4.677	5.350
38	1.009	1.710	2.141	2.518	2.961	3.891	4.661	5.332
39	1.004	1.704	2.133	2.510	2.951	3.878	4.646	5.314
40	0.999	1.697	2.125	2.501	2.941	3.865	4.631	5.298
41	0.994	1.691	2.118	2.493	2.932	3.854	4.617	5.282
42	0.990	1.685	2.111	2.485	2.923	3.842	4.603	5.266
43	0.986	1.680	2.105	2.478	2.914	3.831	4.591	5.252
44	0.982	1.674	2.098	2.470	2.906	3.821	4.578	5.238
45	0.978	1.669	2.092	2.463	2.898	3.811	4.566	5.224
46	0.974	1.664	2.086	2.457	2.890	3.801	4.555	5.211
47	0.971	1.659	2.081	2.450	2.883	3.792	4.544	5.199
48	0.967	1.654	2.075	2.444	2.876	3.783	4.533	5.187
49	0.964	1.650	2.070	2.438	2.869	3.774	4.523	5.175
50	0.960	1.646	2.065	2.432	2.862	3.766	4.513	5.164

Appendix C

Gamma Function Table

TABLE C.1 Values of gamma function $\Gamma(x)$

$$\Gamma(x) = \frac{\Gamma(x+1)}{x}, \quad \Gamma(x) = (x-1)\Gamma(x-1)$$

x	$\Gamma(x)$	x	$\Gamma(x)$	x	$\Gamma(x)$	x	$\Gamma(x)$
1.00	1.000 00	1.25	0.906 40	1.50	0.886 23	1.75	0.919 06
1.01	0.994 33	1.26	0.904 40	1.51	0.885 59	1.76	0.921 37
1.02	0.988 84	1.27	0.902 50	1.52	0.887 04	1.77	0.923 76
1.03	0.983 55	1.28	0.900 72	1.53	0.887 57	1.78	0.926 23
1.04	0.978 44	1.29	0.899 04	1.54	0.888 18	1.79	0.928 77
1.05	0.973 50	1.30	0.897 47	1.55	0.888 87	1.80	0.931 38
1.06	0.968 74	1.31	0.896 00	1.56	0.889 64	1.81	0.934 08
1.07	0.964 15	1.32	0.894 64	1.57	0.890 49	1.82	0.936 85
1.08	0.959 73	1.33	0.893 38	1.58	0.891 42	1.83	0.939 69
1.09	0.955 46	1.34	0.892 22	1.59	0.892 43	1.84	0.942 61
1.10	0.951 35	1.35	0.891 15	1.60	0.893 52	1.85	0.945 61
1.11	0.947 40	1.36	0.890 18	1.61	0.894 68	1.86	0.948 69
1.12	0.943 59	1.37	0.889 31	1.62	0.895 92	1.87	0.951 84
1.13	0.939 93	1.38	0.888 54	1.63	0.897 24	1.88	0.955 07
1.14	0.936 42	1.39	0.887 85	1.64	0.898 64	1.89	0.958 38
1.15	0.933 04	1.40	0.887 26	1.65	0.900 12	1.90	0.961 77
1.16	0.929 80	1.41	0.886 76	1.66	0.901 67	1.91	0.965 23
1.17	0.926 70	1.42	0.886 36	1.67	0.903 30	1.92	0.968 77
1.18	0.923 73	1.43	0.886 04	1.68	0.905 00	1.93	0.972 40
1.19	0.920 89	1.44	0.885 81	1.69	0.906 78	1.94	0.976 10
1.20	0.918 17	1.45	0.885 66	1.70	0.908 64	1.95	0.979 88
1.21	0.915 58	1.46	0.885 60	1.71	0.910 57	1.96	0.983 74
1.22	0.913 11	1.47	0.885 63	1.72	0.912 58	1.97	0.987 68
1.23	0.910 75	1.48	0.885 75	1.73	0.914 67	1.98	0.991 71
1.24	0.908 52	1.49	0.885 95	1.74	0.916 83	1.99	0.995 81
1.25	0.906 40	1.50	0.886 23	1.75	0.919 06	2.00	1.000 00

Appendix D

χ^2 Distribution Tables

TABLE D.1 Pearson's criterion of goodness of fit $P(>\chi^2)$

s'	$P=0.99$	$P=0.98$	$P=0.95$	$P=0.90$	$P=0.80$	$P=0.70$
1	0.000157	0.000628	0.00393	0.0158	0.0642	0.148
2	0.0201	0.0404	0.103	0.211	0.446	0.713
3	0.115	0.185	0.352	0.584	1.005	1.424
4	0.297	0.429	0.711	1.064	1.649	2.195
5	0.554	0.752	1.145	1.610	2.343	3.000
6	0.872	1.134	1.635	2.204	3.070	3.828
7	1.239	1.564	2.167	2.833	3.822	4.671
8	1.646	2.032	2.733	3.490	4.594	5.527
9	2.088	2.532	3.325	4.168	5.380	6.393
10	2.558	3.059	3.940	4.865	6.179	7.267
11	3.053	3.609	4.575	5.578	6.989	8.148
12	3.571	4.178	5.226	6.304	7.807	9.034
13	4.107	4.765	5.892	7.042	8.634	9.926
14	4.660	5.368	6.571	7.790	9.467	10.821
15	5.229	5.985	7.261	8.547	10.307	11.721
16	5.812	6.614	7.962	9.312	11.152	12.624
17	6.408	7.255	8.672	10.085	12.002	13.531
18	7.015	7.906	9.390	10.865	12.857	14.440
19	7.633	8.567	10.117	11.651	13.716	15.352
20	8.260	9.237	10.851	12.443	14.578	16.266
21	8.897	9.915	11.591	13.240	15.445	17.182
22	9.542	10.600	12.338	14.041	16.314	18.101
23	10.196	11.293	13.091	14.848	17.187	19.021
24	10.856	11.992	13.848	15.659	18.062	19.943
25	11.524	12.697	14.611	16.473	18.940	20.867
26	12.198	13.409	15.379	17.292	19.820	21.792
27	12.879	14.125	16.151	18.114	20.703	22.719
28	13.565	14.847	16.928	18.939	21.588	23.647
29	14.256	15.574	17.708	19.768	22.475	24.577
30	14.953	16.306	18.493	20.599	23.364	25.508

TABLE D.1 (*Continued*)

s'	$P = 0.50$	$P = 0.30$	$P = 0.20$	$P = 0.10$	$P = 0.05$	$P = 0.02$	$P = 0.01$
1	0.455	1.074	1.642	2.706	3.841	5.412	6.635
2	1.386	2.408	3.219	4.605	5.991	7.824	9.210
3	2.366	3.665	4.642	6.251	7.815	9.837	11.341
4	3.357	4.878	5.989	7.779	9.488	11.668	13.277
5	4.351	6.064	7.289	9.236	11.070	13.388	15.086
6	5.348	7.231	8.558	10.645	12.592	15.033	16.812
7	6.346	8.383	9.803	12.017	14.067	16.622	18.475
8	7.344	9.524	11.030	13.362	15.507	18.168	20.090
9	8.343	10.656	12.242	14.684	16.919	19.679	21.666
10	9.342	11.781	13.442	15.987	18.307	21.161	23.209
11	10.341	12.899	14.631	17.275	19.675	22.618	24.725
12	11.340	14.011	15.812	18.549	21.026	24.054	26.217
13	12.340	15.119	16.985	19.812	22.362	25.472	27.688
14	13.339	16.222	18.151	21.064	23.685	26.873	29.141
15	14.339	17.322	19.311	22.307	24.996	28.259	30.578
16	15.338	18.418	20.465	23.542	26.296	29.633	32.000
17	16.338	18.511	21.615	24.769	27.587	30.995	33.409
18	17.338	20.601	22.760	25.989	28.869	32.346	34.805
19	18.338	21.689	23.900	27.204	30.144	33.687	36.191
20	19.337	22.775	25.038	28.412	31.410	35.020	37.566
21	20.337	23.858	26.171	29.615	32.671	36.343	38.932
22	21.337	24.939	27.301	30.813	33.924	37.659	40.289
23	22.337	26.018	28.429	32.007	35.172	38.968	41.638
24	23.337	27.096	29.553	33.196	36.415	40.270	42.980
25	24.337	28.172	30.675	34.382	37.652	41.566	44.314
26	25.336	29.246	31.795	35.563	38.885	42.856	45.642
27	26.336	30.319	32.912	36.741	40.113	44.140	46.963
28	27.336	31.391	34.027	37.916	41.337	45.419	48.278
29	28.336	32.461	35.139	39.087	42.557	46.693	49.588
30	29.336	33.530	36.250	40.256	43.773	47.962	50.892

SOURCE: Taken from R. A. Fisher, "Statistical Methods for Research Workers," Oliver & Boyd, Edinburgh.

D. χ^2 Distribution Tables

TABLE D.2 Values of the function $\chi^2_{\alpha, m}$
The function $\chi^2_{\alpha, m}$ is defined by the equation

$$P(\chi^2_m > \chi^2_{\alpha, m}) = \alpha$$

where the random variable χ^2_m has a χ^2 distribution with m degrees of freedom. The probability density function of χ^2_m is equal to

$$P_{\chi^2_m}(x) = \frac{1}{\Gamma(n/2)2^{n/2}} x^{n/2-1} e^{-x/2}, \; x > 0$$

m	\multicolumn{5}{c}{α}				
	0.10	0.05	0.02	0.01	0.005
1	2.7	3.8	5.4	6.6	7.9
2	4.6	6.0	7.8	9.2	11.6
3	6.3	7.8	9.8	11.3	12.8
4	7.8	9.5	11.7	13.3	14.9
5	9.2	11.1	13.4	15.1	16.3
6	10.6	12.6	15.0	16.8	18.6
7	12.0	14.1	16.6	18.5	20.3
8	13.4	15.5	18.2	20.1	21.9
9	14.7	16.9	19.7	21.7	23.6
10	16.0	18.3	21.2	23.2	25.2
11	17.3	19.7	22.6	24.7	26.8
12	18.5	21.0	24.1	26.2	28.3
13	19.8	22.4	25.5	27.7	29.8
14	21.1	23.7	26.9	29.1	31
15	22.3	25.0	28.3	30.6	32.5
16	23.5	26.3	29.6	32.0	34
17	24.8	27.6	31.0	33.4	35.5
18	26.0	28.9	32.3	34.8	37
19	27.2	30.1	33.7	36.2	38.5
20	28.4	31.4	35.0	37.6	40
21	29.6	32.7	36.3	38.9	41.5
22	30.8	33.9	37.7	40.3	42.5
23	32.0	35.2	39.0	41.6	44.0
24	33.2	36.4	40.3	43.0	45.5
25	34.4	37.7	41.6	44.3	47

TABLE D.3 Cumulative χ^2 distribution

$$F(u) = \int_0^u \frac{x^{(n-2)/2} e^{-x/2} dx}{2^{n/2}(n-2)/2!}$$

n	0.005	0.010	0.025	0.050	0.100	0.250	0.500	0.750	0.900	0.950	0.975	0.990	0.995
1	0.0^{4}393	0.0^{3}157	0.0^{3}982	0.0^{2}393	0.0158	0.102	0.455	1.32	2.71	3.84	5.02	6.63	7.88
2	0.0100	0.0201	0.0506	0.103	0.211	0.575	1.39	2.77	4.61	5.99	7.38	9.21	10.6
3	0.0717	0.115	0.216	0.352	0.584	1.21	2.37	4.11	6.25	7.81	9.35	11.3	12.8
4	0.207	0.297	0.484	0.711	1.06	1.92	3.36	5.39	7.78	9.49	11.1	13.3	14.9
5	0.412	0.554	0.831	1.15	1.61	2.67	4.35	6.63	9.24	11.1	12.8	15.1	16.7
6	0.676	0.872	1.24	1.64	2.20	3.45	5.35	7.84	10.6	12.6	14.4	16.8	18.5
7	0.989	1.24	1.69	2.17	2.83	4.25	6.35	9.04	12.0	14.1	16.0	18.5	20.3
8	1.34	1.65	2.18	2.73	3.49	5.07	7.34	10.2	13.4	15.5	17.5	20.1	22.0
9	1.73	2.09	2.70	3.33	4.17	5.90	8.34	11.4	14.7	16.9	19.0	21.7	23.6
10	2.16	2.56	3.25	3.94	4.87	6.74	9.34	12.5	16.0	18.3	20.5	23.2	25.2
11	2.60	3.05	3.82	4.57	5.58	7.58	10.3	13.7	17.3	19.7	21.9	24.7	26.8
12	3.07	3.57	4.40	5.23	6.30	8.44	11.3	14.8	18.5	21.0	23.3	26.2	28.3
13	3.57	4.11	5.01	5.89	7.04	8.30	12.3	16.0	19.8	22.4	24.7	27.7	29.8
14	4.07	4.66	5.63	6.57	7.79	10.2	13.3	17.1	21.1	23.7	26.1	29.1	31.3
15	4.60	5.23	6.26	7.26	8.55	11.0	14.3	18.2	22.3	25.0	27.5	30.6	32.8
16	5.14	5.81	6.91	7.96	9.31	11.9	15.3	19.4	23.5	26.3	28.8	32.0	34.3
17	5.70	6.41	7.56	8.67	10.1	12.8	16.3	20.5	24.8	27.6	30.2	33.4	35.7
18	6.26	7.01	8.23	9.39	10.9	13.7	17.3	21.6	26.0	28.9	31.5	34.8	37.2
19	6.84	7.63	8.91	10.1	11.7	14.6	18.3	22.7	27.2	30.1	32.9	36.2	38.6
20	7.43	8.26	9.59	10.9	12.4	15.5	19.3	23.8	28.4	31.4	34.2	37.6	40.0
21	8.03	8.90	10.3	11.6	13.2	16.3	20.3	24.9	29.6	32.7	35.5	38.9	41.4
22	8.64	9.54	11.0	12.3	14.0	17.2	21.3	26.0	30.8	33.9	36.8	40.3	42.8
23	9.26	10.2	11.7	13.1	14.8	18.1	22.3	27.1	32.0	35.2	38.1	41.6	44.2
24	9.89	10.9	12.4	13.8	15.7	19.0	23.3	28.2	33.2	36.4	39.4	43.0	45.6
25	10.5	11.5	13.1	14.6	16.5	19.9	24.3	29.3	34.4	37.7	40.6	44.3	46.9
26	11.2	12.2	13.8	15.4	17.3	20.8	25.3	30.4	35.6	38.9	41.9	45.6	48.3
27	11.8	12.9	14.6	16.2	18.1	21.7	26.3	31.5	36.7	40.1	43.2	47.0	49.6
28	12.5	13.6	15.3	16.9	18.9	22.7	27.3	32.6	37.9	41.3	44.5	48.3	51.0
29	13.1	14.3	16.0	17.7	19.8	23.6	28.3	33.7	39.1	42.6	45.7	49.6	52.3
30	13.8	15.0	16.8	18.5	20.6	24.5	29.3	34.8	40.3	43.8	47.0	50.9	53.7

SOURCE: This table is abridged from Catherine M. Thompson, Tables of percentage points of the incomplete beta function and of the

TABLE D.4 Percentiles of the χ^2 distribution

ν	\multicolumn{8}{c}{q}							
	0.005	0.010	0.025	0.05	0.10	0.20	0.30	0.40
1	$0.0^4 393$	$0.0^3 157$	$0.0^3 982$	$0.0^2 393$	0.0158	0.0642	0.148	0.275
2	0.0100	0.0201	0.0506	0.103	0.211	0.446	0.713	1.02
3	0.0717	0.115	0.216	0.352	0.584	1.00	1.42	1.87
4	0.207	0.297	0.484	0.711	1.06	1.65	2.19	2.75
5	0.412	0.554	0.831	1.15	1.61	2.34	3.00	3.66
6	0.676	0.872	1.24	1.64	2.20	3.07	3.83	4.57
7	0.989	1.24	1.69	2.17	2.83	3.82	4.67	5.49
8	1.34	1.65	2.18	2.73	3.49	4.59	5.53	6.42
9	1.73	2.09	2.70	3.33	4.17	5.39	6.39	7.36
10	2.16	2.56	3.25	3.94	4.87	6.18	7.27	8.30
11	2.60	3.05	3.82	4.57	5.58	6.99	8.15	9.24
12	3.07	3.57	4.40	5.23	6.30	7.81	9.03	10.2
13	3.57	4.11	5.01	5.89	7.04	8.63	9.93	11.1
14	4.07	4.66	5.63	6.57	7.79	9.47	10.8	12.1
15	4.60	5.23	6.26	7.26	8.55	10.3	11.7	13.0
16	5.14	5.81	6.91	7.96	9.31	11.2	12.6	14.0
17	5.70	6.41	7.56	8.67	10.1	12.0	13.5	14.9
18	6.26	7.01	8.23	9.39	10.9	12.9	14.4	15.9
19	6.84	7.63	8.91	10.1	11.7	13.7	15.4	16.9
20	7.43	8.26	9.59	10.9	12.4	14.6	16.3	17.8
21	8.03	8.90	10.3	11.6	13.2	15.4	17.2	18.8
22	8.64	9.54	11.0	12.3	14.0	16.3	18.1	19.7
23	9.26	10.2	11.7	13.1	14.8	17.2	19.0	20.7
24	9.89	10.9	12.4	13.8	15.7	18.1	19.9	21.7
25	10.5	11.5	13.1	14.6	16.5	18.9	20.9	22.6
26	11.2	12.2	13.8	15.4	17.3	19.8	21.8	23.6
27	11.8	12.9	14.6	16.2	18.1	20.7	22.7	24.5
28	12.5	13.6	15.3	16.9	18.9	21.6	23.6	25.5
29	13.1	14.3	16.0	17.7	19.8	22.5	24.6	26.5
30	13.8	15.0	16.8	18.5	20.6	23.4	25.5	27.4
35	17.2	18.5	20.6	22.5	24.8	27.8	30.2	32.3
40	20.7	22.2	24.4	26.5	29.1	32.3	34.9	37.1
45	24.3	25.9	28.4	30.6	33.4	36.9	39.6	42.0
50	28.0	29.7	32.4	34.8	37.7	41.4	44.3	46.9
75	47.2	49.5	52.9	56.1	59.8	64.5	68.1	71.3
100	67.3	70.1	74.2	77.9	82.4	87.9	92.1	95.8

TABLE D.4 (*Continued*)

					q					
v	0.50	0.60	0.70	0.80	0.90	0.95	0.975	0.990	0.995	0.999
1	0.455	0.708	1.07	1.64	2.71	3.84	5.02	6.63	7.88	10.8
2	1.39	1.83	2.41	3.22	4.61	5.99	7.38	9.21	10.6	13.8
3	2.37	2.95	3.67	4.64	6.25	7.81	9.35	11.3	12.8	16.3
4	3.36	4.04	4.88	5.99	7.78	9.49	11.1	13.3	14.9	18.5
5	4.35	5.13	6.06	7.29	9.24	11.1	12.8	15.1	16.7	20.5
6	5.35	6.21	7.23	8.56	10.6	12.6	14.4	16.8	18.5	22.5
7	6.35	7.28	8.38	9.80	12.0	14.1	16.0	18.5	20.3	24.3
8	7.34	8.35	9.52	11.0	13.4	15.5	17.5	20.1	22.0	26.1
9	8.34	9.41	10.7	12.2	14.7	16.9	19.0	21.7	23.6	27.9
10	9.34	10.5	11.8	13.4	16.0	18.3	20.5	23.2	25.2	29.6
11	10.3	11.5	12.9	14.6	17.3	19.7	21.9	24.7	26.8	31.3
12	11.3	12.6	14.0	15.8	18.5	21.0	23.3	26.2	28.3	32.9
13	12.3	13.6	15.1	17.0	19.8	22.4	24.7	27.7	29.8	34.5
14	13.3	14.7	16.2	18.2	21.1	23.7	26.1	29.1	31.3	36.1
15	14.3	15.7	17.3	19.3	22.3	25.0	27.5	30.6	32.8	37.7
16	15.3	16.8	18.4	20.5	23.5	26.3	28.8	32.0	34.3	39.3
17	16.3	17.8	19.5	21.6	24.8	27.6	30.2	33.4	35.7	40.8
18	17.3	18.9	20.6	22.8	26.0	28.9	31.5	34.8	37.2	42.3
19	18.3	19.9	21.7	23.9	27.2	30.1	32.9	36.2	38.6	43.8
20	19.3	21.0	22.8	25.0	28.4	31.4	34.2	37.6	40.0	45.3
21	20.3	22.0	23.9	26.9	29.6	32.7	35.5	38.9	41.4	46.8
22	21.3	23.0	24.9	27.3	30.8	33.9	36.8	40.3	42.8	48.3
23	22.3	24.1	26.0	28.4	32.0	35.2	38.1	41.6	44.2	49.7
24	23.3	25.1	27.1	29.6	33.2	36.4	39.4	43.0	45.6	51.2
25	24.3	26.1	28.2	30.7	34.4	37.7	40.6	44.3	46.9	52.6
26	25.3	27.2	29.2	31.8	35.6	38.9	41.9	45.6	48.3	54.1
27	26.3	28.2	30.3	32.9	36.7	40.1	43.2	47.0	49.6	55.5
28	27.3	29.2	31.4	34.0	37.9	41.3	44.5	48.3	51.0	56.9
29	28.3	30.3	32.5	35.1	39.1	42.6	45.7	49.6	52.3	58.3
30	29.3	31.3	33.5	36.3	40.3	43.8	47.0	50.9	53.7	59.7
35	34.3	36.5	38.9	41.8	46.1	49.8	53.2	57.3	60.3	66.6
40	39.3	41.6	44.2	47.3	51.8	55.8	59.3	63.7	66.8	73.4
45	44.3	46.8	49.5	52.7	57.5	61.7	65.4	70.0	73.2	80.1
50	49.3	51.9	54.7	58.2	63.2	67.5	71.4	76.2	79.5	86.7
75	74.3	77.5	80.9	85.1	91.1	96.2	100.8	106.4	110.3	118.6
100	99.3	102.9	106.9	111.7	118.5	124.3	129.6	135.6	140.2	149.4

SOURCE: Abridged from A. Hald, "Statistical Tables and Formulas," John Wiley, New York, 1952.

TABLE D.5 Percentiles of the χ^2 distribution

Degrees of freedom ν	α							
	0.005	0.010	0.025	0.05	0.10	0.20	0.30	0.40
1	0.0^{4}393	0.0^{3}157	0.0^{3}982	0.0^{2}393	0.0158	0.0642	0.148	0.275
2	0.0100	0.0201	0.0506	0.103	0.211	0.446	0.713	1.02
3	0.0717	0.115	0.216	0.352	0.584	1.00	1.42	1.87
4	0.207	0.297	0.484	0.711	1.06	1.65	2.19	2.75
5	0.412	0.554	0.831	1.15	1.61	2.34	3.00	3.66
6	0.676	0.872	1.24	1.64	2.20	3.07	3.83	4.57
7	0.989	1.24	1.69	2.17	2.83	3.82	4.67	5.49
8	1.34	1.65	2.18	2.73	3.49	4.59	5.53	6.42
9	1.73	2.09	2.70	3.33	4.17	5.38	6.39	7.36
10	2.16	2.56	3.25	3.94	4.87	6.18	7.27	8.30
11	2.60	3.05	3.82	4.57	5.58	6.99	8.15	9.24
12	3.07	3.57	4.40	5.23	6.30	7.81	9.03	10.2
13	3.57	4.11	5.01	5.89	7.04	8.63	9.93	11.1
14	4.07	4.66	5.63	6.57	7.79	9.47	10.8	12.1
15	4.60	5.23	6.26	7.26	8.55	10.3	11.7	13.0
16	5.14	5.81	6.91	7.96	9.31	11.2	12.6	14.0
17	5.70	6.41	7.56	8.67	10.1	12.0	13.5	14.9
18	6.26	7.01	8.23	9.39	10.9	12.9	14.4	15.9
19	6.84	7.63	8.91	10.1	11.7	13.7	15.4	16.9
20	7.43	8.26	9.59	10.9	12.4	14.6	16.3	17.8
21	8.03	8.90	10.3	11.6	13.2	15.4	17.2	18.8
22	8.64	9.54	11.0	12.3	14.0	16.3	18.1	19.7
23	9.26	10.2	11.7	13.1	14.8	17.2	19.0	20.7
24	9.89	10.9	12.4	13.8	15.7	18.1	19.9	21.7
25	10.5	11.5	13.1	14.6	16.5	18.9	20.9	22.6
26	11.2	12.2	13.8	15.4	17.3	19.8	21.8	23.6
27	11.8	12.9	14.6	16.2	18.1	20.7	22.7	24.5
28	12.5	13.6	15.3	16.9	18.9	21.6	23.6	25.5
29	13.1	14.3	16.0	17.7	19.8	22.5	24.6	26.5
30	13.8	15.0	16.8	18.5	20.6	23.4	25.5	27.4
35	17.2	18.5	20.6	22.5	24.8	27.8	30.2	32.3
40	20.7	22.2	24.4	26.5	29.1	32.3	34.9	37.1
45	24.3	25.9	28.4	30.6	33.4	36.9	39.6	42.0
50	28.0	29.7	32.4	34.8	37.7	41.4	44.3	46.9
75	47.2	49.5	52.9	56.1	59.8	64.5	68.1	71.3
100	67.0	70.1	74.2	77.9	82.4	87.9	92.1	95.8

564   Appendices

TABLE D.6 The Probabilities P for the χ^2 Distribution

Degrees of Freedom, r

χ^2	2	3	4	5	6	7	8	9	10	11	12	13	14	15	16	17	18	19
1	0.6969	0.8013	0.9098	0.9626	0.9856	0.9948	0.9982	0.9994	0.9998	0.9999	1	1	1	1	1	1	1	1
2	0.3679	0.5724	0.7358	0.8491	0.9197	0.9598	0.9810	0.9915	0.9963	0.9985	0.9994	0.9998	0.9999	1	1	1	1	1
3	0.2231	0.3916	0.5578	0.7000	0.8088	0.8850	0.9344	0.9613	0.9814	0.9907	0.9955	0.9979	0.9991	0.9996	0.9998	0.9999	0.9999	1
4	0.1353	0.2615	0.4060	0.5494	0.6767	0.7798	0.8571	0.9114	0.9473	0.9699	0.9831	0.9912	0.9955	0.9977	0.9989	0.9995	0.9998	0.9999
5	0.0821	0.1718	0.2873	0.4159	0.5438	0.6600	0.7576	0.8343	0.8912	0.9312	0.9580	0.9752	0.9858	0.9921	0.9958	0.9978	0.9989	0.9994
6	0.0498	0.1116	0.1991	0.3662	0.4232	0.5398	0.6472	0.7399	0.8153	0.8734	0.9161	0.9462	0.9665	0.9797	0.9881	0.9932	0.9962	0.9979
7	0.0302	0.0719	0.1359	0.2206	0.3208	0.4289	0.5366	0.6371	0.7254	0.7991	0.8576	0.9022	0.9347	0.9576	0.9733	0.9835	0.9901	0.9942
8	0.0183	0.0460	0.0916	0.1562	0.2381	0.3326	0.4335	0.5341	0.6288	0.7133	0.7851	0.8436	0.8893	0.9238	0.9489	0.9665	0.9786	0.9867
9	0.0111	0.0293	0.0611	0.1091	0.1736	0.2527	0.3423	0.4373	0.5321	0.6219	0.7029	0.7729	0.8311	0.8775	0.9134	0.9403	0.9597	0.9735
10	0.0067	0.0186	0.0404	0.0752	0.1247	0.1886	0.2650	0.3505	0.4405	0.5304	0.6160	0.6939	0.7622	0.8197	0.8666	0.9036	0.9319	0.9529
11	0.0041	0.0117	0.0266	0.0514	0.0881	0.1386	0.2017	0.2757	0.3575	0.4433	0.5289	0.6108	0.6860	0.7526	0.8095	0.8566	0.8944	0.9238
12	0.0025	0.0074	0.0174	0.0348	0.0620	0.1006	0.1512	0.2133	0.2851	0.3626	0.4457	0.5276	0.6063	0.6790	0.7440	0.8001	0.8472	0.8856
13	0.0015	0.0046	0.0113	0.0234	0.0430	0.0721	0.1119	0.1626	0.2237	0.2933	0.3690	0.4478	0.5265	0.6023	0.6728	0.7362	0.7916	0.8386
14	0.0009	0.0029	0.0073	0.0156	0.0296	0.0512	0.0818	0.1223	0.1730	0.2330	0.3007	0.3738	0.4497	0.5255	0.5987	0.6671	0.7291	0.7837
15	0.0006	0.0018	0.0047	0.0104	0.0203	0.0360	0.0591	0.0909	0.1321	0.1825	0.2414	0.3074	0.3782	0.4514	0.5246	0.5955	0.6620	0.7226
16	0.0003	0.0011	0.0030	0.0068	0.0138	0.0251	0.0424	0.0669	0.0996	0.1411	0.1912	0.2491	0.3134	0.3821	0.4530	0.5238	0.5925	0.6573
17	0.0002	0.0007	0.0019	0.0045	0.0093	0.0174	0.0391	0.0487	0.0744	0.1079	0.1496	0.1993	0.2562	0.3189	0.3856	0.4544	0.5231	0.5899
18	0.0001	0.0004	0.0012	0.0029	0.0062	0.0120	0.0212	0.0352	0.0550	0.0816	0.1157	0.1575	0.2068	0.2627	0.3239	0.3888	0.4557	0.5224
19	0.0001	0.0003	0.0008	0.0019	0.0042	0.0082	0.0149	0.0252	0.0403	0.0611	0.0885	0.1231	0.1649	0.2137	0.2687	0.3285	0.3918	0.4568
20	0.0000	0.0002	0.0005	0.0013	0.0028	0.0056	0.0103	0.0179	0.0293	0.0453	0.0671	0.0952	0.1301	0.1719	0.2202	0.2742	0.3328	0.3946
21	0.0000	0.0001	0.0003	0.0008	0.0018	0.0038	0.0071	0.0126	0.0211	0.0334	0.0504	0.0729	0.1016	0.1368	0.1785	0.2263	0.2794	0.3368
22	0.0000	0.0001	0.0002	0.0005	0.0012	0.0025	0.0049	0.0089	0.0151	0.0244	0.0375	0.0554	0.0786	0.1078	0.1432	0.1847	0.2320	0.2843
23	0.0000	0.0000	0.0001	0.0003	0.0008	0.0017	0.0034	0.0062	0.0107	0.0177	0.0277	0.0417	0.0603	0.0841	0.1137	0.1493	0.1906	0.2374
24	0.0000	0.0000	0.0001	0.0002	0.0005	0.0011	0.0023	0.0043	0.0076	0.0127	0.0203	0.0311	0.0458	0.0651	0.0895	0.1194	0.1550	0.1962
25	0.0000	0.0000	0.0001	0.0001	0.0003	0.0008	0.0016	0.0030	0.0053	0.0091	0.0148	0.0231	0.0346	0.0499	0.0698	0.0947	0.1249	0.1605
26	0.0000	0.0000	0.0000	0.0001	0.0002	0.0005	0.0010	0.0020	0.0037	0.0065	0.0107	0.0170	0.0259	0.0380	0.0540	0.0745	0.0998	0.1302
27	0.0000	0.0000	0.0000	0.0001	0.0001	0.0003	0.0007	0.0014	0.0026	0.0046	0.0077	0.0124	0.0193	0.0287	0.0415	0.0581	0.0790	0.1047
28	0.0000	0.0000	0.0000	0.0000	0.0001	0.0002	0.0005	0.0010	0.0018	0.0032	0.0055	0.0090	0.0142	0.0216	0.0316	0.0449	0.0621	0.0834
29	0.0000	0.0000	0.0000	0.0000	0.0001	0.0001	0.0003	0.0006	0.0012	0.0027	0.0039	0.0065	0.0104	0.0161	0.0239	0.0345	0.0484	0.0660
30	0.0000	0.0000	0.0000	0.0000	0.0000	0.0001	0.0002	0.0004	0.0009	0.0016	0.0028	0.0047	0.0076	0.0119	0.0180	0.0263	0.0374	0.0518

TABLE D.7 χ^2 values as function of r and p

r	\multicolumn{9}{c}{p}								
	0.95	0.90	0.80	0.70	0.50	0.30	0.20	0.10	0.05
1	0.004	0.016	0.064	0.148	0.455	1.074	1.642	2.71	3.84
2	0.103	0.211	0.446	0.713	1.386	2.41	3.22	4.60	5.99
3	0.352	0.584	1.005	1.424	2.37	3.66	4.64	6.25	7.82
4	0.711	1.064	1.649	2.20	3.36	4.88	5.99	7.78	9.49
5	1.145	1.610	2.34	3.00	4.35	6.06	7.29	9.24	11.07
6	1.635	2.20	3.07	3.83	5.35	7.23	8.56	10.64	12.59
7	2.17	2.83	3.82	4.67	6.35	8.38	9.80	12.02	14.07
8	2.73	3.49	4.59	5.53	7.34	9.52	11.03	13.36	15.51
9	3.32	4.17	5.38	6.39	8.34	10.66	12.24	14.68	16.92
10	3.94	4.86	6.18	7.27	9.34	11.78	13.44	15.99	18.31
11	4.58	5.58	6.99	8.15	10.34	12.90	14.63	17.28	19.68
12	5.23	6.30	7.81	9.03	11.34	14.01	15.81	18.55	21.0
13	5.89	7.04	8.63	9.93	12.34	15.12	16.98	19.81	22.4
14	6.57	7.79	9.47	10.82	13.34	16.22	18.15	21.1	23.7
15	7.26	8.55	10.31	11.72	14.34	17.32	19.31	22.3	25.0
16	7.96	9.31	11.15	12.62	15.34	18.42	20.5	23.5	26.3
17	8.67	10.08	12.00	13.53	16.34	19.51	21.6	24.8	27.6
18	9.39	10.86	12.86	14.44	17.34	20.6	22.8	26.0	28.9
19	10.11	11.65	13.72	15.35	18.34	21.7	23.9	27.2	30.1
20	10.85	12.44	14.58	16.27	19.34	22.8	25.0	28.4	31.4
21	11.59	13.24	15.44	17.18	20.3	23.9	26.2	29.6	32.7
22	12.34	14.04	16.31	18.10	21.3	24.9	27.3	30.8	33.9
23	13.09	14.85	17.19	19.02	22.3	26.0	28.4	32.0	35.2
24	13.85	15.66	18.06	19.94	23.3	27.1	29.6	33.2	36.4
25	14.61	16.47	18.94	20.9	24.3	28.2	30.7	34.4	37.7
26	15.38	17.29	19.82	21.8	25.3	29.2	31.8	35.6	38.9
27	16.15	18.11	20.7	22.7	26.3	30.3	32.9	36.7	40.1
28	16.93	18.94	21.6	23.6	27.3	31.4	34.0	37.9	41.3
29	17.71	19.77	22.5	24.6	28.3	32.5	35.1	39.1	42.6
30	18.49	20.6	23.4	25.5	29.3	33.5	36.2	40.3	43.8

TABLE D.8 χ^2 Distribution†

	Probability										
ν	0.99	0.98	0.95	0.90	0.80	0.20	0.10	0.05	0.02	0.01	0.001
1	0.03157	0.03628	0.00393	0.0158	0.0642	1.642	2.706	3.841	5.412	6.635	10.827
2	0.0201	0.0404	0.103	0.211	0.446	3.219	4.605	5.991	7.824	9.210	13.815
3	0.115	0.185	0.352	0.584	1.005	4.642	6.251	7.815	9.837	11.341	16.268
4	0.297	0.429	0.711	1.064	1.649	5.989	7.779	9.488	11.668	13.277	18.465
5	0.554	0.752	1.145	1.610	2.343	7.289	9.236	11.070	13.388	15.086	20.517
6	0.872	1.134	1.635	2.204	3.070	8.558	10.645	12.592	15.033	16.812	22.457
7	1.239	1.564	2.167	2.833	3.822	9.803	12.017	14.067	16.622	18.475	24.322
8	1.646	2.032	2.733	3.490	4.594	11.030	13.362	15.507	18.168	20.090	26.125
9	2.088	2.532	3.325	4.168	5.380	12.242	14.684	16.919	19.679	21.666	27.877
10	2.558	3.059	3.940	4.865	6.179	13.442	15.987	18.307	21.161	23.209	29.588
11	3.053	3.609	4.575	5.578	6.989	14.631	17.275	19.675	22.618	24.725	31.264
12	3.571	4.178	5.226	6.304	7.807	15.812	18.549	21.026	24.054	26.217	32.909
13	4.107	4.765	5.892	7.042	8.634	16.985	19.812	22.362	25.472	27.688	34.528
14	4.660	5.368	6.571	7.790	9.467	18.151	21.064	23.685	26.873	29.141	36.123
15	5.229	5.985	7.261	8.547	10.307	19.311	22.307	24.996	28.259	30.578	37.697
16	5.812	6.614	7.962	9.312	11.152	20.465	23.542	26.296	29.633	32.000	39.252
17	6.408	7.255	8.672	10.085	12.002	21.615	29.769	27.587	30.995	33.409	40.790
18	7.015	7.906	9.390	10.865	12.857	22.760	25.989	28.869	32.346	34.805	42.312
19	7.633	8.567	10.117	11.651	13.716	23.900	27.204	30.144	33.687	36.191	43.820
20	8.260	9.237	10.851	12.443	14.578	25.038	28.412	31.410	35.020	37.566	45.315
21	8.897	9.915	11.591	13.240	15.445	26.171	29.615	32.671	36.343	38.932	46.797
22	9.542	10.600	12.338	14.041	16.314	27.301	30.813	33.924	37.659	40.289	48.268
23	10.196	11.293	13.091	14.848	17.187	28.429	32.007	35.172	38.968	41.638	49.728
24	10.856	11.992	13.848	15.659	18.062	29.553	33.196	36.415	40.270	42.980	51.179
25	11.524	12.697	14.611	16.473	18.940	30.675	34.382	37.652	41.566	44.314	52.620
26	12.198	13.409	15.379	17.292	19.820	31.795	35.563	38.885	42.856	45.642	54.052
27	12.879	14.125	16.151	18.114	20.703	32.912	36.741	40.113	44.140	46.963	55.476
28	13.565	14.847	16.928	18.939	21.588	34.027	37.916	41.337	45.419	48.278	56.803
29	14.256	15.574	17.708	19.768	22.475	35.139	39.087	42.557	46.693	49.588	58.302
30	14.953	16.306	18.493	20.599	23.364	36.250	40.256	43.773	47.962	50.892	59.703

† For $\gamma > 3$ use the expression $\sqrt{2\chi^2} - \sqrt{2\gamma - 1}$ which is a function inverse to the normal distribution function with unit variance and zero mean.

SOURCE: Fisher, Yates: "Statistical Tables for Biological, Agricultural and Medical Research."

Appendix E

Student Distribution Tables

TABLE E.1 Percentage points of t distribution†

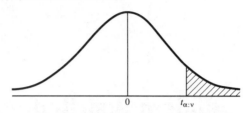

					α					
v	0.40	0.30	0.20	0.10	0.050	0.025	0.010	0.005	0.001	0.0005
1	0.325	0.727	1.376	3.078	6.314	12.71	31.82	63.66	318.3	636.6
2	0.289	0.617	1.061	1.886	2.920	4.303	6.965	9.925	22.33	31.60
3	0.277	0.584	0.978	1.638	2.353	3.182	4.541	5.841	10.22	12.94
4	0.271	0.569	0.941	1.533	2.132	2.776	3.747	4.604	7.173	8.610
5	0.267	0.559	0.920	1.476	2.015	2.571	3.365	4.032	5.893	6.859
6	0.265	0.553	0.906	1.440	1.943	2.447	3.143	3.707	5.208	5.959
7	0.263	0.549	0.896	1.415	1.895	2.365	2.998	3.499	4.785	5.405
8	0.262	0.546	0.889	1.397	1.860	2.306	2.896	3.355	4.501	5.041
9	0.261	0.543	0.883	1.383	1.833	2.262	2.821	3.250	4.297	4.781
10	0.260	0.542	0.879	1.372	1.812	2.228	2.764	3.169	4.144	4.587
11	0.260	0.540	0.876	1.363	1.796	2.201	2.718	3.106	4.025	4.437
12	0.259	0.539	0.873	1.356	1.782	2.179	2.681	3.055	3.930	4.318
13	0.259	0.538	0.870	1.350	1.771	2.160	2.650	3.012	3.852	4.221
14	0.258	0.537	0.868	1.345	1.761	2.145	2.624	2.977	3.787	4.140
15	0.258	0.536	0.866	1.341	1.753	2.131	2.602	2.947	3.733	4.073
16	0.258	0.535	0.865	1.337	1.746	2.120	2.583	2.921	3.686	4.015
17	0.257	0.534	0.863	1.333	1.740	2.110	2.567	2.898	3.646	3.965
18	0.257	0.534	0.862	1.330	1.734	2.101	2.552	2.878	3.611	3.922
19	0.257	0.533	0.861	1.328	1.729	2.093	2.539	2.861	3.579	3.883
20	0.257	0.533	0.860	1.325	1.725	2.086	2.528	2.845	3.552	2.850
21	0.257	0.532	0.859	1.323	1.721	2.080	2.518	2.831	3.527	3.819
22	0.256	0.532	0.858	1.321	1.717	2.074	2.508	2.819	3.505	3.792
23	0.256	0.532	0.858	1.319	1.714	2.069	2.500	2.807	3.485	3.767
24	0.256	0.531	0.857	1.318	1.711	2.064	2.492	2.797	3.467	3.745
25	0.256	0.531	0.856	1.316	1.708	2.060	2.485	2.787	3.450	3.725
26	0.256	0.531	0.856	1.315	1.706	2.056	2.479	2.779	3.435	3.707
27	0.256	0.531	0.855	1.314	1.703	2.052	2.473	2.771	3.421	3.690
28	0.256	0.530	0.855	1.313	1.701	2.048	2.467	2.763	3.408	3.674
29	0.256	0.530	0.854	1.311	1.699	2.045	2.462	2.756	3.396	3.659
30	0.256	0.530	0.854	1.310	1.697	2.042	2.457	2.750	3.385	3.646
40	0.255	0.529	0.851	1.303	1.684	2.021	2.423	2.704	3.307	3.551
50	0.255	0.528	0.849	1.298	1.676	2.009	2.403	2.678	3.262	3.495
60	0.254	0.527	0.848	1.296	1.671	2.000	2.390	2.660	3.232	3.460
80	0.254	0.527	0.846	1.292	1.664	1.990	2.374	2.639	3.195	3.415
100	0.254	0.526	0.845	1.290	1.660	1.984	2.365	2.626	3.174	3.389
200	0.254	0.525	0.843	1.286	1.653	1.972	2.345	2.601	3.131	3.339
500	0.253	0.525	0.842	1.283	1.648	1.965	2.334	2.586	3.106	3.310
∞	0.253	0.524	0.842	1.282	1.645	1.960	2.326	2.576	3.090	3.291

† Table of $t_{\alpha;v}$—the 100α percentage point of t distribution for v degrees of freedom.

SOURCE: This table is reproduced from A. Hald, "Statistical Tables and Formulas," John Wiley, New York, 1952.

E. Student Distribution Tables   569

TABLE E.2 Student's test of significance $2P (> |t|)$

n'	$P = 0.9$	0.8	0.7	0.6	0.5	0.4	0.3	0.2	0.1	0.05	0.02	0.01
1	0.158	0.325	0.510	0.727	1.000	1.376	1.963	3.078	6.314	12.706	31.821	63.657
2	0.142	0.289	0.445	0.617	0.816	1.061	1.386	1.886	2.920	4.303	6.965	9.925
3	0.137	0.277	0.424	0.584	0.765	0.978	1.250	1.638	2.353	3.182	4.541	5.841
4	0.134	0.271	0.414	0.569	0.741	0.941	1.190	1.533	2.132	2.776	3.747	4.604
5	0.132	0.267	0.408	0.559	0.727	0.920	1.156	1.476	2.015	2.571	3.365	4.032
6	0.131	0.265	0.404	0.553	0.718	0.906	1.134	1.440	1.943	2.447	3.143	3.707
7	0.130	0.263	0.402	0.549	0.711	0.896	1.119	1.415	1.895	2.365	2.998	3.499
8	0.130	0.262	0.399	0.546	0.706	0.889	1.108	1.397	1.860	2.306	2.896	3.355
9	0.129	0.261	0.398	0.543	0.703	0.883	1.100	1.383	1.833	2.262	2.821	3.250
10	0.129	0.260	0.397	0.542	0.700	0.879	1.093	1.372	1.812	2.228	2.764	3.169
11	0.129	0.260	0.396	0.540	0.697	0.876	1.088	1.363	1.796	2.201	2.718	3.106
12	0.128	0.259	0.395	0.539	0.695	0.873	1.083	1.356	1.782	2.179	2.681	3.055
13	0.128	0.259	0.394	0.538	0.694	0.870	1.079	1.350	1.771	2.160	2.650	3.012
14	0.128	0.258	0.393	0.537	0.692	0.868	1.076	1.345	1.761	2.145	2.624	2.977
15	0.128	0.258	0.393	0.536	0.691	0.866	1.074	1.341	1.753	2.131	2.602	2.947
16	0.128	0.258	0.392	0.535	0.690	0.865	1.071	1.337	1.746	2.120	2.583	2.921
17	0.128	0.257	0.392	0.534	0.689	0.863	1.069	1.333	1.740	2.110	2.567	2.898
18	0.127	0.257	0.392	0.534	0.688	0.862	1.067	1.330	1.734	2.101	2.552	2.878
19	0.127	0.257	0.391	0.533	0.688	0.861	1.066	1.328	1.729	2.093	2.539	2.861
20	0.127	0.257	0.391	0.533	0.687	0.860	1.064	1.325	1.725	2.086	2.528	2.845
21	0.127	0.257	0.391	0.532	0.686	0.859	1.063	1.323	1.721	2.080	2.518	2.831
22	0.127	0.256	0.390	0.532	0.686	0.858	1.061	1.321	1.717	2.074	2.508	2.819
23	0.127	0.256	0.390	0.532	0.685	0.858	1.060	1.319	1.714	2.069	2.500	2.807
24	0.127	0.256	0.390	0.531	0.685	0.857	1.059	1.318	1.711	2.064	2.492	2.797
25	0.127	0.256	0.390	0.531	0.684	0.856	1.058	1.316	1.708	2.060	2.485	2.787
26	0.127	0.256	0.390	0.531	0.684	0.856	1.058	1.315	1.706	2.056	2.479	2.779
27	0.127	0.256	0.389	0.531	0.684	0.855	1.057	1.314	1.703	2.052	2.473	2.771
28	0.127	0.256	0.389	0.530	0.683	0.855	1.056	1.313	1.701	2.048	2.467	2.763
29	0.127	0.256	0.389	0.530	0.683	0.854	1.055	1.311	1.699	2.045	2.462	2.756
30	0.127	0.256	0.389	0.530	0.683	0.854	1.055	1.310	1.697	2.042	2.457	2.750
∞	0.125 66	0.253 35	0.385 32	0.524 40	0.674 49	0.841 62	1.036 43	1.281 55	1.644 85	1.959 96	2.326 34	2.575 82

SOURCE: Taken from R. A. Fisher, "Statistical Methods for Research Workers," 13th edn, Oliver & Boyd, Edinburgh.

TABLE E.3 Ordinates of the cumulative Student t distribution function $F_T(t)$

ν	\multicolumn{7}{c}{P}
	0.75	0.90	0.95	0.975	0.99	0.995	0.999
1	1.00	3.08	6.31	12.71	31.82	63.66	318.31
2	0.82	1.89	2.92	4.30	6.97	9.93	22.33
3	0.77	1.64	2.35	3.18	4.54	5.84	10.21
4	0.74	1.53	2.13	2.78	3.75	4.60	7.17
5	0.73	1.48	2.02	2.57	3.37	4.03	5.89
6	0.72	1.44	1.94	2.45	3.14	3.71	5.21
7	0.71	1.42	1.90	2.37	3.00	3.50	4.79
8	0.71	1.40	1.86	2.31	2.90	3.36	4.50
9	0.70	1.38	1.83	2.26	2.82	3.26	4.30
10	0.70	1.37	1.81	2.29	2.76	3.17	4.14
11	0.70	1.36	1.80	2.20	2.72	3.11	4.03
12	0.70	1.36	1.78	2.18	2.68	3.06	3.93
13	0.69	1.35	1.77	2.16	2.65	3.01	3.85
14	0.69	1.35	1.76	2.15	2.62	2.98	3.79
15	0.69	1.34	1.75	2.13	2.60	2.95	3.73
16	0.69	1.34	1.75	2.12	2.58	2.92	3.69
17	0.69	1.33	1.74	2.11	2.57	2.90	3.65
18	0.69	1.33	1.73	2.10	2.55	2.89	3.61
19	0.69	1.33	1.73	2.09	2.54	2.86	3.58
20	0.69	1.33	1.73	2.09	2.53	2.85	3.55
21	0.69	1.32	1.72	2.08	2.52	2.83	3.53
22	0.69	1.32	1.72	2.07	2.51	2.82	3.51
23	0.69	1.32	1.71	2.07	2.50	2.81	3.49
24	0.69	1.32	1.71	2.06	2.49	2.80	3.47
25	0.68	1.32	1.71	2.06	2.49	2.79	3.45
26	0.68	1.32	1.71	2.06	2.48	2.78	3.44
27	0.68	1.31	1.70	2.05	2.47	2.77	3.42
28	0.68	1.31	1.70	2.05	2.47	2.76	3.41
29	0.68	1.31	1.70	2.05	2.46	2.76	3.40
30	0.68	1.31	1.70	2.04	2.46	2.75	3.39
40	0.68	1.30	1.68	2.02	2.42	2.70	3.31
50	0.68	1.30	1.68	2.01	2.40	2.68	3.26
60	0.68	1.30	1.67	2.00	2.39	2.66	3.23
70	0.68	1.29	1.67	1.99	2.38	2.65	3.21
80	0.68	1.29	1.66	1.99	2.37	2.64	3.20
90	0.68	1.29	1.66	1.99	2.37	2.63	3.18
100	0.68	1.29	1.66	1.98	2.36	2.63	3.17
150	0.68	1.29	1.66	1.98	2.35	2.61	3.16
∞	0.67	1.28	1.65	1.96	2.33	2.58	3.09

TABLE E.4 Values of the function $t_{\alpha, n}$
The function $t_{\alpha, n}$ is defined by the equation
$$P(\tau_n > t_{\alpha, n}) = \alpha$$
where the random variable τ_n has a Student distribution with n degrees of freedom. The probability density function of τ_n is equal to

$$p_{\tau_n}(x) = \frac{\Gamma((n + 1)/2)}{\Gamma(n/2)\sqrt{\pi n}} \left(1 + \frac{x^2}{n}\right)^{-(n+1)/2}$$

	2α			
n	0.10	0.05	0.02	0.01
5	2.015	2.571	3.365	4.032
6	1.943	2.447	3.143	3.707
7	1.895	2.365	2.998	3.499
8	1.860	2.306	2.896	3.355
9	1.833	2.262	2.821	3.250
10	1.812	2.228	2.764	3.169
12	1.782	2.179	2.681	3.055
14	1.761	2.145	2.624	2.977
16	1.746	2.120	2.583	2.921
18	1.734	2.101	2.552	2.878
20	1.725	2.086	2.528	2.845
22	1.717	2.074	2.508	2.819
30	1.697	2.042	2.457	2.750
∞	1.645	1.960	2.326	2.576

Appendix F

Hermite Polynomials

TABLE F.1 Hermite Polynomials
$H_2(x), H_3(x), H_4(x), H_5(x), H_6(x)$

x	$H_2(x)$	x	$H_2(x)$	x	$H_2(x)$	x	$H_2(x)$
0.00	−1.00000	0.50	−0.7500	1.00	0.0000	1.50	+1.2500
0.01	0.9999	0.51	0.7399	1.01	+0.0201	1.51	1.2801
0.02	0.9996	0.52	0.7296	1.02	0.0404	1.52	1.3104
0.03	0.9991	0.53	0.7191	1.03	0.0609	1.53	1.3409
0.04	0.9984	0.54	0.7084	1.04	0.0816	1.54	1.3716
0.05	0.9975	0.55	0.6975	1.05	0.1025	1.55	1.4025
0.06	0.9964	0.56	0.6864	1.06	0.1236	1.56	1.4336
0.07	0.9951	0.57	0.6751	1.07	0.1449	1.57	1.4649
0.08	0.9936	0.58	0.6636	1.08	0.1664	1.58	1.4964
0.09	0.9919	0.59	0.6519	1.09	0.1881	1.59	1.5281
0.10	0.9900	0.60	0.6400	1.10	0.2100	1.60	1.5600
0.11	0.9879	0.61	0.6279	1.11	0.2321	1.61	1.5921
0.12	0.9856	0.62	0.6156	1.12	0.2544	1.62	1.6244
0.13	0.9831	0.63	0.6031	1.13	0.2769	1.63	1.6569
0.14	0.9804	0.64	0.5904	1.14	0.2996	1.64	1.6896
0.15	0.9775	0.65	0.5775	1.15	0.3225	1.65	1.7225
0.16	0.9744	0.66	0.5644	1.16	0.3456	1.66	1.7556
0.17	0.9711	0.67	0.5511	1.17	0.3689	1.67	1.7889
0.18	0.9676	0.68	0.5376	1.18	0.3924	1.68	1.8224
0.19	0.9639	0.69	0.5239	1.19	0.4161	1.69	1.8561
0.20	0.9600	0.70	0.5100	1.20	0.4400	1.70	1.8900
0.21	0.9559	0.71	0.4959	1.21	0.4641	1.71	1.9241
0.22	0.9516	0.72	0.4816	1.22	0.4884	1.72	1.9584
0.23	0.9471	0.73	0.4671	1.23	0.5129	1.73	1.9929
0.24	0.9424	0.74	0.4524	1.24	0.5376	1.74	2.0276
0.25	0.9375	0.75	0.4375	1.25	0.5625	1.75	2.0625
0.26	0.9324	0.76	0.4224	1.26	0.5876	1.76	2.0976
0.27	0.9271	0.77	0.4071	1.27	0.6129	1.77	2.1329
0.28	0.9216	0.78	0.3916	1.28	0.6384	1.78	2.1684
0.29	0.9159	0.79	0.3759	1.29	0.6641	1.79	2.2041
0.30	0.9100	0.80	0.3600	1.30	0.6900	1.80	2.2400
0.31	0.9039	0.81	0.3439	1.31	0.7161	1.81	2.2761
0.32	0.8976	0.82	0.3276	1.32	0.7424	1.82	2.3124
0.33	0.8911	0.83	0.3111	1.33	0.7689	1.83	2.3489
0.34	0.8844	0.84	0.2944	1.34	0.7956	1.84	2.3856
0.35	0.8775	0.85	0.2775	1.35	0.8225	1.85	2.4225
0.36	0.8704	0.86	0.2604	1.36	0.8496	1.86	2.4596
0.37	0.8631	0.87	0.2431	1.37	0.8769	1.87	2.4969
0.38	0.8556	0.88	0.2256	1.38	0.9044	1.88	2.5344
0.39	0.8479	0.89	0.2079	1.39	0.9321	1.89	2.5721
0.40	0.8400	0.90	0.1900	1.40	0.9600	1.90	2.6100
0.41	0.8319	0.91	0.1719	1.41	0.9881	1.91	2.6481
0.42	0.8236	0.92	0.1536	1.42	1.0164	1.92	2.6864
0.43	0.8151	0.93	0.1351	1.43	1.0449	1.93	2.7249
0.44	0.8064	0.94	0.1164	1.44	1.0736	1.94	2.7636
0.45	0.7975	0.95	0.0975	1.45	1.1025	1.95	2.8025
0.46	0.7884	0.96	0.0784	1.46	1.1316	1.96	2.8416
0.47	0.7791	0.97	0.0591	1.47	1.1609	1.97	2.8809
0.48	0.7696	0.98	0.0396	1.48	1.1904	1.98	2.9204
0.49	0.7599	0.99	−0.0199	1.49	1.2201	1.99	2.9601
0.50	−0.7500	1.00	0.0000	1.50	+1.2500	2.00	+3.0000

TABLE F.1 (Continued)

x	$H_2(x)$	x	$H_2(x)$	x	$H_2(x)$	x	$H_2(x)$
2.00	3.000 0	2.50	+5.250 0	3.00	+8.000 0	3.50	+11.250 0
2.01	3.040 1	2.51	5.300 1	3.01	8.060 1	3.51	11.320 1
2.02	3.080 4	2.52	5.350 4	3.02	8.120 4	3.52	11.390 4
2.03	3.120 9	2.53	5.400 9	3.03	8.180 9	3.53	11.460 9
2.04	3.161 6	2.54	5.451 6	3.04	8.241 6	3.54	11.531 6
2.05	3.202 5	2.55	5.502 5	3.05	8.302 5	3.55	11.602 5
2.06	3.243 6	2.56	5.553 6	3.06	8.363 6	3.56	11.673 6
2.07	3.284 9	2.57	5.604 9	3.07	8.424 9	3.57	11.744 9
2.08	3.326 4	2.58	5.656 4	3.08	8.486 4	3.58	11.816 4
2.09	3.368 1	2.59	5.708 1	3.09	8.548 1	3.59	11.883 1
2.10	3.410 0	2.60	5.760 0	3.10	8.610 0	3.60	11.960 0
2.11	3.452 1	2.61	5.812 1	3.11	8.672 1	3.61	12.032 1
2.12	3.494 4	2.62	5.864 4	3.12	8.734 4	3.62	12.104 4
2.13	3.536 9	2.63	5.916 9	3.13	8.796 9	3.63	12.176 9
2.14	3.579 6	2.64	5.969 6	3.14	8.859 6	3.64	12.249 6
2.15	3.622 5	2.65	6.022 5	3.15	8.922 5	3.65	12.322 5
2.16	3.665 6	2.66	6.075 6	3.16	8.985 6	3.66	12.395 6
2.17	3.708 9	2.67	6.128 9	3.17	9.048 9	3.67	12.468 9
2.18	3.752 4	2.68	6.182 4	3.18	9.112 4	3.68	12.542 4
2.19	3.796 1	2.69	6.236 1	3.19	9.176 1	3.69	12.616 1
2.20	3.840 0	2.70	6.290 0	3.20	9.240 0	3.70	12.690 0
2.21	3.884 1	2.71	6.344 1	3.21	9.304 1	3.71	12.764 1
2.22	3.928 4	2.72	6.398 4	3.22	9.368 4	3.72	12.838 4
2.23	3.972 9	2.73	6.452 9	3.23	9.432 9	3.73	12.912 9
2.24	4.017 6	2.74	6.507 6	3.24	9.497 6	3.74	12.987 6
2.25	4.062 5	2.75	6.562 5	3.25	9.562 5	3.75	13.062 5
2.26	4.107 6	2.76	6.617 6	3.26	9.627 6	3.76	13.137 6
2.27	4.152 9	2.77	6.672 9	3.27	9.692 9	3.77	13.212 9
2.28	4.198 4	2.78	6.728 4	3.28	9.758 4	3.78	13.288 4
2.29	4.244 1	2.79	6.784 1	3.29	9.824 1	3.79	13.364 1
2.30	4.290 0	2.80	6.840 0	3.30	9.890 0	3.80	13.444 0
2.31	4.336 1	2.81	6.896 1	3.31	9.956 1	3.81	13.516 1
2.32	4.382 4	2.82	6.952 4	3.32	10.022 4	3.82	13.592 4
2.33	4.428 9	2.83	7.065 6	3.34	10.088 9	3.83	13.668 9
2.34	4.475 6	2.84	7.065 6	3.34	10.155 6	3.84	13.745 6
2.35	4.522 5	2.85	7.122 5	3.35	10.222 5	3.85	13.822 5
2.36	4.569 6	2.86	7.179 6	3.36	10.289 6	3.86	13.899 6
2.37	4.616 9	2.87	7.236 9	3.37	10.356 9	3.87	13.976 9
2.38	4.664 4	2.88	7.294 4	3.38	10.424 4	3.88	14.054 4
2.39	4.712 1	2.89	7.352 1	3.39	10.492 1	3.89	14.132 1
2.40	4.760 0	2.90	7.410 0	3.40	10.560 0	3.90	14.210 0
2.41	4.808 1	2.91	7.468 1	3.41	10.628 1	3.91	14.288 1
2.42	4.856 4	2.92	7.526 4	3.42	10.696 4	3.92	14.366 4
2.43	4.904 9	2.93	7.584 9	3.43	10.764 9	3.93	14.444 9
2.44	4.953 6	2.94	7.643 6	3.44	10.833 6	3.94	14.523 6
2.45	5.002 5	2.95	7.702 5	3.45	10.902 5	3.95	14.602 5
2.46	5.051 6	2.96	7.761 6	3.46	10.971 6	3.96	14.681 6
2.47	5.100 9	2.97	7.820 9	3.47	11.040 9	3.97	14.760 9
2.48	5.150 4	2.98	7.880 4	3.48	11.110 4	3.98	14.840 4
2.49	5.200 1	2.99	7.940 1	3.49	11.180 1	3.99	14.920 1
2.50	+5.250 0	3.00	+8.000 0	3.50	+11.250 0	4.00	+15.000 0

TABLE F.1 (Continued)

x	$H_3(x)$	x	$H_3(x)$	x	$H_3(x)$	x	$H_3(x)$
0.00	0.000 000	0.50	−1.375 000	1.00	−2.000 000	1.50	−1.125 000
0.01	−0.029 999	0.51	1.397 349	1.01	1.999 699	1.51	1.087 049
0.02	0.059 992	0.52	1.419 392	1.02	1.998 792	1.52	1.048 192
0.03	0.089 973	0.53	1.441 123	1.03	1.997 273	1.53	1.008 423
0.04	0.119 936	0.54	1.462 536	1.04	1.995 136	1.54	0.967 736
0.05	0.149 875	0.55	1.483 625	1.05	1.992 375	1.55	0.926 125
0.06	0.179 784	0.56	1.504 384	1.06	1.988 984	1.56	0.883 584
0.07	0.209 657	0.57	1.524 807	1.07	1.984 957	1.57	0.840 107
0.08	0.239 488	0.58	1.544 888	1.08	1.980 288	1.58	0.795 688
0.09	0.269 271	0.59	1.564 621	1.09	1.974 971	1.59	0.750 321
0.10	0.299 000	0.60	1.584 000	1.10	1.969 000	1.60	0.704 000
0.11	0.328 669	0.61	1.603 019	1.11	1.992 369	1.61	0.656 719
0.12	0.358 272	0.62	1.621 672	1.12	1.955 072	1.62	0.608 472
0.13	0.387 803	0.63	1.639 953	1.13	1.947 103	1.63	0.559 253
0.14	0.417 256	0.64	1.657 856	1.14	1.938 456	1.64	0.509 056
0.15	0.446 625	0.65	1.675 375	1.15	1.929 125	1.65	0.457 875
0.16	0.475 904	0.66	1.692 504	1.16	1.919 104	1.66	0.405 704
0.17	0.505 087	0.67	1.709 237	1.17	1.908 387	1.67	0.352 537
0.18	0.534 168	0.68	1.725 568	1.18	1.896 968	1.68	0.298 368
0.19	0.563 141	0.69	1.741 491	1.19	1.884 841	1.69	0.243 191
0.20	0.592 000	0.70	1.757 000	1.20	1.872 000	1.70	0.187 000
0.21	0.620 739	0.71	1.772 089	1.21	1.858 439	1.71	0.129 789
0.22	0.649 352	0.72	1.786 752	1.22	1.844 152	1.72	0.071 552
0.23	0.677 833	0.73	1.800 983	1.23	1.829 133	1.73	−0.012 283
0.24	0.706 176	0.74	1.814 776	1.24	1.813 376	1.74	+0.048 024
0.25	0.734 375	0.75	1.828 125	1.25	1.796 875	1.75	0.109 375
0.26	0.762 424	0.76	1.841 024	1.26	1.779 624	1.76	0.171 776
0.27	0.790 317	0.77	1.853 467	1.27	1.761 617	1.77	0.235 233
0.28	0.818 048	0.78	1.865 448	1.28	1.742 848	1.78	0.299 752
0.29	0.845 611	0.79	1.876 961	1.29	1.723 311	1.79	0.365 339
0.30	0.873 000	0.80	1.888 000	1.30	1.703 000	1.80	0.432 000
0.31	0.900 209	0.81	1.898 559	1.31	1.681 909	1.81	0.499 741
0.32	0.927 232	0.82	1.908 632	1.32	1.660 032	1.82	0.568 568
0.33	0.954 063	0.83	1.918 213	1.33	1.637 363	1.83	0.638 487
0.34	0.980 696	0.84	1.927 296	1.34	1.613 896	1.84	0.709 504
0.35	1.007 125	0.85	1.935 875	1.35	1.589 625	1.85	0.781 625
0.36	1.033 344	0.86	1.943 944	1.36	1.564 544	1.86	0.854 856
0.37	1.059 347	0.87	1.951 497	1.37	1.538 647	1.87	0.929 203
0.38	1.085 128	0.88	1.958 528	1.38	1.511 928	1.88	1.004 672
0.39	1.110 681	0.89	1.965 031	1.39	1.484 381	1.89	1.081 269
0.40	1.136 000	0.90	1.971 000	1.40	1.456 000	1.90	1.159 000
0.41	1.161 079	0.91	1.976 429	1.41	1.426 779	1.91	1.237 871
0.42	1.185 912	0.92	1.981 312	1.42	1.396 712	1.92	1.317 888
0.43	1.210 493	0.93	1.985 643	1.43	1.365 793	1.93	1.399 057
0.44	1.234 816	0.94	1.989 416	1.44	1.334 016	1.94	1.481 384
0.45	1.258 875	0.95	1.992 625	1.45	1.301 375	1.95	1.564 875
0.46	1.282 664	0.96	1.995 264	1.46	1.267 864	1.96	1.649 536
0.47	1.306 177	0.97	1.997 327	1.47	1.233 477	1.97	1.735 373
0.48	1.329 408	0.98	1.998 808	1.48	1.198 208	1.98	1.822 392
0.49	1.352 351	0.99	1.999 701	1.49	1.162 051	1.99	1.910 599
0.50	−1.375 000	1.00	−2.000 000	1.50	−1.125 000	2.00	+2.000 000

TABLE F.1 (*Continued*)

x	$H_3(x)$	x	$H_3(x)$	x	$H_3(x)$	x	$H_3(x)$
2.00	+2.000 000	2.50	+8.125 000	3.00	+18.000 000	3.50	+32.375 000
2.01	2.090 601	2.51	8.283 251	3.01	18.240 901	3.51	32.713 551
2.02	2.182 408	2.52	8.443 008	3.02	18.483 608	3.52	33.054 208
2.03	2.275 427	2.53	8.604 277	3.03	18.728 127	3.53	33.396 977
2.04	2.369 664	2.54	8.767 064	3.04	18.974 464	3.54	33.741 864
2.05	2.465 125	2.55	8.931 375	3.05	19.222 625	3.55	34.088 875
2.06	2.561 816	2.56	9.097 216	3.06	19.472 616	3.56	34.438 016
2.07	2.659 743	2.57	9.264 593	3.07	19.724 443	3.57	34.789 293
2.08	2.758 912	2.58	9.433 512	3.08	19.978 112	3.58	35.142 712
2.09	2.859 329	2.59	9.603 979	3.09	20.233 629	3.59	35.498 279
2.10	2.961 000	2.60	9.776 000	3.10	20.491 000	3.60	35.856 000
2.11	3.063 931	2.61	9.949 581	3.11	20.750 231	3.61	36.215 881
2.12	3.168 128	2.62	10.124 728	3.12	21.011 328	3.62	36.577 928
2.13	3.273 597	2.63	10.301 447	3.13	21.274 297	3.63	36.942 147
2.14	3.380 344	2.64	10.479 744	3.14	21.539 144	3.64	37.308 544
2.15	3.488 375	2.65	10.659 625	3.15	21.805 875	3.65	37.677 125
2.16	3.597 696	2.66	10.841 096	3.16	22.074 496	3.66	38.047 896
2.17	3.708 313	2.67	11.024 163	3.17	22.345 013	3.67	38.420 863
2.18	3.820 232	2.68	11.208 832	3.18	22.617 432	3.68	38.796 032
2.19	3.933 459	2.69	11.395 109	3.19	22.891 759	3.69	39.173 409
2.20	4.048 000	2.70	11.583 000	3.20	23.168 000	3.70	39.553 000
2.21	4.163 861	2.71	11.772 511	3.21	23.446 161	3.71	39.934 811
2.22	4.281 048	2.72	11.963 648	3.22	23.726 248	3.72	40.318 848
2.23	4.399 567	2.73	12.156 417	3.23	24.008 267	3.73	40.705 117
2.24	4.519 424	2.74	12.350 824	3.24	24.292 224	3.74	41.093 624
2.25	4.640 625	2.75	12.546 875	3.25	24.578 125	3.75	41.484 375
2.26	4.763 176	2.76	12.744 576	3.26	24.865 976	3.76	41.877 376
2.27	4.887 083	2.77	12.943 933	3.27	25.145 783	3.77	42.272 633
2.28	5.012 352	2.78	13.144 952	3.28	25.447 552	3.78	42.670 152
2.29	5.138 989	2.79	13.347 639	3.29	25.741 289	3.79	43.069 939
2.30	5.267 000	2.80	13.552 000	3.30	26.037 000	3.80	43.472 000
2.31	5.396 391	2.81	13.758 041	3.31	26.334 691	3.81	43.876 341
2.32	5.527 168	2.82	13.965 768	3.32	26.634 368	3.82	44.282 968
2.33	5.659 337	2.83	14.175 187	3.33	26.936 037	3.83	44.691 887
2.34	5.792 904	2.84	14.386 304	3.34	27.239 704	3.84	45.103 104
2.35	5.927 875	2.85	14.599 125	3.35	27.545 375	3.85	45.516 625
2.36	6.064 256	2.86	14.813 656	3.36	27.853 056	3.86	45.932 456
2.37	6.202 053	2.87	15.029 903	3.37	28.162 753	3.87	46.350 603
2.38	6.341 272	2.88	15.247 872	3.38	28.474 472	3.88	46.771 072
2.39	6.481 919	2.89	15.467 569	3.39	28.788 219	3.89	47.193 869
2.40	6.624 000	2.90	15.689 000	3.40	29.104 000	3.90	47.619 000
2.41	6.767 521	2.91	15.912 171	3.41	29.421 821	3.91	48.046 471
2.42	6.912 488	2.92	16.137 088	3.42	29.741 688	3.92	48.476 288
2.43	7.058 907	2.93	16.363 757	3.43	30.063 607	3.93	48.908 457
2.44	7.206 784	2.94	16.592 184	3.44	30.387 584	3.94	49.342 984
2.45	7.356 125	2.95	16.822 375	3.45	30.713 625	3.95	49.779 875
2.46	7.506 936	2.96	17.054 336	3.46	31.041 736	3.96	50.219 136
2.47	7.659 223	2.97	17.288 073	3.47	31.371 923	3.97	50.660 773
2.48	7.812 992	2.98	17.523 592	3.48	31.704 192	3.98	51.104 792
2.49	7.968 249	2.99	17.760 899	3.49	32.038 549	3.99	51.551 199
2.50	+8.125 000	3.00	+18.000 000	3.50	+32.375 000	4.00	+52.000 000

TABLE F.1 (Continued)

x	$H_4(x)$	x	$H_4(x)$	x	$H_4(x)$	x	$H_4(x)$
0.00	+3.000 000 00	0.50	+1.562 500 00	1.00	−2.000 000 00	1.50	−5.437 500 00
0.01	2.999 400 01	0.51	1.507 052 01	1.01	2.079 995 99	1.51	5.481 743 99
0.02	2.997 600 16	0.52	1.450 716 16	1.02	2.159 967 84	1.52	5.524 451 84
0.03	2.994 600 81	0.53	1.393 504 81	1.03	2.239 891 19	1.53	5.565 587 19
0.04	2.990 402 56	0.54	1.335 430 56	1.04	2.319 741 44	1.54	5.605 113 44
0.05	2.985 006 25	0.55	1.276 506 25	1.05	2.399 493 75	1.55	5.642 993 75
0.06	2.978 412 96	0.56	1.216 744 96	1.06	2.479 123 04	1.56	5.679 191 04
0.07	2.970 624 01	0.57	1.156 160 01	1.07	2.558 603 99	1.57	5.713 667 99
0.08	2.961 640 96	0.58	1.094 764 96	1.08	2.637 911 04	1.58	5.746 387 04
0.09	2.951 465 61	0.59	1.032 573 61	1.09	2.717 018 39	1.59	5.777 310 39
0.10	2.940 100 00	0.60	0.969 600 00	1.10	2.795 900 00	1.60	5.806 400 00
0.11	2.927 546 41	0.61	0.905 858 41	1.11	2.874 529 59	1.61	5.833 617 59
0.12	2.913 807 36	0.62	0.841 363 36	1.12	2.952 980 64	1.62	5.858 924 64
0.13	2.898 885 61	0.63	0.776 129 61	1.13	3.030 926 39	1.63	5.882 282 39
0.14	2.882 784 16	0.64	0.710 172 16	1.14	3.108 639 84	1.64	5.903 651 84
0.15	2.865 506 25	0.65	0.643 506 25	1.15	3.185 993 75	1.65	5.922 993 75
0.16	2.847 055 36	0.66	0.576 147 36	1.16	3.262 960 64	1.66	5.940 268 64
0.17	2.827 435 21	0.67	0.508 111 21	1.17	3.339 512 79	1.67	5.955 436 79
0.18	2.806 649 76	0.68	0.439 413 76	1.18	3.415 622 24	1.68	5.968 458 24
0.19	2.784 703 21	0.69	0.370 071 21	1.19	3.491 260 79	1.69	5.979 292 79
0.20	2.761 600 00	0.70	0.300 100 00	1.20	3.566 400 00	1.70	5.987 900 00
0.21	2.737 344 81	0.71	0.229 516 81	1.21	3.641 011 19	1.71	5.994 239 19
0.22	2.711 942 56	0.72	0.158 338 56	1.22	3.715 065 44	1.72	5.998 269 44
0.23	2.685 398 41	0.73	0.086 582 41	1.23	3.788 533 59	1.73	5.999 949 59
0.24	2.657 717 76	0.74	+0.014 265 76	1.24	3.861 386 24	1.74	5.999 238 24
0.25	2.628 906 25	0.75	−0.058 593 75	1.25	3.933 593 75	1.75	5.996 093 75
0.26	2.598 969 76	0.76	0.131 978 24	1.26	4.005 126 24	1.76	5.990 474 24
0.27	2.567 914 41	0.77	0.205 869 59	1.27	4.075 953 59	1.77	5.982 337 59
0.28	2.535 746 56	0.78	0.280 249 44	1.28	4.146 045 44	1.78	5.971 641 44
0.29	2.502 472 81	0.79	0.355 099 19	1.29	4.215 371 19	1.79	5.958 343 19
0.30	2.468 100 00	0.80	0.430 400 00	1.30	4.283 900 00	1.80	5.942 400 00
0.31	2.432 635 21	0.81	0.506 132 79	1.31	4.351 600 79	1.81	5.923 768 79
0.32	2.396 085 76	0.82	0.582 278 24	1.32	4.418 442 24	1.82	5.902 406 24
0.33	2.358 459 21	0.83	0.658 816 79	1.33	4.484 392 79	1.83	5.878 268 79
0.34	2.319 763 36	0.84	0.735 728 64	1.34	4.549 420 64	1.84	5.851 312 64
0.35	2.280 006 25	0.85	0.812 993 75	1.35	4.613 493 75	1.85	5.821 493 75
0.36	2.239 196 16	0.86	0.890 591 84	1.36	4.676 579 84	1.86	5.788 767 84
0.37	2.197 341 61	0.87	0.968 502 39	1.37	4.738 646 39	1.87	5.753 090 39
0.38	2.154 451 36	0.88	1.046 704 64	1.38	4.799 660 64	1.88	5.714 416 64
0.39	2.110 534 41	0.89	1.125 177 59	1.39	4.859 589 59	1.89	5.672 701 59
0.40	2.065 600 00	0.90	1.203 900 00	1.40	4.918 400 00	1.90	5.627 900 00
0.41	2.019 657 61	0.91	1.282 850 39	1.41	4.976 058 39	1.91	5.579 966 39
0.42	1.972 716 96	0.92	1.362 007 04	1.42	5.032 531 04	1.92	5.528 855 04
0.43	1.924 788 01	0.93	1.441 347 99	1.43	5.087 783 99	1.93	5.475 519 99
0.44	1.875 880 96	0.94	1.520 851 04	1.44	5.141 783 04	1.94	5.416 915 04
0.45	1.826 006 25	0.95	1.600 493 75	1.45	5.194 493 75	1.95	5.355 993 75
0.46	1.775 174 56	0.96	1.680 253 44	1.46	5.245 881 44	1.96	5.291 709 44
0.47	1.723 396 81	0.97	1.760 107 19	1.47	5.295 911 19	1.97	5.224 015 19
0.48	1.670 684 16	0.98	1.840 031 84	1.48	5.344 547 84	1.98	5.152 863 84
0.49	1.617 048 01	0.99	1.920 003 99	1.49	5.391 755 99	1.99	5.078 207 99
0.50	+1.562 500 00	1.00	−2.000 000 00	1.50	−5.437 500 00	2.00	−5.000 000 00

F. Hermite Polynomials

TABLE F.1 *(Continued)*

x	$H_4(x)$	x	$H_4(x)$	x	$H_4(x)$	x	$H_4(x)$
2.00	−5.000 000 00	2.50	+4.562 500 00	3.00	+30.000 000 00	3.50	+79.562 500 00
2.01	4.918 191 99	2.51	4.890 660 01	3.01	30.724 812 01	3.51	80.864 264 01
2.02	4.832 735 84	2.52	5.225 180 16	3.02	31.459 296 16	3.52	82.179 612 16
2.03	4.743 583 19	2.53	5.566 120 81	3.03	32.203 524 81	3.53	83.508 628 81
2.04	4.650 685 44	2.54	5.913 542 56	3.04	32.957 570 56	3.54	84.851 398 56
2.05	4.553 993 75	2.55	6.267 506 25	3.05	33.721 506 25	3.55	86.208 006 25
2.06	4.453 459 04	2.56	6.628 072 96	3.06	34.495 404 96	3.56	87.578 536 96
2.07	4.349 031 99	2.57	6.995 304 01	3.07	35.279 340 01	3.57	88.963 076 01
2.08	4.240 663 04	2.58	6.369 260 96	3.08	36.073 384 96	3.58	90.361 708 96
2.09	4.128 302 39	2.59	6.750 005 61	3.09	36.877 613 61	3.59	91.774 521 61
2.10	4.011 900 00	2.60	8.137 600 00	3.10	37.692 100 00	3.60	93.201 600 00
2.11	3.891 405 59	2.61	8.532 106 41	3.11	38.516 918 41	3.61	94.643 030 41
2.12	3.766 768 64	2.62	8.933 587 36	3.12	39.352 143 36	3.62	96.098 899 36
2.13	3.637 938 39	2.63	9.342 105 61	3.13	40.197 849 61	3.63	97.569 293 61
2.14	3.504 863 84	2.64	9.757 724 16	3.14	41.054 112 16	3.64	99.054 300 16
2.15	3.367 493 75	2.65	10.180 506 25	3.15	41.921 006 25	3.65	100.554 006 25
2.16	3.225 776 64	2.66	10.610 515 36	3.16	42.798 607 36	3.66	102.068 499 36
2.17	3.079 660 79	2.67	11.047 815 21	3.17	43.686 991 21	3.67	103.597 867 21
2.18	2.929 094 24	2.68	11.492 469 76	3.18	44.586 233 76	3.68	105.142 197 76
2.19	2.774 024 79	2.69	11.944 543 21	3.19	45.496 411 21	3.69	106.701 579 21
2.20	2.614 400 00	2.70	12.404 100 00	3.20	46.417 600 00	3.70	108.276 100 00
2.21	2.450 167 19	2.71	12.871 204 81	3.21	47.349 876 81	3.71	109.865 848 81
2.22	2.281 273 44	2.72	13.345 922 56	3.22	48.293 318 56	3.72	111.470 914 56
2.23	2.107 665 59	2.73	13.828 318 41	3.23	49.248 002 41	3.73	113.091 386 41
2.24	1.929 290 24	2.74	14.318 457 76	3.24	50.214 005 76	3.74	114.727 353 76
2.25	1.746 093 75	2.75	14.816 406 25	3.25	51.191 406 25	3.75	116.378 906 25
2.26	1.558 022 24	2.76	15.322 229 76	3.26	52.180 281 76	3.76	118.046 133 76
2.27	1.365 021 59	2.77	15.835 994 41	3.27	53.180 710 41	3.77	119.729 126 41
2.28	1.167 037 44	2.78	16.357 766 56	3.28	54.192 770 56	3.78	121.427 974 56
2.29	0.964 015 19	2.79	16.887 612 81	3.29	55.216 540 81	3.79	123.142 768 81
2.30	0.755 900 00	2.80	17.425 600 00	3.30	56.252 100 00	3.80	124.873 600 00
2.31	0.542 636 79	2.81	17.971 795 21	3.31	57.299 527 21	3.81	126.620 559 21
2.32	0.324 170 24	2.82	18.526 265 76	3.32	58.358 901 76	3.82	128.383 737 76
2.33	−0.100 444 79	2.83	19.089 079 21	3.33	59.430 303 21	3.83	130.163 227 21
2.34	+0.128 595 36	2.84	19.660 303 36	3.34	60.513 811 36	3.84	131.959 119 36
2.35	0.363 006 25	2.85	20.240 006 25	3.35	61.609 506 25	3.85	133.771 506 25
2.36	0.602 844 16	2.86	20.828 256 16	3.36	62.717 468 16	3.86	135.600 480 16
2.37	0.848 165 61	2.87	21.425 121 61	3.37	63.837 777 61	3.87	137.446 133 61
2.38	1.099 027 36	2.88	22.030 671 36	3.38	64.970 515 36	3.88	139.308 559 36
2.39	1.355 486 41	2.89	22.644 974 41	3.39	66.115 762 41	3.89	141.187 850 41
2.40	1.617 600 00	2.90	23.268 100 00	3.40	67.273 600 00	3.90	143.084 100 00
2.41	1.885 425 61	2.61	23.900 117 61	3.41	68.444 109 61	3.91	144.997 401 61
2.42	2.159 020 96	2.92	24.541 096 96	3.42	69.627 372 96	3.92	146.927 848 96
2.43	2.438 444 01	2.93	25.191 108 01	3.43	70.823 472 01	3.93	148.875 536 01
2.44	2.723 752 96	2.94	25.850 220 96	3.44	72.032 488 96	3.94	150.840 556 96
2.45	3.015 006 25	2.95	26.518 506 25	3.45	73.254 506 25	3.95	152.823 006 25
2.46	3.312 262 56	2.96	27.196 034 56	3.46	74.489 606 56	3.96	154.822 978 56
2.47	3.615 580 81	2.97	27.882 876 81	3.47	75.737 872 81	3.97	156.840 568 81
2.48	3.925 020 16	2.98	28.579 104 16	3.48	76.999 388 16	3.98	158.875 872 16
2.49	4.240 640 01	2.99	29.284 788 01	3.49	78.274 236 01	3.99	160.928 984 01
2.50	+4.562 500 00	3.00	+30.000 000 00	3.50	+79.562 500 00	4.00	+163.000 000 00

TABLE F.1 (Continued)

x	$H_5(x)$	x	$H_5(x)$	x	$H_5(x)$	x	$H_5(x)$
0.00	+0.000 000 00	0.50	+6.281 250 00	1.00	+6.000 000 00	1.50	−3.656 250 00
0.01	0.149 990 00	0.51	6.357 992 53	1.01	5.898 000 05	1.51	3.929 237 42
0.02	0.299 920 00	0.52	6.431 940 40	1.02	5.792 000 80	1.52	4.204 398 80
0.03	0.449 730 02	0.53	6.503 049 55	1.03	5.682 004 07	1.53	4.481 656 40
0.04	0.599 360 10	0.54	6.571 276 50	1.04	5.568 012 90	1.54	4.760 930 70
0.05	0.748 750 31	0.55	6.636 578 44	1.05	5.450 031 56	1.55	5.042 140 31
0.06	0.897 840 78	0.56	6.698 913 18	1.06	5.328 065 58	1.56	5.325 202 02
0.07	1.046 571 68	0.57	6.758 239 21	1.07	5.202 121 73	1.57	5.610 030 74
0.08	1.194 883 28	0.58	6.814 515 68	1.08	5.072 208 08	1.58	5.896 539 52
0.09	1.342 715 90	0.59	6.867 702 43	1.09	4.938 333 95	1.59	6.184 639 52
0.10	1.490 010 00	0.60	6.917 760 00	1.10	4.800 510 00	1.60	6.474 240 00
0.11	1.636 706 11	0.61	6.964 649 63	1.11	4.658 748 16	1.61	6.765 248 32
0.12	1.782 744 88	0.62	7.008 333 28	1.12	4.513 061 68	1.62	7.057 569 92
0.13	1.928 067 13	0.63	7.048 773 65	1.13	4.363 465 18	1.63	7.351 108 30
0.14	2.072 613 78	0.64	7.085 934 18	1.14	4.209 974 58	1.64	7.645 765 02
0.15	2.216 325 94	0.65	7.119 779 06	1.15	4.052 607 19	1.65	7.941 439 69
0.16	2.359 144 86	0.66	7.150 273 26	1.16	3.891 381 66	1.66	8.238 029 94
0.17	2.501 011 99	0.67	7.177 382 51	1.17	3.726 318 04	1.67	8.535 431 44
0.18	2.641 868 96	0.68	7.201 073 36	1.18	3.557 437 76	1.68	8.833 537 84
0.19	2.781 657 61	0.69	7.221 313 13	1.19	3.384 763 66	1.69	9.132 240 82
0.20	2.920 320 00	0.70	7.238 070 00	1.20	3.208 320 00	1.70	9.431 430 00
0.21	3.057 798 41	0.71	7.251 312 94	1.21	3.028 132 46	1.71	9.730 993 01
0.22	3.194 035 36	0.72	7.261 011 76	1.22	2.844 228 16	1.72	10.030 815 44
0.23	3.328 973 63	0.73	7.267 137 16	1.23	2.656 635 68	1.73	10.330 780 79
0.24	3.462 556 26	0.74	7.269 660 66	1.24	2.465 385 06	1.74	10.630 770 54
0.25	3.594 726 56	0.75	7.268 554 69	1.25	2.270 507 81	1.75	10.930 664 06
0.26	3.725 428 14	0.76	7.263 792 54	1.26	2.072 036 94	1.76	11.230 338 66
0.27	3.854 604 89	0.77	7.255 348 42	1.27	1.870 006 94	1.77	11.529 669 53
0.28	3.982 201 04	0.78	7.243 197 44	1.28	1.664 453 84	1.78	11.828 529 76
0.29	4.108 161 11	0.79	7.227 315 64	1.29	1.455 415 16	1.79	12.126 790 31
0.30	4.232 430 00	0.80	7.207 680 00	1.30	1.242 930 00	1.80	12.424 320 00
0.31	4.354 952 92	0.81	7.184 268 44	1.31	1.027 038 97	1.81	12.720 985 51
0.32	4.475 675 44	0.82	7.157 059 84	1.32	0.807 784 24	1.82	13.016 651 36
0.33	4.594 543 54	0.83	7.126 034 06	1.33	0.585 209 59	1.83	13.311 179 89
0.34	4.711 503 54	0.84	7.091 171 94	1.34	0.359 360 34	1.84	13.604 431 26
0.35	4.826 502 19	0.85	7.052 455 31	1.35	+0.130 283 44	1.85	13.896 263 44
0.36	4.939 476 62	0.86	7.009 867 02	1.36	−0.101 972 58	1.86	14.186 532 18
0.37	5.050 404 40	0.87	6.963 390 92	1.37	0.337 357 55	1.87	14.475 091 03
0.38	5.159 203 52	0.88	6.913 011 92	1.38	0.575 819 68	1.88	14.761 791 28
0.39	5.265 832 42	0.89	6.858 715 94	1.39	0.817 305 53	1.89	15.046 482 01
0.40	5.370 240 00	0.90	6.800 490 00	1.40	1.061 760 00	1.90	15.329 010 00
0.41	5.472 375 62	0.91	6.738 322 15	1.41	1.309 126 33	1.91	15.609 219 80
0.42	5.572 189 12	0.92	6.672 201 52	1.42	1.559 346 08	1.92	15.886 953 68
0.43	5.669 630 84	0.93	6.602 118 37	1.43	1.812 359 11	1.93	16.162 051 58
0.44	5.764 651 62	0.94	6.528 064 02	1.44	2.068 103 58	1.94	16.434 351 18
0.45	5.857 202 81	0.95	6.450 030 94	1.45	2.326 515 94	1.95	16.703 687 81
0.46	5.947 236 30	0.96	6.368 012 70	1.46	2.587 530 90	1.96	16.969 894 50
0.47	6.034 704 50	0.97	6.282 004 03	1.47	2.851 081 45	1.97	17.232 801 92
0.48	6.119 560 40	0.98	6.192 000 80	1.48	3.117 098 80	1.98	17.492 238 40
0.49	6.201 757 52	0.99	6.098 000 05	1.49	3.385 512 43	1.99	17.748 029 90
0.50	+6.281 250 00	1.00	+6.000 000 00	1.50	−3.656 250 00	2.00	−18.000 000 00

TABLE F.1 (Continued)

x	$H_5(x)$	x	$H_5(x)$	x	$H_5(x)$	x	$H_5(x)$
2.00	−18.000 000 00	2.50	−21.093 750 00	3.00	+18.000 000 00	3.50	+148.968 750 00
2.01	18.247 969 90	2.51	20.857 447 37	3.01	19.518 080 15	3.51	152.979 362 68
2.02	18.491 758 40	2.52	20.604 578 00	3.02	21.072 642 40	3.52	157.055 402 80
2.03	18.731 181 88	2.53	20.334 822 35	3.03	22.664 172 17	3.53	161.197 551 70
2.04	18.966 054 30	2.54	20.047 857 90	3.04	24.293 158 50	3.54	165.406 494 90
2.05	19.196 187 19	2.55	19.743 359 06	3.05	25.960 094 06	3.55	169.682 922 19
2.06	19.421 389 62	2.56	19.420 997 22	3.06	27.665 475 18	3.56	174.027 527 58
2.07	19.641 468 22	2.57	19.080 440 69	3.07	29.409 801 83	3.57	178.441 009 36
2.08	19.856 227 12	2.58	18.721 354 72	3.08	31.193 577 68	3.58	182.924 070 08
2.09	20.065 468 00	2.59	18.343 401 47	3.09	33.017 310 05	3.59	187.477 416 58
2.10	20.268 990 00	2.60	17.946 240 00	3.10	34.881 510 00	3.60	192.101 760 00
2.11	20.466 589 79	2.61	17.529 526 27	3.11	36.786 692 25	3.61	196.797 815 78
2.12	20.658 061 52	2.62	17.092 913 12	3.12	38.733 375 28	3.62	201.566 303 68
2.13	20.843 196 77	2.63	16.636 050 25	3.13	40.722 081 28	3.63	206.407 947 80
2.14	21.021 784 62	2.64	16.158 584 22	3.14	42.753 336 18	3.64	211.323 476 58
2.15	21.193 611 56	2.65	15.660 158 44	3.15	44.827 669 69	3.65	216.313 622 81
2.16	21.358 461 54	2.66	15.140 413 14	3.16	46.945 615 26	3.66	221.379 123 66
2.17	21.516 115 91	2.67	14.598 985 39	3.17	49.107 710 14	3.67	226.520 720 66
2.18	21.666 353 44	2.68	14.035 509 04	3.18	51.314 495 36	3.68	231.739 159 76
2.19	21.808 950 29	2.69	13.449 614 77	3.19	53.566 515 76	3.69	237.035 191 28
2.20	21.943 680 00	2.70	12.840 930 00	3.20	55.864 320 00	3.70	242.409 570 00
2.21	22.070 313 49	2.71	12.209 078 96	3.21	58.208 460 56	3.71	247.863 055 09
2.22	22.188 619 04	2.72	11.553 682 64	3.22	60.599 493 76	3.72	253.396 410 16
2.23	22.298 362 27	2.73	10.874 358 74	3.23	63.037 979 78	3.73	259.010 403 31
2.24	22.399 306 14	2.74	10.170 721 74	3.24	65.524 482 66	3.74	264.705 807 06
2.25	22.491 210 94	2.75	9.442 382 81	3.25	68.059 570 31	3.75	270.483 398 44
2.26	22.573 834 26	2.76	8.688 949 86	3.26	70.643 814 54	3.76	276.343 958 94
2.27	22.646 931 01	2.77	7.910 027 48	3.27	73.277 791 04	3.77	282.288 274 57
2.28	22.710 253 36	2.78	7.105 216 96	3.28	75.962 079 44	3.78	288.317 135 84
2.29	22.763 550 79	2.79	6.274 116 26	3.29	78.697 263 26	3.79	294.431 337 79
2.30	22.806 570 00	2.80	5.416 320 00	3.30	81.483 930 00	3.80	300.631 680 00
2.31	22.839 054 98	2.81	4.531 419 46	3.31	84.322 671 07	3.81	306.918 966 59
2.32	22.860 746 96	2.82	3.619 002 56	3.32	87.214 081 84	3.82	313.294 006 24
2.33	22.871 384 36	2.83	2.678 653 84	3.33	90.158 761 69	3.83	319.757 612 21
2.34	22.870 702 86	2.84	1.709 954 46	3.34	93.157 313 94	3.84	326.310 602 34
2.35	22.858 435 31	2.85	−0.712 482 19	3.35	96.210 345 94	3.85	332.953 799 06
2.36	22.834 311 78	2.86	+0.314 188 62	3.36	99.318 469 02	3.86	339.688 029 42
2.37	22.798 059 50	2.87	1.370 487 02	3.37	102.482 298 55	3.87	346.514 125 07
2.38	22.742 402 88	2.88	2.456 845 52	3.38	105.702 453 92	3.88	353.432 922 32
2.39	22.688 063 48	2.89	3.573 700 04	3.39	108.979 558 57	3.89	360.445 262 09
2.40	22.613 760 00	2.90	4.721 490 00	3.40	112.314 240 00	3.90	367.551 990 00
2.41	22.526 208 28	2.91	5.900 658 25	3.41	115.707 129 77	3.91	374.753 956 30
2.42	22.425 121 28	2.92	7.111 651 12	3.42	119.158 863 52	3.92	382.052 015 92
2.43	22.310 209 06	2.93	8.354 918 47	3.43	122.670 080 99	3.93	389.447 028 52
2.44	22.181 178 78	2.94	9.630 913 62	3.44	126.241 426 02	3.94	396.939 858 42
2.45	22.037 734 69	2.95	10.940 093 44	3.45	129.873 546 56	3.95	404.531 374 69
2.46	21.879 578 10	2.96	12.282 918 30	3.46	133.567 094 70	3.96	412.222 451 10
2.47	21.706 407 40	2.97	13.659 852 13	3.47	137.322 726 65	3.97	420.013 966 18
2.48	21.517 918 00	2.98	15.071 362 40	3.48	141.141 102 80	3.98	427.906 803 20
2.49	21.313 802 38	2.99	16.517 920 15	3.49	145.022 887 67	3.99	435.901 850 20
2.50	−21.093 750 00	3.00	+18.000 000 00	3.50	+148.968 750 00	4.00	+444.000 000 00

TABLE F.1 (*Continued*)

x	$H_6(x)$	x	$H_6(x)$	x	$H_6(x)$	x	$H_6(x)$
0.00	15.000 000 00	0.50	−4.671 875 00	1.00	+16.000 000 00	1.50	+21.703 125 00
0.01	14.995 500 15	0.51	4.292 683 86	1.01	16.356 960 00	1.51	21.475 571 44
0.02	14.982 002 40	0.52	3.908 971 79	1.02	16.707 680 02	1.52	21.231 573 03
0.03	14.959 512 15	0.53	3.520 907 79	1.03	17.051 920 15	1.53	20.971 001 66
0.04	14.928 038 40	0.54	3.128 663 49	1.04	17.389 440 62	1.54	20.693 733 93
0.05	14.887 593 73	0.55	2.732 413 11	1.05	17.720 001 89	1.55	20.399 651 27
0.06	14.838 194 35	0.56	2.332 333 42	1.06	18.043 364 71	1.56	20.088 640 05
0.07	14.779 860 03	0.57	1.928 603 70	1.07	18.359 290 20	1.57	19.760 591 68
0.08	14.712 614 14	0.58	1.521 405 71	1.08	18.667 539 92	1.58	19.415 402 75
0.09	14.636 483 62	0.59	1.110 923 62	1.09	18.967 875 96	1.59	19.052 975 11
0.10	14.551 499 00	0.60	0.697 344 00	1.10	19.260 061 00	1.60	18.673 216 00
0.11	14.457 694 38	0.61	−0.280 855 78	1.11	19.543 858 40	1.61	18.276 038 15
0.12	14.355 107 41	0.62	+0.138 349 84	1.12	19.819 032 29	1.62	17.861 359 94
0.13	14.243 779 32	0.63	0.560 079 35	1.13	20.085 347 60	1.63	17.429 105 43
0.14	14.123 754 87	0.64	0.984 137 08	1.14	20.342 570 22	1.64	16.979 204 57
0.15	13.995 082 36	0.65	1.410 325 14	1.15	20.590 467 02	1.65	16.511 593 27
0.16	13.857 813 62	0.66	1.838 443 55	1.16	20.828 805 92	1.66	16.026 213 50
0.17	13.712 004 01	0.67	2.268 290 23	1.17	21.057 356 05	1.67	15.523 013 45
0.18	13.557 712 39	0.68	2.699 661 08	1.18	21.275 887 75	1.68	15.001 947 62
0.19	13.395 001 10	0.69	3.132 350 01	1.19	21.484 172 71	1.69	14.462 976 97
0.20	13.223 936 00	0.70	3.566 149 00	1.20	21.681 984 00	1.70	13.906 069 00
0.21	13.044 586 38	0.71	4.000 848 13	1.21	21.869 096 23	1.71	13.331 197 89
0.22	12.857 025 02	0.72	4.436 235 67	1.22	22.045 285 56	1.72	12.738 344 65
0.23	12.661 328 11	0.73	4.872 098 08	1.23	22.210 329 84	1.73	12.127 497 18
0.24	12.457 575 30	0.74	5.308 220 09	1.24	22.364 008 68	1.74	11.498 650 46
0.25	12.245 849 61	0.75	5.744 384 77	1.25	22.506 103 52	1.75	10.851 806 64
0.26	12.026 237 48	0.76	6.180 373 53	1.26	22.636 397 74	1.76	10.186 975 15
0.27	11.798 828 73	0.77	6.615 966 23	1.27	22.754 676 76	1.77	9.504 172 87
0.28	11.563 716 51	0.78	7.050 941 20	1.28	22.860 728 11	1.78	8.803 424 22
0.29	11.320 997 33	0.79	7.485 075 31	1.29	22.954 341 51	1.79	8.084 761 29
0.30	11.070 771 00	0.80	7.918 144 00	1.30	23.035 309 00	1.80	7.348 224 00
0.31	10.813 140 65	0.81	8.349 921 39	1.31	23.103 424 99	1.81	6.593 860 18
0.32	10.548 212 66	0.82	8.780 180 27	1.32	23.158 486 40	1.82	5.821 725 73
0.33	10.276 096 68	0.83	9.208 692 22	1.33	23.200 292 70	1.83	5.031 884 76
0.34	9.996 905 60	0.84	9.635 227 63	1.34	23.228 646 06	1.84	4.224 409 69
0.35	9.710 755 48	0.85	10.059 555 77	1.35	23.243 351 39	1.85	3.399 381 39
0.36	9.417 765 62	0.86	10.481 444 84	1.36	23.244 216 49	1.86	2.556 889 34
0.37	9.118 058 42	0.87	10.900 662 05	1.37	23.231 052 10	1.87	1.697 031 73
0.38	8.811 759 46	0.88	11.316 973 69	1.38	23.203 672 04	1.88	+0.819 915 59
0.39	8.498 997 41	0.89	11.730 145 14	1.39	23.161 893 26	1.89	−0.074 343 04
0.40	8.179 904 00	0.90	12.139 941 00	1.40	23.105 536 00	1.90	0.985 619 00
0.41	7.854 614 05	0.91	12.546 125 10	1.41	23.034 423 82	1.91	1.913 777 88
0.42	7.523 265 37	0.92	12.948 460 60	1.42	22.948 383 77	1.92	2.858 675 86
0.43	7.185 998 79	0.93	13.346 710 03	1.43	22.847 246 43	1.93	3.820 159 60
0.44	6.842 958 09	0.94	13.740 635 38	1.44	22.730 846 05	1.94	4.798 066 08
0.45	6.494 289 98	0.95	14.129 998 14	1.45	22.599 020 64	1.95	5.792 222 48
0.46	6.140 144 10	0.96	14.514 559 39	1.46	22.451 612 08	1.96	6.802 446 02
0.47	5.780 672 93	0.97	14.894 079 85	1.47	22.288 466 22	1.97	7.828 543 84
0.48	5.416 031 81	0.98	15.268 319 98	1.48	22.109 432 97	1.98	8.870 312 84
0.49	5.046 378 86	0.99	15.637 040 00	1.49	21.914 366 44	1.99	9.927 539 55
0.50	−4.671 875 00	1.00	+16.000 000 00	1.50	+21.703 125 00	2.00	−11.000 000 00

TABLE F.1 (*Continued*)

x	$H_6(x)$	x	$H_6(x)$	x	$H_6(x)$	x	$H_6(x)$
2.00	−11.000 000 00	2.50	−75.546 875 00	3.00	−96.000 000 00	3.50	+123.578 125 00
2.01	12.087 459 55	2.51	76.805 492 96	3.01	94.874 638 80	3.51	132.636 242 94
2.02	13.189 672 76	2.52	78.049 437 35	3.02	93.657 100 74	3.52	141.936 957 07
2.03	14.306 383 26	2.53	79.277 704 60	3.03	92.345 182 36	3.53	151.484 213 45
2.04	15.437 323 57	2.54	80.489 271 86	3.04	90.936 650 95	3.54	161.281 999 15
2.05	16.582 214 98	2.55	81.683 096 86	3.05	89.429 244 36	3.55	171.334 342 52
2.06	17.740 767 42	2.56	82.858 117 69	3.06	87.820 670 76	3.56	181.645 313 38
2.07	18.912 679 26	2.57	84.013 252 63	3.07	86.108 608 43	3.57	192.219 023 35
2.08	20.097 637 22	2.58	85.147 399 99	3.08	84.290 705 56	3.58	203.059 626 07
2.09	21.295 316 16	2.59	86.259 437 86	3.09	82.364 579 98	3.59	214.171 317 47
2.10	22.505 379 00	2.60	87.348 224 00	3.10	80.327 819 00	3.60	225.558 836 00
2.11	23.727 476 52	2.61	88.412 595 61	3.11	78.177 979 14	3.61	237.224 962 92
2.12	24.961 247 22	2.62	89.451 369 17	3.12	75.912 585 92	3.62	249.175 522 53
2.13	26.206 317 17	2.63	90.463 340 20	3.13	73.529 133 65	3.63	261.414 382 48
2.14	27.462 299 88	2.64	91.447 283 13	3.14	71.025 085 19	3.64	273.945 953 96
2.15	28.728 796 11	2.65	92.401 951 11	3.15	68.397 871 73	3.65	286.774 692 02
2.16	30.005 393 73	2.66	93.326 075 76	3.16	65.644 892 59	3.66	299.905 095 79
2.17	31.291 667 58	2.67	94.218 367 04	3.17	62.763 514 92	3.67	313.341 708 77
2.18	32.587 179 31	2.68	95.077 513 04	3.18	59.751 073 57	3.68	327.089 119 11
2.19	33.891 477 19	2.69	95.902 179 77	3.19	56.604 870 78	3.69	341.151 959 79
2.20	35.204 096 00	2.70	96.691 011 00	3.20	53.322 176 00	3.70	355.534 909 00
2.21	36.524 556 86	2.71	97.442 628 04	3.21	49.900 225 65	3.71	370.242 690 32
2.22	37.852 367 06	2.72	98.155 629 57	3.22	46.336 222 88	3.72	385.280 073 01
2.23	39.187 019 90	2.73	98.828 591 41	3.23	42.627 337 35	3.73	400.651 872 29
2.24	40.527 994 55	2.74	99.460 066 36	3.24	38.770 704 97	3.74	416.362 949 61
2.25	41.874 755 86	2.75	100.048 583 98	3.25	34.763 427 73	3.75	432.418 212 89
2.26	43.226 754 23	2.76	100.592 650 42	3.26	30.602 573 41	3.76	448.822 616 81
2.27	44.583 425 44	2.77	101.090 748 18	3.27	26.285 175 35	3.77	465.581 163 06
2.28	45.944 190 47	2.78	101.541 335 96	3.28	21.808 232 25	3.78	482.698 900 66
2.29	47.308 455 35	2.79	101.942 848 42	3.29	17.168 707 91	3.79	500.180 926 17
2.30	48.675 611 00	2.80	102.293 696 00	3.30	12.363 531 00	3.80	518.032 384 00
2.31	50.045 033 07	2.81	102.592 264 73	3.31	7.389 594 82	3.81	536.258 466 66
2.32	51.416 081 74	2.82	102.836 916 01	3.32	−2.243 757 08	3.82	554.864 415 05
2.33	52.788 101 61	2.83	103.025 986 41	3.33	+3.077 160 38	3.83	573.855 518 73
2.34	54.160 421 49	2.84	103.157 787 46	3.34	8.576 371 77	3.84	593.237 116 19
2.35	55.532 354 23	2.85	103.230 605 48	3.35	14.257 127 64	3.85	613.014 595 14
2.36	56.903 196 61	2.86	103.242 701 35	3.36	20.122 715 10	3.86	633.193 392 75
2.37	58.272 229 08	2.87	103.192 310 30	3.37	26.176 458 05	3.87	653.778 995 97
2.38	59.638 715 66	2.88	103.077 641 71	3.38	32.421 717 44	3.88	674.776 941 79
2.39	61.001 903 77	2.89	102.896 878 92	3.39	38.861 891 50	3.89	696.192 817 50
2.40	62.361 024 00	2.90	102.648 179 00	3.40	45.500 416 00	3.90	718.032 261 00
2.41	63.715 290 00	2.91	102.329 672 56	3.41	52.340 764 47	3.91	740.300 961 06
2.42	65.063 898 29	2.92	101.939 463 52	3.42	59.386 448 45	3.92	763.004 657 62
2.43	66.406 028 06	2.93	101.475 628 94	3.43	66.641 017 76	3.93	786.149 142 03
2.44	67.740 841 02	2.94	100.936 218 75	3.44	74.108 060 72	3.94	809.740 257 38
2.45	69.067 481 23	2.95	100.319 255 61	3.45	81.791 204 39	3.95	833.783 898 77
2.46	70.385 074 93	2.96	99.622 734 64	3.46	89.694 114 85	3.96	858.286 013 55
2.47	71.692 730 33	2.97	98.844 623 24	3.47	97.820 497 43	3.97	883.252 601 67
2.48	72.989 537 45	2.98	97.982 860 86	3.48	106.174 096 93	3.98	908.689 715 92
2.49	74.274 567 96	2.99	97.035 358 80	3.49	114.758 697 94	3.99	934.603 462 25
2.50	−75.546 875 00	3.00	−96.000 000 00	3.50	+123.578 125 00	4.00	+961.000 000 00

Appendix G

Knots and Coefficients for Numerical Integration Using Hermite Polynomials

TABLE G.1 Knots ξ_k and coefficients A_k for numerical integration using Hermite polynomials

n	k	ξ_k	A_k
1	1	0.000 000 000 000 000	1.772 4 5 385 090 6
2	1	0.707 106 781 186 548	0.886 226 925 452 4
3	1	0.000 000 000 000 000	1.181 6 3 590 060 8
	2	1.224 744 871 391 589	0.295 408 975 150 9
4	1	0.524 647 623 275 290	0.804 914 090 005 5
	2	1.650 680 123 885 785	0.813 128 354 472 5 (-1)
5	1	0.000 000 000 000 000	0.945 308 720 482 9
	2	0.958 572 464 613 819	0.393 619 323 152 2
	3	2.020 182 870 456 086	0.199 532 420 590 5 (-1)
6	1	0.436 077 411 927 617	0.724 629 595 224 4
	2	1.335 849 074 013 697	0.157 067 320 322 9
	3	2.350 604 973 674 492	0.453 000 990 550 9 (-2)
7	1	0.000 000 000 000 000	0.810 264 617 556 8
	2	0.816 287 882 858 965	0.425 607 252 610 1
	3	1.673 551 628 767 471	0.545 155 828 191 3 (-1)
	4	2.651 961 356 835 233	0.971 781 245 099 5 (-3)
8	1	0.381 186 990 207 322	0.661 147 012 558 2
	2	1.157 193 712 446 780	0.207 802 325 814 9
	3	1.981 656 756 695 843	0.170 779 830 074 1 (-1)
	4	2.930 637 420 057 244	0.199 604 072 211 4 (-3)
9	1	0.000 000 000 000 000	0.720 235 215 606 1
	2	0.723 551 018 752 838	0.432 651 559 002 6
	3	1.468 553 289 216 668	0.884 745 273 943 8 (-1)
	4	2.266 580 584 531 843	0.494 362 427 553 7 (-2)
	5	3.190 993 201 781 528	0.396 069 772 632 6 (-4)
10	1	0.342 901 327 223 705	0.610 862 633 735 3
	2	1.036 610 829 789 514	0.240 138 611 082 3
	3	1.756 683 649 299 882	0.338 743 944 554 8 (-1)
	4	2.532 731 674 232 790	0.134 364 574 678 1 (-2)
	5	3.436 159 118 837 738	0.764 043 285 523 3 (-5)

TABLE G.1 (*Continued*)

n	k	ξ_k	A_k
11	1	0.000 000 000 000 000	0.654 759 286 914 6
	2	0.656 809 566 882 100	0.429 359 752 356 1
	3	1.326 557 084 494 933	0.117 227 875 167 7
	4	2.025 948 015 825 755	0.119 113 954 449 1 (-1)
	5	2.783 290 099 781 652	0.346 819 466 323 3 (-3)
	6	3.668 470 846 559 583	0.143 956 039 371 4 (-5)
12	1	0.314 240 376 254 359	0.570 135 236 262 5
	2	0.947 788 391 240 164	0.260 492 310 264 2
	3	1.597 682 635 152 605	0.516 079 856 158 8 (-1)
	4	2.279 507 080 501 060	0.390 539 058 462 9 (-2)
	5	3.020 637 025 120 890	0.857 368 704 358 8 (-4)
	6	3.889 724 897 869 782	0.265 855 168 435 6 (-6)
13	1	0.000 000 000 000 000	0.604 393 187 921 1
	2	0.605 763 879 171 060	0.421 616 296 898 5
	3	1.220 055 036 590 748	0.140 323 320 687 0
	4	1.853 107 651 601 512	0.208 627 752 961 7 (-1)
	5	2.519 735 685 678 238	0.120 745 999 271 9 (-2)
	6	3.246 608 978 372 410	0.204 303 604 027 1 (-4)
	7	4.101 337 596 178 640	0.482 573 185 007 3 (-7)
14	1	0.291 745 510 672 56	0.536 405 909 712 1
	2	0.878 713 787 329 40	0.273 105 609 064 2
	3	1.476 682 731 141 14	0.685 055 342 234 7 (-1)
	4	2.095 183 258 507 72	0.785 005 472 645 8 (-2)
	5	2.748 470 724 985 40	0.355 092 613 551 9 (-3)
	6	3.462 656 933 602 27	0.471 648 435 501 9 (-5)
	7	4.304 448 570 473 63	0.862 859 116 812 5 (-8)
15	1	0.000 000 000 000 00	0.564 100 308 726 4
	2	0.565 069 583 255 58	0.412 028 687 498 9
	3	1.136 115 585 210 92	0.158 488 915 795 9
	4	1.719 992 575 186 49	0.307 800 338 725 5 (-1)
	5	2.325 732 486 173 86	0.277 806 884 291 3 (-2)
	6	2.967 166 927 905 60	0.100 004 441 232 5 (-3)
	7	3.669 950 373 404 45	0.105 911 554 771 1 (-5)
	8	4.499 990 707 309 39	0.152 247 580 425 4 (-8)
16	1	0.273 481 046 138 15	0.507 929 479 016 6
	2	0.822 951 449 144 66	0.280 647 458 528 5
	3	1.380 258 539 198 88	0.838 100 413 989 9 (-1)
	4	1.951 787 990 916 25	0.128 803 115 355 1 (-1)
	5	2.546 202 157 847 48	0.932 284 008 624 2 (-3)
	6	3.176 999 161 979 96	0.271 186 009 253 8 (-4)
	7	3.869 447 904 860 12	0.232 098 084 486 5 (-6)
	8	4.688 738 939 305 82	0.265 480 747 401 1 (-9)
17	1	0.000 000 000 000 0	0.530 917 937 624 9
	2	0.531 633 001 342 7	0.401 826 469 470 4
	3	1.067 648 725 743 5	0.172 648 297 670 1
	4	1.612 924 314 221 2	0.409 200 341 497 6 (-1)
	5	2.173 502 826 666 6	0.506 734 995 762 8 (-2)
	6	2.757 762 915 703 9	0.298 643 286 697 8 (-3)
	7	3.378 932 091 141 5	0.711 228 914 002 1 (-5)
	8	4.061 946 675 875 5	0.497 707 898 163 1 (-7)
	9	4.871 345 193 674 4	0.458 057 893 079 9 (-10)

G. Knots and Coefficients for Numerical Integration Using Hermite Polynomials

TABLE G.1 (*Continued*)

n	k	ξ_k	A_k
18	1	0.258 267 750 519 1	0.483 495 694 725 5
	2	0.776 682 919 267 4	0.284 807 285 670 0
	3	1.300 920 858 389 6	0.973 017 476 413 2 (-1)
	4	1.835 531 604 261 6	0.186 400 423 875 4 (-1)
	5	2.386 299 089 166 7	0.188 852 263 026 8 (-2)
	6	2.961 377 505 531 6	0.918 112 686 792 9 (-4)
	7	3.573 769 068 486 3	0.181 065 448 109 3 (-5)
	8	4.248 117 873 568 1	0.104 672 057 957 9 (-7)
	9	5.048 364 008 874 5	0.782 819 977 211 6 (-11)
19	1	0.000 000 000 000 0	0.502 974 888 276 2
	2	0.503 520 163 423 9	0.391 608 988 613 0
	3	1.010 368 387 134 3	0.183 632 701 307 0
	4	1.524 170 619 393 5	0.508 103 869 090 5 (-1)
	5	2.049 231 709 850 6	0.798 886 677 772 3 (-2)
	6	2.591 133 789 794 5	0.670 877 521 407 2 (-3)
	7	3.157 848 818 347 6	0.272 091 977 631 6 (-4)
	8	3.762 187 351 964 0	0.448 824 314 722 3 (-6)
	9	4.428 532 806 603 8	0.216 305 100 986 4 (-8)
	10	5.220 071 690 537 5	0.132 629 709 449 9 (-11)
20	1	0.245 340 708 300 9	0.462 243 669 600 6
	2	0.737 473 728 545 4	0.286 675 505 362 8
	3	1.234 076 215 395 3	0.109 017 206 020 0
	4	1.738 537 712 116 6	0.248 105 208 874 6 (-1)
	5	2.254 974 002 089 3	0.324 377 334 223 8 (-2)
	6	2.788 806 058 428 1	0.228 338 636 016 3 (-3)
	7	3.347 854 567 383 2	0.780 255 647 853 2 (-5)
	8	3.944 764 040 115 6	0.108 606 937 076 9 (-6)
	9	4.603 682 449 550 7	0.439 934 099 227 3 (-9)
	10	5.387 480 890 011 2	0.222 939 364 553 4 (-12)

SOURCE: H. E. Salzer, R. Zucker, and R. Capuano, Tables of the zeros and weight factors of the first twenty Hermite polynomials, *J. of Research of the National Bureau of Standards*, vol. 48, 1952.

Index

Absorbing barrier(s) 9, 381
Accelerated testing 299
Activation energy 333, 335
Aircrafts 426, 479
Algebra of events 15–23
Amplitude frequency spectrum 197
Arrhenius 334
Availability function 354
Availability index 288, 353, 354
Availability of an item 352
Averaging 176
Axioms of the probability theory 11

Barriers 381
Bathtub curve 273
Bayes formula 39, 40, 82, 83, 303, 304, 308
Bernoulli theorem 87, 145
Bernoulli trials 52
Bertrand's paradox 39
Beta distribution 104
Binomial distribution 18, 52
Boltzmann–Arrhenius equation 334
Brown(ian) motion 352
Brown(ian) process 350
Bueche–Zhurkov formula 334
Buffon's needle problem 38
"Bathtub curve" 273

Campbell theorem 347
Capacity 267
Cars 459
Cauchy distribution 103
Central limit theorem 89, 146
Chain sampling 310, 511
Chaotic phase approximation 444
Chapman–Kolmogorov equation 371
Characteristic functions 223
Characteristics of random variables 117
Chebyshev inequality 146, 270
Chebyshev theorem 145
"Choosy bride" problem 10–11, 31
Clutter 348
Coefficient of variation 81
Combinations 26
Communication channels 221
Communication systems 160
Complex amplitudes 197

Complex frequency characteristic 202, 204
Complex random variables 114
Complex signature 304
Complex spectrum 198
Conditional probability 16, 83
Confidence intervals and levels (limits) 297, 453, 454, 494
Confidence probability 494
Conjugate distributions 295
Continuous random variables 79, 82
Convolution of distributions 122, 138
Correlation coefficient 110
Correlation function 169, 170
Correlation matrix 113
Correlation theory 171
Covariance 109
Crossing the given level 184
Cumulative distribution (function) 79

Degrees of freedom 101
Delphi method 517
Delta correlated process 211
Delta function 182, 382
Demand 267
De Méré's paradox 4
De Moivre–Laplace formula 88
Decision rule 305
Density function 79
Dependability 266, 271, 279, 319
Detection of random signals 355
Deterministic approach xix
Diagnosis 302
Diagnostic matrix 306
Diagnostics 302
Diagnostic signals 302
Diffusion constant 351
Diffusion equation 374
Diffusion flow 375
Dirac delta function 182, 204
Discrete random variables 47
Distribution(s)
    absolute value of a normal random variable 93
    beta 104
    binomial 18, 52
    bivariate 169

canonical form   110
Cauchy   103
chi-   103
chi-squared   101, 103
conjugate   295
cumulative   49
double-exponential   233
Erlang   101, 102, 135, 136
exponential   85
extreme value   227, 232, 234, 248
function   179
gamma   101
geometric   69
hypergeometric   71
initial   342
leptokurtic   51
log(arithmic)-normal   94
Maxwell   378
normal   88, 93, 298
Pearson 101, 495
platykurtic   51
Poisson   60, 282, 339, 352, 367, 535
polynomial   53
Rayleigh   95, 137, 255
Rice   98, 99
Student   105, 456, 500
truncated normal   93
uniform   84
unimodal(single-peak)   50
univariate   169
Weibull   99, 276
Double sampling   510
Drift   375
Dual inputs   219
Duhamel integral   186, 189, 190, 402, 432
Dynamic impedance   203
Dynamic system   219

Earthquakes   483
Equation of renewal   275
Energy   333
Ensemble   176
Entropy   149, 150, 152
Ergodic process   176, 345
Ergodic property   176
Erlang flow   338
Erlang problem   360
Error function   90, 541
Estimate   491
Events   1, 15, 17
Exponential formula of reliability   273
Extreme value statistics   227

Failure   265
Failure flow   275
Failure frequency   273
Failure rate   266, 272, 279
Fatigue   389, 391
Fatigue limit   390

Filtering   355
Flow of events (traffic)   61, 337
  of Poisson type   60, 338
  ordinary   61, 337
  without aftereffect   61, 337
Flows of failures and restorations   352
Fokker–Planck equation   371, 372, 373, 378, 379, 382, 387, 422, 424, 430
Fourier coefficients   196
Fourier integral   197, 201
Fourier integral formula   197
Fourier series   195
Fourier transform   198, 199
Frequency   2
Frequency polygon (table)   48
Frequency response function   203
Frequency spectrum   196
Function of nonfailure   268
Functions of random variables   117
Fundamental principle of counting sample points   23

Gain factor   203
Gambler's ruin problem   7
Gamma distribution   101
Gamma function   100, 555
Gaussian distribution   88
Gaussian noise   350
Gaussian stochastic process   349, 350
Generating functions   72
Geometric distribution   69
Geometric probabilities   34
Geometric tolerance   451
Goodman law   390
Goodness-of-fit
  criteria   280
    Pearson's (chi-squared)   281
    Kolmogorov's   284, 285
  tests   280
Gram–Charlier series   248
Green function   186

Heaviside unit step function   49, 184, 191
Helicopter undercarriage strength   256
Hermite polynomials   246, 573
Histogram   48
Homogeneous process   232
Hypergeometric distribution   70

Image   202
Impedance   203
Impulse response   186
Infant mortality   273
Information   156, 160, 221

Jacobian   121
Jump function   49, 184

Kolmogorov's equation   345

Index

Kolmogorov's backward equation 380, 381
Kolmogorov's forward equation 373 (see Fokker–Planck equation)
Kurtosis 50

Lagrange multipliers 59, 159, 163
Langevin equation 377, 386
Laplace function 59, 89, 301
Laplace theorem 146
Laplace transform 202
Least-squares method 499
Lyapunov's theorem 146
Likelihood relationship 308
Limit theorems 144
Linear dynamic systems 186, 214
Linear filtering 187, 354, 355
Little's formula 360
Log(arithmic)-normal distribution 94
Law of large numbers 145

Maintainability 266
Maintenance 266
Markov chain 341, 344
Markov(ian) process 337, 339, 346, 371
Markov theorem 145
Mathematical expectation 49
Maximum likelihood, method of 306
Mean (value) 49, 80, 169
Mean period 179
Mechanical impedance 202
Moments of random variables 491
Monte Carlo simulation 39, 503–507
MTBF(mean time-between-failures) 266
MTTF(mean time-to-failure) 266
MTTR(mean time-to-repair) 266
Multiserver queue 364, 365

Narrowband process 211, 213
Neumann–Pierson criterion (condition) 314, 316
Noise 354
Nonlinear vibrations 415, 436
Normalization condition 48
  rule 80

Optimal stopping rule 308
Order statistics 228
Ordered sample (series) 48
Ornstein–Uhlenbeck process 378

Palm flow 338, 367
Parseval formula 200, 201
Parseval theorem 201
Periodic impulses 431, 446
Permutations 23
Phase factor 203
Phase(frequency) spectrum 197

Pearson function 395
Poisson flow 338
Poisson impulse process 348
Poisson theorem 145
Poisson process 183, 345, 347
Pollaczek-Khinchin formula 366
Power spectrum 209
Principle of uncertainty 207
Probability 1, 13, 342
  addition rule 13, 19
  axioms of 11
  conditional 16, 83
  fiducial 494
  geometric 34
  of nonfailure 268
  $posterior(i)$ 40, 355
  $prior$ 40
  statistical 2
  threshold 305
  total 21, 40, 476
Probability density 79
Probability integral 54, 89
Probability matrix 342
Process 337
  Brownian 350
  Markov(ian) 339
  normal 230
  Poisson 183, 345
  Wiener 351
  with discrete state 168
  with discrete time 168
  with continuous time 168

Quality control 508
Queues 358

Random fatigue 389
Random function 167
Random initial conditions 410
Random loads and responses 459
Random point 107
Random process 167, 195
  continuous time 168
  discrete time 168
  stationary 175
Random pulse process 348
Random telegraph signal 346
Random variable 47
  centered 49
  complex 114
  continuous 79, 82
  discrete 48
  functions of 117, 167
  statistically independent 109
Random vibrations 399
Random walk 9
Realization 47, 167
Recognition, methods of 303
Reflecting (retaining) barrier 381

Reliability 265, 295
Reliability function 271, 272, 277
Reliability index 269
Reliability indices, choosing of 289
Reliability of repairable items 287
Rice distribution 98, 99
Risk
  buyer's 302
  minimum 313
  seller's 302
Rule of three sigma 91

Safety factor 267, 270
Safety index 269
Sample points, counting 23
Sampling 310, 508, 510, 511
Set theory 11
Shannon formula 221
Ships 137, 237, 249, 260, 465
Shot noise 347
Signal 354
Signature 302, 304
Significant value 214
Single sampling 508
Single-server queue 359, 362, 363
Skewness 50
Slamming of ships 260
Smoluchowski's equation 371
Smoothening of random signals 355
Spectral density 198
Spectrum of a function (process) 198
Standard (mean square) deviation 50, 81
Static impedance 203
Stationary process 175, 232
Statistical decisions, methods of 311
Statistical independence 109
Statistical linearization 416
Statistics
  Boltzmann 30, 333
  Bose–Einstein 30
  Fermi–Dirac 30
Stirling's formula 26, 32
Stochastic differential equations 376, 386

Stochastic instability 440
Stochastic phase approximation 444
Stochastic process 167
Strength 267
  fatigue 316
  ultimate 316
Structural reliability 316
Summation of damages 391
Systematic flow 375
System(s)
  elastic 402
  linear 186
  of many random variables 113
  of two random variables 107
  communications 160

Technical diagnostics 302
Tolerance 451
Transfer function 202
Transformation of random processes 186
Transition function 371
Transition matrix 342
Transition probabilities 345
Tyhonov formula 378, 385

Unbiased estimator 491
Urn model 14, 52

Variance 50, 80, 169, 274
Venn diagram 12
von Karman's equations 400

Wald's method 308, 310, 311, 512
Weibull distribution 99
Weighting function 186
"White noise" 211, 376, 385, 387, 408, 409, 418
Wiener filtering 357
Wiener–Hopf equation 354, 357
Wiener–Khinchin formulas 208
Wiener–Levy process 351
Wiener process 211, 213
Wöhler curve 389

About the Author

Ephraim Suhir is Distinguished Member of Technical Staff and Principal Investigator at Bell Laboratories, Lucent Technologies, Basic Research Area, Physical Sciences and Engineering Research Division, in Murray Hill, New Jersey, USA. He conducts research-and-engineering studies in materials science, applied mechanics, applied physics, and mechanical, reliability, microelectronics, photonics, and manufacturing engineering. Dr. Suhir is Fellow of the American Society of Mechanical Engineers, Institute of Electrical and Electronic Engineers, and the Society of Plastics Engineers, and Senior Member of the American Physical Society. He is the co-founder and Technical Editor of the ASME Journal of Electronic Packaging, member of the Editorial Board of the International Journal of Solids and Structures, and an organizing committee member for various national and international conferences and symposia. Dr. Suhir received numerous best paper awards, as well as many awards for his pioneering contributions to the engineering profession.